BARRON'S

THE TRUSTED NAME IN TEST PREP

2025

AP®

Statistics

PREMIUM

Martin Sternstein, Ph.D.

Acknowledgments

Thanks to my brother, Allan, my sister-in-law, Marilyn, my sons, Jonathan and Jeremy, my daughters-in-law, Asia and Cheryl, and my grandchildren, Jaiden, Jordan, Josiah, Luna, Jayme, and Layla, for their heartfelt love and support. Most of all, thanks are due to my wife, Faith, whose love, warm encouragement, and always calm and optimistic perspective on life provide a home environment in which deadlines can be met and goals easily achieved. My sincere appreciation goes to the participants who have attended my AP Statistics workshops for teaching me just as much as I have taught them, and special thanks for their many useful suggestions are due to the following exceptional teachers:

David Bock	Ithaca High School, Ithaca, NY (retired)
Paul Buckley	Gonzaga College High School, Washington, DC
Dawn Dentato	Somers High School, Lincolndale, NY (retired)
Jared Derksen	Rancho Cucamonga High School, Rancho Cucamonga, CA
Sarah Johnson	Grand Blanc High School, Grand Blanc, MI
Lee Kucera	University of California, Irvine, CA
Laura Marshall	Phillips Exeter Academy, Exeter, NH
Brendan Murphy	John Bapst Memorial High School, Bangor, ME
Leigh Nataro	Moravian College, Bethlehem, PA
Penny Smeltzer	Austin Peace Academy, Austin, TX
Daren Starnes	The Lawrenceville School, Lawrenceville, NJ
Josh Tabor	Canyon del Oro High School, Oro Valley, Arizona
Doug Tyson	Central York High School, York, PA
Jane Viau	Frederick Douglass Academy, New York, NY
Dawn White	Silver Lake Regional High School, Kingston, MA
Luke Wilcox	East Kentwood High School, Kentwood, MI
Dash Young-Saver	KIPP University Prep High School, San Antonio, TX

Martin Sternstein
Ithaca College

Published by Kaplan North America, LLC, d/b/a Barron's Educational Series
1515 West Cypress Creek Road
Fort Lauderdale, Florida 33309
www.barronseduc.com

ISBN: 978-1-5062-9197-0

10 9 8 7 6 5 4 3 2 1

Kaplan North America, LLC, d/b/a Barron's Educational Series print books are available at special quantity discounts to use for sales promotions, employee premiums, or educational purposes. For more information or to purchase books, please call the Simon & Schuster special sales department at 866-506-1949.

About the Author

Dr. Martin Sternstein, Professor Emeritus at Ithaca College, was honored by Princeton Review as one of the nation's "300 Best College Professors." He is a long-time College Board consultant and has been a Reader and Table Leader for the AP Statistics exam for many years. He has strong interests in national educational and social issues concerning equal access to math education for all. For two years, he was a Fulbright Professor in Liberia, West Africa, after which he developed a popular "Math in Africa" course, and he is the only mathematician to have given a presentation at the annual Conference on African Linguistics. He also taught the first U.S. course for college credit in chess theory.

Table of Contents

PART 1: DIAGNOSTIC TEST

PART 2: UNITS REVIEW

PART 3: FINAL REVIEW

PART 4: PRACTICE TESTS

PART 5: APPENDICES

VISIT BARRON'S ONLINE LEARNING HUB FOR FOUR MORE FULL-LENGTH PRACTICE TESTS.

How to Use This Book

This book provides comprehensive review and extensive practice for the latest AP Statistics course and exam.

About the Exam

Start with the Exam Overview, which outlines the exam format. Familiarize yourself with all of the units covered on this test, review the different question types, and learn how the exam will be scored.

Review and Practice

Study all nine units in Part 2, which are organized according to the nine units of AP Statistics, and cover the topics recommended by the AP Statistics Development Committee. Every chapter includes Learning Objectives that will be covered, a review of each topic, dozens of figures and tables that illustrate key concepts, and end-of-chapter summaries. Interspersed among these units are 29 quizzes (mini-AP exams with both multiple-choice and free-response questions), which should be used as progress checks.

Then, consult Part 3, the Final Review, which has several invaluable sections.

- "Selecting an Appropriate Inference Procedure" offers hints on inference recognition followed by two quizzes on naming the procedure to use, defining parameters, listing conditions to be checked, and stating hypotheses, if appropriate.
- "Statistical Insights into Social Issues" includes two quizzes of comprehensive review questions that cover the whole AP Statistics curriculum. These quizzes aim to give an appreciation of the power of statistics and show how this subject gives insights into some of society's most pressing issues.
- "The Investigative Task" helps you prepare for free-response Question 6, which counts for one-eighth of your total grade on the exam; there are three illustrative examples followed by a quiz with seven practice investigative tasks.
- "50 Misconceptions," "50 Common Errors on the AP Exam," and "50 AP Exam Hints, Advice, and Reminders" provide tips to remember and pitfalls to avoid on test day.

Diagnostic Test

When you are ready for final review, take the full-length diagnostic test in Part 1 to determine which topics you know well and which ones you may want to brush up on. Complete the entire test, and then check all of the answer explanations, especially for any questions you may have missed. Then, consult the Study Guide for units in the book that you should focus on in your review.

Practice Tests

There are four full-length practice tests in the book that mirror the actual exam in format, content, and level of difficulty. Each test is followed by detailed answers and explanations for all questions.

Appendices

The end of this book consists of a series of helpful appendices, including the answers and explanations for all quizzes, important formulas to know, templates, a guide to inference, and much more. Be sure to go over these sections before completing your final review.

Online Practice

There are also four additional full-length practice tests online where all questions are answered and explained. You may take these tests in practice (untimed) mode or in timed mode.

For Students

This book is intended both as a topical review during the year and as a final review in the weeks before the AP exam. Study the text and illustrative examples carefully, and try to complete the practice quiz problems before referring to the solutions. Simply reading the detailed explanations without first striving to work through the questions on your own is not the best approach. Remember, *mathematics is not a spectator sport!* Use the practice quizzes at appropriate times throughout the school year, and you will develop confidence and a deeper understanding of the material.

A good piece of advice is to develop critical practices (like checking assumptions and conditions), acquire strong technical skills, and always write clear and thorough, yet to the point, interpretations and conclusions in context. Final answers to most problems should not be numbers but, rather, sentences explaining and analyzing numerical results. To help develop skills and insights to tackle AP free-response questions (which often choose contexts students haven't seen before), read newspapers and magazines, and figure out how to apply what you are learning to better understand articles that reference numbers, graphs, and studies.

On the day of the exam, eat a healthy meal before the test. Bring a watch (not a cell phone or smartwatch) to help pace yourself. Do not bring a ruler, white out, or highlighters. Bring extra batteries for your calculator and more than one sharp pencil with good erasers. Know that scoring at least 40% correct on the exam should be enough for at least a 3 or higher, so don't panic if you can't answer a question. And scoring 70% should earn you a 5! Furthermore, no matter what your score, plan to take more statistics classes in college.

The AP Statistics course is one of the fastest growing and most important courses offered in the high school curriculum. If you work with your teacher and study hard, you will find this to be an enjoyable course, you will do well on the AP exam, and you will develop into a more thoughtful citizen of this world! One day, when you're a data scientist, an engineer, a doctor or nurse, a manager, or whatever it is you do, you may look back on the AP exam as only a distant memory, but you'll remember the truly important things you learned in your AP Statistics class. You'll remember how to think critically and compassionately about the world around you and how to communicate your knowledge, insights, and discoveries; these are skills you'll carry with you for the rest of your life.

For Teachers

This book is fully aligned with the nine units and exam format outlined in the AP Statistics Course and Exam Description (the CED) and elaborated upon in AP Classroom. Ideally, each individual unit review, paired with practice quiz problems, should be assigned after the unit has been covered in class. The full-length diagnostic test and practice tests should be reserved for final review shortly before the AP exam. These tests are at the same level of difficulty as the actual exam, have the number of multiple-choice questions for each topic as specified by the College Board, and have the specific topic free-response questions as itemized by the College Board. While this review book is not designed to substitute for an in-class experience, students have found it to be a valuable resource both for learning the essential concepts and for preparing for the AP exam.

BARRON'S ESSENTIAL 5

As you review the content in this book and work toward earning that **5** on your AP STATISTICS exam, here are five things that you **MUST** know:

Graders want to give you credit—help them! Make them understand *what* you are doing, *why* you are doing it, and *how* you are doing it. Don't make the reader guess at what you are doing.

- **Communication** is just as important as statistical knowledge!

- Be sure you understand **exactly what you are being asked to do or find or explain** and **approach each problem systematically**.

- Some problems look scary on first reading but are not overly difficult and are surprisingly straightforward. Questions that take you beyond the scope of the AP curriculum will be phrased in ways that you should be able to answer them based on what you have learned in your AP Statistics class.

- *Naked* or *bald answers* will receive little or **no** credit. You must show where answers come from.

- On the other hand, don't give more than one solution to the same problem—you will receive credit only for the weaker one.

Describing and comparing distributions is fundamental in descriptive statistics.

- Reference shape, center, variability, and unusual features such as outliers, gaps, and clusters when describing one-variable quantitative data. Don't forget context!

- Use comparative language, rather than simply making separate lists, when comparing two distributions. Don't forget context!

- Be able to analyze displays such as dotplots, histograms, boxplots, and stemplots.

- Be able to analyze parallel boxplots and back-to-back stemplots.

- Reference form (linear or nonlinear), direction (positive or negative), and strength (weak, moderate, or strong) when describing bivariate data. Don't forget context!

Data collection is the first step in the data analysis process.

- Understand the difference between observational studies and experiments, and know the strengths and weaknesses of each.

- Understand that random sampling, the use of chance in selecting a sample from a population, is critical in being able to generalize from a sample to a population.

- Be able to describe how to implement sampling methods, including simple random sampling, stratified sampling, cluster sampling, and systematic sampling.

- Understand that random assignment of subjects to treatments in experiments is critical in minimizing the effect of possible confounding variables.

- Be able to describe how to set up an experiment using random assignment and possibly blinding or blocking.

Distributions describe variability, and variability is the most fundamental concept in statistics. Understand the terminology:

- *population distribution* (variability in an entire population),

- *sample distribution* (variability within a particular sample), and

- *sampling distribution* (variability between samples).

- The larger the sample size, the more the **sample distribution** looks like the population distribution.

- Central limit theorem: the larger the sample size, the more the **sampling distribution** (probability distribution of the sample means) looks like a normal distribution.

Choosing the correct procedure and performing proper checks are critical.

- Categorical variables lead to proportions or chi-square procedures, while quantitative variables lead to means or linear regression.

- Estimating a quantity indicates a confidence interval, while looking for evidence to test a claim indicates a hypothesis test.

- Know the proper checks for each procedure and state them correctly. (Listing wrong conditions will lose points.)

- Verifying assumptions and conditions means more than simply listing them with little check marks—you must show work or give some reason to confirm verification.

Exam Overview

Exam Format

The exam consists of two parts: a 90-minute section with 40 multiple-choice questions and a 90-minute free-response section with five open-ended questions and one investigative task to complete. During grading, the two sections of the exam are given equal weight. Students have remarked that the first section involves "lots of reading" while the second section involves "lots of writing." The percentage of questions from each content area is approximately 25% data analysis, 15% experimental design, 25% probability, and 35% inference.

Multiple-Choice Section

In the multiple-choice section, the questions are much more conceptual than computational, and thus use of the calculator is minimal. The multiple-choice section can be broken down as follows: Exploring One-Variable Data (6–9 questions); Exploring Two-Variable Data (2–3 questions); Collecting Data (5–6 questions); Probability, Random Variables, and Probability Distributions (4–8 questions); Sampling Distributions (3–5 questions); Inference for Categorical Data: Proportions (5–6 questions); Inference for Quantitative Data: Means (4–7 questions); Inference for Categorical Data: Chi-Square (1–2 questions); and Inference for Quantitative Data: Slopes (1–2 questions).

Good strategies for working multiple-choice questions include the following:

- Read carefully so you are clear what is being asked for.
- Underline or circle key words or numbers in the question.
- Make notes in the margin as to what numbers represent.
- Cross out answer choices you know are incorrect.
- Cross out answer choices that may be true but don't relate to the specific question asked.
- Use this process of elimination and then narrow down to the BEST answer.
- Answer every question.

Free-Response Section

In the free-response section, the first five open-ended questions can be broken down as follows:

- 1 multipart question with a primary focus on collecting data
- 1 multipart question with a primary focus on exploring data
- 1 multipart question with a primary focus on probability and sampling distributions
- 1 question with a primary focus on inference
- 1 question that combines two or more skill categories

The investigative task, the sixth question in the free-response section, assesses multiple skills and content in a nonroutine way.

Good strategies and guidance for working free-response questions include the following:

- Be sure you understand exactly what you are being asked to do, find, or explain.
- Underline key words, phrases, and numbers.
- Make notes in the margin as to what numbers represent.
- Use proper terminology: words like bias, correlation, normal, power, range, skew, and even statistic have specific statistical meanings.
- Indicate your methods clearly, as problems will be graded on the correctness of the methods as well as the accuracy of the results and explanations.

- One-variable distributions: don't forget context as well as shape, center, spread, and unusual features.
- Linear regressions: be able to interpret the slope, the *y*-intercept, and the coefficient of determination in context.
- Sampling methods or experimental design: be able to explain how you will pick a random sample or randomly assign subjects to treatment groups.
- Probability: name the distribution, give the parameters, note boundary values, and give the calculated answer.
- Inference: name procedures, define parameters, include confirmation of underlying assumptions, perform calculations (use software if quicker), and state conclusions in context, not just as numbers.
- Describe both an advantage of one thing and a disadvantage of the other if asked why one thing is better than another.

Calculator Usage

On the AP Statistics exam, you will be furnished with a list of formulas (from (I) Descriptive Statistics, (II) Probability and Distributions, and (III) Inferential Statistics) and tables (including standard normal probabilities, *t*-distribution critical values, and χ^2 critical values). While you will be expected to bring a graphing calculator with statistics capabilities to the exam, it is not recommended leaving answers in terms of calculator syntax. Furthermore, many students have commented that calculator usage was less than they had anticipated. However, even though the calculator is a tool, to be used sparingly and as needed, you need to be proficient with this technology. You also must be comfortable with reading generic computer output.

Scoring

The score on the multiple-choice section is based on the number of correct answers, with no points deducted for incorrect answers. So don't leave any blank answers!

Free-response questions are scored on a 0 to 4 scale with 1 point for a *minimal* response, 2 points for a *developing* response, 3 points for a *substantial* response, and 4 points for a *complete* response. Individual parts of these questions are scored as E for *essentially* correct, P for *partially* correct, and I for *incorrect*. Note that *essentially* correct does not mean *perfect*. Work is graded *holistically*—that is, a student's complete response is considered as a whole whenever scores do not fall precisely on an integer value on the 0 to 4 scale.

Each of the first five open-ended questions counts as 15% of the total free-response score, and the investigative task counts as 25% of the free-response score. The first open-ended question is typically the most straightforward. After doing this one to build confidence, students might consider looking at the investigative task since it counts more.

Each completed AP exam will receive a grade based on a 5-point scale, with 5 being the highest score and 1 being the lowest score. Most colleges and universities accept a grade of 3 or better for credit, advanced placement, or both. Over the years, average cut scores, together with the approximate percent of students receiving each score, are as in the following table.

AP Score	Total Points	Students (%)
5	70–100	15%
4	55–69	20%
3	40–54	25%
2	30–39	20%
1	0–29	20%

PART 1
Diagnostic Test

ANSWER SHEET
Diagnostic Test

1.	Ⓐ Ⓑ Ⓒ Ⓓ Ⓔ	11.	Ⓐ Ⓑ Ⓒ Ⓓ Ⓔ	21.	Ⓐ Ⓑ Ⓒ Ⓓ Ⓔ	31.	Ⓐ Ⓑ Ⓒ Ⓓ Ⓔ
2.	Ⓐ Ⓑ Ⓒ Ⓓ Ⓔ	12.	Ⓐ Ⓑ Ⓒ Ⓓ Ⓔ	22.	Ⓐ Ⓑ Ⓒ Ⓓ Ⓔ	32.	Ⓐ Ⓑ Ⓒ Ⓓ Ⓔ
3.	Ⓐ Ⓑ Ⓒ Ⓓ Ⓔ	13.	Ⓐ Ⓑ Ⓒ Ⓓ Ⓔ	23.	Ⓐ Ⓑ Ⓒ Ⓓ Ⓔ	33.	Ⓐ Ⓑ Ⓒ Ⓓ Ⓔ
4.	Ⓐ Ⓑ Ⓒ Ⓓ Ⓔ	14.	Ⓐ Ⓑ Ⓒ Ⓓ Ⓔ	24.	Ⓐ Ⓑ Ⓒ Ⓓ Ⓔ	34.	Ⓐ Ⓑ Ⓒ Ⓓ Ⓔ
5.	Ⓐ Ⓑ Ⓒ Ⓓ Ⓔ	15.	Ⓐ Ⓑ Ⓒ Ⓓ Ⓔ	25.	Ⓐ Ⓑ Ⓒ Ⓓ Ⓔ	35.	Ⓐ Ⓑ Ⓒ Ⓓ Ⓔ
6.	Ⓐ Ⓑ Ⓒ Ⓓ Ⓔ	16.	Ⓐ Ⓑ Ⓒ Ⓓ Ⓔ	26.	Ⓐ Ⓑ Ⓒ Ⓓ Ⓔ	36.	Ⓐ Ⓑ Ⓒ Ⓓ Ⓔ
7.	Ⓐ Ⓑ Ⓒ Ⓓ Ⓔ	17.	Ⓐ Ⓑ Ⓒ Ⓓ Ⓔ	27.	Ⓐ Ⓑ Ⓒ Ⓓ Ⓔ	37.	Ⓐ Ⓑ Ⓒ Ⓓ Ⓔ
8.	Ⓐ Ⓑ Ⓒ Ⓓ Ⓔ	18.	Ⓐ Ⓑ Ⓒ Ⓓ Ⓔ	28.	Ⓐ Ⓑ Ⓒ Ⓓ Ⓔ	38.	Ⓐ Ⓑ Ⓒ Ⓓ Ⓔ
9.	Ⓐ Ⓑ Ⓒ Ⓓ Ⓔ	19.	Ⓐ Ⓑ Ⓒ Ⓓ Ⓔ	29.	Ⓐ Ⓑ Ⓒ Ⓓ Ⓔ	39.	Ⓐ Ⓑ Ⓒ Ⓓ Ⓔ
10.	Ⓐ Ⓑ Ⓒ Ⓓ Ⓔ	20.	Ⓐ Ⓑ Ⓒ Ⓓ Ⓔ	30.	Ⓐ Ⓑ Ⓒ Ⓓ Ⓔ	40.	Ⓐ Ⓑ Ⓒ Ⓓ Ⓔ

Diagnostic Test

Section I: Questions 1–40

SPEND 90 MINUTES ON THIS PART OF THE EXAM.

> **DIRECTIONS:** The questions or incomplete statements that follow are each followed by five suggested answers or completions. Choose the response that best answers the question or completes the statement.

1. It is estimated that 70 percent of Americans have pets. However, favorite pet of choice differs by geographic location. A random sample of pet owners is cross-classified by geographic location and pet of choice. The results are summarized in the following segmented bar chart.

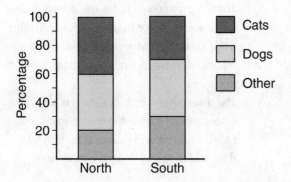

   Which of the following is an *incorrect* conclusion?

   (A) More pet owners in the South than in the North answered "Other."
   (B) Twice as many pet owners in the North answered "Dogs" than answered "Other."
   (C) The same number of pet owners in the South answered "Cats" as answered "Other."
   (D) In both the North and South, the same proportion of pet owners answered "Dogs."
   (E) A greater proportion of pet owners in the North than in the South answered "Cats."

2. Is there a linear relationship between calories and sodium content in beef hot dogs? A random sample of 20 beef hot dogs gives the following regression output:

```
Dependent variable is: Sodium
Predictor     Coef  SE Coef      T      P
Constant   -228.33    77.97  -2.93  0.009
Calories    4.0133   0.4922   8.15  0.000
S = 48.5799  R-Sq = 78.7%  R-Sq(adj) = 77.5%
```

   Which of the following gives a 99% confidence interval for the slope of the regression line?

   (A) $4.0133 \pm 2.861\left(\dfrac{0.4922}{\sqrt{20}}\right)$
   (B) $4.0133 \pm (2.861)(0.4922)$
   (C) $4.0133 \pm (2.878)(0.4922)$
   (D) $4.0133 \pm 2.861\left(\dfrac{48.5799}{\sqrt{20}}\right)$
   (E) $4.0133 \pm 2.878\left(\dfrac{48.5799}{\sqrt{20}}\right)$

3. In tossing a fair coin, which of the following sequences is most likely to appear?

   (A) HHHHH
   (B) HTHTHT
   (C) HTHHTTH
   (D) TTHTHHTH
   (E) All are equally likely.

4. There have been growing numbers of news stories about White Americans calling the police on people of color whose behavior is completely normal. A criminologist hypothesizes that the mean number of such incidents across the country is 2 per day. A sociologist believes the true mean is greater than 2 per day and plans a hypothesis test at the 5% significance level on a random sample of 50 days. For which of the following possible true values of μ will the power of the test be greatest?

(A) 1.5
(B) 1.85
(C) 2.0
(D) 2.15
(E) 2.4

5. A simple random sample is defined by

(A) the method of selection.
(B) how representative the sample is of the population.
(C) whether or not a random number generator is used.
(D) the assignment of different numbers associated with the outcomes of some chance situation.
(E) examination of the outcome.

6. Can shoe size be predicted from height? In a random sample of 50 teenagers, the standard deviation in heights was 8.7 cm, while the standard deviation in shoe size was 2.3. The least squares regression equation was:

Predicted shoe size $= -33.6 + 0.25(\text{Height in cm})$

What was r, the correlation coefficient?

(A) $\dfrac{(0.25)(8.7)}{2.3}$

(B) $\dfrac{(0.25)(2.3)}{8.7}$

(C) $\dfrac{2.3}{\left(\dfrac{8.7}{\sqrt{50}}\right)}$

(D) $\dfrac{8.7}{\left(\dfrac{2.3}{\sqrt{50}}\right)}$

(E) There is not enough information to calculate the correlation coefficient.

Questions 7–9 refer to the following situation:

A researcher would like to show that a new oral diabetes medication she developed helps control blood sugar level better than insulin injection. She plans to run a hypothesis test at the 5% significance level.

7. What would be a Type I error?

(A) The researcher concludes she has sufficient evidence that her new medication helps more than insulin injection, and her medication really is better than insulin injection.
(B) The researcher concludes she has sufficient evidence that her new medication helps more than insulin injection when in reality her medication is not better than insulin injection.
(C) The researcher concludes she does not have sufficient evidence that her new medication helps more than insulin injection, and her medication really is not better than insulin injection.
(D) The researcher concludes she does not have sufficient evidence that her new medication helps more than insulin injection when in reality her medication is better than insulin injection.
(E) The researcher concludes she has sufficient evidence that her new medication controls blood sugar level the same as insulin injection, and in reality there is a difference.

8. What would be a Type II error?

   (A) The researcher concludes she has sufficient evidence that her new medication helps more than insulin injection, and her medication really is better than insulin injection.
   (B) The researcher concludes she has sufficient evidence that her new medication helps more than insulin injection when in reality her medication is not better than insulin injection.
   (C) The researcher concludes she does not have sufficient evidence that her new medication helps more than insulin injection, and her medication really is not better than insulin injection.
   (D) The researcher concludes she does not have sufficient evidence that her new medication helps more than insulin injection when in reality her medication is better than insulin injection.
   (E) The researcher concludes she has sufficient evidence that her new medication controls blood sugar level the same as insulin injection, and in reality there is a difference.

9. The researcher thinks she can improve her chances by running five identical significance tests, each using a different group of volunteers with diabetes, hoping that at least one of the tests will show that her new oral diabetes medication helps control blood sugar level better than insulin injection. What is the probability of committing at least one Type I error?

   (A) 0.05
   (B) $5(0.05)(0.95)^4$
   (C) $1 - (0.95)^5$
   (D) $(0.95)^5$
   (E) 0.95

10. A financial analyst determines the yearly research and development investments for 50 blue chip companies. She notes that the distribution is distinctly not bell-shaped. If the 50 dollar amounts are converted to z-scores, what can be said about the standard deviation of the 50 z-scores?

    (A) It is less than the standard deviation of the raw scores.
    (B) It is greater than the standard deviation of the raw scores.
    (C) It is equal to the standard deviation of the raw scores.
    (D) It equals $\frac{\sigma}{\sqrt{50}}$ where σ is the population standard deviation of the raw scores.
    (E) It equals 1.

11. A coin is weighted so that the probability of heads is 0.6. The coin is tossed 20 times, and the number of heads is noted. This procedure is repeated a total of 200 times, and the number of heads is recorded each time. What kind of distribution has been simulated?

    (A) The sampling distribution of the sample proportion with $n = 20$ and $p = 0.6$
    (B) The sampling distribution of the sample proportion with $n = 200$ and $p = 0.6$
    (C) The sampling distribution of the sample proportion with $\bar{x} = (20)(0.6)$ and $\sigma = \sqrt{20(0.6)(0.4)}$
    (D) The binomial distribution with $n = 20$ and $p = 0.6$
    (E) The binomial distribution with $n = 200$ and $p = 0.6$

12. A 100-question multiple-choice history exam is graded as number correct minus $\frac{1}{4}$ number incorrect, so scores can take values from -25 to $+100$. Suppose the standard deviation for one class's results is reported to be -3.14. What is the proper conclusion?

    (A) More students received negative scores than positive scores.
    (B) At least half the class received negative scores.
    (C) Some students must have received negative scores.
    (D) Some students must have received positive scores.
    (E) An error was made in calculating the standard deviation.

13. Of the 423 seniors graduating this year from a city high school, 322 plan to go on to college. When the principal asks an AP student to calculate a 95% confidence interval for the proportion of this year's graduates who plan to go to college, the student says that this would be inappropriate. Why?

    (A) The independence assumption may have been violated (students tend to do what their friends do).
    (B) There is no evidence that the data come from a normal or nearly normal population (GPAs help determine college admission and may be skewed).
    (C) Randomization was not used.
    (D) There is a difference between a confidence interval and a hypothesis test with regard to the proportion of graduates planning on college.
    (E) The population proportion is known, so a confidence interval has no meaning.

14. An AP Statistics student in a large high school plans to survey his fellow students with regard to their preferences between using a laptop or using a tablet. Which of the following survey methods is unbiased?

    (A) The student comes to school early and surveys the first 50 students who arrive.
    (B) The student passes a survey card to every student with instructions to fill it out at home and drop the filled-out card in a box by the school entrance the next day.
    (C) The student creates an online survey and asks everyone to respond.
    (D) The student goes to all of the high school sports events for a week, hands out the survey, and waits for each student to fill it out and hand it back.
    (E) None of the above sampling methods are unbiased.

15. In a random sample of 500 students, it was reported that test grades went up an average of at least 10 points for 70 percent of the students when usage of cell phones was banned during the school day. What was the degree of confidence if the margin of error was ± 2.5 percent?

    (A) $P(-0.025 < z < 0.025)$
    (B) $P(0.675 < z < 0.725)$
    (C) $P\left(-\dfrac{0.025}{\sqrt{(0.5)(0.5)/500}} < z < \dfrac{0.025}{\sqrt{(0.5)(0.5)/500}}\right)$
    (D) $P\left(-\dfrac{0.025}{\sqrt{(0.7)(0.3)/500}} < z < \dfrac{0.025}{\sqrt{(0.7)(0.3)/500}}\right)$
    (E) $P\left(\dfrac{0.675}{\sqrt{(0.7)(0.3)/500}} < z < \dfrac{0.725}{\sqrt{(0.7)(0.3)/500}}\right)$

16. In a random sample of 10 insects of a newly discovered species, an entomologist measures an average life expectancy of 17.3 days with a standard deviation of 2.3 days. Assuming all conditions for inference are met, what is a 95% confidence interval for the mean life expectancy for insects of this species?

    (A) $17.3 \pm 1.96\left(\dfrac{2.3}{\sqrt{9}}\right)$
    (B) $17.3 \pm 1.96\left(\dfrac{2.3}{\sqrt{10}}\right)$
    (C) $17.3 \pm 2.228\left(\dfrac{2.3}{\sqrt{9}}\right)$
    (D) $17.3 \pm 2.228\left(\dfrac{2.3}{\sqrt{10}}\right)$
    (E) $17.3 \pm 2.262\left(\dfrac{2.3}{\sqrt{10}}\right)$

17. Time management and procrastination are difficult problems for college students. The distribution for the percentage of study time that occurs in the 24 hours prior to a final exam is approximately normal with mean 44 and standard deviation 21. Consider two different random samples taken from the population of college students, one of size 10 and one of size 100. Which of the following is true about the sampling distributions of the sample mean for the two sample sizes?

    (A) Both distributions are approximately normal with mean 44 and standard deviation 21.
    (B) Both distributions are approximately normal. Both the mean and the standard deviation for the $n = 10$ sampling distribution are greater than for the $n = 100$ distribution.
    (C) Both distributions are approximately normal with the same mean. The standard deviation for the $n = 10$ sampling distribution is greater than for the $n = 100$ distribution.
    (D) Only the $n = 100$ sampling distribution is approximately normal. Both distributions have mean 44 and standard deviation 10.
    (E) Only the $n = 100$ sampling distribution is approximately normal. Both distributions have the same mean. The standard deviation for the $n = 10$ sampling distribution is greater than for the $n = 100$ distribution.

18. One of the primary objectives of street lighting is the prevention of personal and property crime. One study of 30 randomly chosen metropolitan areas measured street lighting and crime in each area, each measured on a 10-point scale. Among the following, which is the best statistical test to use in analyzing this data?

    (A) Two-sample t-test of population means
    (B) Linear regression t-test
    (C) Chi-square test of independence
    (D) Chi-square test of homogeneity
    (E) Chi-square test of goodness of fit

19. A study of gun ownership and homicide rates in developed countries resulted in the following scatterplot and regression line.

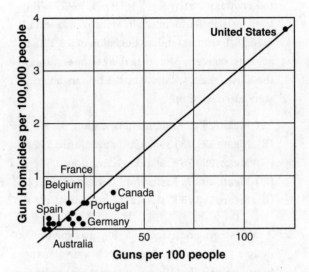

Which of the following is true about the point corresponding to the United States?

    (A) It has high leverage and is a regression outlier.
    (B) It has high leverage but is not a regression outlier.
    (C) It is a regression outlier but does not have high leverage.
    (D) It is not a regression outlier and does not have high leverage.
    (E) It has high leverage, but whether it is a regression outlier cannot be determined.

20. A campus has 55% male and 45% female students. Suppose 30% of the male students pick basketball as their favorite sport compared with 20% of the females. If a randomly chosen student picks basketball as the student's favorite sport, what is the probability the student is male?

    (A) $\dfrac{0.30}{0.30 + 0.20}$

    (B) $\dfrac{0.55}{0.30 + 0.20}$

    (C) $\dfrac{0.30}{(0.55)(0.30) + (0.45)(0.20)}$

    (D) $\dfrac{0.55}{(0.55)(0.30) + (0.45)(0.20)}$

    (E) $\dfrac{(0.55)(0.30)}{(0.55)(0.30) + (0.45)(0.20)}$

21. The kelvin is a unit of measurement for temperature; 0 K is absolute zero, the temperature at which all thermal motion ceases. Conversion from Fahrenheit to Kelvin is given by $K = \frac{5}{9} \times (F - 32) + 273$. The average daily temperature in Monrovia, Liberia, is 78.35°F with a standard deviation of 6.3°F. If a scientist converts Monrovia daily temperatures to the Kelvin scale, what will be the new mean and standard deviation?

   (A) Mean, 25.75 K; standard deviation, 3.5 K
   (B) Mean, 231.75 K; standard deviation, 3.5 K
   (C) Mean, 298.75 K; standard deviation, 3.5 K
   (D) Mean, 298.75 K; standard deviation, 258.72 K
   (E) Mean, 298.75 K; standard deviation, 276.5 K

22. A cattle veterinarian is considering two experimental designs to compare two sources of bovine growth hormone, or BVH, to spur increased milk production in Guernsey cattle. Design 1 involves flipping a coin as each cow enters the stockade, and if *heads*, giving it BVH from bovine cadavers, and if *tails*, giving it BVH from engineered *E. coli*. Design 2 involves flipping a coin as each cow enters the stockade, and if *heads*, giving it BVH from bovine cadavers for a specified period of time and then switching to BVH from engineered *E. coli* for the same period of time, and if *tails*, the order is reversed. With both designs, daily milk production is noted. Which of the following is accurate?

   (A) Neither design uses randomization since there is no indication that cows will be randomly picked from the population of all Guernsey cattle.
   (B) Design 1 is a completely randomized design, while Design 2 is a block design.
   (C) Both designs use double-blinding, but neither uses a placebo.
   (D) In the second design, BVH from bovine cadavers and BVH from engineered *E. coli* are confounded.
   (E) One of the two designs is actually an observational study, while the other is an experiment.

23. The purpose of the linear regression *t*-test is

   (A) to determine if there is a linear association between two numerical variables.
   (B) to find a confidence interval for the slope of a regression line.
   (C) to find the *y*-intercept of a regression line.
   (D) to be able to calculate residuals.
   (E) to be able to determine the consequences of Type I and Type II errors.

24. A fair die is tossed 12 times, and the number of 3's is noted. This is repeated 200 times. Which of the following distributions is the most likely to occur?

   (A)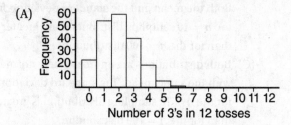

   (B)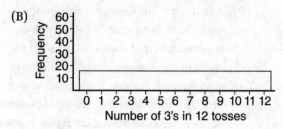

   (C)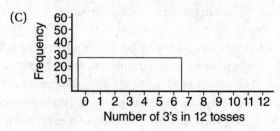

   (D)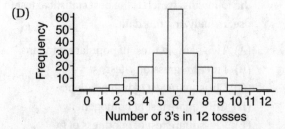

   (E)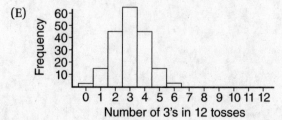

25. Which of the following is a true statement about sampling?

    (A) If the sample is random, the size of the sample usually doesn't matter.
    (B) If the sample is random, the size of the population usually doesn't matter.
    (C) A sample of less than 1% of the population is too small for statistical inference.
    (D) A sample of more than 10% of the population is too large for statistical inference.
    (E) All of the above are true statements.

26. Suppose, in a study of mated pairs of soldier beetles, it is found that the measure of the elytron (hardened forewing) length is always 0.5 millimeters longer in the female. What is the correlation between elytron lengths of mated females and males?

    (A) −1
    (B) −0.5
    (C) 0
    (D) 0.5
    (E) 1

27. A random sample of 100 individuals who were singled out at an international airport security checkpoint is reviewed, and the individuals are classified according to region of origin:

Region of Origin	United States	Europe	Arabic	Asia, non-Arabic	Other
Number singled out	41	19	15	13	12

The proportion of travelers in each category who use this airport follows:

Region of Origin	United States	Europe	Arabic	Asia, non-Arabic	Other
Proportion	0.64	0.12	0.08	0.09	0.07

We wish to test whether the distribution of people singled out is the same as the distribution of people who use the airport with regard to region of origin. What is the appropriate χ^2 statistic?

(A) $\frac{(41-64)^2}{64} + \frac{(19-12)^2}{12} + \frac{(15-8)^2}{8} + \frac{(13-9)^2}{9} + \frac{(12-7)^2}{7}$

(B) $\frac{(41-64)^2}{41} + \frac{(19-12)^2}{19} + \frac{(15-8)^2}{15} + \frac{(13-9)^2}{13} + \frac{(12-7)^2}{12}$

(C) $\frac{(0.41-0.64)^2}{0.64} + \frac{(0.19-0.12)^2}{0.12} + \frac{(0.15-0.08)^2}{0.08} + \frac{(0.13-0.09)^2}{0.09} + \frac{(0.12-0.07)^2}{0.07}$

(D) $\frac{(0.41-0.64)^2}{0.41} + \frac{(0.19-0.12)^2}{0.19} + \frac{(0.15-0.08)^2}{0.15} + \frac{(0.13-0.09)^2}{0.13} + \frac{(0.12-0.07)^2}{0.12}$

(E) $\frac{(41-64)^2}{20} + \frac{(19-12)^2}{20} + \frac{(15-8)^2}{20} + \frac{(13-9)^2}{20} + \frac{(12-7)^2}{20}$

28. The age distribution for a particular debilitating disease has a mean greater than the median. Which of the following graphs most likely illustrates this distribution?

(A)
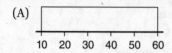

(B)
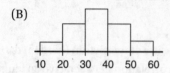

(C)
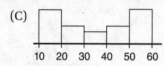

(D)
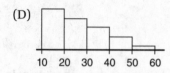

(E)
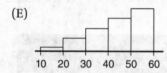

29. In a random sample of 10 people, the probability that k people show a particular genetic abnormality is given by $\binom{10}{k}(0.38)^k (0.62)^{10-k}$ for $k = 0, \ldots, 10$.

What is the mean of the associated random variable?

(A) 0.38
(B) 0.62
(C) 3.8
(D) 5.0
(E) 6.2

30. It is hypothesized that high school varsity pitchers throw fastballs at an average of 80 mph. A random sample of varsity pitchers is timed with radar guns resulting in a 95% confidence interval of (74.5, 80.5). Which of the following is a correct statement?

(A) There is a 95% chance that the mean fastball speed of all varsity pitchers is 80 mph.
(B) There is a 95% chance that the mean fastball speed of all varsity pitchers is 77.5 mph.
(C) Most of the interval is below 80, so there is evidence at the 5% significance level that the mean of all varsity pitchers is something other than 80 mph.
(D) The test $H_0: \mu = 80$, $H_a: \mu \neq 80$ is not significant at the 5% significance level, but it would be at the 1% level.
(E) It is likely that the true mean fastball speed of all varsity pitchers is within 3 mph of the sample mean fastball speed.

31. A recent study noted prices and battery lives of 10 top-selling tablet computers. The data follow:

	1	2	3	4	5	6	7	8	9	10
Cost	303	450	260	480	540	390	350	400	600	450
Battery life (hr)	8.5	10	7	11	10	9	8	9.5	11	9.5

The residual plot of the least squares model is

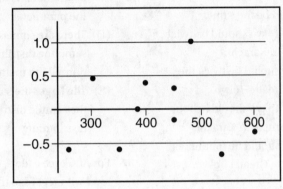

What is the model's predicted battery life for the tablet computer costing $480?

(A) 10 hr

(B) 10.5 hr

(C) 11 hr

(D) 11.5 hr

(E) 12 hr

32. Should college athletes be required to give their coaches their social media account user names and passwords? A survey of student-athletes is to be taken. The statistician believes that Division I, II, and III players may differ in their views, so she selects a random sample of athletes from each division to survey. This is a

(A) simple random sample.

(B) stratified sample.

(C) cluster sample.

(D) systematic sample.

(E) convenience sample.

33. In a random sample of 1,500 college students, a pollster found 45% prefer a female president and 42% prefer a male president. To calculate a 95 percent confidence interval for the difference in the proportion of college students who prefer a female president over a male president, he uses

$(0.45 - 0.42) \pm 1.96 \sqrt{\dfrac{(0.45)(0.55)}{1,500} + \dfrac{(0.42)(0.58)}{1,500}}$. Were conditions for inference met?

(A) Yes, because there was a random sample, 1,500 is less than 10% of all college students, and 1,500(0.45), 1,500(0.55), 1,500(0.42), and 1,500(0.58) are all ≥ 10.

(B) No, because there was no random assignment between female and male college students.

(C) No, because 1,500 is not greater than 10% of all college students.

(D) No, because the independence assumption is violated.

(E) No, because the random sample may not be truly representative of the population.

34. The population of the Greater Tokyo area is 34,400,000 and of Karachi is 17,200,000. A random sample of citizens is to be taken in each city, and 95% confidence intervals for the mean age in each city will be calculated. Assuming roughly equal sample standard deviations, to obtain the same margin of error for each confidence interval,

    (A) the sample sizes should be the same.
    (B) the sample in Greater Tokyo should be twice the size of the sample in Karachi.
    (C) the sample in Karachi should be twice the size of the sample in Greater Tokyo.
    (D) the sample in Greater Tokyo should be four times the size of the sample in Karachi.
    (E) the sample in Karachi should be four times the size of the sample in Greater Tokyo.

35. A truant officer determines the mean and standard deviation of the number of student absences for all school days during an academic year. Which of the following is the best description of the standard deviation?

    (A) Approximately the median difference between the number of students absent on individual days and the median number of absences on all days
    (B) Approximately the mean difference between the number of students absent on individual days and the mean number of absences on all days
    (C) The difference between the greatest number and the least number of absences among all days during the year
    (D) The difference between the greatest number and the mean number of absences among all days during the year
    (E) The difference between the greatest number and the least number of absences among the middle 50 percent of all the daily absences during the year

36. Which of the following is an *incorrect* statement?

    (A) Statistics are random variables with their own probability distributions.
    (B) The standard error does not depend on the size of the population.
    (C) Bias means that, on average, our estimate of a parameter is different from the true value of the parameter.
    (D) There are some statistics for which the sampling distribution is not approximately normal, no matter how large the sample size.
    (E) The larger the sample size, the closer the sample distribution is to a normal distribution.

37. For Air Force cadets, the recommended fitness level with regard to the number of push-ups is 34. In a test whether or not current classes of recruits can meet this standard, a *t*-test of $H_0: \mu = 34$ against $H_a: \mu < 34$ gives a *P*-value of 0.068. Using this data, among the following, which is the largest level of confidence for a two-sided confidence interval that does not contain 34?

    (A) 85%
    (B) 90%
    (C) 92%
    (D) 95%
    (E) 96%

38. A particular car is tested for stopping distance in feet on wet pavement at 30 mph using tires with one tread design and then tires with another tread design. For each set of tires, the test is repeated 30 times, and the following parallel boxplots give a comparison of the resulting five-number summaries.

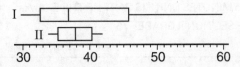

Which of the following is a reasonable conclusion?

(A) Distribution I is skewed right, while distribution II is bell-shaped.
(B) Distribution I is skewed left, while distribution II is a normal distribution.
(C) The mean of distribution I is greater than the mean of distribution II.
(D) The range of distribution I is approximately $46 - 33 = 13$.
(E) The upper 50% of the values in distribution I are all greater than the lower 50% of the values in distribution II.

39. E and F are events on the same probability space with $P(E) = 0.3$ and $P(F) = 0.4$. What is the relationship between $P(E \cap F)$ and 0.12?

(A) $P(E \cap F) > 0.12$
(B) $P(E \cap F) < 0.12$
(C) $P(E \cap F) = 0.12$
(D) $P(E \cap F) \neq 0.12$
(E) There is not enough information to determine a relationship between $P(E \cap F)$ and 0.12.

40. Do middle school and high school students have different views on what makes someone popular? Random samples of 100 middle school and 100 high school students yield the following counts with regard to three choices: lots of money, good at sports, and good looks:

What makes a Student Popular			
	Money	Sports	Looks
Middle school students	22	48	30
High school students	36	24	40

A chi-square test of homogeneity yields which of the following test statistics?

(A) $\dfrac{(22-29)^2}{22} + \dfrac{(48-36)^2}{48} + \dfrac{(30-35)^2}{30} + \dfrac{(36-29)^2}{36} + \dfrac{(24-36)^2}{24} + \dfrac{(40-35)^2}{40}$

(B) $\dfrac{(22-29)^2}{29} + \dfrac{(48-36)^2}{36} + \dfrac{(30-35)^2}{35} + \dfrac{(36-29)^2}{29} + \dfrac{(24-36)^2}{36} + \dfrac{(40-35)^2}{35}$

(C) $\left(\dfrac{22}{29}\right)^2 + \left(\dfrac{48}{36}\right)^2 + \left(\dfrac{30}{35}\right)^2 + \left(\dfrac{36}{29}\right)^2 + \left(\dfrac{24}{36}\right) + \left(\dfrac{40}{35}\right)^2$

(D) $\dfrac{(22)(29)}{58} + \dfrac{(48)(36)}{72} + \dfrac{(30)(35)}{70} + \dfrac{(36)(29)}{58} + \dfrac{(24)(36)}{72} + \dfrac{(40)(35)}{70}$

(E) $\sqrt{(22-29)^2 + (48-36)^2 + (30-35)^2 + (36-29)^2 + (24-36)^2 + (40-35)^2}$

STOP If there is still time remaining, you may check your work on this section.

Section II: Part A

Questions 1–5

SPEND ABOUT 65 MINUTES ON THIS PART OF THE EXAM.

PERCENTAGE OF SECTION II GRADE—75

> **DIRECTIONS:** You must show all work and indicate the methods you use. You will be graded on the correctness of your methods and on the accuracy of your results and explanations.

1. A horticulturist plans a study on the use of compost tea for plant disease management. She obtains 16 identical beds, each containing a random selection of five minipink rose plants. She plans to use two different composting times (two and five days), two different compost preparations (aerobic and anaerobic), and two different spraying techniques (with and without adjuvants). Midway into the growing season she will check all plants for rose powdery mildew disease.

    (a) List the complete set of treatments.

    (b) Describe a completely randomized design for the treatments above.

    (c) Explain the advantage of using only minipink roses in this experiment.

    (d) Explain a disadvantage of using only minipink roses in this experiment.

2. A top-100, 7.0-rated tennis pro wishes to compare a new racket against his current model. He is interested in whether there is statistical evidence that his hitting speed with the new racket is different than that with his old racket. He strings the new racket with the same type of strings at 60 pounds tension that he uses on his old racket. From past testing, he knows that the average forehand crosscourt volley with his old racket is 82 miles per hour (mph). On an indoor court, using a ball machine set at 70 mph, which is the same speed he had his old racket tested against, he takes 47 swings with the new racket. An associate with a speed gun records an average of 83.5 mph with a standard deviation of 3.4 mph.

    (a) Was random sampling, random assignment, both, or neither involved in this study? Explain.

    (b) What is the parameter of interest, and what are the appropriate hypotheses for testing whether his hitting speed with the new racket is different than that with his old racket?

    (c) What inference procedure should be used?

    (d) Is the normality condition satisfied? Explain.

    (e) If all conditions are assumed to be satisfied, what is the test statistic and the P-value?

    (f) At a significance level of 0.01, state an appropriate conclusion in context.

    (g) A 99% confidence interval based on this data is (82.167, 84.833). Interpret this interval in context.

    (h) Is this confidence interval consistent with the test decision in part (f)? Explain.

3. On the social media platform Snapchat, users who contact each other once per day develop a Snapstreak. The Snapstreaks of students at a large suburban high school have an approximately normal distribution with mean 26.4 and standard deviation 8.2.

    (a) What is the probability that a given Snapstreak is over 30?

    (b) In a random sample of five independent Snapstreaks of students at this school, what is the probability that a majority are over 30?

    (c) What is the probability that the mean of the five independent Snapstreaks is over 30?

4. Two hundred fifty randomly chosen people raised in the United States were asked their expression for "soft drink" and the geographic region where they were raised. The results are summarized in the following two-way table.

Expression for "soft drink"	Geographic Region within the U.S.			
	Northeast	Midwest	West	South
Coke	10	5	5	30
Soda	50	20	40	5
Pop	20	50	10	5

(a) Of the people calling soft drinks "soda," what proportion were from the Northeast?

(b) Of the people from the West, what proportion called soft drinks "pop"?

(c) The mosaic plot below displays the distribution of soft drink expressions given by people from different geographic locations. Describe what this plot reveals about the association between these two variables for the 250 people in the study.

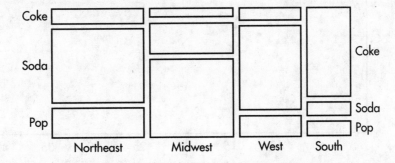

5. Researchers want to study whether a capuchin monkey named Rafiki can correctly predict (better than guess-work) whether the Dow Jones Industrial Average (DJIA) will go up or down on any given day. In a random sample of 10 days, Rafiki correctly predicted the rise or fall of the DJIA 7 of the 10 days (by choosing to eat a banana from a box with an "up" arrow or a box with a "down" arrow).

(a) What are the null and alternative hypotheses and the parameter of interest?

(b) If you conduct a simulation to investigate whether the observed result provides strong evidence that Rafiki can correctly predict the rise or fall of the DJIA, what would you use for the probability of success, for the sample size, and for the number of samples?

(c) Which of the following graphs could reasonably have come from the simulation?

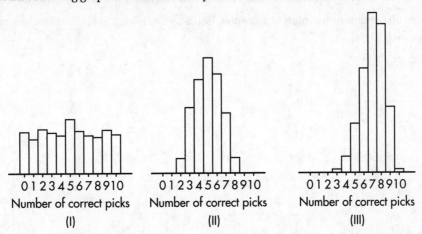

(d) Which of 6.5, 0.65, 0.15, and 0.015 is closest to the P-value? Interpret it in context.

(e) Make a conclusion based on the simulation and P-value.

Section II: Part B

Question 6

SPEND ABOUT 25 MINUTES ON THIS PART OF THE EXAM.
PERCENTAGE OF SECTION II GRADE—25

6. A college counselor is interested in whether or not the number of hours a student studies per week has a statistically significant linear association to the student's GPA. She takes a random sample of 25 students and, for each, records weekly study hours versus GPA on an anonymous survey. The resulting scatterplot along with regression output for a linear model is shown below.

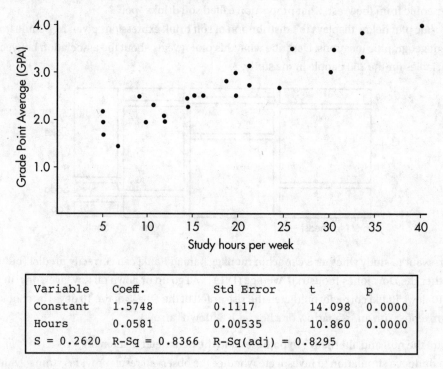

Variable	Coeff.	Std Error	t	p
Constant	1.5748	0.1117	14.098	0.0000
Hours	0.0581	0.00535	10.860	0.0000

S = 0.2620   R-Sq = 0.8366   R-Sq(adj) = 0.8295

(a) Interpret the *y*-intercept in context.

(b) Interpret the slope in context.

(c) What proportion of the variation in GPAs is not accounted for by the linear regression model?

(d) Assuming all conditions for inference are met, find a 95% confidence interval for the slope, and interpret this in context.

A math professor further analyzes the data and notes that 10 of the students took AP Statistics in high school. The modified scatterplot showing this additional information is shown below.

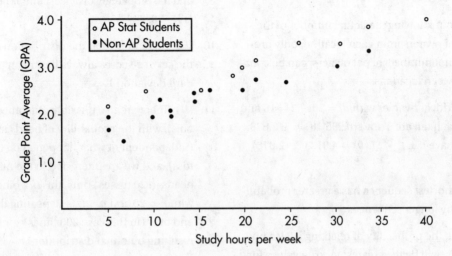

Using the linear regression model from the original analysis, the professor calculates that the average residual from the students who took AP Statistics is 0.2414 while the average residual from the students who did not take AP Statistics is −0.1609.

(e) Use the residual calculations to estimate how much greater the GPA for a student who takes AP Statistics in high school would be, on average, than the GPA for a student who does not take AP Statistics if the students study the same number of hours per week.

The professor then creates two regression models, one for the students who take AP Statistics and one for the other students. The resulting regression equations are shown below.

Linear Fit for Students who take AP Statistics in High School:

    Predicted GPA = 1.9427 + 0.0524(Hours)

Linear Fit for Students who do not take AP Statistics in High School:

    Predicted GPA = 1.5396 + 0.0502(Hours)

(f) A 95% confidence interval for the true difference in the two slopes is 0.0022 ± 0.0134. Based on this interval, is there a significant difference in the two slopes? Explain.

(g) A 95% confidence interval for the true difference in the two y-intercepts is 0.4031 ± 0.2767. Based on this interval, is there a significant difference in the two y-intercepts? Explain.

STOP   If there is still time remaining, you may check your work on this section.

Answer Explanations

Section I

1. **(A)** Without knowing the actual number of the sample pet owners from each location, only proportions, not numbers of pet owners, can be compared between locations.

2. **(C)** The critical t-values with $df = 20 - 2 = 18$ and 0.005 in each tail are $\pm \text{invT}(0.995,18) = \pm 2.878$. Thus, we have $b \pm t^* \times SE(b) = 4.0133 \pm (2.878)(0.4922)$.

3. **(A)** The shortest sequence has a greater probability than any longer sequence.

4. **(E)** Power, the probability of rejecting a false null hypothesis, will be the greatest for true values furthest from the hypothesized value in the direction of the alternative hypothesis. Here, $H_a: \mu > 2.0$, and 2.4 is furthest from 2.0 among the value choices that are greater than 2.0.

5. **(A)** A simple random sample may or may not be representative of the population. It is a method of selection in which every possible sample of the desired size has an equal chance of being selected.

6. **(A)** A formula relating the given statistics is $b = r\dfrac{S_y}{S_x}$, which in this case gives $0.25 = r\dfrac{2.3}{8.7}$ and thus $r = \dfrac{(0.25)(8.7)}{2.3}$.

7. **(B)** The null hypothesis is that the new medication is no better than insulin injection, while the alternative hypothesis is that the new medication is better. A Type I error means a mistaken rejection of a true null hypothesis.

8. **(D)** A Type II error means a mistaken failure to reject a false null hypothesis. A false null hypothesis here means that her medication really is better than insulin injection, and failure to realize this means she does not have sufficient evidence that it is better.

9. **(C)** Running a hypothesis test at the 5% significance level means that the probability of committing a Type I error is 0.05. Then the probability of not committing a Type I error is 0.95. Assuming the tests are independent, the probability of not committing a Type I error on any of the five tests is $(0.95)^5$,

and the probability of at least one Type I error is $1 - (0.95)^5$.

10. **(E)** No matter what the distribution of raw scores, the set of z-scores always has mean 0 and standard deviation 1.

11. **(D)** There are two possible outcomes (heads and tails), with the probability of heads always 0.6 (independent of what happened on the previous toss), and we are interested in the number of heads in 20 tosses. Thus, this is a binomial model with $n = 20$ and $p = 0.6$. Repeating this over and over (in this case 200 times) simulates the resulting binomial distribution.

12. **(E)** The standard deviation can never be negative.

13. **(E)** When a complete census is taken (all 423 seniors were in the study), the population proportion is known and a confidence interval has no meaning.

14. **(E)** The method described in (A) is a convenience sample, (B) and (C) are voluntary response surveys, and (D) suffers from undercoverage bias.

15. **(D)** $SE(\hat{p}) = \sqrt{(0.7)(0.3)/500}$. Then $z\, SE(\hat{p}) = 0.025$ gives critical z-scores of $\pm\dfrac{0.025}{SE(\hat{p})}$. The degree of confidence is the probability between these two values.

16. **(E)** With $df = n - 1 = 10 - 1 = 9$ and 95% confidence, the critical t-values are $\pm\text{invT}(0.975, 9) = \pm 2.262$. $SE(\bar{x}) = \dfrac{s}{\sqrt{n}} = \dfrac{2.3}{\sqrt{10}}$. The confidence interval is $\bar{x} \pm t^*SE(\bar{x}) = 17.3 \pm 2.262\left(\dfrac{2.3}{\sqrt{10}}\right)$.

17. **(C)** Because the population is approximately normal, normality can be assumed for both sampling distributions. Both will have mean 44. The standard deviation for the $n = 10$ sampling distribution is $\dfrac{21}{\sqrt{10}}$, which is greater than the standard deviation for the $n = 100$ sampling distribution, which is $\dfrac{21}{\sqrt{100}}$.

18. **(B)** The chi-square tests all involve counts, and comparing means doesn't make sense in this context. We can plot (street lighting score, crime score) for the 30 metropolitan areas, look for a pattern in the scatterplot, and perform a linear regression t-test.

19. **(B)** The point has high leverage because its x-value is far from the mean of the x-values. It is not a regression outlier because it has a small residual when compared with the other residuals.

20. **(E)**

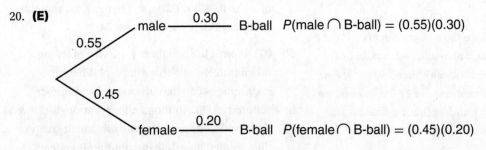

$$P(\text{B-ball}) = (0.55)(0.30) + (0.45)(0.20)$$

$$P(\text{male} \mid \text{B-ball}) = \frac{P(\text{male} \cap \text{B-ball})}{P(\text{B-ball})} = \frac{(0.55)(0.30)}{(0.55)(0.30) + (0.45)(0.20)}$$

21. **(C)** Adding the same constant to every value in a set adds the same constant to the mean but leaves the standard deviation unchanged. Multiplying every value in a set by the same constant multiplies the mean and standard deviation by that constant. So the new mean is $\frac{5}{9} \times (78.35 - 32) + 273 = 298.75$, and the new standard deviation is $\frac{5}{9} \times 6.3 = 3.5$.

22. **(B)** Design 2 is an example of a matched pairs design, a special case of a block design; here, each subject is compared with itself with respect to the two treatments. Both designs definitely use randomization with regard to assignment of treatments, but since they do not use randomization in selecting subjects from the general population, care must be taken in generalizing any conclusions. It's not clear whether or not the researchers who do the observations and measurements know which treatment individual cows are receiving, so there is no way to conclude if there is or is not blinding. The two sources of BVH are different treatments, and so they are not being confounded. In both designs, treatments are randomly applied, so neither is an observational study.

23. **(A)** The linear regression t-test generally has null hypothesis H_0: $\beta = 0$ that there is no linear relationship; if the P-value is small enough, there is evidence of a linear association; that is, there is evidence that $\beta \neq 0$.

24. **(A)** We have a binomial distribution with $n = 12$ and $p = \frac{1}{6}$. With mean $= np = 12\left(\frac{1}{6}\right) = 2$, choice (A) is the only reasonable choice.

25. **(B)** The size of the sample always matters: the larger the sample, the greater the power of statistical tests. One percent of a large population is large. Larger samples are better, but if the sample is greater than 10% of the population, the best statistical techniques are not those covered in the AP curriculum. While not in the AP curriculum, if the population is small, its size may matter as shown by the finite population correction factor.

26. **(E)** The points on the scatterplot all fall on the straight line:

$$\text{Female length} = \text{Male length} + 0.5$$

27. **(A)** $\chi^2 = \sum \dfrac{(\text{observed} - \text{expected})^2}{\text{expected}}$. Expected values are found by multiplying the proportions times the sample size of 100.

28. **(D)** The distributions in (A), (B), and (C) appear roughly symmetric, so the mean and median will be roughly the same. The distribution in (D) is skewed to the right, so the mean will most likely be greater than the median, while the distribution in (E) is skewed to the left, so the mean will most likely be less than the median.

29. **(C)** This is a binomial with $n = 10$ and $p = 0.38$, so the mean is $np = 10(0.38) = 3.8$.

30. **(E)** Choices (A), (B), and (C) are common misconceptions. Since the 95% confidence interval contains 80, a two-sided test would not be significant at the 5% significance level or lower. The interval can be expressed as 77.5 ± 3. In other words, we are 95% confident that the true mean fastball speed is within 3 mph of 77.5 mph.

31. **(A)** Residual = observed − predicted, so $1.0 = 11 - $ predicted *and* predicted $= 10$.

32. **(B)** Stratified sampling is when the population is divided into homogeneous groups (the three divisions in this example) and a random sample of individuals is chosen from each group.

33. **(D)** The two-sample z-interval for a difference in proportions is inappropriate here, because the two proportions do not come from *independent* samples.

34. **(A)** The margin of error, $\pm t^* \frac{s}{\sqrt{n}}$, depends on n, the sample size, not the population size.

35. **(B)** The standard deviation is an approximate average measure of how far individual values of a set differ from the mean of the set.

36. **(E)** The larger the sample size, the closer the sample distribution is to the population distribution. The central limit theorem roughly says that if multiple samples of size n are drawn randomly and independently from a population, the histogram of the means of those samples will be approximately normal. Statistics have probability distributions called sampling distributions. The standard error is based on the spread (variability) of the sample and on the sample size. The central limit theorem does not apply to all statistics as it does to sample means. Many sampling distributions are not normal; for example, the sampling distribution of the sample max is not a normal distribution. An estimator of a parameter is unbiased if we have a method that, through repeated samples, is on average the same value as the parameter.

37. **(A)** With 0.068 in a tail, the confidence interval with 34 at one end would have a confidence level of $1 - 2(0.068) = 0.864$, so anything higher than 86.4% confidence will contain 34.

38. **(C)** From a boxplot there is no way of telling if a distribution is bell-shaped. (Very different distributions can have the same five-number summary.) Distribution I appears strongly skewed right, and so its mean is probably much greater than its median, while distribution II appears roughly symmetric, and so its mean is probably close to its median. The interquartile range, not the range, in distribution I is 13.

39. **(E)** If E and F are independent, then it would follow that $P(E \cap F) = (0.3)(0.4) = 0.12$, but without further information, all that can be said is that $P(E \cap F)$ is between 0 and 0.3.

40. **(B)** $\chi^2 = \sum \frac{(\text{observed} - \text{expected})^2}{\text{expected}}$ and cell calculations [expected value of a cell equals (row total)(column total)/(table total)] or χ^2-test on a calculator such as the TI-84, Casio Prizm, or HP Prime will yield expected cells of 29, 36, 35, 29, 36, and 35.

Section II: Part A

1. (a) There are $2 \times 2 \times 2 = 8$ different treatments:

   Two-day, aerobic, with adjuvant
   Two-day, aerobic, without adjuvant
   Two-day, anaerobic, with adjuvant
   Two-day, anaerobic, without adjuvant
   Five-day, aerobic, with adjuvant
   Five-day, aerobic, without adjuvant
   Five-day, anaerobic, with adjuvant
   Five-day, anaerobic, without adjuvant

   (b) We must randomly assign the treatment combinations to the beds. (Roses have already been randomly assigned to the beds.) With 8 treatments and 16 beds, each treatment should be assigned to 2 beds. For example, give each bed a random number between 1 and 16 (use a computer to generate 16 random integers between 1 and 16 *without replacement*), and then assign the first treatment in the above list to the beds with the numbers 1 and 2, assign the second treatment in the above list to the beds with the numbers 3 and 4, and so on.

   (c) Using only minipink roses in this experiment gives reduced variability and increases the likelihood of determining differences among the treatments.

   (d) Using only minipink roses in this experiment limits the scope and makes it difficult to generalize the results to other species of roses.

2. (a) There is no indication that any kind of randomization was used. (For example, he could have taken swings on a random sampling of days to take into account variation in his strength or concentration on different days.)

   (b) The parameter of interest is $\mu =$ his true mean hitting speed with the new racket. The hypotheses are $H_0: \mu = 82$ and $H_a: \mu \neq 82$.

   (c) One-sample t-test for a population mean.

   (d) Yes, the normality condition is satisfied because the sample size $n = 47 \geq 30$. This is sufficiently large for the central limit theorem to apply, which gives approximate normality for the sampling distribution of the sample mean.

   (e) Calculator software gives $t = 3.0246$ and $P = 0.00406$.

   (f) With this small of a P-value, $0.00406 < 0.01$, there is sufficient evidence to reject the null hypothesis, that is, there is convincing evidence that his hitting speed with the new racket is different than that with his old racket.

   (g) We are 99% confident that the true hitting speed with the new racket is between 82.167 and 84.833 mph.

   (h) Yes, this is consistent with the answer in part (f) because 82 is not in the interval (82.167, 84.833).

> **TIP**
>
> Graders want to give you credit. Help them! Make them understand *what* you are doing, *why* you are doing it, and *how* you are doing it. Don't make the reader guess at what you are doing. *Communication* is just as important as statistical knowledge!

SCORING

This question is scored in three sections. Section 1 consists of parts (a), (b), (c), and (d). Section 2 consists of parts (e) and (f). Section 3 consists of parts (g) and (h).

Section 1 is essentially correct for (1) noting there is no randomization, (2) the correct parameter, (3) the correct hypotheses, (4) the correct procedure, and (5) justifying the normality condition. Section 1 is partially correct if the response satisfies 3 or 4 of these 5 components.

Section 2 is essentially correct for (1) the correct test statistic, (2) the correct P-value, (3) the correct decision to reject the null hypothesis, (4) linkage to the P-value in the conclusion, and (5) giving the conclusion in context with reference to the population, not the sample. Section 2 is partially correct if the response satisfies 3 or 4 of these 5 components.

Section 3 is essentially correct for (1) a correct interpretation of the confidence interval and (2) answering in the affirmative about consistency with reference to 82 not being in the confidence interval. Section 3 is partially correct if the response satisfies 1 of these 2 components.

4	**Complete Answer**	All three sections essentially correct.
3	**Substantial Answer**	Two sections essentially correct and one section partially correct.
2	**Developing Answer**	Two sections essentially correct OR one section essentially correct and one or two sections partially correct OR all three sections partially correct.
1	**Minimal Answer**	One section essentially correct OR two sections partially correct.

3. (a) Normal distribution with $\mu = 26.4$ and $\sigma = 8.2$.

$$P(x > 30) = P\left(z > \frac{30 - 26.4}{8.2}\right) = P(z > 0.4390) = 0.3303$$

TIP

Full credit would also be given in 3(a) for: Normal, $\mu = 26.4$, $\sigma = 8.2$, $P(x > 30) = 0.330$.

(b) $B(n = 5, p = 0.3303)$, and $1 - \texttt{binomcdf(5, 0.3303, 2)} = 0.2054$
[or P(at least 3 out of 5 are > 30) $= 10(0.3303)^3(0.6697)^2 + 5(0.3303)^4(0.6697) + (0.3303)^5 = 0.2054$]

(c) The distribution of $\bar{x}$ is normal with mean $\mu_{\bar{x}} = 26.4$ and standard deviation

$$\sigma_{\bar{x}} = \frac{8.2}{\sqrt{5}} = 3.667. \text{ So, } P(\bar{x} > 30) = P\left(z > \frac{30 - 26.4}{3.667}\right) = P(z > 0.9817) = 0.1631.$$

TIP

Full credit would also be given in 3(b) for: Binomial, $n = 5$, $p = 0.3303$, $P(x \geq 3) = 0.2054$.

TIP

Full credit would also be given in 3(c) for: Normal, $\mu_{\bar{x}} = 26.4$, $\sigma_{\bar{x}} = \frac{8.2}{\sqrt{5}}$, 3.667, $P(\bar{x} > 30) = 0.1631$.

SCORING

Part (a) is essentially correct if the distribution is named (normal), the parameters are given (mean and standard deviation), the cutoff with direction is indicated (>30), and the correct probability is calculated. Simply writing normalcdf(30,∞,26.4,8.2) = 0.3303 is a partially correct response; however, writing normalcdf(lower bound = 30, upper bound = ∞, mean = 26.4, standard deviation = 8.2) = 0.3303 is essentially correct.

Part (b) is essentially correct if the correct probability is calculated and the derivation is clear. Part (b) is partially correct for indicating a binomial with $n = 5$ and $p =$ answer from (a) but calculating incorrectly. Simply writing 1 = binomcdf(5,0.3303,2) = 0.2054 without designating the parameters is also a partially correct response.

Part (c) is essentially correct for specifying a normal distribution and both $\mu_{\bar{x}}$ and $\sigma_{\bar{x}}$ and for correctly calculating the probability. Part (c) is partially correct for specifying a normal distribution and both $\mu_{\bar{x}}$ and $\sigma_{\bar{x}}$ but incorrectly calculating the probability, or for failing to specify a normal distribution and both $\mu_{\bar{x}}$ and $\sigma_{\bar{x}}$ but correctly calculating the probability.

4	**Complete Answer**	All three parts essentially correct.
3	**Substantial Answer**	Two parts essentially correct and one part partially correct.
2	**Developing Answer**	Two parts essentially correct OR one part essentially correct and one or two parts partially correct OR all three parts partially correct.
1	**Minimal Answer**	One part essentially correct OR two parts partially correct.

4. (a) $\dfrac{50}{50 + 20 + 40 + 5} = \dfrac{50}{115}$

    (b) $\dfrac{10}{5 + 40 + 10} = \dfrac{10}{55}$

    (c)   Based upon the mosaic plot, there is an association between the geographic location where someone is raised and what they call a soft drink. In this study, most people from the Northeast called soft drinks "soda," most people from the Midwest called soft drinks "pop," most people from the West called soft drinks "soda," and most people from the South called soft drinks "Coke."

SCORING

Part (a) is essentially correct for the expression $\dfrac{50}{115}$ and partially correct if only 0.435 without showing where this came from.

Part (b) is essentially correct for the expression $\dfrac{10}{55}$ and partially correct if only 0.182 without showing where this came from.

Part (c) is essentially correct for correctly noting the observed preferences for each geographic location and partially correct for noting only two or three of the preferences.

4	Complete Answer	All three parts essentially correct.
3	Substantial Answer	Two parts essentially correct and one part partially correct.
2	Developing Answer	Two parts essentially correct OR one part essentially correct and one or two parts partially correct OR all three parts partially correct.
1	Minimal Answer	One part essentially correct OR two parts partially correct.

5. (a) H_0: Rafiki can do no better than guesswork in predicting winners, H_0: $p = 0.5$.
        H_a: Rafiki can correctly predict winners better than guesswork, H_a: $p > 0.5$. The parameter of interest is $p =$ the true proportion of times Rafiki can correctly predict whether the DJIA will go up or down on any given day.

    (b) $p = 0.5$, $n = 10$, and a large number for number of samples.

    (c) Graph (II). The simulation should be centered at the value based on the null hypothesis probability, not the observed value.

    (d) Looking at areas under (II), the correct graph, among the given choices, 7 or greater correct picks seems to have a probability closest to 0.15. If the null hypothesis were true, that is, if Rafiki can do no better than guesswork, the estimated probability of Rafiki correctly predicting 7 or more out of 10 rises or falls of the DJIA is about 0.15.

    (e) With this large of a P-value, $0.15 > 0.05$, there is not sufficient evidence to reject H_0; that is, there is not convincing evidence that Rafiki can pick the rise and fall of the DJIA better than guesswork.

SCORING

This question is scored in four sections. Section 1 consists of part (a). Section 2 consists of parts (b) and (c). Section 3 consists of part (d). Section 4 consists of part (e).

Section 1 is essentially correct for correctly stating the hypotheses and correctly naming the parameter and is partially correct for giving one of these two components correctly.

Section 2 is essentially correct for $p = 0.5$, $n = 10$, any large number for number of samples, and picking Graph (II), and is partially correct for giving two or three of these four components correctly.

Section 3 is essentially correct for correctly picking 0.15 and giving a correct interpretation of the P-value and is partially correct for correctly picking 0.15 but giving a weak interpretation of the P-value.

Section 4 is essentially correct for a correct conclusion in context with linkage to the simulated P-value and is partially correct if only missing linkage or only missing context.

Count partially correct answers as one-half an essentially correct answer.

4	**Complete Answer**	Four essentially correct answers.
3	**Substantial Answer**	Three essentially correct answers.
2	**Developing Answer**	Two essentially correct answers.
1	**Minimal Answer**	One essentially correct answer.

Use a holistic approach to decide a score totaling between two numbers.

Section II: Part B

6. (a) **Think:** Remember, this has to be referred to as a predicted value or an average value when $x = 0$, where x in this problem is weekly study hours.

   Answer: The y-intercept, 1.5748, refers to a student who studies 0 hours per week. Thus, a student who doesn't study any hours at all is predicted to have a GPA of 1.5748.

   (b) **Think:** Again, you have to refer to a predicted or an average increase in the y-value for each unit increase in the x-value.

   Answer: The slope is 0.0581. This means that for each additional hour of study per week, the predicted GPA increases by 0.0581. In other words, this model predicts that the GPA increases by an average of 0.0581 for each additional hour of study per week.

   (c) **Think:** The coefficient of determination, R-Sq, gives the proportion of the variation in the y-variable that *is* accounted for by the linear regression model. So the proportion *not* accounted for by the regression model must be $1 -$ R-Sq.

   Answer: $1 - 0.8366 = 0.1634$

   (d) **Think:** This is the easiest confidence interval to find because the computer printout gives the standard error of the sample slope as well as the sample slope. Remember that for linear regression inference, $df = n - 2$.

   Answer: With $df = 25 - 2 = 23$ and 0.025 in each tail, the critical t-values are $\pm$invT(0.975, 23) $= \pm 2.0687$. The 95% confidence interval of the true slope is $b \pm ts_b = 0.0581 \pm 2.0687(0.00535) = 0.0581 \pm 0.0111$. We are 95% confident that for every additional study hour per week, the average increase in GPA is between 0.0470 and 0.0692.

   (e) **Think:** The residual averages suggest that the regression line tends to underestimate the GPA of students who take AP Statistics in high school by 0.2414 and to overestimate the GPA of students who do not take AP Statistics in high school by 0.1609. The difference between these two residual averages is what we are looking for!

   Answer: We calculate $0.2414 - (-0.1609) = 0.4023$. Thus for two students, one of whom had taken AP Statistics in high school, who study the same number of hours weekly, the student who had taken AP Statistics in high school would be estimated to have a GPA 0.4023 higher than the student who had not taken AP Statistics in high school.

   (f) **Think:** If the interval is entirely above 0 or entirely below 0, there is statistical significance, but if 0 is in the interval, there is not. We also note that a 95 percent confidence level is equivalent to a 5 percent significance level.

   Answer: 0.0022 ± 0.0134 gives the interval $(-0.0112, 0.0156)$, which does contain 0. Thus, no, the confidence interval does not indicate a significant difference (at the 5 percent significance level) between the two slopes.

(g) **Think:** Just as in part (f), this depends on whether or not 0 is in the interval.

  Answer: 0.4031 ± 0.2767 gives the interval $(0.1264, 0.6798)$, which is entirely positive and does not contain 0. Thus, yes, the confidence interval does indicate a significant difference (at the 5 percent significance level) between the two y-intercepts.

SCORING

This question is scored in three sections. Section 1 consists of parts (a), (b) and (c). Section 2 consists of parts (d) and (e). Section 3 consists of parts (f) and (g).

Section 1 is essentially correct for correct interpretations of the y-intercept in (a) and slope in context in (b) and a correct calculation in (c). Section 1 is partially correct if two of these three components are correct.

Section 2 is essentially correct for a correct confidence interval, a correct conclusion in context, a correct calculation of the difference of the average residuals, and a correct interpretation of this difference using nondeterministic language. Section 2 is partially correct if three of these four components are correct.

Section 3 is essentially correct when correct answers with explanations are given in both (f) and (g). Section 3 is partially correct for correct conclusions with weak explanations or for a completely correct statement for only one of (f) and (g).

4	**Complete Answer**	All three sections essentially correct.
3	**Substantial Answer**	Two sections essentially correct and one section partially correct.
2	**Developing Answer**	Two sections essentially correct OR one section essentially correct and one or two sections partially correct OR all three sections partially correct.
1	**Minimal Answer**	One section essentially correct OR two sections partially correct.

AP Score for the Diagnostic Test

Multiple-choice section (40 questions)

Number correct × 1.25 = _____

Free-response section (5 open-ended questions plus an investigative task)

Question 1 _____ × 1.875 = _____
out of 4

Question 2 _____ × 1.875 = _____
out of 4

Question 3 _____ × 1.875 = _____
out of 4

Question 4 _____ × 1.875 = _____
out of 4

Question 5 _____ × 1.875 = _____
out of 4

Question 6 _____ × 3.125 = _____
out of 4

Total points from multiple-choice and free-response sections = _____

Conversion chart based on a recent AP exam:

Total Points	AP Score
70–100	5
55–69	4
40–54	3
30–39	2
0–29	1

Study Guide for the Diagnostic Test Multiple-Choice Questions

Circle all of the questions you answered incorrectly, and focus your review on the unit or units that contain the most circled questions.

Unit 1: Exploring One-Variable Data

Questions 1, 10, 12, 19, 21, 28, 35, 38

Unit 2: Exploring Two-Variable Data

Questions 6, 19, 26, 31

Unit 3: Collecting Data

Questions 5, 13, 14, 22, 25, 32

Unit 4: Probability, Random Variables, and Probability Distributions

Questions 3, 9, 11, 20, 24, 29, 39

Unit 5: Sampling Distributions

Questions 17, 24, 36

Unit 6: Inference for Categorical Data: Proportions

Questions 7, 8, 9, 13, 15, 33

Unit 7: Inference for Quantitative Data: Means

Questions 4, 7, 8, 9, 16, 30, 34, 37

Unit 8: Inference for Categorical Data: Chi-Square

Questions 27, 40

Unit 9: Inference for Quantitative Data: Slopes

Questions 2, 18, 23

PART 2
Units Review

1

Exploring One-Variable Data

(6–9 multiple-choice questions on the AP exam)

Learning Objectives

In this unit, you will learn how

→ To be able to distinguish between categorical variables and quantitative variables.

→ To be able to create and interpret a bar graph for categorical data given as a frequency or relative frequency table.

→ To be able to calculate summary statistics for univariate (one-variable) quantitative data including center (mean and median), variability (range, interquartile range, variance, and standard deviation), and the five-number summary (minimum, Q_1, median, Q_3, and maximum).

→ To be able to describe univariate quantitative data including shape (such as bell-shaped, symmetric, right or left skewed, uniform, unimodal, or bimodal), center, spread (variability), and unusual features (such as gaps, clusters, modes, and outliers), and always remembering to do so in context (including units).

→ To be able to create displays for quantitative data including dotplots, stemplots, histograms, and boxplots.

→ To be able to read and interpret specific information from the above displays and from cumulative frequency graphs.

→ To be able to compare two distributions, being sure to use comparative terminology rather than simply making two separate lists, and again not forgetting context.

→ To be able to calculate using the empirical rule applied to normal distributions.

The first two units involve descriptive statistics, taking one-variable and two-variable data, displaying them, drawing conclusions from the displays and basic calculations, and communicating these observations. As an introduction to what is expected on the exam, consider the following example.

> **Example 1.1**

An association between life expectancy and educational attainment is illustrated in the following graphs.

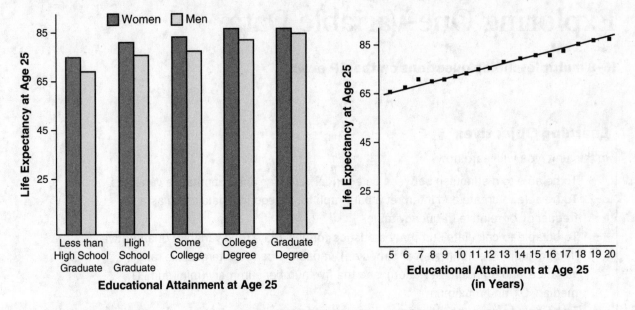

(a)   What does this say about women's versus men's life expectancies? Explain.

(b)   Does this say that attaining a higher educational level will give a person greater life expectancy? Explain.

(c)   Could a pie chart be used to display these data? Explain.

(d)   Do the points in the second graph seem to follow a pattern that is linear or nonlinear? Is this pattern in the positive or negative direction? Is it strongly, moderately, or weakly aligned?

(e)   Two points on the second graph are (12, 72) and (20, 84). They correspond to 25-year-olds who graduated from high school and graduate school, respectively. What information does the slope of a line between these points predict?

✓ **Solution**

(a)   At each educational attainment level, women's life expectancy is higher than men's.

(b)   This does not indicate a cause-and-effect relationship. Instead, it only shows an association. For example, it could be that more intelligent people attain higher educational levels and also take better care of their health and so have greater life expectancies. If this was the case, intelligence, not educational level, would lead to greater life expectancy.

(c)   No, a pie chart could not be used to display these data because the life expectancy for each educational attainment level does not represent parts of a single whole.

(d)   Visually, the points appear to follow a linear, positive, strong pattern because the points are tightly clustered around an upward-sloping, straight line.

(e)   The slope is $(84 − 72)/(20 − 12) = 12/8 = 1.5$. This predicts a 1.5-year rise in life expectancy for each additional year of education.

Categorical Variables

A categorical (or qualitative) variable takes on values that are category names or group labels. The values can be organized into frequency tables or relative frequency tables or can be represented graphically by displays such as bar graphs (bar charts), dotplots, and pie charts.

> **Example 1.2**

During the first week of 2024, the results of a survey revealed that 1,100 parents wanted to keep the school year to the current 180 days, 300 wanted to shorten it to 160 days, 500 wanted to extend it to 200 days, and 100 expressed no opinion. (Noting that there were 2,000 parents surveyed, percentages can be calculated.)

A table showing frequencies and relative frequencies:

Desired School Length	Number of Parents (frequency)	Relative Frequency	Percent of Parents
180 days	1,100	1,100/2,000 = 0.55	55%
160 days	300	300/2,000 = 0.15	15%
200 days	500	500/2,000 = 0.25	25%
No opinion	100	100/2,000 = 0.05	5%

> **NOTE**
>
> Frequency gives the number of cases falling into each category, while relative frequency gives the proportion or percent of cases falling into each category.

Bar graphs showing frequencies and relative frequencies:

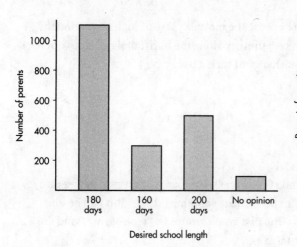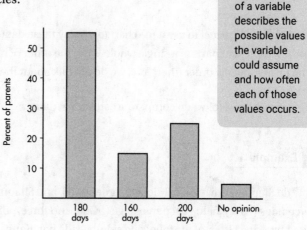

> The distribution of a variable describes the possible values the variable could assume and how often each of those values occurs.

Note that, as above, graphs should always have appropriate labeling and scaling.

> **Example 1.3**

When asked to choose their favorite hip-hop artist, 8 students chose Eminem, 5 picked 2Pac, 6 picked Notorious B.I.G., 3 picked Nas, and 3 picked Jay-Z. These data can be displayed in the following dotplot and pie chart.

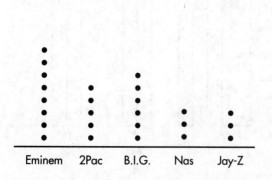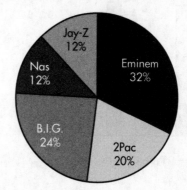

> **NOTE**
>
> 8 + 5 + 6 + 3 + 3 = 25, and 8/25 = 32%, 5/25 = 20%, 6/25 = 24%, and 3/25 = 12%.

Among those students who did not choose Eminem, what proportion chose Jay-Z?

(continued)

✓ Solution

Because $25 - 8 = 17$ students did not choose Eminem and 3 of these 17 students chose Jay-Z, the proportion is $\frac{3}{17}$.

❯ Example 1.4

The first years of the Covid-19 pandemic illustrated stark disparities in the health outcomes for people of color. The following bar chart shows 2020–2022 mortality data (cumulative mortality rates per 100,000 population) by race/ethnicity.

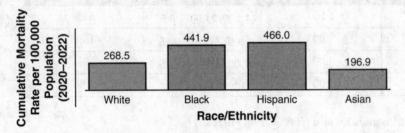

It would not be correct to use a pie chart to display these data because the mortality rate of each race/ethnicity does not represent parts of a single whole. Also, the order of race/ethnicity along the horizontal axis does not need to be in any specific order. Therefore, we do not talk about the "shape" of such a bar graph.

Bar graphs are useful when comparing categorical data.

❯ Example 1.5

How do Stephen Curry (SC), Michael Jordan (MJ), and Shaquille O'Neal (SO) compare with regard to free throw percentage (FT%), field goal percentage (FG%), and three-point percentage shooting (3P%)? This can be illustrated by using either of the following side-by-side bar graphs. The first graph groups by "type of shot," and the second graph groups by "player" along the respective horizontal axes.

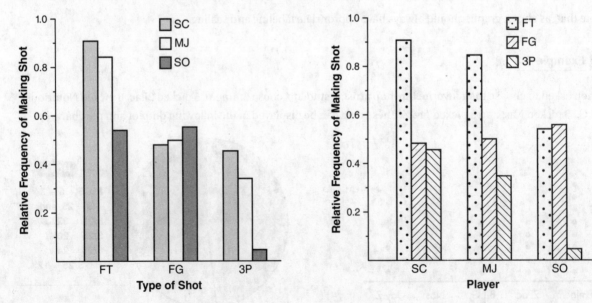

Representing a Quantitative Variable with Tables and Graphs

A quantitative variable takes on numerical values for a measured or counted quantity. The values can be organized into frequency tables or relative frequency tables or can be represented graphically by displays such as dotplots, histograms, stemplots, cumulative relative frequency plots, or boxplots.

A quantitative variable can be categorized as either *discrete* or *continuous*. A discrete quantitative variable takes on a finite or countable number of values. There are "gaps" between each of the values. For example, the number of teachers in different schools throughout a state is a countable or finite amount in each school. No school has 100.7 or 252½ teachers. There is a "gap" between 100 and 101 and between 252 and 253. Other examples of discrete quantitative variables include the SAT scores of students in a school or the amount of money in bank accounts in a city. A continuous quantitative variable can take on uncountable or infinite values with no gaps, such as heights and weights of students. Time is a continuous quantitative variable because, for example, you can measure someone's age as 18 years, 6 months, 17 days, 4 hours, 19 seconds, 39 milliseconds, … and so on.

> **TIP**
>
> To help decide if a variable is a quantitative variable, ask if it makes sense to find an average. For example, you can find the average "height of students" but not the average "eye color."

❯ Example 1.6

The *dotplot* below shows the lengths of stay (in days) for all patients admitted to a rural hospital during the first week of January 2024.

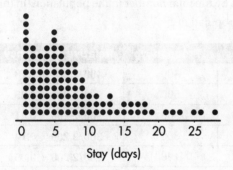

Stay (days)

> **TIP**
>
> Typically, discrete quantitative variables represent something that is being counted, whereas continuous quantitative variables represent something that is being measured.

Histograms, which are useful for large data sets involving quantitative variables, show counts or percents falling either at certain values or between certain values. Although the AP Statistics exam does not stress the construction of histograms, there are often questions that involve interpreting given histograms.

> **Example 1.7**

Suppose there are 2,200 seniors in a city's six high schools. Four hundred of the seniors are taking no AP classes, 500 are taking one, 900 are taking two, 300 are taking three, and 100 are taking four. These data can be displayed in the following histogram:

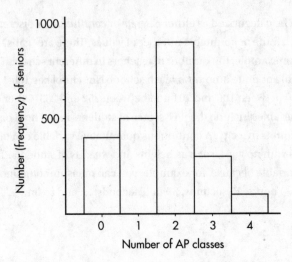

Sometimes, instead of labeling the vertical axis with frequencies, it is more convenient or more meaningful to use *relative frequencies*, that is, frequencies divided by the total number in the population. In this example, divide the "number of seniors" by the total number in the population.

Number of AP Classes	Frequency	Relative Frequency
0	400	400/2,200 = 0.18
1	500	500/2,200 = 0.23
2	900	900/2,200 = 0.41
3	300	300/2,200 = 0.14
4	100	100/2,200 = 0.05

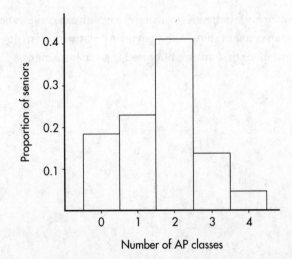

Note that the shape of the histogram is the same whether the vertical axis is labeled with frequencies or with relative frequencies.

> ### Example 1.8

Consider the following histogram of exam scores, where the vertical axis has not been labeled.

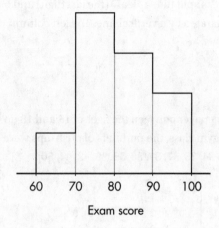

Exam score

What can we learn from this histogram?

✓ Solution

It is impossible to determine the actual frequencies; that is, we have no idea if there were 25 students, 100 students, or any particular number of students who took the exam. However, we can determine the relative frequencies by noting the fraction of the total area that is over any interval.

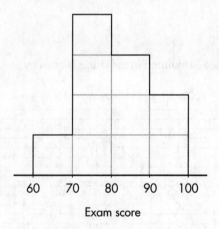

Exam score

We can divide this area into 10 equal portions and then note that $\frac{1}{10}$, or 10%, of the area is between 60 and 70, so 10% of the students scored between 60 and 70. Similarly, 40% scored between 70 and 80, 30% scored between 80 and 90, and 20% scored between 90 and 100.

Although it is usually not possible to divide histograms so nicely into 10 equal areas, the principle of relative frequencies corresponding to relative areas still applies. Also note how this example shows the number of exam scores falling *between* certain values, whereas the previous example showed the number of AP classes taken for *each* value. Note also that it is impossible to say exactly what the minimum value is (only that it is between 60 and 70) or exactly what the maximum value is (only that it is between 90 and 100).

TIP

Relative frequencies are the usual choice when comparing distributions of different-size populations.

Although a histogram may show how many scores fall into each grouping or interval, the exact values of individual scores are lost. An alternative pictorial display, called a *stemplot* (also called a stem and leaf display), retains this individual information and is useful for giving a quick overview of a distribution, displaying the relative density and shape of the data. Each data value is split into a "leaf" (the last digit) and a "stem" (the first digit or digits). A stemplot contains two columns separated by a vertical line. The left column contains the stems, and the right column contains the leaves.

> ### Example 1.9

How many nonstop push-ups can a teenager between the ages of 15 and 18 do? In one study in a mixed-gender high school gym class, the numbers of push-ups were {2, 5, 7, 10, 12, 12, 14, 16, 16, 18, 19, 20, 21, 29, 32, 34, 35, 37, 37, 38, 39, 39, 42, 44, 50}.

```
0 | 2 5 7
1 | 0 2 2 4 6 6 8 9
2 | 0 1 9
3 | 2 4 5 7 7 8 9 9
4 | 2 4
5 | 0
```

Number of push-ups
(5|0 means 50 push-ups)

Sometimes we sum frequencies and show the result visually in a cumulative relative frequency plot (also known as an *ogive*).

> ### Example 1.10

The following graph shows 2024 school enrollment in the United States by age.

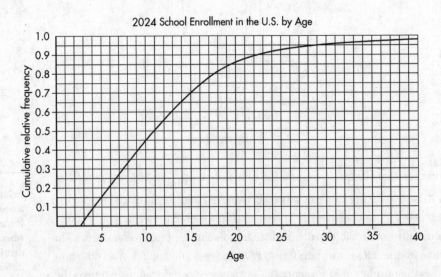

2024 School Enrollment in the U.S. by Age

(continued)

What can we learn from this cumulative relative frequency plot? *Possible answers:* Place a finger on age 5 and slide it vertically along the grid line until your finger hits the graph. Then slide your finger horizontally toward the vertical axis. The value on the vertical axis is 0.15, and therefore 15% of the school enrollment is below age 5. Similarly, going over to the graph from 0.5 on the vertical axis, we see that 50% of the school enrollment is below and 50% is above a middle age of 11. Going up from age 29, we see that 0.95, or 95%, of the enrollment is below age 29, and thus 5% is above age 29. Going over from 0.25 and 0.75 on the vertical axis, we see that the middle 50% of school enrollment is between ages 6 and 7 at the lower end and age 16 at the upper end.

Describing the Distribution of a Quantitative Variable

When describing a distribution, always comment on <u>S</u>hape, <u>O</u>utliers, <u>C</u>enter, and <u>S</u>pread (SOCS). Alternatively, you can use the acronym <u>C</u>enter, <u>U</u>nusual features, <u>S</u>hape, and <u>S</u>pread (CUSS). Always mention context somewhere in the description. Important aspects that should be addressed include the following.

1. *Shape:* Although distributions come in an endless variety of shapes, certain common patterns deserve special mention.
   - A *symmetric* distribution is one in which the two halves are mirror images of each other. For example, the weights of all people in some organizations fall into symmetric distributions with two mirror image bumps, one for men's weights and one for women's weights.
   - A distribution is *skewed to the right* if it spreads far and thinly toward the higher values. For example, ages of nonagenarians (people in their 90s) is a distribution with sharply decreasing numbers as one moves from 90-year-olds to 99-year-olds.
   - A distribution is *skewed to the left* if it spreads far and thinly toward the lower values. For example, track meet times often show a distribution bunched at a higher end with a few low values.
   - A *bell-shaped* distribution is symmetric with a center mound and two sloping tails. For example, the distribution of IQ scores across the general population is roughly symmetric with a center mound at 100 and two sloping tails.
   - A distribution is *uniform* if its histogram is a horizontal line. For example, tossing a fair die and noting how many dots (pips) appear on top yields a roughly uniform distribution because 1 through 6 are all equally likely.

> **TIP**
>
> In the real world, distributions are rarely perfectly symmetric, perfectly bell-shaped, or perfectly uniform. So we usually say "roughly" or "approximately" symmetric, bell-shaped, or uniform.

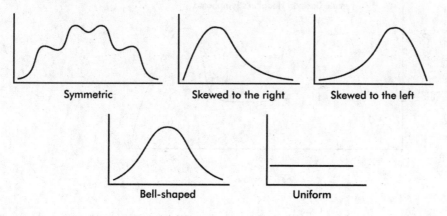

Symmetric   Skewed to the right   Skewed to the left

Bell-shaped   Uniform

2. *Center:* The center either separates the values (or the area under the curve in a histogram) roughly in half (the *median*) or is calculated as an average (the *mean*).

3. *Spread:* The spread or variability is the scope of the values from smallest to largest. More specifically, the *range, interquartile range, variance,* and *standard deviation* are discussed depending on the distribution.

4. *Outliers:* Outliers are values falling far outside the overall pattern. (For example, most people can hold their breath between 20 and 60 seconds; 5 seconds for a person with emphysema and 5 minutes for a free diver would be outliers in a breath-holding distribution.) Two specific mathematical definitions of outliers will be discussed later.

5. *Modes:* Modes show major peaks. (For example, the numbers of siblings of each student in most high schools tends to be *unimodal* with a peak at 1 sibling, while the duration of Old Faithful's eruptions is *bimodal* with peaks around both 1.9 and 4.5 minutes.) However, every little bump in the data should not be called a mode!

6. *Clusters:* Clusters show natural subgroups into which the values fall. (For example, the salaries of teachers in Ithaca, NY, fall into three overlapping clusters: one for public school teachers, a higher one for Ithaca College professors, and an even higher one for Cornell University professors.)

7. *Gaps:* Gaps show holes where no values fall. (For example, the Office of the Dean sends letters to students being put on the honor roll and to those being put on academic warning for low grades. Thus, the GPA distribution of students receiving letters from the dean has a huge gap in the middle.)

You have already seen several of these on the previous pages:

- The dotplot in Example 1.6 illustrates a skewed right distribution with a center around a hospital stay of 6 days and a spread from 0 to 28 days.
- The histogram of Example 1.7 illustrates a roughly bell-shaped distribution with a center of 2 AP classes and a spread from 0 to 4 AP classes.
- The histogram of Example 1.8 illustrates a unimodal distribution with a center at an exam score of 80 and a spread from about 60 to about 100.
- The stemplot of Example 1.9 illustrates a bimodal distribution with a center of 21 push-ups, a spread from 2 to 50 push-ups, and a gap between 21 and 29 push-ups.

> **Example 1.11**

Hodgkin's lymphoma is a cancer of the lymphatic system, the system that drains excess fluid from the blood and protects against infection. Consider the following histogram:

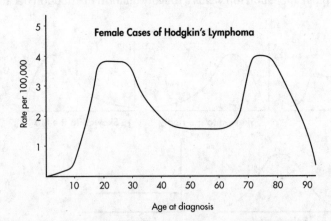

The distribution of ages of diagnosis of female cases of Hodgkin's lymphoma is bimodal, roughly symmetric, with a center around 50, a spread from 0 to around 93, and two distinct clusters centered around 25 and 75 years old.

❯ Example 1.12

The histogram below shows employee computer usage (number accessing the Internet) during given times at a company's main office.

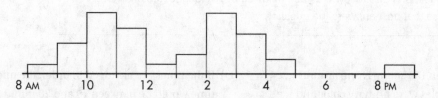

NOTE

In plots such as this, there is no way to know the exact minimum or maximum values, only intervals in which these values fall.

The distribution of employee computer usage (number accessing the Internet) during given times is bimodal, with a center around 1 P.M., a spread from around 8 A.M. to around 9 P.M., a gap between 5 P.M. and 8 P.M., and an outlier or outliers between 8 P.M. and 9 P.M. Computer usage at this company appears heaviest at midmorning and midafternoon with a dip in usage during the noon lunch hour. You should look for reasons behind outliers. In this case, the evening usage could indicate employees returning after dinner or possibly a hacker using company computer power for free!

A distribution skewed to the left has a cumulative frequency plot that rises slowly at first and then steeply later. In contrast, a distribution skewed to the right has a cumulative frequency plot that rises steeply at first and then slowly later.

❯ Example 1.13

Consider the essay-grading policies of three teachers: Mr. Abrams gives very low scores; Mrs. Brown gives equal numbers of low and high scores; and Ms. Connors gives very high scores. Histograms of the grades (with 1 representing the lowest score and 4 representing the highest score) are as follows:

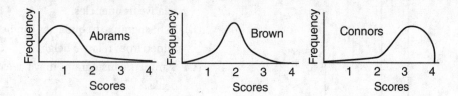

The distribution of Mr. Abrams's scores is skewed right, of Mrs. Brown's scores is roughly bell-shaped, and of Ms. Connors's scores is skewed left. All three distributions are unimodal and have roughly the same spread from around 0 to around 4. The center of Mr. Abrams's scores is the lowest, while the center of Ms. Connors's scores is the highest. These distributions translate into the following cumulative frequency plots:

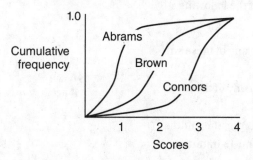

Quiz 1

Multiple-Choice Questions

> **DIRECTIONS:** The questions that follow are each followed by five suggested answers. Choose the response that best answers the question.

1. Data were collected on 100 stamps from around the world. Two variables were primary color and width of stamp. Are these variables categorical or quantitative?

   (A) Both are categorical.
   (B) Both are quantitative.
   (C) Primary color is categorical and width is quantitative.
   (D) Primary color is quantitative and width is categorical.
   (E) This cannot be answered without further information.

2. The times, in minutes, taken by 155 high school students to complete a level of the video game *Angry Birds* are summarized in the histogram below.

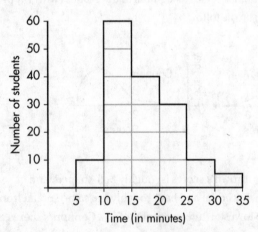

   Based on the histogram, which of the following must be true?

   (A) The minimum time taken by any of these students was 5 minutes.
   (B) The maximum time taken by any of these students was 35 minutes.
   (C) If the times are arranged in order, the middle time would be between 10 and 15 minutes.
   (D) If the times are arranged in order, the middle time would be between 15 and 20 minutes.
   (E) The same number of students took less than 15 minutes as took over 20 minutes.

3. Which of the following is a true statement?

   (A) Stemplots are useful for both quantitative and categorical data sets.
   (B) Stemplots are equally useful for small and very large data sets.
   (C) Stemplots can show symmetry, gaps, clusters, and outliers.
   (D) Stemplots may or may not show individual values.
   (E) Stems may be skipped if there is no data value for a particular stem.

4. Which of the following is an *incorrect* statement?

   (A) In histograms, relative areas correspond to relative frequencies.
   (B) In histograms, frequencies can be determined from relative heights.
   (C) Symmetric histograms may have multiple peaks.
   (D) Two students working with the same set of data may come up with histograms that look different.
   (E) Displaying outliers may be more problematic when using histograms than when using stemplots.

5. A sales rep asked 50 Generation X, 75 Millennial, and 100 Generation Z respondents their preference between Company A and Company B with regard to sound quality of their smart speakers. The results are shown in the following table.

	Company A	Company B
Generation X	19	31
Millennials	43	32
Generation Z	68	32

Which of the following statements is supported by the table?

(A) The same percentage of Millennials and Generation Z preferred Company B.

(B) The percentage of Generation X who preferred Company B is greater than the percentage of Millennials who preferred Company B.

(C) Less than half the non-Millennials preferred Company A.

(D) Of those who preferred Company A, more than half were either Generation X or Millennials.

(E) Of those who preferred Company B, less than half were not Generation Z.

6. In a favorite movie survey of 20 AP Statistics students, 4 chose Comedy, 5 chose Action, 6 chose Romance, 1 chose Drama, and 4 chose Sci-Fi. These data are illustrated in the following pie chart.

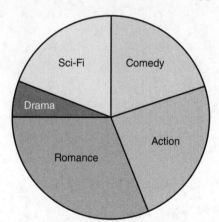

Favorite Type of Movie

Among those not choosing Comedy or Romance, what proportion chose Sci-Fi or Action?

(A) $\dfrac{4+5}{4+6}$

(B) $\dfrac{4+5}{20-(4+6)}$

(C) $\dfrac{4+5}{20-(4+5)}$

(D) $\dfrac{4+5}{20}$

(E) $1-\dfrac{4+5}{20}$

Question 7 refers to the following five histograms:

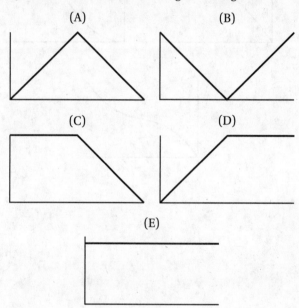

7. To which of the above five histograms does the following cumulative relative frequency plot correspond?

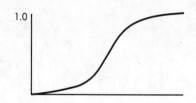

(A) A

(B) B

(C) C

(D) D

(E) E

Question 8 refers to the following five cumulative relative frequency plots:

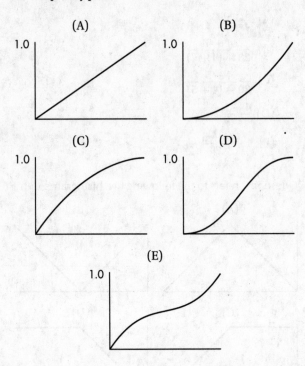

(A)

(B)

(C)

(D)

(E)

8. To which of the above cumulative relative frequency plots does the following histogram correspond?

(A) A
(B) B
(C) C
(D) D
(E) E

Free-Response Questions

> **DIRECTIONS:** You must show all work and indicate the methods you use. You will be graded on the correctness of your methods and on the accuracy of your final answers.

1. The miles run by 25 randomly selected runners during a given week are given below.

Miles run (x)	$5 \leq x < 10$	$10 \leq x < 15$	$15 \leq x < 20$	$20 \leq x < 25$
Frequency	3	11	7	4

   (a) Construct a relative frequency histogram.
   (b) What percent of the runners ran under 15 miles?
   (c) If each of the four intervals is divided in half and the corresponding numbers given, what percent would be under 15 miles?

2. The dotplot below shows the numbers of goals scored by the 20 teams playing in a city's high school soccer games on a particular day.

Goals scored by each team

   (a) Describe the distribution.
   (b) One superstar scored six goals, but his team still lost. What are all possible final scores for that game? Explain.
   (c) Is it possible that each of the teams that scored exactly two goals were the winners of that game? Explain.

3. The winning percentages for a major league baseball team over the past 22 years are shown in the following stemplot:

```
46 | 0 8
47 | 1 4 7 9
48 | 5 8 8 9
49 | 3 4 7
50 |
51 |
52 | 5 8
53 | 2 5 6
54 | 4 8 9
55 | 6
```

(55 | 6   means 55.6%)

(a) Interpret the lowest value.
(b) Describe the distribution.
(c) Give a reason that one might argue that the team is more likely to lose a given game than win it.
(d) Give a reason that one might argue that the team is more likely to win a given game than lose it.

4. A college basketball team keeps records of career average points per game of players playing at least 75% of team games during their college careers. The cumulative relative frequency plot below summarizes statistics of players graduating over the past 10 years.

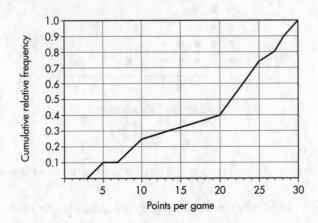

(a) Interpret the point (20, 0.4) in context.
(b) Interpret the intersection of the plot with the horizontal axis in context.
(c) Interpret the horizontal section of the plot from 5 to 7 points per game in context.
(d) The players with the top 10% of the career average points per game will be listed on a plaque. What is the cutoff score for being included on the plaque?
(e) What proportion of the players averaged between 10 and 20 points per game?

The answers for this quiz can be found beginning on page 475.

Summary Statistics for a Quantitative Variable

Given a raw set of data, often we can detect no overall pattern. Perhaps some values occur more frequently, a few extreme values may stand out, and the range of values is usually apparent. The presentation of data, including summarizations and descriptions, and involving such concepts as representative or average values, measures of variability, positions of various values, and the shape of a distribution, falls under the broad topic of *descriptive statistics*. This aspect of statistics is in contrast to *inferential statistics*, the process of drawing inferences from limited data, a subject discussed in later units.

In the following paragraphs, we consider the two primary ways of denoting the "center" of a distribution:

1. The *median,* which is the middle number of a set of numbers arranged in numerical order.
2. The *mean,* which is found by summing items in a set and dividing by the number of items.

> ## Example 1.14

Consider the following set of home run distances (in feet) to center field in 13 ballparks: {387, 400, 400, 410, 410, 410, 414, 415, 420, 420, 421, 457, 461}. What are the median and mean home run distances to center field for this set of 13 ballparks?

✓ Solution

The median is 414 feet (there are six values below 414 and six values above), while the mean is

$$\frac{387 + 400 + 400 + 410 + 410 + \cdots + 457 + 461}{13} = 417.3 \text{ feet}$$

The word *median* is derived from the Latin *medius,* which means "middle." The values under consideration are arranged in ascending or descending order. If there is an odd number of values, the median is the middle one. If there is an even number of values, the median is found by adding the two middle values and dividing by 2. Thus the median of a set has the same number of elements above it as below it.

REMEMBER

Don't forget to put the data in order before finding the median.

The median is not affected by exactly how large the larger values are or by exactly how small the smaller values are. Thus, it is a particularly useful measurement when the extreme values, *outliers,* are in some way suspicious or when you want to diminish their effect. For example, if 10 mice try to solve a maze and nine succeed in less than 15 minutes while one is still trying after 24 hours, the most representative value is the median (not the mean, which is over 2 hours). Similarly, if the salaries of four executives are each between $240,000 and $245,000 while a fifth is paid less than $20,000, again the most representative value is the median (the mean is under $200,000). It is often said that the median is "resistant" to extreme values.

Although the median is often useful in descriptive statistics, the *mean,* or more accurately, the *arithmetic mean,* is most important for statistical inference and analysis. Also, for the layperson, the "average" is usually understood to be the mean.

The mean of a *whole population* (the complete set of items of interest) is denoted by the Greek letter μ (mu), while the mean of a *sample* (a part of a population) is denoted by $\bar{x}$. For example, the mean value of the set of all houses in the United States might be $\mu = \$156,400$, while the mean value of 100 randomly chosen houses might be $\bar{x} = \$152,100$ or perhaps $\bar{x} = \$163,800$ or even $\bar{x} = \$224,000$. (Each sample will probably contain different dollar amounts than other samples and therefore will likely have a different mean than other samples.) In statistics we learn how to estimate a population mean from a sample mean.

NOTE

A parameter, such as μ, is a numerical summary of a population, while a statistic, such as $\bar{x}$, is a numerical summary of sample data.

Mathematically, the mean $= \frac{\Sigma x}{n}$, where Σx represents the sum of all the elements of the set under consideration and n is the actual number of elements. Σ is the uppercase Greek letter sigma.

> **Example 1.15**

Suppose the salaries of the six part-time employees at a computer security company are $3,000; $7,000; $15,000; $22,000; $23,000; and $38,000.

(a)   What are the mean and median salaries?

(b)   What will the new mean and median salaries be if everyone receives a $3,000 increase?

(c)   What will the new mean and median salaries be if only the person with the highest salary receives a $3,000 increase?

(d)   What will be the new mean and median if everyone receives a 10% raise?

✓ **Solutions**

(a)   The mean is $\dfrac{3,000 + 7,000 + 15,000 + 22,000 + 23,000 + 38,000}{6} = \$18,000$.

The median is $\dfrac{15,000 + 22,000}{2} = \$18,500$.

(b)   The mean is $\dfrac{6,000 + 10,000 + 18,000 + 25,000 + 26,000 + 41,000}{6} = \$21,000$.

The median is $\dfrac{18,000 + 25,000}{2} = \$21,500$.

Note that adding the same constant to each value increases both the mean and the median by that same constant ($18,000 + $3,000 = $21,000 and $18,500 + $3,000 = $21,500).

(c)   The mean is $\dfrac{3,000 + 7,000 + 15,000 + 22,000 + 23,000 + 41,000}{6} = \$18,500$.

The median is $\dfrac{15,000 + 22,000}{2} = \$18,500$.

Note that the mean, unlike the median, is sensitive to a change in any value.

(d)   The mean is $\dfrac{3,300 + 7,700 + 16,500 + 24,200 + 25,300 + 41,800}{6} = \$19,800$.

The median is $\dfrac{16,500 + 24,200}{2} = \$20,350$.

Note that multiplying each value by the same constant multiplies both the mean and the median by that same constant (110% of $18,000 is $19,800 and 110% of $18,500 is $20,350).

In describing a set of numbers, not only is it useful to designate a central value (mean or median), but also it is important to be able to indicate the *variability* of the measurements. An explosion engineer in mining operations aims for small variability—it would not be good for the 30-minute fuses actually to have a range of 10–50 minutes before detonation. On the other hand, a teacher interested in distinguishing better students from those who have difficulty with the material aims to design exams with large variability in results—it would not be helpful if all the students scored exactly the same. The players on two basketball teams may have the same average height, but this observation doesn't tell the whole story. If the dispersions are quite different, one team may have a 7-foot player, whereas the other has no one over 6 feet tall. Two Mediterranean holiday cruises may advertise the same average age for their passengers. One, however, may have passengers only between 20 and 25 years old, while the other has only middle-aged parents in their forties together with their children under age 10.

To understand variability, we either calculate the mean and standard deviation (defined below) or calculate the following *five-number summary* of a data set:

- The minimum.
- The 25th percentile, Q_1, is the median of the lower half of the data set. In other words, about 25% of the numbers in the data set lie below Q_1 and about 75% lie above Q_1.

TIP

Variability is the single most fundamental concept in statistics and is the key to understanding statistics.

- The median.
- The 75th percentile, Q_3, is the median of the upper half of the data set. In other words, about 75% of the numbers in the data set lie below Q_3 and about 25% lie above Q_3.
- The maximum.

There are four primary ways of describing variability, or dispersion:

1. *Range:* the difference between the largest and smallest values (or maximum minus minimum).
   - Although the range gives a measure of spread, it is entirely dependent on the two extreme values and is insensitive to the values in the middle.
   - Some quality control techniques involve taking periodic small samples and basing further action on the range found in such samples.
2. *Interquartile range (IQR):* the difference between the largest and smallest values after removing the lower and upper quartiles.
   - IQR = range of middle 50% = $Q_3 - Q_1$ = 75th percentile minus 25th percentile.
   - This removes the influence of extreme values.
3. *Variance:* the average of the squared differences from the mean.
   - The variance of a complete population is $\sigma^2 = \dfrac{\Sigma(x - \mu)^2}{n}$.
   - The variance of a sample is $s^2 = \dfrac{\Sigma(x - \bar{x})^2}{n - 1}$.
4. *Standard deviation (SD):* the square root of the variance.
   - The standard deviation of a complete population is $\sigma = \sqrt{\dfrac{\Sigma(x - \mu)^2}{n}}$.
   - The standard deviation of a sample is $s = \sqrt{\dfrac{\Sigma(x - \bar{x})^2}{n - 1}}$.
   - The standard deviation gives a typical distance that each value is away from the mean. In other words, it is a representative value for the variability of a population or sample.
   - Although variance is measured in square units, the standard deviation is measured in the same units as are the data.

There are two numerical rules for designating outliers:

1. The more common rule is that outliers are any value less than $Q_1 - 1.5(\text{IQR})$ or greater than $Q_3 + 1.5(\text{IQR})$.
2. An acceptable alternative rule is that outliers are all values located two or more standard deviations above, or below, the mean.

❯ Example 1.16

In a random sample of 40 NBA players, the points scored by each player on a particular game day are summarized in the following table and are shown in the dotplot.

Number of points	4	6	7	10	15	21	26	32
Number of players	2	5	5	11	9	6	1	1

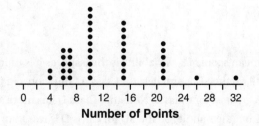

(continued)

(a)   What is the five-number summary?
(b)   What are the range and interquartile range? Show your work.
(c)   Using the 1.5(IQR) rule, what values are outliers, if any? Show your work.
(d)   The mean number of points scored is 12.55 with a standard deviation of 6.46. Using the two standard deviations rule, what values are outliers, if any? Show your work.

✓ Solution

(a)   The minimum number of points scored is 4. With $n = 40$, count 10 values to obtain $Q_1 = 7$. Counting 20–21 values shows that the median is 10. Counting 10 values from the high side gives $Q_3 = 15$. The maximum is 32. The five-number summary is {4, 7, 10, 15, 32}.

(b)   The range is Max minus Min $= 32 - 4 = 28$.
The interquartile range is IQR $= Q_3 - Q_1 = 15 - 7 = 8$.

(c) $Q_1 - 1.5(IQR) = 7 - 1.5(8) = -5$ and no values are below this.
$Q_3 + 1.5(IQR) = 15 + 1.5(8) = 27$ and there is one value, 32, above this.
Using the 1.5(IQR) rule, there is one outlier, 32.

(d) $\bar{x} - 2s = 12.55 - 2(6.46) = -0.37$ and no values are below this.
$\bar{x} + 2s = 12.55 + 2(6.46) = 25.47$ and there are two values, 26 and 32, above this.
Using the two standard deviations rule, there are two outliers, 26 and 32.

❯ Example 1.17

Consider the following dotplots of student scores on an AP exam of four AP Calculus classes, each with 10 students. Without making any calculations, arrange the four classes in order from smallest SD to largest SD.

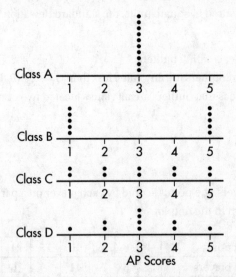

✓ Solution

Although all four classes have a mean score of 3, variability is the issue, as it is unique for each class. The scores in Class A have no variability at all, so the SD is smallest for Class A. The scores in Class B are all as far as possible away from the mean, so the SD is greatest for Class B. Note that Class D has more scores near the middle, while Class C has more scores further from the middle. So the SD for Class D is less than the SD for Class C. The correct ordering of the classes from smallest SD to largest SD is: A-D-C-B.

We have seen ways of choosing a value to represent the center of a distribution. We also need to be able to talk about the *position* of other values. There are three important, recognized procedures for designating position:

1. *Simple ranking*, which involves arranging the elements in some order and noting where in that order a particular value falls;

2. *Percentile ranking*, which indicates what percentage of all values fall at or below the value under consideration;

3. The z-*score*, which states very specifically the number of standard deviations a particular value is above or below the mean.

❯ Example 1.18

It is recommended that the "good cholesterol," high-density lipoprotein (HDL), be present in the blood at levels of at least 40 mg/dL. Suppose all 50 members of a high school football team are tested, and the resulting HDL levels are as follows: {53, 26, 45, 33, 64, 29, 73, 29, 21, 58, 70, 41, 48, 55, 55, 39, 57, 48, 9, 59, 56, 39, 68, 50, 65, 30, 38, 54, 49, 35, 56, 70, 43, 86, 52, 40, 28, 40, 67, 50, 47, 54, 59, 29, 29, 42, 45, 37, 51, 40}. The mean of this list is 47.22 mg/dL with a standard deviation of 15.05 mg/dL. Given that there are 31 higher HDL levels than 41, what is the position of the HDL score of 41 as measured in simple ranking, percentile ranking, and z-score?

> **NOTE**
>
> A negative z-score indicates that a value is less than the mean.

✓ Solution

Since there are 31 higher HDL levels on the list, the 41 has a simple ranking of 32 out of 50. Of all the HDL levels, 19 are at or lower than 41, so the percentile ranking is $\frac{19}{50} = 38\%$. The HDL score of 41 has a z-score of $\frac{41 - 47.22}{15.05} = -0.413$.

Simple ranking is easily calculated and easily understood. We know what it means for someone to graduate second in a class of 435 or for a player from a team of size 30 to have the seventh-best batting average. Simple ranking is useful even when no numerical values are associated with the elements. For example, detergents can be ranked according to relative cleansing ability without any numerical measurements of strength.

Percentile ranking, the percent of the data that is less than or equal to a value, is helpful in comparing positions with different bases. We can more easily compare a rank of 176 out of 704 with a rank of 187 out of 935 by noting that the first has a rank of 75% and the second a rank of 80%. Percentile rank is also useful when the exact population size is not known or is irrelevant. For example, it is more meaningful to say that Jennifer scored in the 90th percentile on a national exam rather than trying to determine her exact ranking among some large number of test takers.

> **NOTE**
>
> More formally, the first quartile, Q_1, is the median of the half of the ordered data set from the minimum to the position of the median. The third quartile, Q_3, is the median of the half of the ordered data set from the position of the median to the maximum.

The *quartiles*, Q_1 and Q_3, lie one-quarter and three-quarters of the way up a list, respectively. Their percentile ranks are 25% and 75%, respectively. The interquartile range defined earlier can also be defined as $Q_3 - Q_1$. The *deciles* lie one-tenth and nine-tenths of the way up a list, respectively, and have percentile ranks of 10% and 90%.

The z-*score* is a measure of position that takes into account both the center and the dispersion of the distribution. More specifically, the z-score of a value tells how many standard deviations the value is from the mean. Mathematically, $x - \mu$ gives the raw distance from μ to x; dividing by σ converts this to number of standard deviations. Thus $z = \frac{x - \mu}{\sigma}$, where x is the raw score, μ is the mean, and σ is the standard deviation. If the score x is greater than the mean μ, z is positive; if x is less than μ, z is negative.

Given a z-score, we can reverse the procedure and find the corresponding raw score. Solving for x gives $x = \mu + z\sigma$.

❯ Example 1.19

Because of hurricanes, refinery outages, and Middle East tensions, the mean price of gasoline in one northeast city hit \$3.80 per gallon with a standard deviation of \$0.05. Then \$3.90 has a z-score of $\frac{3.90 - 3.80}{0.05} = +2$, while \$3.65 has a z-score of $\frac{3.65 - 3.80}{0.05} = -3$. Alternatively, a z-score of $+2.2$ corresponds to a raw score of $3.80 + 2.2(0.05) = 3.80 + 0.11 = 3.91$, while a z-score of -1.6 corresponds to $3.80 - 1.6(0.05) = 3.72$.

It is often useful to portray integer z-scores and the corresponding raw scores as follows:

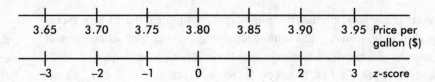

One purpose of z-scores is to act as a common "measuring stick" with which to compare values from different populations. For example, a high school senior might have a higher GPA than a high school sophomore, but the senior's GPA might be less far above the average GPA of seniors than the sophomore's GPA is above the average GPA of sophomores. We can make a comparison using z-scores.

Changing units, for example, from dollars to rubles or from miles to kilometers, is common in a world that seems to become smaller all the time. It is instructive to note how measures of center, variability, and, by extension, position are affected by such changes.

Adding the same constant to every value increases the mean and median by that same constant; however, the distances between the increased values stay the same, and so the range and standard deviation remain unchanged. Multiplying every value by the same constant multiplies the mean, median, range, and standard deviation by that same constant. However, none of these changes will affect a simple ranking, a percentile ranking, or even a z-score.

NOTE

Another consequence is that no matter what the distribution of raw scores, the set of z-scores always has mean 0 and standard deviation 1.

❯ Example 1.20

A set of experimental measurements of the freezing point of an unknown liquid yield a mean of 25.32 degrees Celsius with a standard deviation of 1.47 degrees Celsius.

(a)   If all the measurements are converted to the Kelvin scale, what are the new mean and standard deviation? (Kelvins are equivalent to degrees Celsius plus 273.16.)

(b)   What is the z-score of 27 degrees Celsius?

✓ Solution

(a)   The new mean is $25.32 + 273.16 = 298.48$ Kelvins. The standard deviation, however, remains numerically the same, 1.47 Kelvins. Graphically, you should picture the whole distribution moving over by the constant 273.16; the mean moves, but the standard deviation (which measures variability) doesn't change.

(b)   For 27 degrees Celsius, $z = (27 - 25.32)/1.47 = 1.14$. It is instructive to note that 27 degrees Celsius corresponds to $27 + 273.16 = 300.16$ Kelvins with again a z-score of $z = (300.16 - 298.48)/1.47 = 1.14$.

> **Example 1.21**

Measurements of the sizes of farms in an upstate New York county yield a mean of 59.2 hectares with a standard deviation of 11.2 hectares.

(a)   If all the measurements are converted from hectares (metric system) to acres (one acre was originally the area a yoke of oxen could plow in one day), what are the new mean and standard deviation? (One hectare is equivalent to 2.471 acres.)

(b)   What is the z-score of 55 hectares?

✓ **Solution**

(a)   The new mean is $2.471 \times 59.2 = 146.3$ acres with a standard deviation of $2.471 \times 11.2 = 27.7$ acres. Graphically, multiplying each value by the constant 2.471 both moves and spreads out the distribution.

(b) $z = (55 - 59.2)/11.2 = -0.375$. It is instructive to note that 55 hectares equals $2.471(55) = 135.905$ acres, and again $z = (135.905 - 146.3)/27.7 = -0.375$.

Graphical Representations of Summary Statistics

> **Example 1.22**

Below is a histogram of the home prices (in $1,000) of a Midwestern urban city.

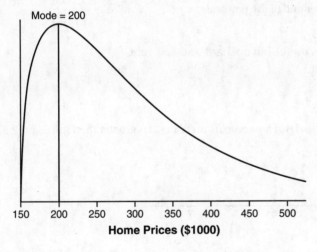

The distribution of home prices is unimodal, is skewed right, has a mode of $200,000, and varies from $150,000 to about $525,000.

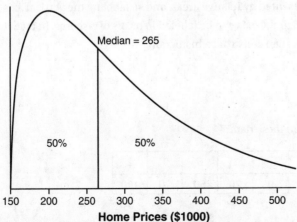

The median, $265,000, divides the distribution in half, so it can be illustrated by a line that divides the area of the histogram in half.

(continued)

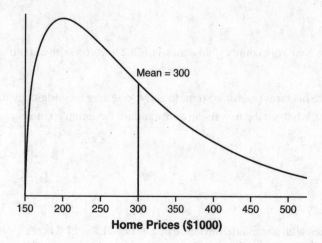

Mean = 300

Home Prices ($1000)

The mean, $300,000, is affected by the spacing of all the values. If the histogram is considered a solid region, the mean corresponds to a line passing through the center of gravity, or balance point.

> **Example 1.23**

Suppose that the faculty salaries at a college have a median of $82,500 and a mean of $88,700. What does this indicate about the shape of the distribution of the salaries?

✓ **Solution**

The median is less than the mean, and so the salaries are probably skewed to the right. There are a few highly paid professors, with the bulk of the faculty at the lower end of the pay scale.

It should be noted that the above principle is a useful, but not hard-and-fast, rule.

> **Example 1.24**

The set given by the dotplot below is skewed to the right; however, its median (3) is greater than its mean (2.97).

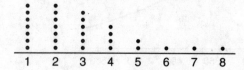

We have seen that relative frequencies are represented by relative areas, and so labeling the vertical axis is not crucial. If we know the standard deviation, the horizontal axis can be labeled in terms of z-scores. In fact, if we are given the percentile rankings of various z-scores, we can construct a histogram.

> **Example 1.25**

Suppose we are asked to construct a histogram from these data:

z-score	−2	−1	0	1	2
Percentile ranking	0	20	60	70	100

(continued)

We note that the entire area is less than z-score $+2$ and greater than z-score -2. Also, 20% of the area is between z-scores -2 and -1, 40% is between -1 and 0, 10% is between 0 and 1, and 30% is between 1 and 2. Thus the histogram is as follows.

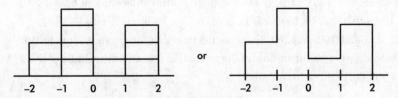

or

Now suppose we are given four in-between z-scores as well:

z-score	-2.0	-1.5	-1.0	-0.5	0.0	0.5	1.0	1.5	2.0
Percentile ranking	0	5	20	30	60	65	70	80	100

Then we have:

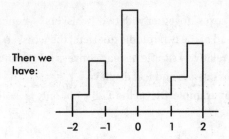

With 1,000 z-scores perhaps the histogram would look like:

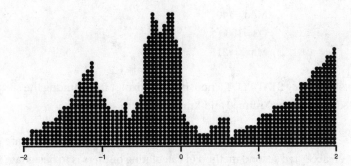

The height at any point is meaningless; what is important is relative areas.

(a)   In the final diagram above, what percentage of the area is between z-scores of $+1$ and $+2$?

(b)   What percent is to the left of 0?

✓ Solutions

(a)   Still 30%.

(b)   Still 60%.

A *boxplot* (also called a *box and whisker display*) is a visual representation of dispersion that shows the smallest value, the largest value, the middle (median), the middle of the bottom half of the set (Q_1), and the middle of the top half of the set (Q_3). The five numbers that are identified in a boxplot are called the five-number summary.

> ## Example 1.26

After an AP Statistics teacher hears that every one of her 27 students received a 3 or higher on the AP exam, she treats the class to a game of bowling. The individual student bowling scores are 210, 130, 150, 140, 150, 210, 150, 125, 85, 200, 70, 150, 75, 90, 150, 115, 120, 125, 160, 140, 100, 95, 100, 215, 130, 160, and 200. The students note that the greatest score is 215, the smallest is 70, the middle is 140, the middle of the top half is 160, and the middle of the bottom half is 100. A boxplot of these five numbers is

TIP

The IQR is the *length* of the box, not the box itself. So, the median is in the box, or is between Q_1 and Q_3, but is not *in* the IQR.

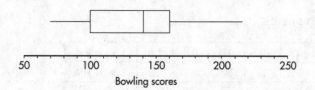

Bowling scores

Be careful about describing the shape of a distribution when all that you have is a boxplot. For example, "approximately normal" is never a possible conclusion.

Note that the display consists of two "boxes" together with two "whiskers"—hence the alternative name. The boxes show the spread of the two middle quarters; the whiskers show the spread of the two outer quarters. This relatively simple display conveys information not immediately available from histograms or stem and leaf displays.

Putting the above data into a list, for example, L1, on the TI-84, not only gives the five-number summary

TIP

Note that a boxplot gives one measure of center (the median) and two measures of variability (the range and the IQR). A boxplot does not give information as to the mean or standard deviation.

$$1\text{-Var Stats}$$
$$\text{minX}=70$$
$$\text{Q1}=100$$
$$\text{Med}=140$$
$$\text{Q3}=160$$
$$\text{MaxX}=215$$

but also gives the boxplot itself using STAT PLOT, choosing the boxplot from among the six type choices, and then using ZoomStat or in WINDOW, letting Xmin=0 and Xmax=225.

Values more than $1.5 \times \text{IQR}$ (1.5 times the interquartile range) outside the two boxes are plotted separately as outliers. As noted earlier, an accepted second method of calculating outliers is to designate as outliers all values located two or more standard deviations above, or below, the mean.

> ## Example 1.27

Inputting the lengths of words in a selection of Shakespeare's plays results in the following summary statistics and histogram:

n	mean	SD	min	Q_1	med	Q_3	max
498	4.114	2.085	1	3	4	5	12

(continued)

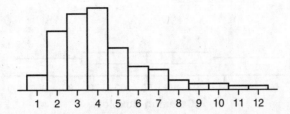

(a) Find all outliers, if any.

(b) Draw a boxplot.

✓ **Solutions**

(a) Outliers consist of any word lengths less than $Q_1 - 1.5(IQR) = 3 - 1.5(5 - 3) = 0$ or greater than $Q_3 + 1.5(IQR) = 5 + 1.5(5 - 3) = 8$. (The second method of outlier calculation gives all values less than $4.114 - 2(2.085) = -0.056$ or greater than $4.114 + 2(2.085) = 8.284$.) Word lengths of 9, 10, 11, and 12 are outliers.

(b)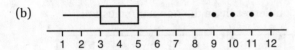

Note: In general, boxplots cannot show overall shape; however, they may suggest skewness.

It should be noted that different sets can have the same five-number summary and thus the same boxplots but have dramatically different distributions.

> **Example 1.28**

Let $A = \{0, 5, 10, 15, 25, 30, 35, 40, 45, 50, 71, 72, 73, 74, 75, 76, 77, 78, 100\}$, $B = \{0, 22, 23, 24, 25, 26, 27, 28, 29, 50, 55, 60, 65, 70, 75, 85, 90, 95, 100\}$, and $C = \{0, 10, 20, 25, 25, 30, 35, 40, 50, 60, 65, 70, 75, 75, 80, 90, 100\}$. Simple inspection indicates very different distributions; however, the TI-84 gives identical boxplots with Min $= 0$, $Q_1 = 25$, Med $= 50$, $Q_3 = 75$, and Max $= 100$ for each. The dotplots are shown below.

Draw the common boxplot.

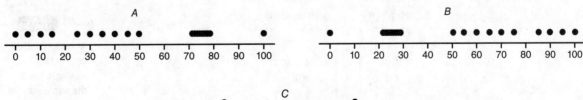

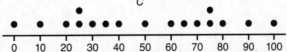

(continued)

✓ Solution

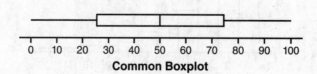

Common Boxplot

⟩ Example 1.29

Among the eleven new hires at a computer security company, two started with salaries at $45,000, two at $48,000, three at $50,000, one at $53,000, and two at $54,000. The eleventh new hire was the CEO's nephew. His starting salary was $475,000. Summary statistics are as follows:

n	mean	SD	min	Q_1	med	Q_3	max	IQR
11	88,364	128,271	45,000	48,000	50,000	54,000	475,000	6,000

Someone applying to this company who is interested in a typical starting salary would probably want to delete the CEO's nephew's salary from consideration. Removing this outlier results in the summary statistics:

n	mean	SD	min	Q_1	med	Q_3	max	IQR
10	49,700	3302	45,000	48,000	50,000	53,000	54,000	5,000

Note that the mean and standard deviation, both sensitive to outliers, changed dramatically when the outlier was removed. However, the median and IQR, as well as Q_1 and Q_3, all resistant to outliers, changed only slightly, if at all, when the outlier was removed.

Comparing Distributions of a Quantitative Variable

Many studies involve comparing groups. Following are four examples showing comparisons involving back-to-back stemplots, side-by-side histograms, parallel boxplots, and cumulative frequency plots.

⟩ Example 1.30

The numbers of wins for the 30 NBA teams at the end of a recent season are shown in the following *back-to-back stemplot*.

> **NOTE**
> The leaves radiate out from a common stem.

> **NOTE**
> Although both sets in this example are of the same size, in general you do not have to have the same number of data points on each side of the plot.

```
        Eastern          Western
       Conference       Conference
          9 7    | 1 | 9
          9 2    | 2 |
          9 9 2  | 3 | 3 3 3 6 7 9
    9 8 2 2 1    | 4 | 8 8 9
          8 1    | 5 | 0 3 3 4 7
            0    | 6 |
```

7 | 1 | 9 represents Eastern and
Western Conference teams with
17 and 19 wins, respectively.

Compare the two distributions.

(continued)

✓ **Solution**

Shape: The distribution of wins in the Eastern Conference (EC) is roughly bell-shaped, while the distribution of wins in the Western Conference (WC) is roughly uniform with a low outlier.

Center: Counting values (8th out of 15) gives medians of $m_{EC} = 41$ and $m_{WC} = 48$. Thus, the WC distribution of wins has the greater center.

Spread: The range of the EC distribution of wins is $60 - 17 = 43$, while the range of the WC distribution of wins is $57 - 19 = 38$. Thus, the EC distribution of wins has the greater spread.

> **Important**
>
> When asked for a comparison, don't forget to address Shape, Unusual features (Outliers, gaps, clusters), Center, and Spread (SOCS or CUSS) and to refer to context. You must use *comparative* words—that is, you must state which center and which spread is larger (or if they are approximately the same). Simply describing each distribution separately is not enough and will be penalized.

Unusual features: The WC distribution has an apparent outlier at 19 and a gap between 19 and 33, which is different than the EC distribution that has no apparent outliers or gaps.

⟩ **Example 1.31**

Two surveys, one of high school students and one of college students, asked students how many hours they sleep per night. The following histograms summarize the distributions.

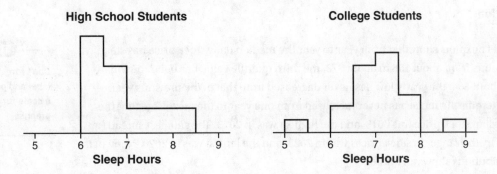

Compare the two distributions.

✓ **Solution**

Shape: The distribution of sleep hours in the high school student distribution is skewed right, while the distribution of sleep hours in the college student distribution is unimodal and roughly symmetric.

Center: The median sleep hours for the high school students (between 6.5 and 7) is less than the median sleep hours for the college students (between 7 and 7.5).

Spread: The range of the college student sleep hour distribution is greater than the range of the high school student sleep hour distribution.

Unusual features: The college student sleep hour distribution has two distinct gaps, 5.5 to 6 and 8 to 8.5, and possible low and high outliers, while the high school student sleep hour distribution doesn't clearly show possible gaps or outliers.

> **Example 1.32**

The following are parallel boxplots showing the daily price fluctuations of a particular common stock over the course of 5 years. What trends do the boxplots show?

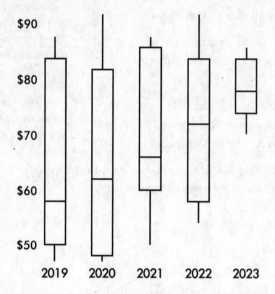

✓ **Solution**

The parallel boxplots show that from year to year the median daily stock price has steadily risen 20 points from about $58 to about $78, the third quartile value has been roughly stable at about $84, the yearly low has never decreased from that of the previous year, and the interquartile range has never increased from one year to the next. Note that the lowest median stock price was 2019, and the highest was in 2023. The smallest spread (as measured by the range) in stock prices was in 2023, and the largest was in 2020. None of the price distributions shows an outlier.

TIP

Don't forget to label and provide a scale for all graphs!

> **Example 1.33**

The graph below compares cumulative frequency plotted against age for the U.S. population in 1860 and in 1980.

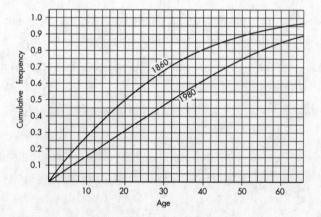

How do the medians and interquartile ranges compare?

(continued)

✓ **Solution**

Looking across from 0.5 on the vertical axis, we see that in 1860 half the population was under the age of 20, while in 1980 all the way up to age 32 must be included to encompass half the population. Looking across from 0.25 and 0.75 on the vertical axis, we see that for 1860, $Q_1 = 9$ and $Q_3 = 35$ and so the interquartile range is $35 - 9 = 26$ years, while for 1980, $Q_1 = 16$ and $Q_3 = 50$ and so the interquartile range is $50 - 16 = 34$ years. Thus, both the median and the interquartile range were greater in 1980 than in 1860.

Quiz 2

Multiple-Choice Questions

> **DIRECTIONS:** The questions that follow are each followed by five suggested answers. Choose the response that best answers the question.

1. The graph below shows household income in Laguna Woods, California.

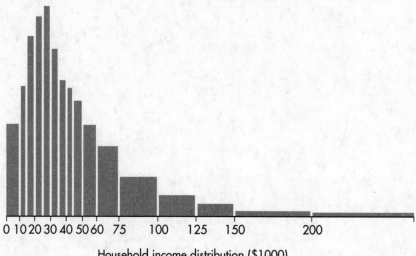

Household income distribution ($1000)

What can be said about the ratio $\dfrac{\text{Mean household income}}{\text{Median household income}}$?

(A) Approximately zero

(B) Less than one, but definitely above zero

(C) Approximately one

(D) Greater than one

(E) Cannot be answered without knowing the standard deviation

2. Students in an algebra class were timed in seconds while solving a series of mathematical brainteasers. One student's time had a standardized score of $z = 2.40$. If the times are all changed to minutes, what will then be the student's standardized score?

(A) $z = 0.04$

(B) $z = 0.4$

(C) $z = 1.80$

(D) $z = 2.40$

(E) The new standardized score cannot be determined without knowing the class mean.

3. Dieticians are concerned about sugar consumption in teenagers' diets (a 12-ounce can of soft drink typically has 10 teaspoons of sugar). In a random sample of 55 students, the number of teaspoons of sugar consumed for each student on a randomly selected day is tabulated. Summary statistics are noted below:

   Min = 10   Max = 60   First quartile = 25   Third quartile = 38

   Median = 31   Mean = 31.4 $n = 55$ $s = 11.6$

   Which of the following is a true statement?

   (A) None of the values are outliers.
   (B) The value 10 is an outlier, and there can be no others.
   (C) The value 60 is an outlier, and there can be no others.
   (D) Both 10 and 60 are outliers, and there may be others.
   (E) The value 60 is an outlier, and there may be others at the high end of the data set.

4. Below is a boxplot of CO_2 levels (in grams per kilometer) for a sampling of year 2024 vehicles.

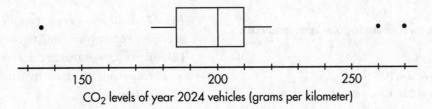

   CO₂ levels of year 2024 vehicles (grams per kilometer)

   Suppose follow-up testing determines that the low outlier should be 10 grams per kilometer less and the two high outliers should each be 5 grams per kilometer greater. What effect, if any, will these changes have on the mean and median CO_2 levels?

   (A) Both the mean and median will be unchanged.
   (B) The median will be unchanged, but the mean will increase.
   (C) The median will be unchanged, but the mean will decrease.
   (D) The mean will be unchanged, but the median will increase.
   (E) Both the mean and median will change.

5. The graph below shows cumulative proportions plotted against grade point averages for a large public high school.

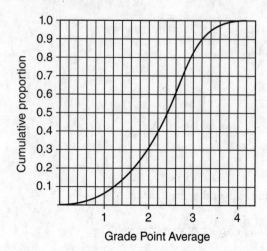

What are the median grade point average and the IQR?

(A) Median = 0.8, IQR = 1.8
(B) Median = 2.0, IQR = 2.8
(C) Median = 2.4, IQR = 1.0
(D) Median = 2.5, IQR = 1.0
(E) Median = 2.6, IQR = 1.8

6. Can either the standard deviation or the inter-quartile range ever be negative?

(A) Standard deviation: never
     Interquartile range: never
(B) Standard deviation: never
     Interquartile range: sometimes
(C) Standard deviation: sometimes
     Interquartile range: never
(D) Standard deviation: sometimes
     Interquartile range: sometimes
(E) This cannot be answered without knowing the context.

7. A teacher is teaching two AP Statistics classes. On the final exam, the 20 students in the first class averaged 92, while the 25 students in the second class averaged only 83. If the teacher combines the classes, what will the average final exam score be?

(A) 87
(B) 87.5
(C) 88
(D) None of the above.
(E) More information is needed to make this calculation.

8. Allan famously quipped that when he moved from Pennsylvania to California, the average IQ dropped in both states. What would have had to been true for this to happen?

(A) Allan's IQ was greater than the average IQ in both Pennsylvania and California.
(B) Allan's IQ was greater than the average IQ in Pennsylvania and less than the average IQ in California.
(C) Allan's IQ was less than the average IQ in Pennsylvania and greater than the average IQ in California.
(D) Allan's IQ was less than the average IQ in both Pennsylvania and California.
(E) There is no way for Allan's statement to be true.

Questions 9 and 10 refer to the following: 250 students are taking a college history course. There is one class of size 150 and four classes of size 25.

9. What is the average class size among the five classes?

(A) 35
(B) 50
(C) 100
(D) 125
(E) 137.5

10. Among the 250 students, what is the average size, per student, of their history class?

(A) 35
(B) 50
(C) 100
(D) 125
(E) 137.5

Free-Response Questions

DIRECTIONS: You must show all work and indicate the methods you use. You will be graded on the correctness of your methods and on the accuracy of your final answers.

1. Victims spend from 5 to 5,840 hours repairing the damage caused by identity theft with a mean of 330 hours and a standard deviation of 245 hours.

   (a) What would be the mean, range, standard deviation, and variance for hours spent repairing the damage caused by identity theft if each of the victims spent an additional 10 hours?
   (b) What would be the mean, range, standard deviation, and variance for hours spent repairing the damage caused by identity theft if each of the victims' hours spent increased by 10%?

2. Are women better than men at multitasking? Suppose in one study of multitasking a random sample of 200 female and 200 male high school students were assigned several tasks at the same time, such as solving simple mathematics problems, reading maps, and answering simple questions while talking on a telephone. Total times taken to complete all the tasks are given in the histograms below.

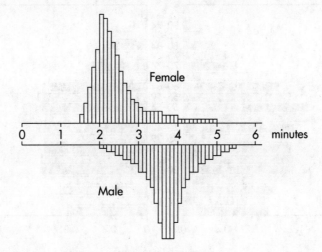

   Write a few sentences comparing the distributions of times to complete all tasks by females and by males.

3. Suppose a distribution has mean 300 and standard deviation 25. If the z-score of Q_1 is -0.7 and the z-score of Q_3 is 0.7, what values would be considered outliers?

4. Below are population pyramids from Liberia and Canada.

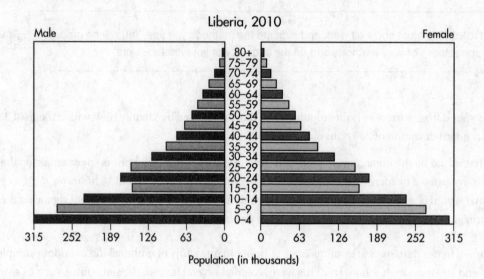

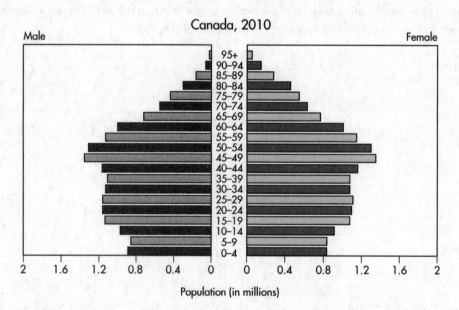

(a) The approximate median age of the Liberian population falls in which of these intervals: 0–4, 15–19, 30–34, 40–44? Explain.

(b) Explain why it is impossible to calculate the mean age of either population.

(c) Which country has more children younger than 10 years of age? Explain.

(d) Does the population pyramid indicate that Canadian men or Canadian women live longer? Explain.

(e) In 2010, Liberia had recently come out of a civil war with the extensive use of child soldiers. How is this visible in the population pyramid?

5. Cumulative frequency plots of the heights (in centimeters) of males from (A) Sudan and (B) Rwanda are given below.

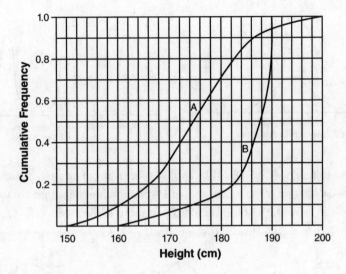

Compare the distributions of heights of males from Sudan and Rwanda.

The answers for this quiz can be found beginning on page 476.

The Normal Distribution

One bell-shaped distribution, called the normal distribution, is valuable for providing a useful model in describing various natural phenomena. However, we will see that the real importance of the normal distribution in statistics is that it can be used to describe the results of many sampling procedures.

The normal distribution curve is bell-shaped and symmetric and has an infinite base. Long, flat-looking tails cover many values but only a small proportion of the area.

The mean of a normal distribution is equal to the median and is located at the center. We want a unit of measurement that applies equally well to any normal distribution, and we choose a unit that arises naturally out of the curve's shape. There is a point on each side where the slope is steepest. These two points are called *points of inflection*, and the distance from the mean to either point is precisely equal to one standard deviation. Thus, it is convenient to measure distances under the normal curve in terms of *z*-scores. (Recall from earlier in this unit that *z*-scores represent the number of standard deviations a particular value is above or below the mean.)

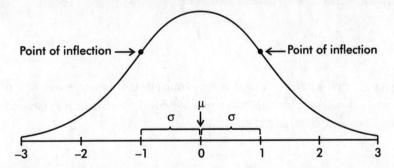

Every normal distribution is completely determined by two parameters: its mean μ and its standard deviation σ.

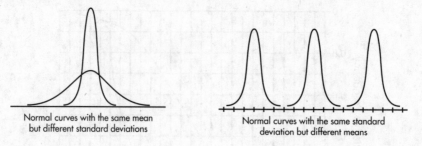

Normal curves with the same mean
but different standard deviations

Normal curves with the same standard
deviation but different means

The *empirical rule* (also called the *68-95-99.7 rule*) applies specifically to normal distributions. About 68% of the values lie within 1 standard deviation of the mean, about 95% of the values lie within 2 standard deviations of the mean, and about 99.7% of the values lie within 3 standard deviations of the mean.

In the following figures, the horizontal axes show z-scores:

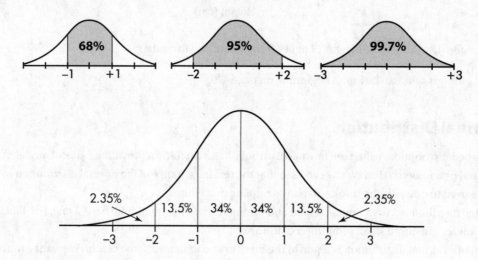

> **Example 1.34**

Suppose that taxicabs in New York City are driven an average of 75,000 miles per year with a standard deviation of 12,000 miles. What information does the empirical rule give us?

✓ **Solution**

Assuming that the distribution is roughly normal, we can conclude that approximately 68% of the taxis are driven between 63,000 and 87,000 miles per year, approximately 95% are driven between 51,000 and 99,000 miles, and virtually all are driven between 39,000 and 111,000 miles.

The empirical rule also gives a useful quick estimate of the standard deviation in terms of the range. We can see in the figure on the previous page that 95% of the data fall within a span of 4 standard deviations (from -2 to $+2$ on the z-score line) and 99.7% of the data fall within 6 standard deviations (from -3 to $+3$ on the z-score line). It is therefore reasonable to conclude that for these data the standard deviation is roughly between one-fourth and one-sixth of the range of a *finite* set. Since we can find the range of a finite set almost immediately, the empirical rule technique for estimating the standard deviation is often helpful in pointing out gross arithmetic errors.

> ### Example 1.35

A financial adviser randomly samples 110 recent college graduates and asks about their monetary debt upon graduation. The resulting data are displayed in the following dotplot:

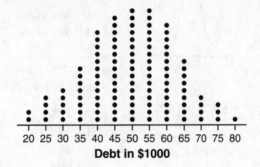

Debt in $1000

Which of the following is the best estimate of the standard deviation of these debts?

(A) 5
(B) 7.5
(C) 12.5
(D) 17.5
(E) 20

✓ Solution

The range of this data is $80 - 20 = 60$. The distribution is roughly bell-shaped, so by the empirical rule, the standard deviation is expected to be between $\frac{1}{6}(60) = 10$ and $\frac{1}{4}(60) = 15$. Among the answer choices, 12.5 is the only one in the interval from 10 to 15, so choose (C).

Don't be confused into thinking that z-scores mean normality. A z-score can be calculated whenever the mean and standard deviation are known, no matter what the distribution. Only in the very special case of nearly normal distributions can one move easily from z-scores to percentiles and probabilities.

Quiz 3

Multiple-Choice Questions

> **DIRECTIONS:** The questions that follow are each followed by five suggested answers.
> Choose the response that best answers the question.

1. Consider the following two normal curves.

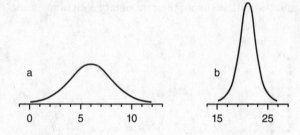

Which has the smaller mean, and which has the smaller standard deviation?

(A) Smaller mean, *a*; smaller standard deviation, *a*

(B) Smaller mean, *a*; smaller standard deviation, *b*

(C) Smaller mean, *b*; smaller standard deviation, *a*

(D) Smaller mean, *b*; smaller standard deviation, *b*

(E) Smaller mean, *a*; same standard deviation

Questions 2–4 refer to the following. Suppose an average life expectancy in a particular country is 74.3 years with a standard deviation of 9.7 years. Assuming a roughly normal distribution, consider the following graph of the distribution.

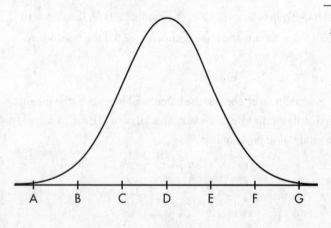

2. Point E on this normal curve corresponds to

(A) 54.9 years.

(B) 64.6 years.

(C) 74.3 years.

(D) 84.0 years.

(E) 93.7 years.

3. About what percent of the years are between 54.9 and 93.7 (points B and F)?

(A) 17%

(B) 34%

(C) 68%

(D) 95%

(E) 99.7%

4. About what percent of the years are less than 64.6 (point C)?

(A) 17%

(B) 34%

(C) 68%

(D) 95%

(E) 99.7%

5. The distribution of body fat percentages of adult males is approximately normal with a standard deviation of 2.5%. A male whose body fat percentage is over 25% is considered obese. If 16% of males are obese, what is the average body fat percentage?

(A) 16%

(B) 17.5%

(C) 20%

(D) 22.5%

(E) 25%

6. The following dotplot shows the speeds (in mph) of 100 fastballs thrown by a major league pitcher.

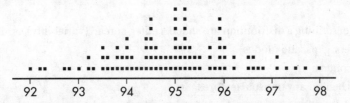

Which of the following is the best estimate of the standard deviation of these speeds?

(A) 0.5 mph
(B) 1.1 mph
(C) 1.6 mph
(D) 2.2 mph
(E) 6.0 mph

Free-Response Questions

DIRECTIONS: You must show all work and indicate the methods you use. You will be graded on the correctness of your methods and on the accuracy of your final answers.

1. It is estimated that 23% of American teen spending goes toward food. Assume that yearly amounts spent on food can be approximated by a normal model with a mean of $600 and a standard deviation of $120.

   (a) A teen spending $720 would be at what percentile?

   (b) What percent of American teens spend between $360 and $840 per year on food?

   (c) 99.7% of American teens spend between what two dollar amounts on food per year?

2. A high school guidance counselor tallied the number of colleges that seniors applied to in Fall 2023 and created the following descriptive statistics for the data set.

Variable	n	Mean	Median	StDev	Min	Max	Q_1	Q_3
Colleges	50	6.4	5	4.80	0	15	2	10

Do the numbers of colleges applied to by this school's seniors in Fall 2023 seem to roughly follow a normal model?

The answers for this quiz can be found beginning on page 478.

SUMMARY

- The four keys to describing a distribution are shape, center, spread (variability), and unusual features (such as outliers, clusters, gaps, and modes).
- Always provide context.
- Look for reasons behind any unusual features.
- A few common shapes that arise are symmetric, skewed to the right, skewed to the left, bell-shaped, and uniform distributions.
- Categorical variables are summarized by counts (frequency) and proportions (relative frequency) and visually displayed by dotplots and bar graphs.
- For quantitative data, dotplots, histograms, cumulative relative frequency plots (ogives), stemplots, and boxplots, give useful displays.
- In a histogram, relative area corresponds to relative frequency.
- The two principal measurements of the center of a distribution are the mean and the median.
- The principal measurements of the spread (variability) of a distribution are the range (maximum value minus minimum value), the interquartile range ($IQR = Q_3 - Q_1$), the variance, and the standard deviation.
- Adding the same constant to every value in a set adds the same constant to the mean and median but leaves all the aforementioned measures of variability unchanged.
- Multiplying every value in a set by the same constant multiplies the mean, median, range, IQR, and standard deviation by that constant.
- The mean, range, variance, and standard deviation are sensitive to extreme values, while the median and interquartile range are not.
- The principal measurements of position are simple ranking, percentile ranking, and the z-score (which measures the number of standard deviations a particular value is above or below the mean).
- No matter what the distribution of raw scores, the set of z-scores always has mean 0 and standard deviation 1.
- The empirical rule (the 68-95-99.7 rule) applies specifically to normal distributions.
- In skewed left data, the mean is usually less than the median, while in skewed right data, the mean is usually greater than the median.
- Boxplots show the five-number summary: the minimum value, the first quartile (Q_1), the median, the third quartile (Q_3), and the maximum value; they indicate outliers as distinct points. In general, boxplots cannot show overall shape; however, they may suggest skewness.
- Two sets can have the same five-number summary and thus the same boxplots but have dramatically different distributions.
- Outliers are any values less than $Q_1 - 1.5(IQR)$ or greater than $Q_3 + 1.5(IQR)$. (An accepted second method of calculating outliers is to designate as outliers all values located two or more standard deviations above, or below, the mean.)
- When asked for a comparison of two data sets, in addition to comparing shape and unusual features, you must state which center and which spread is greater (two separate calculations are not a comparison).
- Advantages and disadvantages of four principal plots or graphical displays:
  - *Dotplot:* Advantages—shows every value, and it is easy to see the shape of the distribution. Disadvantage—difficult to plot for very large data sets.
  - *Stemplot:* Advantages—shows every value, and it is easy to see the shape of the distribution. Disadvantage—difficult to plot for very large data sets.
  - *Histogram:* Advantages—easy to make for large data sets, and it is easy to see the shape of the distribution. Disadvantage—does not show individual values.
  - *Boxplot:* Advantages—quickly shows the five-number summary and outliers while splitting the data into quartiles. Disadvantages—does not show individual values and can hide features of the distribution, such as clusters, gaps, and even the overall shape.

2

Exploring Two-Variable Data

(2–3 multiple-choice questions on the AP exam)

Learning Objectives

In this unit, you will learn how

→ To be able to calculate frequencies and conditional frequencies from a two-way table.

→ To be able to interpret side-by-side bar charts, segmented bar charts, and mosaic plots.

→ To be able to describe bivariate data from a scatterplot using the terminology direction (positive or negative), form (linear or nonlinear), strength (such as weak, moderate, or strong), and any unusual points, and always in context.

→ To be able to find the equation of a least squares regression line from generic computer regression output.

→ To be able to calculate the slope and y-intercept of the least squares regression line using relevant summary statistics (means, standard deviations, and correlation).

→ To be able to interpret the slope and y-intercept of a linear regression line in context, using nondeterministic language.

→ To be able to calculate and interpret a residual.

→ To be able to roughly estimate a correlation coefficient from a scatterplot.

→ To be able to interpret the coefficient of determination in context.

→ To be able to identify outliers, influential points, and points of high leverage in a scatterplot.

In this unit, you will work with two-way tables, together with graphs and calculations, to explore relationships in two-variable categorical data sets. To understand association between two quantitative variables, you will analyze scatterplots and generic computer output, assess correlation, interpret slopes and intercepts in context, investigate residuals, and consider nonlinearity. The principal objectives of data analysis with two variables are to explore whether or not there is an association between two variables and to describe the nature of any association among variables.

Two Categorical Variables

Qualitative data often encompass two categorical variables that may or may not have a dependent relationship. These data can be displayed in a *two-way table* (also called a *contingency table*).

> **Example 2.1**

The Cuteness Factor: A Japanese study had 250 volunteers look at pictures of cute baby animals, adult animals, or tasty-looking foods before testing their level of focus in solving puzzles.

		Level of focus		
		Low	Medium	High	*Total*
Pictures viewed	Baby animals	5	20	40	*65*
	Adult animals	30	40	15	*85*
	Tasty foods	55	35	10	*100*
	Total	*90*	*95*	*65*	*250*

"Pictures viewed" is the *row variable*, and "level of focus" is the *column variable*. In analyzing tables, it is usually helpful to sum rows and columns as is done above. These totals are called *marginal frequencies* and can be put in the form of proportions or percentages called *relative frequencies*. The *relative frequency distributions* for level of focus and pictures viewed are

Level of focus

Low: $\frac{90}{250} = 0.36 = 36\%$

Medium: $\frac{95}{250} = 0.38 = 38\%$

High: $\frac{65}{250} = 0.26 = 26\%$

Pictures viewed

Baby animals: $\frac{65}{250} = 0.26 = 26\%$

Adult animals: $\frac{85}{250} = 0.34 = 34\%$

Tasty foods: $\frac{100}{250} = 0.40 = 40\%$

The distributions can be displayed in bar graphs as follows:

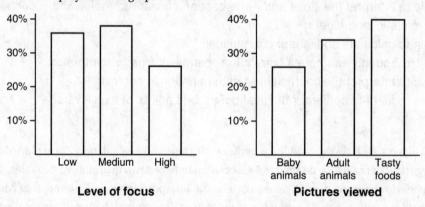

The relative frequency distributions described and calculated above do not describe or measure the relationship between the two categorical variables. For this we must consider the information in the body of the table, not just the sums in the margins.

> **Example 2.2**

We are interested in predicting the level of focus from the pictures viewed, and so we look at *conditional relative frequencies* for each row separately. For instance, in Example 2.1, what proportion or percentage of the participants who viewed baby animals then had each of the levels of focus?

Low: $\frac{5}{65} = 0.077 = 7.7\%$

Medium: $\frac{20}{65} = 0.308 = 30.8\%$

High: $\frac{40}{65} = 0.615 = 61.5\%$

> **NOTE**
>
> *Conditional relative frequencies* are the relative frequencies for a specific part of the table, for example, the cell frequencies in a row divided by the total for that row or the cell frequencies in a column divided by the total for that column.

(continued)

This *conditional relative frequency distribution* can be displayed either with groupings of bars or by a *segmented bar chart* where each segment has a length corresponding to its relative frequency:

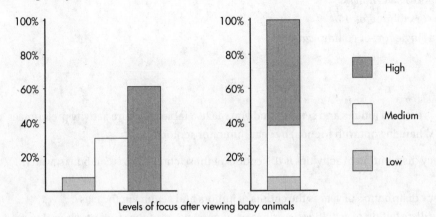

Levels of focus after viewing baby animals

Similarly, the level of focus conditional relative frequency distributions for the participants who viewed adult animals and who viewed tasty foods can be calculated.

Level of focus

Participants who viewed adult animals	Low:	$\frac{30}{85} = 0.353 = 35.3\%$	Participants who viewed tasty foods	Low:	$\frac{55}{100} = 0.55 = 55\%$
	Medium:	$\frac{40}{85} = 0.471 = 47.1\%$		Medium:	$\frac{35}{100} = 0.35 = 35\%$
	High:	$\frac{15}{85} = 0.176 = 17.6\%$		High:	$\frac{10}{100} = 0.10 = 10\%$

Both of the following bar charts give good visual representations of the data:

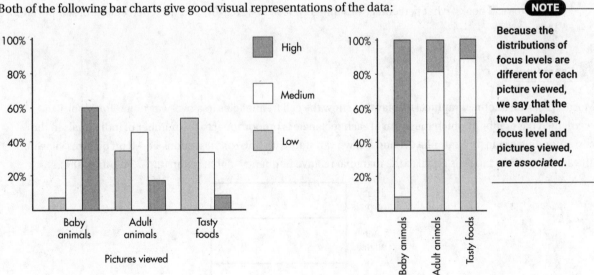

The groupings of bars and the segmented bar charts both clearly indicate that viewing pictures of baby animals tends to lead to higher levels of focus, viewing tasty foods tends to lead to lower levels of focus, and viewing adult animals tends to lead to medium levels of focus.

In summary:

- *Frequency* refers to *counts.*
- *Relative frequency* refers to *percentages.*
- *Marginal* refers to row or column *totals.*
- *Conditional* refers to a single row or column *category.*

❯ Example 2.3

Suppose your English class is a mix of juniors and seniors, and you make a table of leisure activity preferences (reading books, watching TV, hanging out with friends) by year (junior or senior).

- The marginal frequency distribution of activities is the counts of those choosing to read books, watch TV, and hang out.
- The marginal frequency distribution of year is the counts of juniors and seniors in the class.
- The relative frequency distribution of activities describes what percentage of the students choose to read books, what percentage choose to watch TV, and what percentage choose to hang out.
- The relative frequency distribution of year describes what percentage of the class are juniors and what percentage are seniors.
- The conditional frequency distribution among those choosing to read books shows how many of the juniors choose to read books and how many of the seniors choose to read books.
- The conditional frequency distribution of activities among juniors shows how many of the juniors choose to read books, how many juniors choose to watch TV, and how many juniors choose to hang out.
- The conditional relative frequency distribution among those choosing to watch TV describes what percentage of those choosing to watch TV are juniors and what percentage choosing to watch TV are seniors.
- The conditional relative frequency distribution of activities for seniors describes what percentage of seniors choose to read books, what percentage of seniors choose to watch TV, and what percentage of seniors choose to hang out.

❯ Example 2.4

Mosaic plots are another graphical display that show the cell frequencies in a two-way table. They look like segmented bar graphs in which the width of each rectangle is proportional to the number of individuals in the corresponding group. In Examples 2.1 and 2.2, we can start with horizontal sections based on pictures viewed. The area of each rectangle is equal to the marginal relative frequency of that group (26%, 34%, and 40%).

(continued)

Then, each horizontal section is split into sections based on focus level. The area of each smaller rectangle is equal to the conditional relative frequency for each specific group as calculated in Example 2.2 (for example, 7.7%, 30.8%, and 61.5% for baby animals).

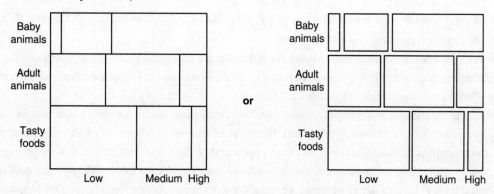

Alternatively, we could have first started with vertical sections based on level of focus.

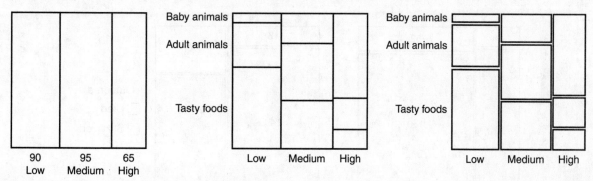

> Example 2.5

Suppose you need heart surgery and are trying to decide between two surgeons, Dr. Fixit and Dr. Patch. You find out that each operated 250 times last year with the following results:

	Dr. F	Dr. P
Died	60	50
Survived	190	200

Which surgeon would you choose? Among Dr. Fixit's 250 patients, 190 survived, for a survival rate of $\frac{190}{250} = 0.76$ or 76%, while among Dr. Patch's 250 patients, 200 survived, for a survival rate of $\frac{200}{250} = 0.80$ or 80%. Your choice seems clear.

However, everything may not be so clear-cut. Suppose that on further investigation you determine that the surgeons operated on patients who were in either good or poor condition with the following results:

Good condition	Dr. F	Dr. P
Died	8	17
Survived	60	120

Poor condition	Dr. F	Dr. P
Died	52	33
Survived	130	80

Note that adding corresponding boxes from these two tables gives the original table above.

(continued)

How do the surgeons compare when operating on patients in good health? In poor health?

Dr. Fixit's 68 patients in good condition have a survival rate of $\frac{60}{68} = 0.882$, or 88.2%, while Dr. Patch's 137 patients in good condition have a survival rate of $\frac{120}{137} = 0.876$, or 87.6%. Similarly, we note that Dr. Fixit's 182 patients in poor condition have a survival rate of $\frac{130}{182} = 0.714$ or 71.4%, while Dr. Patch's 113 patients in poor condition have a survival rate of $\frac{80}{113} = 0.708$ or 70.8%.

Thus, Dr. Fixit does better with patients in good condition (88.2% versus Dr. Patch's 87.6%) and also does better with patients in poor condition (71.4% versus Dr. Patch's 70.8%). However, Dr. Fixit has a lower overall patient survival rate (76% versus Dr. Patch's 80%)! How can this be?

This problem is an example of *Simpson's paradox*, where a conjecture can be reversed when several different groups of data are combined to form a single group. The effect of another variable is masked when the groups are combined. In this particular example, closer scrutiny reveals that Dr. Fixit operates on many more patients in poor condition than Dr. Patch, and these patients in poor condition are precisely the ones with lower survival rates. Thus, even though Dr. Fixit does better with each group of patients, his overall rating is lower. Our original table hid the effect of the variable related to the condition of the patients. Mosaic plots help clarify the paradox:

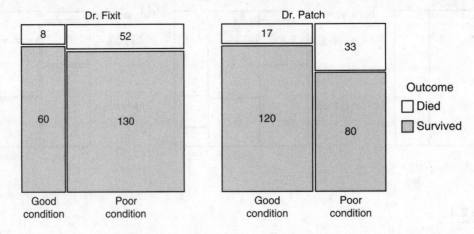

Note how the above visually illustrates that Dr. Fixit operates on more patients in poor condition than in good condition, while Dr. Patch operates on more patients in good condition than in poor condition. It also illustrates that Dr. Fixit's patients, in both good condition and poor condition, survive at higher rates than Dr. Patch's patients.

Quiz 4

Multiple-Choice Questions

DIRECTIONS: The questions that follow are each followed by five suggested answers. Choose the response that best answers the question.

Questions 1–5 are based on the following: Students, teachers, and administrators were asked which of three teacher characteristics (challenging, enthusiastic, strict) they considered most important for a successful classroom experience. Five hundred people in a high school community were surveyed with the following results:

	Challenging	Enthusiastic	Strict	Total
Student	50	150	50	250
Teacher	125	50	25	200
Administrator	15	10	25	50
Total	190	210	100	500

1. What percentage of those surveyed were students?

   (A) 10%
   (B) 20%
   (C) 30%
   (D) 40%
   (E) 50%

2. What percentage of those surveyed picked challenging as most important and were teachers?

   (A) 25%
   (B) 38%
   (C) 40%
   (D) 62.5%
   (E) 65.8%

3. What percentage of administrators picked strict as most important?

   (A) 5%
   (B) 10%
   (C) 20%
   (D) 25%
   (E) 50%

4. What percentage of those picking enthusiastic as most important were students?

   (A) 30%
   (B) 42%
   (C) 50%
   (D) 60%
   (E) 71.4%

5. Which group of people were most likely to pick strict as most important?

   (A) Student
   (B) Teacher
   (C) Administrator
   (D) Teacher and administrator, equally
   (E) Student, teacher, and administrator, equally

Questions 6–7 are based on the following. A study of music preferences in three geographic locations resulted in the following segmented bar chart:

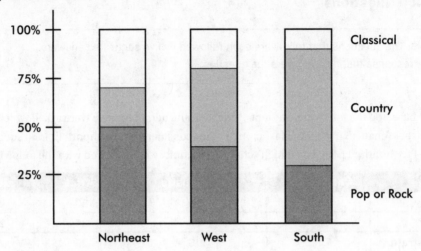

6. Which of the following is greatest?

(A) The percentage of those from the Northeast who prefer classical is greatest.
(B) The percentage of those from the West who prefer country is greatest.
(C) The percentage of those from the South who prefer pop or rock is greatest.
(D) The above are all equal.
(E) It is impossible to determine the answer without knowing the actual numbers of people involved.

7. Which of the following is greatest?

(A) The number of people in the Northeast who prefer pop or rock is greatest.
(B) The number of people in the West who prefer classical is greatest.
(C) The number of people in the South who prefer country is greatest.
(D) The above are all equal.
(E) It is impossible to determine the answer without knowing the actual numbers of people involved.

---

8. High school students were surveyed as to whether or not their GPA was 3.0 or higher and what was the most severe punishment they had ever received from their parents. The results are displayed in the following mosaic plot.

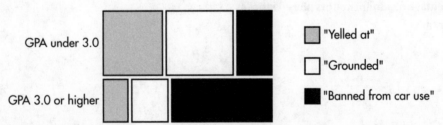

Based on the above plot, which of the following is a true statement?

(A) The number of students with GPAs under 3.0 who were banned from car use is greater than the number of students with GPAs 3.0 or higher who were banned from car use.
(B) More students had a GPA 3.0 or higher and were either yelled at or grounded than had a GPA under 3.0 and were yelled at.
(C) Of the students who were grounded, a greater proportion had GPAs 3.0 or higher than GPAs under 3.0.
(D) More students had a GPA under 3.0 than a GPA 3.0 or higher.
(E) More students were yelled at than were grounded.

Free-Response Questions

> **DIRECTIONS:** You must show all work and indicate the methods you use. You will be graded on the correctness of your methods and on the accuracy of your final answers.

1. For one study of racial disparities in traffic stops, 225 random records for drivers pulled over for speeding at night (when it is difficult to distinguish race/ethnicity at a distance) were analyzed. The following two-way table summarizes one aspect of the data.

	White drivers	Black drivers	Latino drivers	*Total*
Ticket	36	30	27	93
Warning	84	30	18	132
Total	120	60	45	225

   (a) Of the drivers pulled over for speeding who were given a ticket, what proportion were Latino?

   (b) Of the drivers pulled over for speeding who were White, what percent were given a warning?

   (c) The mosaic plot displays the distribution of tickets versus warnings given by race/ethnicity. Describe what this graph reveals about the association between these two variables for the 225 drivers pulled over for speeding.

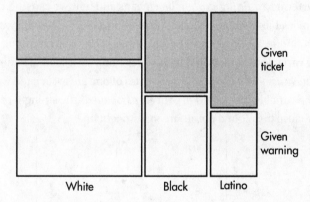

2. In 1973, a question arose concerning possible gender discrimination in admissions to a particular graduate school. The graduate school analyzed admissions to the largest graduate programs, and below are partial results of that study:

Program	Men Accepted	Men Rejected	Women Accepted	Women Rejected
A	511	314	89	19
E	53	138	95	298

   (a) Find the percentage of men and the percentage of women accepted by each of the two programs. Comment on any pattern or bias you see.

   (b) Find the percentage of men and the percentage of women accepted overall by these two programs. Does this appear to contradict the results from part (a)?

   (c) If you worked in the graduate admissions office, what would you say to an inquiring reporter who is investigating gender bias in graduate admissions?

The answers for this quiz can be found beginning on page 479.

Two Quantitative Variables

Many important applications of statistics involve examining whether two or more quantitative (numerical) variables are related to one another. For example, is there a relationship between the smoking histories of pregnant women and the birth weights of their children? Between SAT scores and success in college? Between amount of fertilizer used and amount of crop harvested?

Two questions immediately arise. First, how can the strength of an apparent relationship be measured? Second, how can an observed relationship be put into functional terms? For example, a real estate broker might not only wish to determine whether a relationship exists between the prime rate and the number of new homes sold in a month but might also find useful an expression with which to predict the number of home sales, given a particular value of the prime rate.

A graphical display, called a scatterplot, gives an immediate visual impression of a possible relationship between two variables, while a numerical measurement, called the *correlation coefficient* (or simply the *correlation*), is often used as a quantitative value of the strength of a linear relationship. In either case, evidence of a relationship is not evidence of causation.

Suppose a relationship is perceived between two quantitative variables, X and Y. We are interested in the strength and direction of this relationship and in any deviation from the basic pattern of this relationship. We graph the pairs (x, y). In this topic, we examine whether the relationship, as illustrated by the scatterplot, can be reasonably explained in terms of a linear function, that is, one whose graph is a straight line.

> **NOTE**
>
> No matter whether the variables are positively or negatively associated, for any two points, anything is possible, and either point could be higher or lower than the other.

We need to know what the term *best-fitting straight line* means and how we can find this line. Furthermore, we want to be able to gauge whether the relationship between the variables is strong enough so that finding and making use of this straight line is meaningful.

When larger values of one variable are associated with larger values of a second variable, the variables are called *positively associated*. When larger values of one are associated with smaller values of the other, the variables are called *negatively associated*. The strength of the association is gauged by how close the plotted points are to a straight line.

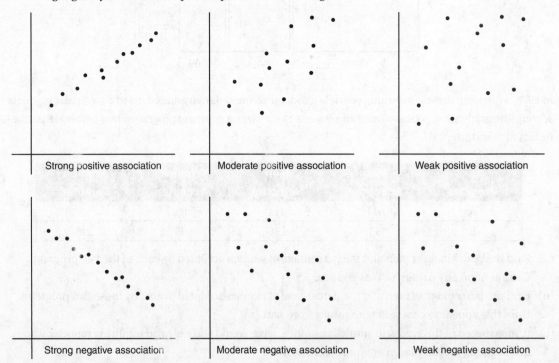

| Strong positive association | Moderate positive association | Weak positive association |
| Strong negative association | Moderate negative association | Weak negative association |

Sometimes different dots in a scatterplot are labeled with different symbols or different colors to show a categorical variable. The resulting labeled scatterplot might distinguish, for example, between men and women, or between stocks and bonds, and so on.

> **Example 2.6**

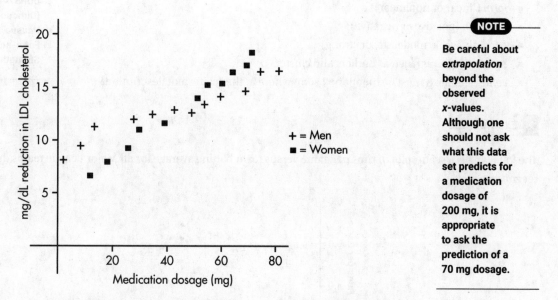

> **NOTE**
>
> Be careful about *extrapolation* beyond the observed *x*-values. Although one should not ask what this data set predicts for a medication dosage of 200 mg, it is appropriate to ask the prediction of a 70 mg dosage.

This diagram is a labeled scatterplot distinguishing men with plus signs and women with square dots to show a categorical variable, gender. The plot indicates that higher doses of the medication lead to greater reductions in LDL (bad cholesterol) for both men and women. However, lower doses of the medication help men more than women in lowering LDL, while higher doses help women more than men.

When analyzing the overall pattern in a scatterplot, it is also important to note *clusters* and *outliers*.

> **Example 2.7**

An experiment was conducted to determine the effect of temperature and light on the potency of a particular antibiotic. One set of vials of the antibiotic was stored under different temperatures but under the same lighting, while a second set of vials was stored under different lightings but under the same temperature.

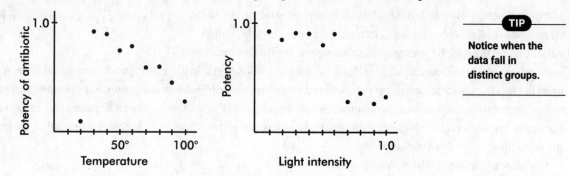

> **TIP**
>
> Notice when the data fall in distinct groups.

In the first scatterplot, notice the roughly linear pattern with one outlier far outside this pattern. A possible explanation is that the antibiotic is more potent at lower temperatures but only down to a certain temperature, at which it drastically loses potency.

In the second scatterplot, notice the two clusters. It appears that below a certain light intensity the potency is one value, while above that intensity it is another value. In each cluster, there seems to be no association between intensity and potency.

As noted in the above examples, when asked to describe a scatterplot, you must consider:

- *form* (linear or nonlinear)
- *direction* (positive or negative)
- *strength* (weak, moderate, or strong)
- *unusual features* (such as outliers and clusters)
- *context* (context must be mentioned somewhere in the scatterplot description)

TIP

DUFS + context (Direction, Unusual features, Form, and Strength)

> Example 2.8

The following is a scatterplot of runs per game versus team batting average for all Major League teams during a recent season.

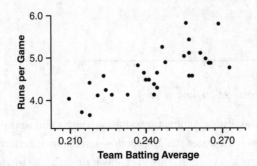

Describe the scatterplot.

✓ Solution

The scatterplot of runs per game versus team batting average for all Major League teams during this season has form: linear; direction: positive; and strength: moderate.

Correlation

Although a scatter diagram usually gives an intuitive visual indication when a linear relationship is strong, in most cases it is quite difficult to visually judge the specific strength of a relationship. For this reason, there is a mathematical measure called *correlation* (or the *correlation coefficient*). Important as correlation is, we always need to keep in mind that significant correlation does not necessarily indicate *causation*. For example, there is a very strong correlation between ice cream sales and frequency of forest fires! (As ice cream sales rise, in general so does temperature, and with higher temperatures, there are more forest fires.) Furthermore, correlation measures the strength of only a *linear* relationship.

Important

Correlation does not imply causation!

Correlation, designated by r, has the formula

$$r = \frac{1}{n-1} \sum \left(\frac{x_i - \bar{x}}{s_x} \right) \left(\frac{y_i - \bar{y}}{s_y} \right)$$

in terms of the means and standard deviations of the two variables. We note that the formula is actually the sum of the products of the corresponding z-scores divided by 1 less than the sample size. However, you should be able to calculate correlation quickly using the statistical package on your calculator. (Examining the formula helps you understand where correlation is coming from, but you will NOT have to use the formula to calculate r.)

The correlation formula gives the following:

- The formula is based on standardized scores (z-scores), and so changing units does not change the correlation r.
- Since the formula does not distinguish between which variable is called x and which is called y, interchanging the variables (on the scatterplot) does not change the value of the correlation r.
- The division in the formula gives that the correlation r is unit-free.

The value of r always falls between -1 and $+1$, with -1 indicating perfect negative correlation and $+1$ indicating perfect positive correlation. It should be stressed that a correlation at or near zero does not mean there is not a relationship between the variables; there may still be a strong *nonlinear* relationship. Additionally, a correlation close to -1 or $+1$ does not necessarily mean that a linear model is the *most* appropriate model.

> **NOTE**
>
> **You should establish a reasonable case for linearity (eyeball the scatterplot) before calculating r. Linearity is a prerequisite, not a conclusion.**

> **NOTE**
>
> **The correlation r is not a measure of linearity. It measures strength and gives the direction of a linear relationship.**

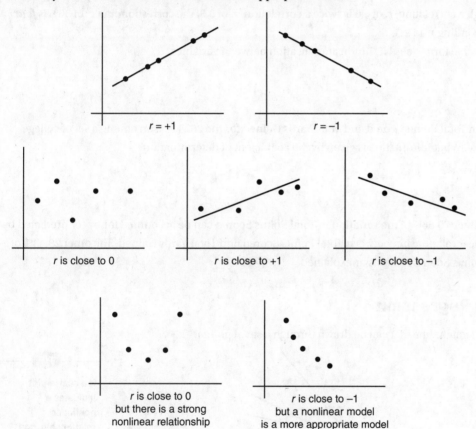

Very roughly, for the purpose of the AP exam, one might say that a correlation with absolute value above 0.8 is "high," from 0.5 to 0.8 is "moderate," and below 0.5 is "low." However, in the real world, everything depends on context. For example, a doctor might feel that an indicator for the severity of cancer spread that has a correlation of 0.9 is not good enough if the result determines major differences in treatment. On the other hand, a financial advisor who discovers a variable with a correlation of 0.1 with stock performance might feel that this is enough to recommend major investments.

The *coefficient of determination*, r^2, measures the percentage of variability in the y-values that can be explained by the linear regression model. That is, $r^2 = \dfrac{\text{variability explained by the model}}{\text{total variability in the } y\text{-values}}$. Important points are as follows:

- The coefficient of determination takes on values between 0.0 and 1.0, or equivalently between 0% and 100%.
- If $r^2 = 100\%$, then all the data points fall on a straight line, and thus all the variation in the response variable y is predictable by variation in the explanatory variable x.

- As a formula, $r^2 = \dfrac{\sum(\hat{y}_i - \bar{y})^2}{\sum(y_i - \bar{y})^2} = 1 - \dfrac{\sum(y_i - \hat{y}_i)^2}{\sum(y_i - \bar{y})^2}$, but you will not have to use this formula as r^2 will almost always be given in generic computer output, usually as "R-Sq =."

- When asked to interpret r^2, fill in: "___% of the variation in (the y-variable in context) is accounted for by the linear model relating (the y-variable in context) and (the x-variable in context)."

- Alternatively, you can say "___% of the variation in (the y-variable in context) is explained by the variation in (the x-variable in context) in the linear model."

- When calculating r from r^2, the square root function on your calculator always gives the positive root; however, r may be positive or negative. To decide whether r is positive or negative, remember that r will always take the same sign as the slope.

- In generic computer output, do not confuse "R-Sq =," which gives our desired r^2, with "R-Sq(adj) =," which refers to a multiple regression topic that is not covered in this course. You will be marked wrong if you use "R-Sq(adj)."

- Note that although a correlation, r, of 0.6 is twice a correlation, r, of 0.3, the corresponding r^2 of 36% is *four* times the corresponding r^2 of 9%.

- Note that a high value for r^2 does NOT indicate causation between variables.

❯ Example 2.9

The correlation between Total Points Scored and Total Yards Gained for the 2023 season among a set of college football teams is $r = 0.84$. What information is given by the coefficient of determination?

✓ Solution

$r^2 = (0.84)^2 = 0.7056$. Thus, 70.56% of the variation in Total Points Scored can be accounted for by (or predicted by or explained by) the linear relationship between Total Points Scored and Total Yards Gained. The other 29.44% of the variation in Total Points Scored remains unexplained.

Least Squares Regression

What is the best-fitting straight line that can be drawn through a set of points?

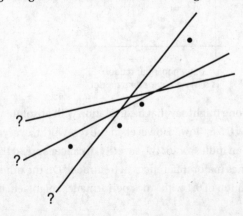

TIP

If a scatterplot indicates a nonlinear relationship, don't try to force a straight-line fit.

On the basis of our experience with measuring variances, by *best-fitting straight line* we mean the straight line that minimizes the sum of the squares of the vertical differences between the observed values and the values predicted by the line.

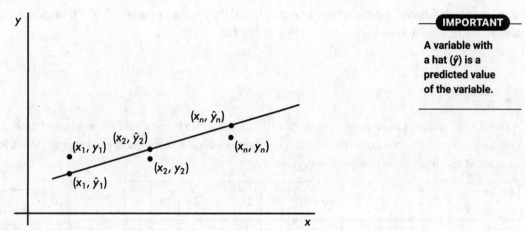

IMPORTANT

A variable with a hat ($\hat{y}$) is a predicted value of the variable.

That is, in this figure, we wish to minimize

$$(y_1 - \hat{y}_1)^2 + (y_2 - \hat{y}_2)^2 + \cdots + (y_n - \hat{y}_n)^2$$

It is reasonable, intuitive, and correct that the best-fitting line will pass through $(\bar{x}, \bar{y})$, where $\bar{x}$ and $\bar{y}$ are the means of the variables X and Y. Then, from the basic expression for a line with a given slope through a given point, we have:

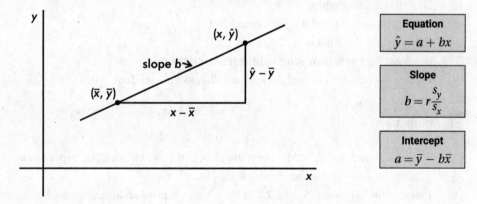

Equation
$\hat{y} = a + bx$

Slope
$b = r\dfrac{s_y}{s_x}$

Intercept
$a = \bar{y} - b\bar{x}$

The slope is the amount that the predicted y-value changes for every unit increase in x. Similarly, the y-intercept is the predicted y-value for $x = 0$.

The slope b can be determined from the formula

$$b = r\frac{s_y}{s_x} = \frac{rs_y}{s_x}$$

where r is the correlation and s_x and s_y are the standard deviations of the two sets. That is, each standard deviation change in x results in a change of r standard deviations in $\hat{y}$. If you graph z-scores for the y-variable against z-scores for the x-variable, the slope of the regression line is precisely r, and in fact, the linear equation becomes $z_y = r z_x$.

This best-fitting straight line, that is, the line that minimizes the sum of the squares of the differences between the observed values and the values predicted by the line, is called the *least squares regression line*, or simply the *regression line*. It can be calculated directly by entering the two data sets and using the statistics package on your calculator.

❯ Example 2.10

Consider the three points (2, 11), (3, 17), and (4, 29). Given any straight line, we can calculate the sum of the squares of the three vertical distances from these points to the line. What is the smallest possible value this sum can be?

(continued)

✓ Solution

Calculator software gives the regression line to be $\hat{y} = -8 + 9x$. This regression line minimizes the sum of the squares of the vertical distances between the points and the line. In this case $(2, 10)$, $(3, 19)$, and $(4, 28)$ are on the line, so the minimum sum is $(10 - 11)^2 + (19 - 17)^2 + (28 - 29)^2 = 6$.

❯ Example 2.11

A sociologist conducts a survey of 15 teens. The number of "close friends" and the number of times Facebook is checked every evening are noted for each student. Letting X and Y represent the number of close friends and the number of Facebook checks, respectively, gives:

Close friends: X	25	23	30	25	20	33	18	21	22	30	26	26	27	29	20
Facebook checks: Y	10	11	14	12	8	18	9	10	10	15	11	15	12	14	11

(a) Identify the explanatory and response variables.
(b) Draw a scatterplot.
(c) Describe the scatterplot.
(d) What is the equation of the regression line?
(e) Interpret the slope in context.
(f) Interpret the coefficient of determination in context.
(g) Predict the number of evening Facebook checks for a student with 24 close friends.

✓ Solutions

(a) The explanatory variable, X, is the number of close friends, and the response variable, Y, is the number of evening Facebook checks.
(b) Plotting the 15 points $(25, 10)$, $(23, 11)$, . . . , $(20, 11)$ gives an intuitive visual impression of the relationship:

> **TIP**
> Be sure to label axes and show number scales whenever possible.

> **NOTE**
> Although the overall pattern is *positive*, given any two points, it is still possible for the first point to have a larger *x*-value and a smaller *y*-value than the second point.

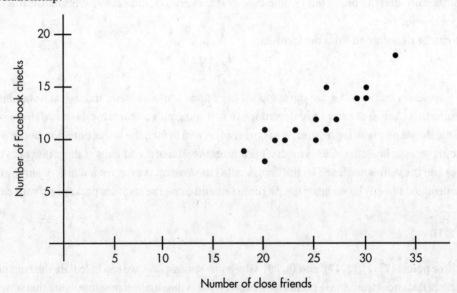

(continued)

(c)   The relationship between the number of close friends and the number of evening Facebook checks appears to be linear, positive, and strong.

(d)   Calculator software gives $\hat{y} = -1.73 + 0.5492x$, where x is the number of close friends and y is the number of evening Facebook checks. We can instead write the following:

$$\text{Predicted Facebook checks} = -1.73 + 0.5492(\text{Close friends}) \text{ or}$$

$$\widehat{\text{Facebook checks}} = -1.73 + 0.5492(\text{Close friends})$$

NOTE

The "hat" above a variable refers to the *predicted* value of that variable.

(e)   The slope is 0.5492. Each additional close friend leads to an average of 0.5492 more evening Facebook checks.

(f)   Calculator software gives $r = 0.8836$, so $r^2 = 0.78$. Thus, 78% of the variation in the number of evening Facebook checks is explained by the linear relationship between the number of evening Facebook checks and the number of close friends.

(g) $-1.73 + 0.5492(24) = 11.45$ evening Facebook checks. Students with 24 close friends will average 11.45 evening Facebook checks.

NOTE

We can also say that 22% of the variation in number of evening Facebook checks is *not* accounted for by the linear regression model.

⟩ Example 2.12

By using a respirometer while subjects exercise, scientists can calculate calories burned. The following are walking times (at 4 mph on a treadmill) and calories burned for a sample of six people weighing between 120 and 140 pounds.

Time (minutes): x	30	90	90	75	60	50
Calories burned: y	185	630	585	500	430	400

(a)   Predict how many calories are burned by a 130-pound person walking at 4 mph for 44 minutes.

(b)   Predict how long a 130-pound person has to walk to burn off the calories in a 449-calorie slice of pizza (one-quarter of a large pizza)?

✓ Solutions

(a)   Calculator software gives

$$\widehat{\text{Calories}} = 23.8 + 6.55(\text{Time})$$

and $23.8 + 6.55(44) = 312$. We say that the *mean* (or predicted) number of calories burned by a 130-pound person walking at 4 mph for 44 minutes is 312.

(b)   Note that the roles of x and y are reversed here, but we cannot algebraically solve from the above equation because x and $\hat{y}$ are not the same as y and $\hat{x}$. Calculator software gives the equation of this regression line to be:

$$\widehat{\text{Time}} = -0.829 + 0.1465(\text{Calories})$$

And $-0.829 + 0.1465(449) = 64.9$ minutes. We say that the *mean* (or predicted) number of minutes of walking at 4 mph to burn off a 449-calorie slice of pizza is 64.9 minutes.

NOTE

The numbers 23.8 and 6.55 are statistics, which estimate the true y-intercept and true slope, respectively. We do not know the values of the parameters that these statistics estimate.

As seen above, when we use the regression line to predict a y-value for a given x-value, we are actually predicting the mean y-value for that given x-value. For any given x-value, there are many possible y-values, and we are predicting their mean.

> ### Example 2.13

For those under 30, RNA viral diseases are no fun, but are not typically deadly. However, above age 70, RNA viral diseases are very dangerous. One study produced the following statistics:

NOTE

Ages of RNA viral disease victims: $\bar{x} = 52.5$, $s_x = 29.96$
Probability of dying from the RNA viral disease: $\bar{y} = 0.0465$, $s_y = 0.0580$
Correlation $r = 0.817$

While the slope of the regression line will always have some meaning in context, the y-intercept may or may not be meaningful.

(a)   Find the slope of the least squares regression line and interpret it in context.
(b)   Find the y-intercept of the least squares regression line and interpret it in context.
(c)   What is the equation of the least squares regression line with age as the explanatory variable?
(d)   Based on this study, what is the predicted probability of dying for a 70-year-old with an RNA viral disease?

✓ Solutions

(a)   The slope is $b = r\dfrac{s_y}{s_x} = 0.817\left(\dfrac{0.0580}{29.96}\right) = 0.001582$. As age increases by one year, the model predicts that the probability of dying from an RNA viral disease will increase by an average of 0.001582.

(b)   The y-intercept $= \bar{y} - b\bar{x} = 0.0465 - 0.001582(52.5) = -0.03656$. This refers to a predicted probability when $x = 0$ and is meaningless in this context. Probabilities cannot be negative, and it is unclear what an age of 0 years might refer to.

(c)   Predicted probability of dying $= -0.03656 + 0.001582(\text{Age})$

(d)   The predicted probability of dying for a 70-year old with an RNA viral disease is $-0.03656 + 0.001582(70) = 0.074$.

Residuals

The difference between an observed and a predicted value is called the *residual*. When the regression line is graphed on the scatterplot, the residual of a point is the vertical distance the point is from the regression line.

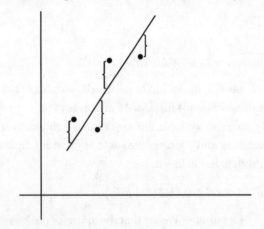

Important points about *residuals* are as follows:

- Order is important: residual equals *observed* minus *predicted*.
- When the data point is above the regression line, the residual is positive; a data point below the regression line gives a negative residual.
- A positive residual means the linear model *underestimated* the actual response value, while a negative residual means the linear model *overestimated* the actual response value.
- The sum, and thus the mean, of the residuals is always zero.
- The regression line is the line that minimizes the sum of the *squares* of the residuals.

- The standard deviation of the residuals gives a measure of how the points are spread around the regression line. In computer output, the standard deviation of the residuals is typically given as "S =."
- To obtain a residual plot, the residuals can be plotted against either the x-values or the $\hat{y}$-values.
- A residual plot with a definite pattern is an indication that a nonlinear model will show a better fit to the data than the straight regression line.
- Understand that although a linear model may be appropriate, it may be weak with a low r^2-value. Alternatively, a linear model may not be the best model (as evidenced by the residual plot), but it still might be a very good model with high r^2-value.

› Example 2.14

What do each of the following residual plots indicate about the linear model?

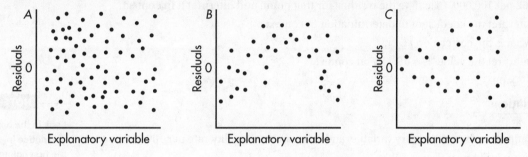

✓ Solution

In A, there is no apparent pattern in the residual plot, indicating that a linear model is appropriate.

In B, the residual plot shows a strong pattern, indicating that a nonlinear model will be a better fit than a straight-line model.

In C, the residual plot shows "fanning," indicating that the linear model gives a stronger fit for smaller x-values than for larger x-values.

The ability to interpret computer output is important not only to do well on the AP Statistics exam but also to understand statistical reports in the business and scientific worlds.

› Example 2.15

Does strength of the sun, as indicated by latitude, have an effect on skin cancer mortality? A study looked at 49 randomly selected regions in North America and noted the skin cancer mortality rate (deaths per 100,000 population) and latitude of the region. Calculating a least squares regression line gives the following computer output:

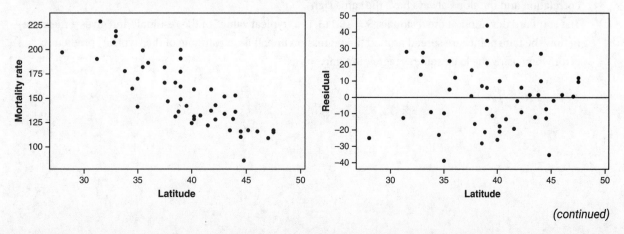

(continued)

Predictor	Coef	SE Coef	T	P
Constant	389.19	23.81	16.34	0.000
Latitude	−5.9776	0.5984	−9.99	0.000
S = 19.115	R-Sq = 68.0%		R-Sq(adj) = 67.3%	

(a)   Identify the explanatory and response variables.

(b)   Describe the relationship given by the scatterplot.

(c)   Does it appear that a line is an appropriate model for the data? Explain.

(d)   What is the slope of the regression line, and what does it signify?

(e)   Is it reasonable to conclude a cause-and-effect relationship between skin cancer mortality and latitude (probably due to strength of the sun at different latitudes)?

(f)   One of the regions was a location in Maryland with a latitude of 39 and a skin cancer mortality rate of 162 per 100,000. Calculate the residual for that point, and interpret it in context.

(g)   Interpret the coefficient of determination in context.

(h)   What is the correlation r?

(i)   Interpret the value "S = 19.115" in context.

✓ Solutions

NOTE

Look in the box and notice in the first column the words "Constant" and "Latitude." In the second column, to the right of "Constant," is the number 389.19, which is the y-intercept of the regression line; to the right of "Latitude" (the explanatory variable) is the number −5.9776, which is the slope of the regression line.

(a)   Latitude is the explanatory variable, and skin cancer mortality rate per 100,000 population is the response variable.

(b)   There is a moderate, negative, linear relationship between latitude and skin cancer mortality rate.

(c)   Yes, a line appears to be an appropriate model between latitude and skin cancer mortality rate because the scatterplot looks linear and there is no apparent pattern in the residual plot.

(d)   The slope of the regression line is −5.9776, signifying that each unit increase in latitude is associated with a 5.9776 decrease in the skin cancer mortality rate per 100,000 population, on average.

(e)   No, correlation does not imply causation.

(f)   For a latitude of 39, the predicted skin cancer mortality rate is $389.19 + (−5.9776)(39) = 156.06$. The residual is actual minus predicted $= 162 − 156.06 = +5.94$. The observed skin cancer mortality rate is 5.94 more than what is predicted by the regression model.

(g)   68.0% of the variation in skin cancer mortality rate is accounted for by the linear model relating skin cancer mortality rate and latitude.

(h) $r = \pm\sqrt{r^2} = -\sqrt{0.680} = -0.825$ where the negative square root is used because the slope is negative (the correlation and the slope always have the same sign).

(i)   The standard deviation of the residuals, S = 19.115, is a "typical value" of the residuals and gives a measure of how the data points are spread around the regression line. It is an estimate of the "typical" prediction error when using the least squares regression line.

Outliers, Influential Points, and Leverage

In a scatterplot, *regression outliers* are indicated by points falling far away from the overall pattern. That is, outliers are points with relatively large discrepancies between the value of the response variable and a predicted value for the response variable. In terms of residuals, a point is an outlier if its residual is an outlier in the set of residuals.

❯ Example 2.16

A scatterplot of grade point average (GPA) versus weekly television time for a group of high school seniors is as follows:

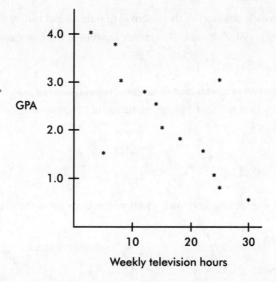

By direct observation of the scatterplot, we note that there are two outliers: one person who watches 5 hours of television weekly yet has only a 1.5 GPA and another person who watches 25 hours weekly yet has a 3.0 GPA. Note also that although the value of 30 weekly hours of television may be considered an outlier for the television hours variable and the 0.5 GPA may be considered an outlier for the GPA variable, the point (30, 0.5) is *not* an outlier in the regression context because it does not fall off the straight-line pattern.

Scores whose removal would sharply change the regression line are called *influential scores*. Sometimes this description is restricted to points with extreme *x*-values. An influential score may have a small residual but still have a greater effect on the regression line than scores with possibly larger residuals but average *x*-values.

> **NOTE**
>
> An outlier is determined by a distance measurement, whereas an influential point is determined by a change measurement.

❯ Example 2.17

Consider the following scatterplot of six points and the regression line:

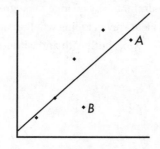

(continued)

The heavy line in the scatterplot on the left below shows what happens when point *A* is removed, and the heavy line in the scatterplot on the right below shows what happens when point *B* is removed.

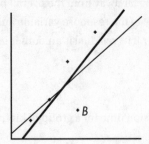 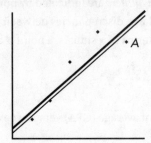

Note that the regression line is greatly affected by the removal of point *A* but not by the removal of point *B*. Thus, point *A* is an *influential score*, while point *B* is not. This is true in spite of the fact that point *A* is closer to the original regression line than point *B*.

There are many ways in which a point can be influential. Removal of the point may markedly change the slope, the *y*-intercept, or the correlation. If you claim that a point is influential, you should also state which statistic (slope, intercept, or correlation) it influences.

> ### Example 2.18

Consider the effect of outliers in the *x*, the *y*, and both *x* and *y* directions on *b*, the slope of the regression line, and on *r*, the correlation:

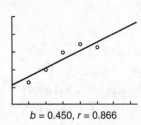 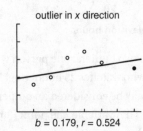 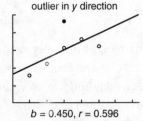 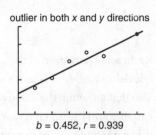

	outlier in *x* direction	outlier in *y* direction	outlier in both *x* and *y* directions
$b = 0.450, r = 0.866$	$b = 0.179, r = 0.524$	$b = 0.450, r = 0.596$	$b = 0.452, r = 0.939$

Note how an outlier in the *x* direction can dramatically affect both the slope and the correlation. Also note how an outlier in both the *x* and *y* directions, if maintaining the same pattern, will result in roughly the same slope but a much increased correlation (in absolute value).

A point is said to have *high leverage* if its *x*-value is far from the mean of the *x*-values. Such a point has the strong potential to change the regression line. If it happens to line up with the pattern of the other points, its inclusion might not influence the equation of the regression line, but it could well strengthen the correlation and r^2, the coefficient of determination. However, if the point does not line up, its inclusion can dramatically change the regression line; in this case, it is also an influential point. The only clear way of telling whether or not a point with high leverage is an influential point is to calculate the linear model twice, both with and without the point in question.

> **NOTE**
>
> **The only "regression outlier" is found in the third graph where the solid point has a residual far greater than any other residual.**

> Example 2.19

Consider the four scatterplots below, each with a cluster of points and one additional point separated from the cluster.

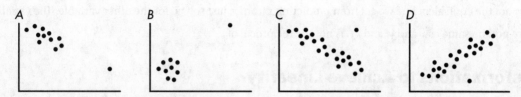

In *A*, the additional point has high leverage (its *x*-value is much greater than the mean *x*-value), has a small residual (it fits the overall pattern) so is not an outlier, and is somewhat influential (while its removal would have little effect on the regression equation, removal would lower the correlation in absolute value).

In *B*, the additional point has high leverage (its *x*-value is much greater than the mean *x*-value), probably has a small residual (the regression line would pass close to it), and is very influential (removing it would dramatically change the slope and the correlation, reducing both to close to 0).

In *C*, the additional point has some leverage (its *x*-value is greater than the mean *x*-value but not very much greater), has a large residual compared with other residuals (so it's a regression outlier), and is somewhat influential (its removal would change the slope to more negative).

In *D*, the additional point has no leverage (its *x*-value appears to be close to the mean *x*-value), has a large residual compared with other residuals (so it's a regression outlier), and is not influential (its removal would increase the *y*-intercept very slightly and would have very little if any effect on the slope).

TIP

Outliers and high leverage points are often influential, but not always.

> Example 2.20

Imagine a scatterplot of (GPA, IQ) for 150 high school students and the corresponding regression line. Now add one student from the chess club who has an average GPA but a very high IQ. She may raise the average IQ a little, shifting the line up marginally and lowering the correlation slightly. Because of her average GPA, though, she has no leverage, and the slope remains the same. She's an outlier in the (GPA, IQ) relationship but not influential.

More on Regression

Much can be noted from a scatterplot:

- If an oval is drawn around the points in a scatterplot, a rounder oval indicates *r* is closer to 0 while a skinnier oval indicates *r* is closer to ± 1.
- An unusual point in a scatterplot can strongly affect *r*, just as an outlier in one-variable data can strongly affect $\bar{x}$ and *s*.

The regression equation $\hat{y} = a + bx = a + \left(r\frac{s_y}{s_x} \right)x$ has two important implications.

First, if the correlation $r = +1$, then $\hat{y} = a + \left(\frac{s_y}{s_x} \right)x$. For each standard deviation s_x increase in *x*, the predicted *y*-value increases by s_y. If the correlation $r = +0.4$, for example, then $\hat{y} = a + \left(0.4\frac{s_y}{s_x} \right)x = \left(\frac{0.4s_y}{s_x} \right)x$. For each standard deviation s_x increase in *x*, the predicted *y*-value increases by $0.4s_y$. Thus, unless $r = +1$, we don't predict *y* to be the same number of SDs from the mean as *x*. In fact, we reduce that prediction by a factor of *r*, predicting *y* to be closer to the mean than *x* was.

For example, the mean heights of sons of tall fathers tend to be less than the mean heights of the fathers, and the mean heights of sons of short fathers tend to be greater than the mean heights of the fathers. We call this *regression to the mean*.

Second, the regression equation for predicting x from y has the slope $r\frac{s_x}{s_y}$. Thus, the line we get from predicting y from x is not the equivalent line we get from predicting x from y, just solved for the other variable. The variables $\hat{y}$ and x are not the same as $\hat{x}$ and y, and $r\frac{s_y}{s_x}$ is not the reciprocal of $r\frac{s_x}{s_y}$.

Transformations to Achieve Linearity

Often a straight-line pattern is not the best model for depicting a relationship between two variables. A clear indication of this problem is when the scatterplot shows a distinctive curved pattern. Another indication is when the residuals show a distinctive pattern rather than a random scattering. In such a case, the nonlinear model can sometimes be revealed by transforming one or both of the variables and then noting a linear relationship. Useful transformations often result from using the *log* or *ln* buttons on your calculator to create new variables.

› Example 2.21

Consider the following years and corresponding populations:

Year, x:	1980	1990	2000	2010	2020
Population (1000s), y:	44	65	101	150	230

The linear model is $\hat{y} = -9{,}022 + 4.57x$ with $r^2 = 0.943$.

So, 94.3% of the variability in population is accounted for by the linear model. However, the scatterplot and residual plot indicate that a nonlinear relationship would be an even stronger model.

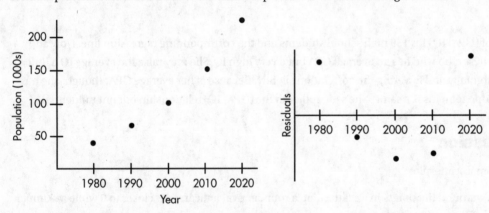

Linear fits to $(x, \log y)$ and $(\log x, \log y)$ result in the following two residual plots:

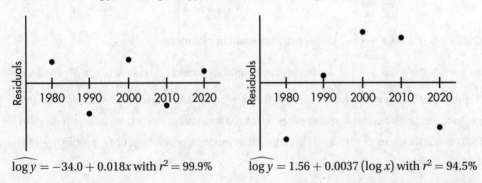

$\widehat{\log y} = -34.0 + 0.018x$ with $r^2 = 99.9\%$ $\widehat{\log y} = 1.56 + 0.0037 (\log x)$ with $r^2 = 94.5\%$

TIP

After transformation of data, either increased randomness in the residual plot or an increase in r^2, both offer evidence that the least squares regression line for the transformed data is a more appropriate model than the regression line for the untransformed data.

Two of the three residual plots have distinct curved patterns. The residual plot coming from log y versus x illustrates a more random pattern in the residual plot as well as the greatest r^2. The best model is thus $\widehat{\log y} = -34.0 + 0.018x$. In context we have $\widehat{\log{(\text{Pop})}} = -34.0 + 0.018(\text{Year})$. So, for example, the population predicted for the year 2025 would be calculated $\widehat{\log{(\text{Pop})}} = -34.0 + 0.018(2025) = 2.45$, and so $\widehat{\text{Pop}} = 10^{2.45} = 282$ thousand or 282,000.

There are many useful transformations. For example:

Log y as a linear function of x, log $y = a + bx$, reexpresses as an *exponential*:
$$y = 10^{a+bx}, \text{ or } y = c\,10^{bx}, \text{ where } c = 10^a$$

Log y as a linear function of log x, log $y = a + b\log x$, reexpresses as a *power*:
$$y = 10^{a+b\log x}, \text{ or } y = cx^b, \text{ where } c = 10^a$$

$\sqrt{y}$ as a linear function of x, $\sqrt{y} = a + bx$, reexpresses as a *quadratic*:
$$y = (a + bx^2)$$

$\frac{1}{y}$ as a linear function of x, $\frac{1}{y} = a + bx$, reexpresses as a *reciprocal*:
$$y = \frac{1}{a + bx}$$

y as a linear function of log x, $y = a + b\log x$, is a *logarithmic* function.

> These very brief reexpressions are shown here for completeness and background, but you will not be responsible for performing these transformations. You do need to be able to recognize the need for a transformation, justify its appropriateness (residuals plot), and use the model to make predictions.

Quiz 5

Multiple-Choice Questions

> **DIRECTIONS:** The questions that follow are each followed by five suggested answers. Choose the response that best answers the question.

1. A rural college is considering constructing a windmill to generate electricity but is concerned over noise levels. A study is performed measuring noise levels (in decibels) at various distances (in feet) from the campus library, and a least squares regression line is calculated with a correlation of 0.74. Which of the following is a proper and most informative conclusion for an observation with a negative residual?

   (A) The measured noise level is 0.74 times the predicted noise level.
   (B) The predicted noise level is 0.74 times the measured noise level.
   (C) The measured noise level is greater than the predicted noise level.
   (D) The predicted noise level is greater than the measured noise level.
   (E) The slope of the regression line at that point must also be negative.

2. A college baseball coach tabulates the number of practice hours in a batting cage and the number of hits over the course of a season for a random sample of players. When the least squares regression analysis is performed, the correlation is 0.67. Which of the following is the correct way to label the correlation?

   (A) 0.67
   (B) 0.67 hits per hour
   (C) 0.67 hours per hit
   (D) 0.67 hit hours
   (E) 0.67 hour hits

Questions 3–5 refer to the following:

The relationship between winning game proportions when facing the sun and when the sun is on one's back is analyzed for a random sample of 10 professional tennis players. The computer printout for regression is below:

Predictor	Coef	SE Coef	T	P
Constant	0.05590	0.02368	2.36	0.046
Facing	0.92003	0.03902	23.58	0.000

S = 0.0242922   R-Sq = 98.6%   R-Sq(adj) = 98.4%

3. What is the equation of the regression line, where "facing" and "back" are the winning game proportions when facing the sun and with back to the sun, respectively?

   (A) $\widehat{facing} = 0.056 + 0.920(back)$
   (B) $\widehat{back} = 0.056 + 0.920(facing)$
   (C) $\widehat{facing} = 0.920 + 0.056(back)$
   (D) $\widehat{back} = 0.920 + 0.056(facing)$
   (E) $\widehat{facing} = 0.024 + 0.039(back)$

4. What is the correlation?

   (A) −0.986
   (B) −0.984
   (C) 0.984
   (D) 0.986
   (E) 0.993

5. For one player, the winning game proportions were 0.55 and 0.59 for "facing" and "back," respectively. What was the associated residual?

   (A) −0.0488
   (B) −0.028
   (C) 0.028
   (D) 0.0488
   (E) 0.3608

6. Data are obtained for a group of college freshmen examining their total SAT scores from their senior year of high school and their GPAs during their first year of college. The resulting regression equation is

   $$\widehat{\text{GPA}} = 0.55 + 0.00161(\text{SAT score}) \text{ with } r = 0.632$$

   What percentage of the variation in GPAs can be accounted for by looking at the linear relationship between GPAs and SAT scores?

   (A) 0.161%
   (B) 16.1%
   (C) 39.9%
   (D) 63.2%
   (E) This value cannot be computed from the information given.

7. Suppose the correlation between two variables is $r = 0.23$. What will the new correlation be if 0.14 is added to all values of the x-variable, every value of the y-variable is doubled, and the two variables are interchanged?

   (A) 0.74
   (B) 0.37
   (C) 0.23
   (D) −0.23
   (E) −0.74

8. Consider the following scatterplot and regression analysis of 15 data points:

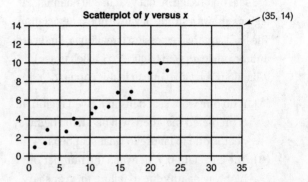

   Slope $b = 0.377$
   Coefficient of determination $r^2 = 95.0\%$
   Standard deviation of the residuals $s = 0.640$

   With the addition of a data point at (35, 14), which one of the following choices gives the most likely new regression statistics?

   (A) Slope $b = 0.377$
       Coefficient of determination $r^2 = 97.0\%$
       Standard deviation of the residuals $s = 0.663$
   (B) Slope $b = 0.377$
       Coefficient of determination $r^2 = 97.0\%$
       Standard deviation of the residuals $s = 0.617$
   (C) Slope $b = 0.377$
       Coefficient of determination $r^2 = 93.0\%$
       Standard deviation of the residuals $s = 0.640$
   (D) Slope $b = 0.377$
       Coefficient of determination $r^2 = 93.0\%$
       Standard deviation of the residuals $s = 0.663$
   (E) Slope $b = 0.397$
       Coefficient of determination $r^2 = 95.0\%$
       Standard deviation of the residuals $s = 0.640$

9. A study of the number of highway deaths due to failure to wear seat belts in a year versus percentage seat belt usage in that year in 300 national regions shows a strong negative linear association. The least squares regression equation is: Predicted number of deaths = 11,100 − 305.1(Belt usage). What does "negative" mean in this context?

   (A) If no one wore seat belts in a given region, it is predicted that there will be 11,100 highway deaths due to failure to wear seat belts.

   (B) The correlation, r, between the number of highway deaths due to failure to wear seat belts and the percentage of seat belt usage is negative.

   (C) If a given region has a lower percentage of seat belt usage than a second region, the given region will have a higher number of highway deaths due to failure to wear seat belts than the second region.

   (D) Regions with a higher percentage of seat belt usage tend to have lower numbers of highway deaths due to failure to wear seat belts.

   (E) If a region has a 1 percent gain in seat belt usage, then it will have a reduction of 305 highway deaths due to failure to wear seat belts, on average.

10. A simple random sample of 35 world-ranked chess players provides the following statistics:

    Number of hours of study per day: $\bar{x} = 6.2$, $s_x = 1.3$
    Yearly winnings: $\bar{y} = \$208,000$, $s_y = \$42,000$
    Correlation $r = 0.15$

    Based on these data, what is the resulting linear regression equation?

    (A) $\widehat{\text{Winning}} = 178,000 + 4,850(\text{Hours})$
    (B) $\widehat{\text{Winning}} = 169,000 + 6,300(\text{Hours})$
    (C) $\widehat{\text{Winning}} = 14,550 + 31,200(\text{Hours})$
    (D) $\widehat{\text{Winning}} = 7,750 + 32,300(\text{Hours})$
    (E) $\widehat{\text{Winning}} = -52,400 + 42,000(\text{Hours})$

11. Consider the set of points {(2, 5), (3, 7), (4, 9), (5, 12), (10, n)}. What should n be so that the correlation between the x- and y-values is 1?

    (A) 21
    (B) 24
    (C) 25
    (D) A value different from any of the above.
    (E) No value for n can make $r = 1$.

12. The scatterplot below has one point labeled X. Does this point have high leverage, a large residual, both, or neither?

    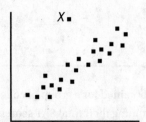

    (A) High leverage and a large residual
    (B) High leverage and a small residual
    (C) Low leverage and a large residual
    (D) Low leverage and a small residual
    (E) This cannot be answered without calculating the regression line.

13. The following are the graphs of three scatterplots, *X*, *Y*, *Z*, and of three residual plots, 1, 2, 3.

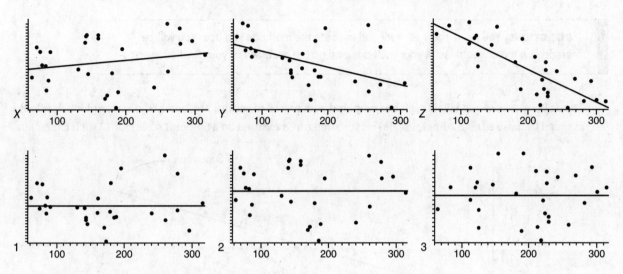

Which scatterplot corresponds to which residual plot?

(A) *X* and 1, *Y* and 2, *Z* and 3
(B) *X* and 1, *Y* and 3, *Z* and 2
(C) *X* and 2, *Y* and 1, *Z* and 3
(D) *X* and 2, *Y* and 3, *Z* and 1
(E) *X* and 3, *Y* and 1, *Z* and 2

Free-Response Questions

DIRECTIONS: You must show all work and indicate the methods you use. You will be graded on the correctness of your methods and on the accuracy of your final answers.

1. Comic book heroes and villains can be compared with regard to various attributes. The scatterplot below looks at speed (measured on a 20-point scale) versus strength (measured on a 50-point scale) for 17 such characters.

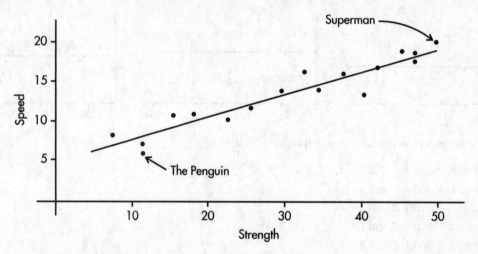

Computer software showing the results of fitting a straight line to the data by the method of least squares gives:

$$\text{Predicted speed} = 4.626 + 0.283(\text{Strength}) \qquad \text{R-sq} = 84.28\%$$

(a) Find the correlation for the linear relationship between speed and strength.

(b) What is the slope of the regression line, and what does it signify?

(c) Given that the mean of the strength values is 30.47, what is the mean of the speed values?

(d) Given that the standard deviation of the strength values is 13.72, what is the standard deviation of the speed values?

(e) Thanos has a strength score of 45 and a speed score of 18. Does the linear model underestimate or overestimate his speed and by how much?

(f) How will the correlation change if the point (42, 16.5), which corresponds to Supergirl, is removed? Explain.

(g) If the points corresponding to both Superman and the Penguin are removed, will the slope increase, decrease, or remain unchanged? Explain.

2. An outlier can have a striking effect on the correlation r. Consider the following three scatterplots:

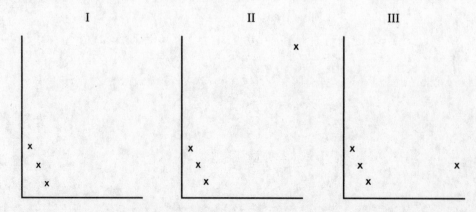

(a) How do the outliers in II and III affect the correlation from I?

(b) Insert a fourth point in I such that the correlation doesn't change.

3. A continuing political debate concerns how to address the large student debt carried by college graduates. Graph (I) is a scatterplot showing the amount of student debt of 25 randomly selected college graduates and their stress level as self-reported on a 1–100 scale.

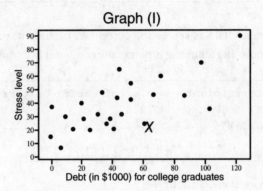

Computer output from linear regression gives:

Predictor	Coef	SE Coef
Constant	19.376	4.782
Debt	0.42640	0.08906

(a) Describe the association between stress level and debt for this sample of 25 college graduates.

(b) The point labeled X represents a student with $60,000 debt and a stress level of 25. Calculate and interpret the residual of that point.

Two other possible explanatory variables with regard to stress level are income and health as measured on a 3 to 8 HRQoL (health-related quality of life) index. Graph (II) is a scatterplot of health plotted with the corresponding residuals from the regression of stress on debt. Graph (III) is a scatterplot of income plotted with the corresponding residuals from the regression of stress on debt.

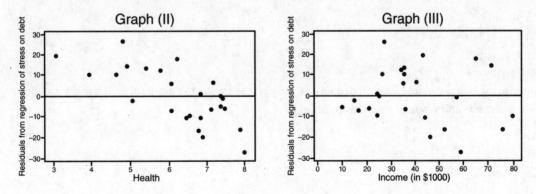

(c) Which variable, health or income, in addition to debt should be used to improve the prediction of stress level? Explain.

The answers for this quiz can be found beginning on page 480.

Quiz 6

Multiple-Choice Questions

DIRECTIONS: The questions that follow are each followed by five suggested answers. Choose the response that best answers the question.

1. A study collects data on average total SAT scores and percentage of students who took the exam at 100 randomly selected high schools. The following is part of the computer printout for regression:

```
Variable        Coefficient      s.e. of coeff      t-ratio      prob
Constant        1176.32          12.65              92.99        ≤ 0.0001
SAT             -2.84276         0.2461             -11.55       ≤ 0.0001

R-squared = 76.5%          R-squared (adj) = 76.1%
```

Which of the following is a correct conclusion?

(A) "SAT" in the variable column indicates that SAT score is the dependent (response) variable.
(B) The correlation is 0.875.
(C) The y-intercept indicates the mean total SAT score if percent of students taking the exam has no effect on total SAT scores.
(D) The r^2-value indicates that the residual plot does not show a strong pattern.
(E) Schools with lower percentages of students taking the exam tend to have higher average total SAT scores.

2. Suppose the correlation is negative. Given two points from the scatterplot, which of the following is possible?

   I. The first point has a larger x-value and a smaller y-value than the second point.
   II. The first point has a larger x-value and a larger y-value than the second point.
   III. The first point has a smaller x-value and a larger y-value than the second point.

(A) I only
(B) II only
(C) III only
(D) I and III only
(E) I, II, and III

3. Suppose the regression line for a set of data, $a + 3x$, passes through the point $(2, 5)$. If $\bar{x}$ and $\bar{y}$ are the sample means of the x- and y-values, respectively, then $\bar{y} =$

(A) $\bar{x}$
(B) $-2 + \bar{x}$
(C) $5 + \bar{x}$
(D) $3\bar{x}$
(E) $-1 + 3\bar{x}$

4. Which of the following statements about correlation r is true?

   (A) A correlation of 0.2 means that 20% of the points are highly correlated.
   (B) Perfect correlation, that is, when the points lie exactly on a straight line, results in $r = 0$.
   (C) Correlation is not affected by which variable is called x and which is called y.
   (D) Correlation is not affected by extreme values.
   (E) A correlation of 0.75 indicates a relationship that is 3 times as linear as one for which the correlation is only 0.25.

5. Which of the following statements about residuals from the least squares line are true?

   I. Residuals may be positive, negative, or zero.
   II. The linear regression line maximizes the sum of the squares of the residuals.
   III. A definite pattern in the residual plot is an indication that a nonlinear model will show a better fit to the data than the straight regression line.

   (A) I and II only
   (B) I and III only
   (C) II and III only
   (D) I, II, and III
   (E) None of the above gives the complete set of true responses.

6. In a study of winning percentage in home games versus average home attendance for professional baseball teams, the resulting regression line is

   Predicted winning percentage = 44 + 0.0003(Attendance)

   What is the residual if a team has a winning percentage of 55% with an average attendance of 34,000?

   (A) −11.0
   (B) −0.8
   (C) 0.8
   (D) 11.0
   (E) 23.0

7. Suppose the correlation between two variables is −0.57. If each of the y-scores is multiplied by −1, which of the following is true about the new scatterplot?

   (A) It slopes up to the right, and the correlation is −0.57.
   (B) It slopes up to the right, and the correlation is +0.57.
   (C) It slopes down to the right, and the correlation is −0.57.
   (D) It slopes down to the right, and the correlation is +0.57.
   (E) None of the above is true.

8. A study of selling prices of homes in a southern California community (in $1,000) versus size of the homes (in 1,000s of square feet) shows a moderate positive linear association. The least squares regression equation is Predicted selling price = 35.3 + 214.1(Size). What does "linear" mean in this context?

   (A) The points in the scatterplot line up in a straight line.
   (B) There is no distinct pattern in the residual plot.
   (C) The coefficient of determination, r^2, is large (close to 1).
   (D) As home size increases by 1,000 square feet, the selling price tends to change by a constant amount, on average.
   (E) Each increase of 1,000 square feet in home size gives an increase of 214.1($1,000) in selling price.

9. Does lower sun exposure, as measured by greater distance from the equator, result in lower incidence of skin cancer? A study of skin cancer mortality versus latitude of 100 northern hemisphere locales shows a strong negative linear association. What does "strong" mean in this context?

(A) More sun exposure causes greater numbers of skin cancer deaths.

(B) More sun exposure is associated with greater numbers of skin cancer deaths.

(C) A locale's incidence of skin cancer deaths has a linear association with its latitude.

(D) A least squares model predicts that the greater the latitude, the lower the average incidence of skin cancer.

(E) The actual incidence of skin cancer at a given latitude will be very close to what is predicted by a least squares model.

10. Consider n pairs of numbers. Suppose $\bar{x} = 2$, $s_x = 3$, $\bar{y} = 4$, and $s_y = 5$. Of the following, which could be the least squares line?

(A) $\hat{y} = -2 + x$

(B) $\hat{y} = 2x$

(C) $\hat{y} = -2 + 3x$

(D) $\hat{y} = \frac{5}{3} - x$

(E) $\hat{y} = 6 - x$

11. With regard to a least squares regression line, the sum of the residuals is always zero. Does it follow that the mean or the median of the residuals also must be zero?

(A) Mean of the residuals: Yes, must be zero
Median of the residuals: Yes, must be zero

(B) Mean of the residuals: Yes, must be zero
Median of the residuals: No, may or may not be zero

(C) Mean of the residuals: No, may or may not be zero
Median of the residuals: Yes, must be zero

(D) Mean of the residuals: No, may or may not be zero
Median of the residuals: No, may or may not be zero

(E) The answer depends on the context.

12. The scatterplot below has one point labeled X. Does this point have high leverage, a large residual, both, or neither?

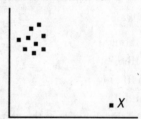

(A) High leverage and a large residual

(B) High leverage and a small residual

(C) Low leverage and a large residual

(D) Low leverage and a small residual

(E) This cannot be answered without calculating the regression line.

13. The following are the graphs of three scatterplots, *X*, *Y*, *Z*, and of three residual plots, 1, 2, 3.

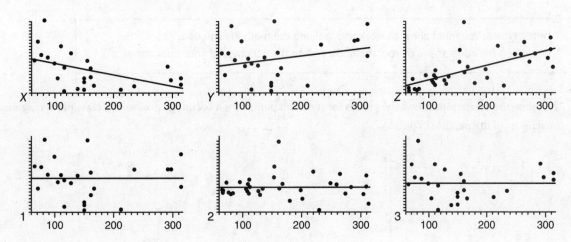

Which scatterplot corresponds to which residual plot?

(A) *X* and 1, *Y* and 2, *Z* and 3

(B) *X* and 1, *Y* and 3, *Z* and 2

(C) *X* and 2, *Y* and 1, *Z* and 3

(D) *X* and 2, *Y* and 3, *Z* and 1

(E) *X* and 3, *Y* and 1, *Z* and 2

Free-Response Questions

> **DIRECTIONS:** You must show all work and indicate the methods you use. You will be graded on the correctness of your methods and on the accuracy of your final answers.

1. The following scatterplot shows the grades for research papers for a sociology professor's class plotted against the lengths of the papers (in pages).

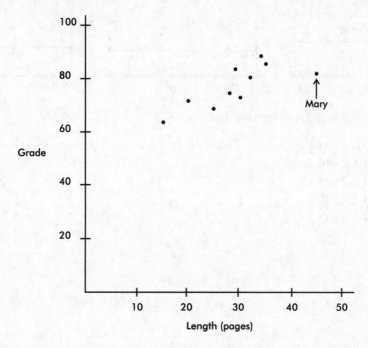

Mary turned in her paper late and was told by the professor that her grade would have been higher if she had turned it in on time. A computer printout fitting a straight line to the data (not including Mary's score) by the method of least squares gives:

```
Grade = 46.51 + 1.106 Length
R-sq = 74.6%
```

   (a) Find the correlation coefficient for the relationship between grade and length of paper based on these data (excluding Mary's paper).
   (b) What is the slope of the regression line and what does it signify?
   (c) How will the correlation coefficient change if Mary's paper is included? Explain your answer.
   (d) How will the slope of the regression line change if Mary's paper is included? Explain your answer.

2. A scatterplot of the number of accidents per day on a particular interstate highway during a 30-day month is as follows:

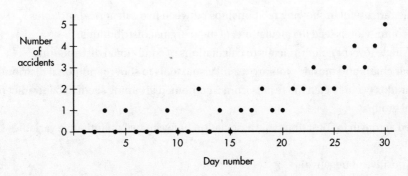

(a) Draw a histogram of the frequencies of the number of accidents.

(b) Name a feature apparent in the scatterplot but not in the histogram.

(c) Name a feature clearly shown by the histogram but not as obvious in the scatterplot.

3. In a study of how exam scores are influenced by hours of sleep the night before the exam, a random sample of 13 students was outfitted with wristbands that tracked sleep time. The scatterplot of the resulting data follows.

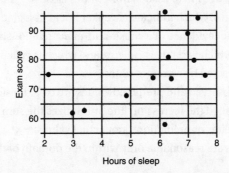

(a) Describe the association between exam score and hours of sleep for these 13 students.

(b) Describe the association between exam score and hours of sleep for the 8 students who received at least 6 hours of sleep the night before the exam.

(c) Explain how the appropriateness of using linear regression is critically influenced by one student's data.

The answers for this quiz can be found beginning on page 481.

SUMMARY

- Two-way tables are useful in showing relationships between two categorical variables.
- The row and column totals lead to calculations of the marginal distributions.
- Focusing on single rows or columns leads to calculations of conditional distributions.
- Segmented bar charts and mosaic plots are useful visual tools to show conditional distributions.
- Simpson's paradox occurs when the results from a combined grouping seem to contradict the results from the individual groups.
- One suggested acronym for describing the association between two quantitative variables shown in a scatterplot is DUFS:
  - Direction (positive or negative)
  - Unusual features (outliers, influential points, clusters)
  - Form (linear or not)
  - Strength (weak, moderate, strong)
- If the relationship appears roughly linear, the correlation coefficient, r, and especially the coefficient of determination, r^2, are useful measurements.
- The value of r is always between -1 and $+1$, with positive values indicating positive linear association and negative values indicating negative linear association; values close to -1 or $+1$ indicate a stronger linear association than values close to 0, which indicate a weaker linear association.
- Evidence of an association is not evidence of a cause-and-effect relationship!
- Correlation is not affected by which variable is called x and which is called y, or by changing units.
- Correlation can be strongly affected by extreme values.
- The differences between the observed and predicted values are called residuals.
- The best-fitting straight line, called the regression line, minimizes the sum of the squares of the residuals.
- The regression line gives estimates or predictions, not actual values.
- The regression line predictions are reasonable only within the domain of the data; extrapolation outside the domain is unreliable.
- The y-intercept a and the slope b of the least squares regression line, $\hat{y} = a + bx$, can be calculated from summary statistics using $b = r\dfrac{s_y}{s_x}$ and $a = \bar{y} - b\bar{x}$.
- The slope b describes the *average* or *predicted* increase or decrease in the y-variable for each one unit increase in the x-variable.
- The y-intercept a describes the *average* or *predicted* outcome of y if $x = 0$.
- For the linear regression model, the mean of the residuals is always 0.
- A definite pattern in the residual plot indicates that a nonlinear model may fit the data better than the straight regression line.
- The coefficient of determination, r^2, gives the percentage of variation in y that is accounted for by a linear relationship between x and y. Alternatively, $1 - r^2$ gives how much of the variability in y is *unaccountable* for by the regression line.
- Outliers are points whose residuals in absolute value are large when compared with those of other residuals.
- Influential points are points whose removal would sharply change the slope, correlation, and/or y-intercept of the regression line.
- High leverage points are those whose x-values are far from the mean of the x-values.
- Nonlinear models can sometimes be studied by transforming one or both variables and then noting a linear relationship.
- It is very important to be able to interpret generic computer output.

3

Collecting Data

(5–6 multiple-choice questions on the AP exam)

Learning Objectives

In this unit, you will learn how

→ To be able to distinguish between a sample and a population and between a statistic and a parameter.

→ To be able to explain whether an observational study is retrospective or prospective.

→ To be able to distinguish between a convenience sample, a voluntary response sample, and a random sample.

→ To be able to identify, describe, and implement different sampling methods including a simple random sample, a cluster sample, a stratified sample, and a systematic sample.

→ To be able to describe the benefits and drawbacks of various sampling methods.

→ To be able to explain a particular bias and how it is likely to result in the sample underestimating or overestimating a population parameter.

→ To be able to explain why a particular study is observational or an experiment.

→ To be able to design an experiment using one or more of the following: different levels of an explanatory variable, a placebo, blinding or double-blinding, blocking, and matched pairs design.

→ To be able to explain why or why not a cause-and-effect conclusion is justified.

→ To understand the difference between, and how to perform, a random sampling and a random assignment.

→ To be able to explain in context why a particular variable might lead to confounding.

→ To be able to explain why or why not the results of a study can be generalized to a larger population.

> **TIP**
>
> **You should always be aware to what population you can reasonably generalize the results of your study.**

In this unit, you will discover that data collection is the critical first step in the data analysis process, you will learn principles of sampling and experimental design, you will see how bias can result, and you will understand how to control for confounding. You will also be able to distinguish between retrospective versus prospective studies, observational versus experimental studies, random sampling versus random assignment, stratified versus cluster sampling, stratification versus blocking, experimental units versus treatment groups, blocks versus treatment groups, and completely randomized design versus randomized block design. Then, finally, you will begin to understand the meaning of statistical significance and what kind of conclusions can result from well-designed experiments.

There is a key question that you will begin to reflect upon in this unit and that will be more fully answered in the later units on inference. That is, suppose we measure some difference between two groups: what are the possible explanations? The difference could have resulted simply by chance, or there could be some unknown confounding variable that is causing the difference in measure, or there could be a causal association whereby something in one group causes the difference in measure.

In the real world, time and cost considerations usually make it impossible to analyze an entire population. Do companies like Apple question every potential consumer before designing a new smartphone? Does a television producer check every household's viewing preferences before deciding whether a pilot program will be continued? By studying statistics, we learn how to estimate a population characteristic (called a population *parameter*) by considering a sample measurement (called a sample *statistic*). Later in this review book we will see how to estimate population proportions, means, and slopes by looking at sample proportions, means, and slopes.

To derive conclusions about the larger population, we need to be confident that the sample we have chosen represents the population fairly. Analyzing the data with computers is often easier than gathering the data, but the frequently quoted "garbage in, garbage out" applies here. If data are poorly collected, then the measurements we calculate will be meaningless. Nothing can help if the data are collected poorly. Unfortunately, many of the statistics that we are bombarded with in newspapers, on the radio, and on television are based on poorly designed data collection procedures.

> **NOTE**
>
> MYTH: A random sample will *always* be representative of the population.
>
> FACT: A random sample, *simply by chance*, might turn out not to be very representative.

The following questions should always be asked when collecting data:

- How are the data being collected?
- Does the collection method reasonably lead to a sample representative of the population or not?
- How might the individuals in our sample differ from those not in our sample?
- If our sample is not representative of the intended population, how might this lead to an underestimate or overestimate of the measurement of interest?

There are four important principles in picking a representative sample in order to be able to generalize our findings to the whole population:

1. The use of *randomization* in sample selection increases the chance that the resulting sample exhibits all the known and unknown features of the population.
2. The *sample size*, not the fraction of the population surveyed, is key. Larger samples give better information about the population than smaller samples.
3. A sample is generalizable only to the population from which the sample was selected.
4. It is not possible to conclude a cause-and-effect relationship between variables using data collected from an observational study.

There are four important principles in designing an experiment in order to be able to make a cause-and-effect conclusion:

1. There should be at least two treatment groups, one of which can be a control group.
2. The use of *randomization* in assignment to treatment groups increases the chance that treatment groups are as similar as possible other than which treatment each group receives.
3. The size of treatment groups, that is, the number of subjects in each treatment group, is key. Larger numbers of subjects give better information than smaller numbers.
4. If there is an important difference among the subjects, similar subjects should be grouped into blocks, and treatments should be randomized within the blocks.

Retrospective Versus Prospective Observational Studies

Observational studies aim to gather information about a population without disturbing the population. *Retrospective studies* look backward, examining existing data, while *prospective studies* watch for outcomes, tracking individuals into the future.

> **Example 3.1**

Retrospective studies of the 2014–2016 Ebola epidemic in West Africa have looked at the timing, numbers, and locations of reported cases and have tried to understand how the outbreak spread. There is now much better understanding of transmission through contact with bodily fluids of infected people. Several prospective studies involve ongoing surveillance to see how experience and tools to rapidly identify cases will now limit future epidemics.

Advantages and Disadvantages

Retrospective studies tend to be smaller scale, quicker to complete, and less expensive. In cases such as addressing diseases with low incidence, the study begins right from the start with people who have already been identified. However, researchers have much less control, usually having to rely on past recordkeeping of others. Furthermore, the existing data will often have been gathered for other purposes than the topic of interest. Then there is the problem of subjects' inaccurate memories and possible biases.

Prospective studies usually have greater accuracy in data collection and are less susceptible to recall error from subjects. Researchers do their own recordkeeping and can monitor exactly what variables they are interested in. However, these studies can be very expensive and time-consuming, as they often follow large numbers of subjects for a long time.

Bias

The one thing that most quickly invalidates a sample and makes obtaining useful information and drawing meaningful conclusions impossible is *bias*. **A sampling method is biased if in some critical way it consistently results in samples that do not represent the population.** This typically leads to certain responses being repeatedly favored over others. Sampling bias is a property of the sampling method, not of any one sample generated by the method.

> The way data are obtained is crucial. A large sample size cannot make up for a poor survey design.

Voluntary response surveys are based on individuals who choose to participate, typically give too much emphasis to people with strong opinions, and undersample people who don't care much about a topic. For example, radio call-in programs about controversial topics do not produce meaningful data on what proportion of the population favor or oppose related issues. Online surveys posted to websites are a common source of voluntary response bias. Although a voluntary response survey may be easy and inexpensive to conduct, the sample is likely to be composed of strongly opinionated people, especially those with negative opinions on a subject.

> **TIP**
>
> **Think about potential biases before collecting data!**

Convenience surveys, like interviews at shopping malls, are based on choosing individuals who are easy to reach. These surveys tend to produce data highly unrepresentative of the entire population. For example, door-to-door household surveys miss people who do not happen to be at home and typically miss groups such as college students, prison inmates, and the homeless. Although convenience samples may allow a researcher to conduct a survey quickly and inexpensively, generalizing from the sample to a population is almost impossible.

There are many other sources of sampling bias. Although it is not necessary to know every possible kind of bias by name, it is important to recognize when a sample design is flawed, explain why it is flawed, and understand the direction of the bias. You will need to be able to decide if the bias will likely cause an underestimate or overestimate of a population parameter.

> **TIP**
>
> **Unless specifically asked to name a source of bias, it is best to simply describe the flaw that is caused by the bias *and state the direction of bias in context*. A wrong name will be penalized.**

Undercoverage bias happens when there is inadequate representation, and thus some groups in the population are left out of the process of choosing the sample. For example, telephone surveys ignore those who don't have telephones. *Response bias* occurs when the question itself can lead to misleading results because people don't want to be perceived as having unpopular or unsavory views or don't want to admit to having committed crimes. *Nonresponse bias*, where there are low response rates, occurs when individuals chosen for the sample can't be contacted or refuse to participate, and it is often unclear which part of the population is responding. *Quota sampling bias*, where interviewers are given free choice in picking people in the (problematic, if not impossible) attempt to pick representatively with no randomization, is a recipe for disaster. *Question wording bias* can occur when nonneutral or poorly worded questions lead to very unrepresentative responses or can occur even when the order in which questions are asked makes a difference.

Example 3.2

The *Military Times*, in collaboration with the Institute for Veterans and Military Families at Syracuse University, conducted a voluntary and confidential online survey of U.S. service members who were readers of the *Military Times*. Their military status was verified through official Defense Department email addresses. What were possible sources of bias?

Solution

First, *voluntary online surveys* are very suspect because they typically overcount strongly opinionated people. Second, *undercoverage bias* is likely because only readers of the *Military Times* took part in the survey. Note that *response bias* was probably not a problem because the survey was confidential.

When given a data collection method and asked to describe a bias that could occur, you need to address the following:

- Why is there bias? (For example, "The survey was voluntary response," or "This was a convenience survey," or "Only certain people had a chance to be selected," or "People might be untruthful because they don't want to be perceived as having an unpopular view.")
- How is it biased? (For example, "Those choosing to answer might feel stronger about the issue than the general population," or "Those picked for the sample might be less likely than the general population to have full-time jobs.")
- What will this do to the estimate of the population parameter of interest? (For example, "This would likely result in an underestimate of the population proportion of ...," or "This would likely result in an overestimate of the population mean of ....")

In some questions, you can receive full credit arguing either direction (a likely underestimate or overestimate) as long as there is a clear direction and a reasonable explanation tied to the data collection method that supports the argument.

Sampling Methods

From politics to business, from science to sociology, and from sports to fighting crime, statisticians use available data to understand and make predictions about entire populations.

Collecting data from every individual in a population, called a *census*, might seem to provide complete information, but unless the population is small, this method would be prohibitively expensive and time-consuming. Furthermore, it could be ridiculous, as, for example, tasting every single pizza a company makes would leave nothing left to sell. Finally, if a questionnaire is poorly worded, answers, even from an entire population, might be meaningless.

Given that it is almost always impractical, if not impossible, to examine a whole population, we must settle for examining a sample. The list of all individuals from which the sample is drawn is called the *sampling frame*. Of primary importance is how to choose a representative sample from the sampling frame.

How can we increase our chance of choosing a representative sample? One technique is to write the name of each member of the population on a card, mix the cards thoroughly in a large box, and pull out a specified number of cards. This method gives everyone in the population an equal chance of being selected as part of the sample. Unfortunately, this method is usually too time-consuming, and bias might still creep in if the mixing is not thorough. A *simple random sample* (SRS), **one in which every possible sample of the desired size has an equal chance of being selected**, can be more easily obtained by assigning a number to everyone in the population and using a random digit table or having a computer generate random numbers to indicate choices.

> **NOTE**
>
> In an SRS, it is also true that every individual has an equal chance of being selected.

> Example 3.3

Suppose 80 students are taking an AP Statistics course and the teacher wants to pick a sample of 10 students randomly to try out a practice exam. She can make use of a random number generator on a computer as follows. Assign the students numbers 1, 2, 3, . . . , 80. Use a computer to generate 10 random integers between 1 and 80 without replacement, that is, throw out repeats. The sample consists of the students with assigned numbers corresponding to the 10 unique computer-generated numbers.

An alternative solution is to first assign the students numbers 01, 02, 03, . . . , 80. While reading off two digits at a time from a random digit table, the teacher ignores any numbers over 80, ignores 00, and ignores repeats, stopping when she has a set of 10. If the table began 75425 56573 90420 48642 27537 61036 15074 84675, she would choose the students numbered 75, 42, 55, 65, 73, 04, 27, 53, 76, and 10. Note that 90 and 86 are ignored because they are over 80. Note that the second and third occurrences of 42 are ignored because they are repeats.

> **Important**
>
> To receive full credit, there are 3 steps:
>
> 1. assigning integers,
> 2. using a computer random number generator to generate distinct integers in a given range, and
> 3. linking selected integers with corresponding individuals selected for the sample.

> **TIP**
>
> When using a random digit table, it is important that each label has the same number of digits. It would be wrong to assign the digits 1, 2, 3, . . . , 80 rather than 01, 02, 03, . . . , 80. (This is not necessary when using a random number generator.)

Advantages of simple random sampling include the following:

- The simplicity of simple random sampling makes it relatively easy to interpret data collected.
- This method requires minimal advance knowledge of the population other than knowing the complete sampling frame.
- Simple random sampling allows us to make generalizations (i.e., statistical inferences) from the sample to the population.
- Among the major parameters with which we will work, simple random sampling tends to be *unbiased*; that is, when repeated many times, it gives sample statistics that are centered around the true parameter value.

Disadvantages of simple random sampling include the following:

- The need for a list of all potential subjects can be a formidable task.
- Although simple random sampling is a straightforward procedure to understand, it can be difficult to execute, especially if the population is large. For example, how would you obtain a simple random sample of students from the population of all high school students in the United States?
- The need to repeatedly contact nonrespondents can be very time-consuming.
- Important groups may be inadvertently left out of the sample when using this method.

There are other sampling methods available. Each has its own set of advantages and disadvantages.

Stratified sampling involves dividing the population into homogeneous groups called *strata*, then picking random samples from each of the strata, and finally combining these individual samples into what is called a *stratified random sample*. For example, we can stratify by age, gender, income level, or race/ethnicity; pick a sample of people from each stratum; and combine to form the final sample. For instance, a political pollster might want to be sure to include respondents from various minority groups. Stratifying by religion or race/ethnicity and including people from each group would ensure this diversity in the final sample.

> **NOTE**
> All individuals in a given stratum have a characteristic in common.

Advantages of stratified sampling include the following:

- This ensures that every stratum is adequately represented and removes the chance of any significant segment of the population being entirely neglected.
- Looking at subgroups within a population can make the work of collecting data more manageable.
- This method may speed up the time it takes to conduct a survey and so is more cost-effective.
- Samples taken within a stratum have reduced variability, which means the resulting estimates are more precise than when using other sampling methods.
- More information, insights, and important differences among groups can become apparent.

> **TIP**
> You should be able to explain why you think that stratification should be used in a given problem.

Disadvantages of stratified sampling include the following:

- It necessitates prior knowledge about the population and the variables used to stratify it, which may not be always available or accurate.
- It may be difficult to classify every member of a population into one of the stratum.
- It may necessitate knowing the size of the population in each stratum.
- Evaluating the results may be more difficult, more complex, than with simple random sampling.
- Like an SRS, this method might be difficult to implement with large populations.
- Significant planning may be necessary to ensure that the subdivisions are logical in the context of the study. Forcing subdivisions when none really exist is meaningless.

We could further do *proportional sampling*, where the sizes of the random samples from each stratum depend on the proportion of the total population represented by the stratum.

Cluster sampling involves dividing the population into heterogeneous groups called *clusters* and then picking everyone or everything in a random selection of one or more of the clusters. For example, to survey high school seniors, we could randomly pick several senior class homerooms in which to conduct our study and sample all students in those selected homerooms.

> **NOTE**
> Each cluster should resemble the entire population.

Advantages of cluster sampling include the following:

- Clusters are often chosen for ease, convenience, and quickness.
- With limited fixed funds, cluster sampling usually allows for a larger sample size than do other sampling methods.
- This method is especially appropriate when the construction of a complete list of the population is impossible, expensive, or too difficult to organize.

Disadvantages of cluster sampling include the following:

- Although clusters should mirror the full diversity of the population, realistically this is difficult to accomplish.
- Difficulties in analysis can result if the clusters are of very different sizes.
- With a given sample size, cluster sampling usually provides less precision than either an SRS or a stratified sample provides.
- If the population doesn't have natural clusters and the designated clusters are not representative of the population, selection could easily result in a biased sample.

The differences between strata and clusters include the following:

- Although all strata are represented in the sample, only a subset of clusters are in the sample.
- Stratified sampling is accurate when each stratum consists of *homogeneous* elements, while cluster sampling is accurate when each cluster consists of *heterogeneous* elements.
- Randomization plays a key role in both methods! With stratified sampling, the randomization takes place within each stratum. With cluster sampling, the randomization takes place when choosing which clusters to include.

Systematic sampling is a relatively simple and quick method. It involves listing the population in some order (for example, alphabetically), choosing a random point to start, and then picking every tenth (or hundredth, or thousandth, or kth) person from the list. This gives a reasonable sample as long as the original order of the list is not in any way related to the variables under consideration.

TIP
Depending upon how the random starting point is chosen, it may be necessary to wind the list around to the beginning.

Advantages of systematic sampling include the following:

- Systematic sampling is simpler and quicker to implement than the other methods discussed. Subjects don't have to be numbered. For example, this is a nice option for situations like exit polling or surveying people as they leave a stadium.
- It is a practical, straightforward, reasonably random method especially useful for investigators unfamiliar with advanced statistics.
- If similar members in the list are grouped together, we can end up with a kind of stratified sample, only more easily implemented.

Disadvantages of systematic sampling include the following:

- If randomization is a top priority, this method is weak as the only randomization in systematic sampling occurs in choosing the initial subject.
- If the population size isn't roughly known, determining the fixed interval for sample selection is problematic and can lead to inaccurate results.
- If the list happens to have a periodic structure similar to the period of the sampling, a very biased sample could result. For example, picking every 12th seat in an airplane when there are 6 seats in a row would result in all aisle or all middle or all window seats in the sample.

> **Example 3.4**

Suppose a sample of 100 high school students from a Chicago school of size 5,000 is to be chosen to determine their views on whether they think the Cubs will win another World Series this century. One method would be to have each student write his or her name on a card, put the cards into a box, have the principal reach in and pull out 100 cards, and then choose the names on those cards to be the sample. However, questions could arise regarding how well the cards are mixed. For example, how might the outcome be affected if all students in one PE class toss their names in at the same time so that their cards are clumped together? A second method would be to number the students 1 through 5,000 and use a random number generator to pick 100 unique (throw out repeats) integers between 1 and 5,000. The sample is then the students whose numbers correspond to the 100 generated numbers. A third method would be to assign each student a number from 0001 to 5,000 and use a random digit table, picking out four digits at a time, ignoring repeats, 0000, and numbers over 5,000, until a unique set of 100 numbers are picked. Then choose the students corresponding to the selected 100 numbers. What are alternative procedures?

✓ **Solution**

From a list of the students, the surveyor could choose a random starting name and then simply note every 50th name (systematic sampling). Since students in each class have certain characteristics in common, the surveyor could use a random selection method to pick 25 students from each of the separate lists of freshmen, sophomores, juniors, and seniors (stratified sampling). If each homeroom has a random mix of 20 students from across all grade levels, the surveyor could randomly pick five homerooms with the sample consisting of all the students in these five rooms (cluster sampling).

> **NOTE**
> Understand that these other procedures are still proper random samples even if they are not simple random samples.

Sometimes we randomly pick a single sample, cross-categorize on two variables, and then test to see if there is an association between the variables. However, sometimes we instead randomly pick a sample from each of two or more populations and then compare the distribution of some variable among the different populations. This second method is called *sampling for homogeneity*.

> **NOTE**
> We'll revisit sampling for homogeneity later when developing chi-square inference, but this choice of sampling method is really a design issue and worth recognizing at this time.

Sampling Variability

No matter how well designed and well conducted a survey is, it still gives a sample *statistic* as an estimate for a population *parameter*. Different samples give different sample statistics, all of which are estimates for the same population parameter. So *sampling variability*, also called *sampling error*, is naturally present. This variability can be described using probability; that is, we can say how likely we are to have a certain size error. Generally, the associated probabilities are smaller when the sample size is larger. However, **whenever a sample is taken, sampling variability (sampling error) will be present**. This is not an error that someone is committing but is simply sample-to-sample variation.

One shouldn't confuse **accuracy** (centered at the right place) with **precision** (low variability).

> ## Example 3.5

Suppose we are trying to estimate the mean age of high school teachers and have four methods of choosing samples. We choose 10 samples using each method. Later we find the true mean, $\mu = 42$. Plots of the results of each sampling method are given below.

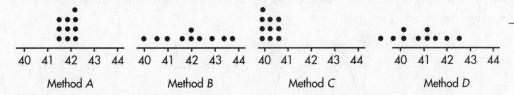

Method A   Method B   Method C   Method D

Method *A* exhibits high accuracy and high precision. Method *B* exhibits high accuracy and low precision. Method *C* exhibits low accuracy and high precision. Method *D* exhibits low accuracy and low precision. Low accuracy (not centered at the right place) indicates probable bias in the selection method. Note that shape and variability in distributions are completely irrelevant to the issue of sampling bias. Sampling bias is focused on the *center* of the distribution.

It needs to be emphasized that *bias* is a question of sampling methodology, not a characteristic of an individual sample. It's possible for a biased methodology to produce a sample that characterizes the population well, but that's just sampling variability resulting in a lucky break. With bias, samples will collectively misrepresent the population, but that does not mean that all of them will do so. With an unbiased methodology, samples will collectively represent the population; however, because of sampling variability, any given sample still can be a "bad" sample. In this case, it would be wrong to say that the sample is biased, but rather it's simply an example of sampling variability.

> ## Example 3.6

Pesticide residue on store-bought fruit can be a health problem. Ms. D has two AP Statistics classes with 30 students in each. For a study asking high school students whether or not they wash fruit before eating them, each student in the first class obtained a random sample of size 10 to question, while each student in the second class obtained a random sample of size 50 to question. The data obtained are displayed in two dotplots:

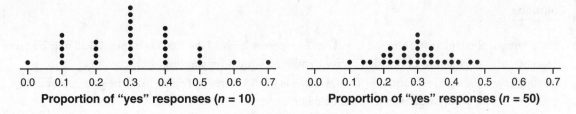

Proportion of "yes" responses (*n* = 10)   Proportion of "yes" responses (*n* = 50)

Note the relationship between sampling variability and sample size. Estimates from larger samples produce less sampling variability; that is, they are more tightly clustered around the presumed true parameter value.

⟩ **Example 3.7**

Air Quality Index (AQI) is the EPA's index for reporting air quality. An AQI value of 50 or below represents good air quality, while an AQI value over 150 represents unhealthy air quality. Suppose that the distribution of AQI scores in the 500 highest-income counties is roughly normal with a mean of 50, while the distribution of AQI scores in the 500 lowest-income counties is roughly normal with a mean of 100. The figure below represents the distributions of AQI scores in these high- and low-income counties.

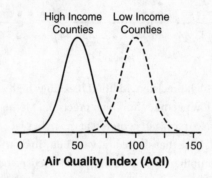

For an environmental study, 110 of these 1,000 counties will be randomly selected. Two methods are under consideration.

   Method I: Take an SRS from among the 1,000 low- and high-income counties.

   Method II: Randomly select either the low- or high-income counties and take an SRS of those 500 counties.

(a)   Will a sample obtained using Method II be representative of the 1,000 counties? Explain.

(b)   Which of the two sampling methods will result in a sample with less variability in AQI? Explain.

(c)   The figure below represents the AQI of 110 counties obtained by using one of the two methods. Explain which method was most likely used.

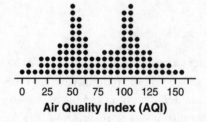

✓ **Solutions**

(a)   No, a sample obtained using Method II will not be representative of the 1,000 counties. It will only be representative of either the low- or the high-income counties, whichever was randomly chosen.

(b)   Method II will result in less variability in AQI because all the scores will come from the one randomly selected income group of 500 similar-income counties.

(c)   Method I was most likely used to select this sample. The bimodal shape of the sample distribution indicates that counties were likely selected from both high- and low-income groups, which happens with Method I. Method II would most likely result in a unimodal distribution centered around either 50 or 100.

Quiz 7

Multiple-Choice Questions

> **DIRECTIONS:** The questions that follow are each followed by five suggested answers. Choose the response that best answers the question.

1. In September 2020, medical researchers analyzed the summer records of hospitalized Covid-19 patients, collecting data on disease severity and whether or not the patients had been taking statin medications for high cholesterol within 30 days prior to hospitalization. The researchers found that statin use was associated with a lesser severity of the disease. What kind of research was this?

   (A) A retrospective observational study
   (B) A prospective observational study
   (C) A voluntary response survey
   (D) A convenience survey
   (E) An experiment

2. An increasing number of U.S. public schools are adopting school uniforms. Two possible wordings for a question on whether or not students should have to wear school uniforms are as follows:

   I. Many educators believe in creating a level playing field to reduce socioeconomic disparities. Do you believe that students should have to wear school uniforms?

   II. Many sociologists believe that students have a right to express their individuality. Do you believe that students should have to wear school uniforms?

   One of these questions showed that 18% of the population favors school uniforms, while the other question showed that 23% of the population favors school uniforms. Which question probably produced which result and why?

   (A) The first question probably showed 23% of the population favors school uniforms, and the second question probably showed 18% because of the lack of randomization in the choice of pro-uniform and anti-uniform arguments as evidenced by the wording of the questions.

   (B) The first question probably showed 18% and the second question probably showed 23% because of stratification in the wording of the questions.

   (C) The first question probably showed 23% and the second question probably showed 18% because of the lack of a neutral cluster in the sample.

   (D) The first question probably showed 18% and the second question probably showed 23% because of response bias due to the wording of the questions.

   (E) The first question probably showed 23% and the second question probably showed 18% because of response bias due to the wording of the questions.

3. Which of the following is a true statement?

   (A) If bias is present in a sampling procedure, it can be overcome by dramatically increasing the sample size.
   (B) There is no such thing as a "bad sample."
   (C) Sampling techniques that use probability techniques effectively eliminate bias.
   (D) Convenience samples often lead to undercoverage bias.
   (E) Voluntary response samples often underrepresent people with strong opinions.

4. In a study of successes and failures in adopting new academic standards, a random sample of high school principals will be selected from each of the 50 states. Selected individuals will be asked a series of evaluative questions. Why is stratification used here?

   (A) To minimize response bias
   (B) To minimize nonresponse bias
   (C) To minimize voluntary response bias
   (D) Because each state is roughly representative of the U.S. population as a whole
   (E) To obtain higher statistical precision because variability of responses within a state is less likely than variability of responses found in the overall population

5. Which of the following is a true statement about sampling variability (sampling error)?

   (A) Sampling variability can be eliminated only if a survey is both extremely well designed and extremely well conducted.
   (B) Sampling variability reflects natural variation among samples, is always present, and can be described using probability.
   (C) Sampling variability is generally larger when the sample size is larger.
   (D) Sampling variability implies an error, possibly very small, but still an error on the part of the surveyor.
   (E) Sampling variability is higher when bias is present.

6. Each of the 30 MLB teams has 25 active-roster players. A sample of 60 players is to be chosen as follows. Each team will be asked to place 25 cards with its players' names into a hat and randomly draw out two names. The two names from each team will be combined to make up the sample. Will this method result in a simple random sample of the 750 baseball players?

   (A) Yes, because each player has the same chance of being selected.
   (B) Yes, because each team is equally represented.
   (C) Yes, because this is an example of stratified sampling, which is a special case of simple random sampling.

   (D) No, because the teams are not chosen randomly.
   (E) No, because not each group of 60 players has the same chance of being selected.

7. A researcher plans a study to examine the depth of belief in God among the adult population. He obtains a simple random sample of 100 adults as they leave church one Sunday morning. All but one of them agree to participate in the survey. Which of the following is a true statement?

   (A) Proper use of chance as evidenced by the simple random sample makes this a well-designed survey.
   (B) The high response rate makes this a well-designed survey.
   (C) Selection bias makes this a poorly designed survey.
   (D) The validity of this survey depends on whether or not the adults attending this church are representative of all churches.
   (E) The validity of this survey depends upon whether or not similar numbers of those surveyed are male and female.

8. A sports statistician plans to survey Division 1 college football programs as to graduation rates of their student athletes. The statistician will use a cluster sample with the NCAA Division 1 football conferences as clusters. A random sample of two conferences will be selected and their football programs analyzed. What is one disadvantage to selecting a cluster sample to investigate Division 1 football program graduation rates?

   (A) With two full conference football programs, the sample might be too large to effectively analyze.
   (B) The conferences in the sample might not be representative of the Division 1 population.
   (C) Cluster sampling is generally much more expensive than other sampling techniques.
   (D) The graduation rates of student athletes might not be homogeneous among schools in a given conference.
   (E) There is no disadvantage to using the proposed cluster sample.

9. Political advisors want to survey registered voters as to their top concerns. Which of the following sampling methods does NOT have voluntary response bias as a potential source of error? In each case, assume all 100 chosen people answer the survey.

(A) Give the survey to 100 randomly chosen people from among callers to a radio station who identify themselves as registered voters.

(B) Give the survey to 100 randomly chosen people from official voter registration lists.

(C) Give the survey to 100 randomly chosen people who respond to a newspaper ad asking about registered voters' concerns.

(D) Give the survey to 100 randomly chosen people from among those who approach and agree to answer questions of a surveyor standing on a busy street corner.

(E) Give the survey to 100 randomly chosen people from among those giving feedback to the question posted on a neutral online website.

10. A survey is to be taken as to whether or not people believe the moon landing conspiracy theory, which claims the six crewed landings (1969–1972) were faked. Which of the following describes a design in which *nonresponse bias* is likely present?

(A) Surveys were completed by the first 100 people responding to the question posed on a radio call-in program.

(B) Surveys were completed by 100 of the 200 people to whom the survey was mailed.

(C) Surveys were completed by 100 people chosen at random.

(D) Surveys were completed by all 100 professors teaching in a university science program.

(E) Surveys were completed by all 100 students who were sent the survey at a small liberal arts college with 850 total students.

Free-Response Questions

DIRECTIONS: You must show all work and indicate the methods you use. You will be graded on the correctness of your methods and on the accuracy of your final answers.

1. A student is interested in estimating the average length of words in a 600-page textbook and plans the following three-stage sampling procedure:

I. After noting that each of the 12 chapters has a different author, the student decides to obtain a sample of words from each chapter.

II. Each chapter is approximately 50 pages long. The student uses a random number generator to pick three pages from each chapter and use all words from these pages to find the average length of words for that author.

III. On each chosen page, the student notes the length of every tenth word.

(a) The first stage above represents what kind of sampling procedure? Give an advantage in using it in this context.

(b) The second stage above represents what kind of sampling procedure? Give an advantage in using it in this context.

(c) The third stage above represents what kind of sampling procedure? Give a *disadvantage* to using it dependent upon an author's writing style.

2. Pelicans are considered an endangered species, and there is a need to protect them from becoming extinct. Unfortunately, the U.S. Fish & Wildlife Service is unable to protect all pelican colonies; however, it plans to protect those colonies identified as most important. One criterion in determining the importance of a geographical region is the density of nesting sites within it.

Given an aerial view of the nests in a geographical region used for pelican nesting, the service plans to divide the rectangular picture into 150 equal-size plots as below:

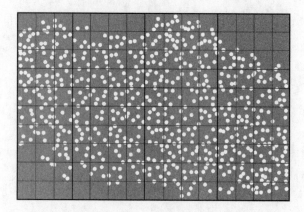

The service will randomly pick 15 plots (10% of the total), count the nests in each of those plots, sum and multiply by 10 to estimate the total number of nests in the region, and then calculate density. Two methods for counting the number of nests are under consideration.

Method I: Randomly pick one of the above rows and use all the 15 plots in that row.

Method II: Randomly pick one of the plots in each of the 15 columns to obtain the 15 plots.

(a) What sampling procedure is each method striving for, and how can each method be implemented?

(b) Which of these two methods is preferable here? Explain.

3. For each of the following survey designs, identify the problem and the effect it would have on the estimate of interest.

(a) Should spending be increased for the athletic program? A school official decides to sample spectators randomly at the next home basketball game.

(b) Do people believe that eating organic will lower the risk of cancer? A health magazine sends a questionnaire to all their subscribers.

(c) A supermarket quality control inspector opens boxes of strawberries. If no mold is seen among the visible strawberries, he accepts the box.

(d) Are the patients following their diet instructions? A family physician asks each of her patients during their office visits.

(e) Parents are asked to write to a newspaper with a yes or no response to whether or not they are happy with the education their children are receiving in the local public schools.

The answers for this quiz can be found beginning on page 484.

Experiments Versus Observational Studies

In an experiment, we impose some change or treatment and measure the response. In an observational study, we simply observe and measure something that has taken place or is taking place while trying not to cause any changes by our presence. Understanding the differences between observational studies and experiments is one of the most important ideas to come away with from your AP Statistics class.

Observational Studies	Experiments
Observe and measure without influencing	Impose *treatments* and measure responses
Can only show associations	Can suggest *cause-and-effect* relationships
Use of *random sampling* in order to be able to generalize to a population	Use of *random assignment* to minimize the effect of confounding variables
Usually less expensive and less time-consuming	Can be expensive and time-consuming
Use of *strata,* and randomization within strata, for greater accuracy and to reduce variability	Use of *blocks*, and randomization within blocks, to control for certain variables
	Possible ethical concerns over imposing certain treatments
	Use of a control group for a baseline comparison
	Use of blinding and double-blinding

› Example 3.8

A study is to be designed to determine whether a particular commercial review course actually helps raise SAT scores among students attending a particular high school. How could an observational study be performed? An experiment? Which is more appropriate here?

✓ Solution

An observational study would interview students at the school who have taken the SAT exam, asking whether or not they took the review course, and then the SAT results of those who have and have not taken the review course would be compared.

An experiment performed on students at the school who are planning to take the SAT exam would use chance to randomly assign some students to take the review course while others to not take the review course and would then compare the SAT results of those who have and have not taken the review course.

An argument in favor of the experimental approach is that with the observational study, there could be many explanations for any SAT score difference noted between those who took the course and those who didn't. For example, students who took the course might be the more serious high school students. Higher score results might be due to their more serious outlook and have nothing to do with taking the course. The experiment tries to control for confounding variables by using random assignment to make the groups taking and not taking the course as similar as possible except for whether or not the individuals take the course.

> **TIP**
>
> **It is important to identify the *response variable* (SAT scores in this example) right away. This is critical in answering any experimental design question.**

(continued)

An argument in favor of the observational study approach is that there are ethical considerations in forcing some students to take the review course who might not have wanted to and in withholding the review course from some students who might have wanted to take it.

The Language of Experiments

An experiment is performed on objects called *experimental units*. If the units are people, they are called *subjects*. Experiments involve *explanatory variables*, called *factors*, that are believed to have an effect on *response variables*. A group is intentionally treated with some *level* of the explanatory variable, and the outcome of the response variable is measured.

Several primary principles deal with the proper planning and conducting of experiments:

- Possible confounding variables are controlled for by making conditions (other than the treatments) similar for all treatment groups.
- Chance should be used in the random assignment of subjects to treatment groups.
- Natural variation in outcomes can be lessened by using more subjects.
- You should always be asking whether or not you can reasonably draw cause-and-effect connections between explanatory and response variables.
- To draw cause-and-effect connections, experiments **may** use *control groups* to have something to compare to. Depending on the experiment, a control group may be given an inactive treatment (a *placebo*), an established active treatment, or no treatment at all.
- Depending on the experiment, there may be ethical issues on whether or not a treatment can be used.

❯ Example 3.9

In an experiment to test exercise and blood pressure reduction, volunteers are randomly assigned to participate in 0, 1, or 2 hours of exercise per day for 5 days per week over the next 6 months. What is the explanatory variable with the corresponding levels, and what is the response variable?

✓ Solution

The explanatory variable, hours of exercise, is being implemented at three levels: 0, 1, and 2 hours per day. The response variable is not specified but could be the measured change in either systolic or diastolic blood pressure readings after 6 months.

Suppose the volunteers were further randomly assigned to follow either the DASH (Dietary Approaches to Stop Hypertension) or the TLC (Therapeutic Lifestyle Changes) diet for the 6 months. There would then be two factors, hours of exercise with three levels and diet plan with two levels, and a total of six treatments (DASH diet with 0 hours daily exercise, DASH with 1 hour exercise, DASH with 2 hours exercise, TLC diet with 0 hours daily exercise, TLC with 1 hour exercise, and TLC with 2 hours exercise).

Two variables are *confounded* when both are associated with a response variable and are also associated with each other. Their effects on the response variable cannot be distinguished, and it may be impossible to tell which explanatory variable truly causes the response. For example:

- In a random survey of adults, it is noted that people who regularly brush their teeth have fewer cavities. However, people who regularly brush may also choose better diets, and it might be unclear whether brushing or diet leads to fewer cavities. In this case, brushing and diet would be confounded.

TIP

If asked to list all the treatments, don't create ones that aren't there!

NOTE

When there is more than one factor, the number of treatments is the product of the number of levels for each factor.

- In an experiment to determine if different fertilizers cause differences in plant growth, a horticulturist might decide to have many test plots using one or the other of the fertilizers, with equal numbers of sunny and shady plots for each fertilizer, so that fertilizer and sun are not confounded. However, if two fertilizers require different amounts of watering, it might be difficult to determine if the difference in fertilizers or the difference in watering is the real cause of observed differences in plant growth. In this case, fertilizer and water would be confounded.

› Example 3.10

Many are concerned that vaping produces the so-called gateway effect. Studies have shown that young people who vape are more likely to go on to smoke cigarettes than young people who do not vape.

(a)   What are the explanatory and response variables?
(b)   Are these observational studies or experiments?

Some researchers suggest that a confounding variable might be that many people feel that smoking in general is "cool."

(c)   Explain what this means to be a confounding variable in the above context.
(d)   Explain how this confounding variable might lead researchers to conclude a higher or lower proportion of people become cigarette smokers because they vape than actually become cigarette smokers because they vape.

✓ Solutions

(a)   The explanatory variable is whether or not a young adult vapes, and the response variable is whether or not the person becomes a cigarette smoker later in life.
(b)   These are observational studies. It would have been highly unethical to randomly select young adults and instruct them to vape as part of an experiment to see if they go on to cigarette smoking.
(c)   People who think smoking is "cool" are more likely to vape as young adults and are more likely to smoke cigarettes later in life. So, it may be that it's not vaping that leads to cigarette smoking but, rather, that thinking smoking is "cool" leads both to vaping as a young adult and to smoking cigarettes later in life.
(d)   This confounding variable would lead researchers to conclude a higher proportion of people become cigarette smokers because they vape than actually become cigarette smokers because they vape, because some become cigarette smokers because they feel that smoking is "cool," not because they vape.

In an experiment, there is often a *control group* to determine if the treatment of interest has an effect. There are several types of control groups. A control group can be a collection of experimental units not given any treatment, or given the current treatment, or given a treatment with an inactive substance (a placebo). When a control group receives the current treatment or a placebo, these count as treatments if you are asked to enumerate the treatments.

Random assignment to the various treatment groups, including to the control group, can help reduce the problem posed by confounding variables. The beauty of random assignment is that it works on variables that you "know" to be confounders, variables that you "suspect" to be confounders, and variables you are clueless about or don't even know they exist. Random assignment to treatments, called *completely randomized design*, is critical in good experimental design. This is especially true if the subjects are not randomly selected, as is the case in most medical and drug experiments.

> Example 3.11

Sixty patients, ages 5 to 12, all with common warts are enrolled in a study to determine if application of duct tape is as effective as cryotherapy in the treatment of warts. Subjects will receive either cryotherapy (liquid nitrogen applied to each wart for 10 seconds every 2 weeks) for 6 treatments or duct tape occlusion (applied directly to the wart) for 2 months. Describe a completely randomized design.

NOTE

Random assignment *minimizes*, but can't *eliminate*, the effect of possible confounding variables.

✓ Solution

Assign each patient an integer from 1 to 60. Use a random integer generator on a calculator to pick integers between 1 and 60, throwing away repeats, until 30 unique integers have been selected. (Or numbering the patients with two-digit numbers from 01 to 60, use a random number table, reading off two digits at a time, ignoring repeats, 00, and numbers over 60, until 30 unique numbers have been selected.) The 30 patients corresponding to the 30 selected integers will be given the cryotherapy treatment. (A third design would be to put the 60 names on identical slips of paper, put the slips in a hat, mix them well, and then pick out 30 slips, without replacement, with the corresponding names given cryotherapy.) The remaining 30 patients will receive the duct tape treatment. At the end of the treatment schedules, compare the proportion of each group that had complete resolution of the warts being studied.

Random assignment allows us to make a valid comparison of treatments.

It is a fact that many people respond to any kind of perceived treatment. This is called the *placebo effect*. For example, when given a sugar pill after surgery but told that it is a strong pain reliever, many people feel immediate relief from their pain. *Blinding* occurs when the subjects don't know which of the different treatments (such as placebos) they are receiving. *Double-blinding* is when neither the subjects nor the *response evaluators* know who is receiving which treatment.

The goal is to control every variable we can so that all treatment groups have the same experience, save the one variable we are manipulating. Blinding and double-blinding are useful tactics to accomplish that, and a placebo is a useful tool in blinding.

> Example 3.12

There is a pressure point on the wrist that some doctors believe can be used to help control the nausea experienced following certain medical procedures. The idea is to place a band containing a small marble firmly on a patient's wrist so that the marble is located directly over the pressure point. Describe how a double-blind experiment might be run on 50 postoperative patients.

✓ Solution

Assign each patient an integer from 1 to 50. Use a random integer generator on a calculator to pick integers between 1 and 50, ignoring repeats, until 25 unique integers have been selected. (Or numbering the patients with two-digit numbers from 01 to 50, from a random number table read off two digits at a time, throwing away repeats, 00, and numbers over 50, until 25 unique numbers have been selected.) Put wristbands with marbles over the pressure point on the patients with these assigned numbers. (A third experimental design would be to put the 50 names on identical slips of paper, put the slips in a hat, mix them well, and then pick out 25 slips, without replacement, with the corresponding names given wristbands with marbles over the pressure point.)

NOTE

Blinding and placebos in experiments are important but are not always feasible. You can still have "experiments" without using these.

(continued)

Put wristbands with marbles on the remaining patients also but *not* over the pressure point. Have a researcher check by telephone with all 50 patients at designated time intervals to determine the degree of nausea being experienced. Neither the patients nor the researcher on the telephone should know which patients have the marbles over the correct pressure point.

NOTE

Of course, with double-blinding, someone (the person directing the study) must know who is receiving which treatment!

A *matched pairs design* (also called a *paired comparison design*) is when two treatments are compared based on the responses of paired subjects, one of whom receives one treatment while the other receives the second treatment. Often the paired subjects are really single subjects who are given both treatments, one at a time in random order.

❯ Example 3.13

Does seeing pictures of accidents caused by drunk drivers influence one's opinion on penalties for drunk drivers? How could a comparison test be designed for 100 subjects?

✓ Solution

The subjects could be asked questions about drunk driving penalties before and then again after seeing the pictures, and any change in answers could be noted. This would be a poor design because there is no control group, there is no use of randomization, and subjects might well change their answers because they realize that that is what is expected of them after seeing the pictures.

A better design is to use randomization to split the subjects into two groups, half of whom simply answer the questions while the other half first see the pictures and then answer the questions. For example, put the 100 names on identical slips of paper, put the slips in a hat, mix them well, and then pick out 50 slips (without replacement) with the corresponding names designating the subjects who see the pictures.

Another possibility is to use a group of 50 biological twins as subjects. One of each set of twins is randomly picked (e.g., based on choosing an odd or even digit from a random number table) to answer the questions without seeing the pictures, while the other first sees the pictures and then answers the questions. The answers could be compared from each set of twins. This is a paired comparison test that might help minimize confounding variables due to family environment, heredity, and so on.

Just as stratification, in sampling, first divides the population into representative groups called strata, *blocking,* in experimental design, first divides the subjects into representative groups called *blocks*. Subjects within *each* block are randomly assigned treatments. One can think of blocking as running parallel experiments before combining the results. This technique helps reduce variability as well as helps control certain possible confounding variables by bringing them directly into the picture. The paired comparison (matched pairs) design is a special case of blocking in which each pair can be considered a block of size 2.

TIP

Use proper terminology! The language of experiments is different from the language of sampling. Don't mix up *blocking* and *stratification*.

❯ Example 3.14

There is a rising trend for star college athletes to turn professional without finishing their degrees. A study is performed to assess whether reading an article about professional salaries has an impact on such decisions. Randomization can be used to split the subjects into two groups, and those in one group are given the article before answering questions. How can a block design be incorporated into the design of this experiment?

(continued)

✓ **Solution**

For example, the subjects can be split into two blocks, underclass and upperclass, before using randomization to assign some to read the article before questioning. With this design, the impact of the salary article on freshmen and sophomores can be distinguished from the impact on juniors and seniors.

Similarly, blocking can be used to separately analyze men and women, those with high GPAs and those with low GPAs, those in different sports, those with different majors, and so on.

Completely Randomized Design

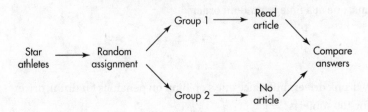

Block Design

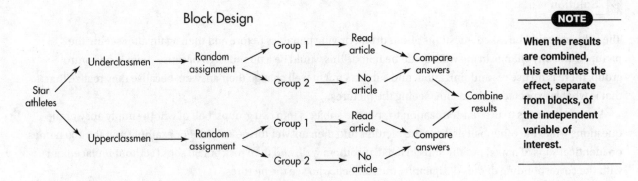

> **NOTE**
>
> When the results are combined, this estimates the effect, separate from blocks, of the independent variable of interest.

The following emphasizes some points with regard to experimental design:

- The goal of random assignment is to create treatment groups that are similar in all respects except for the treatment imposed.
- It is not enough to say "use random assignment." You must be able to carefully explain how the random assignment of treatments to subjects is to be accomplished. Don't forget to name which treatment is given to which randomly assigned group.
- It is not enough simply to say that two variables are confounded. You must be able to explain the possible incorrect conclusion in context.
- It is important to be able to describe how to use random assignment for different experimental designs (completely randomized, randomized blocked, or matched pairs):
  - In completely randomized design, the experimental units are assigned to the treatments (or the treatments to units) completely at random.
  - In randomized block design, the randomization occurs *within blocks*. Blocks are *not* formed at random, and treatments are not randomly assigned to blocks.
  - In matched pairs design, the randomized block design occurs with blocks of size 2.
- When blinding is used, subjects in the treatment and control groups experience the placebo effect equally.

Replication and Generalizability of Results

When differences are observed in a comparison test, the researcher must decide whether these differences are *statistically significant* or whether they can be explained by natural variation. One important consideration is the size of the sample: the larger the sample, the more significant the observation. This is the principle of *replication*. In other words, the treatment should be repeated on a sufficient number of subjects so that real response differences are more apparent.

Randomization, in the form of random assignment, is critical to minimize the effect of confounding variables. However, in order to generalize experimental results to a larger population (as we try to do in sample surveys), it would also be necessary that the group of subjects used in the experiment be randomly selected from the population. For example, it is hard to generalize from the effect a television commercial has on students at a private midwestern high school to the effect the same commercial has on retired senior citizens in Florida. If the experiment was performed on those high school students, it can be generalized only to the population of all students at private midwestern high schools.

NOTE

Replication refers to having more than one experimental unit in each treatment group, not multiple trials of the same experiment.

Inference and Experiments

What kind of conclusions result from well-designed experiments? Can we conclude that observed differences in the treatment groups are statistically significant; that is, are the differences so large as to be unlikely to have occurred by chance? How can we decide whether the differences are large enough?

If the treatments make no difference, just by chance there probably will be some variation. In later units, you will learn mathematically how to decide if something is different enough to be considered statistically significant. However, for now, you should understand that something is statistically significant if the probability of it happening just by chance is so small that you're convinced there must be another explanation.

TIP

Most experiments are not performed on random samples from a population. So, don't use *random sample* terminology when you mean *random assignment*.

› Example 3.15

For a fair pair of dice, the probability of throwing "7 or 11" is approximately 22.2 percent. We would expect for the sum to come up 7 or 11 an average of 22.2 times for every 100 tosses if the dice are fair. Suppose we throw a pair of dice 100 times, and 7 or 11 comes up 26 times. This would appear to be within the bounds of random fluctuations, and we would have no reason to suspect the dice of being unfair. However, if 7 or 11 came up 55 times, that would raise our suspicions. While it would still be possible that the dice were fair, we could comfortably say that we had convincing evidence to believe that they were unfair. However, what if 7 or 11 came up 30 times? Is that different enough from 22.2 to raise our suspicions? We'll learn how to answer this question later after studying probability distributions.

› Example 3.16

Sixty students volunteered to participate in an experiment comparing the effects of coffee, caffeinated cola, and herbal tea on pulse rates. Twenty students are randomly assigned to each of the three treatments. For each student, the change in pulse rate was measured after consuming eight ounces of the treatment beverage. The results are summarized with the parallel boxplots below.

(continued)

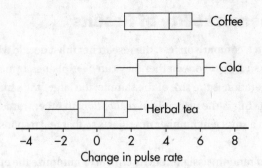

What are reasonable conclusions?

✓ Solution

The median change in pulse rate for the cola drinkers was higher than that for the coffee drinkers; however, looking at the overall spreads, that observed difference does not seem significant. The difference between the coffee and caffeinated cola drinkers with respect to change in pulse rate is likely due to random chance. Now compare the coffee and caffeinated cola drinkers' results to that of the herbal tea drinkers. Although there is some overlap, there is not much. It seems reasonable to conclude the difference is statistically significant; that is, drinking coffee or caffeinated cola results in a greater rise in pulse rate than drinking herbal tea.

〉 Example 3.17

Harmane is a potent neurotoxin linked to human disease and is found in a variety of cooked meats. Chicken in fast-food restaurants appears to be a particular problem. A fast-food chain claims that their cooked chicken has an 8 ng/g (nanograms per gram) concentration, but a nutritionist believes the true figure is greater. She randomly samples chicken nuggets on ten different days and obtains a mean harmane concentration of 8.48 ng/g. Is this statistically significant? The chain claims that this is possible just by chance sample variability. The chain's scientists simulate 100 such samples of size 10 from a laboratory setting where the known mean concentration is 8 ng/g in order to show that 8.48 ng/g is possible and not unusual. The resulting 100 means are summarized in the following dotplot:

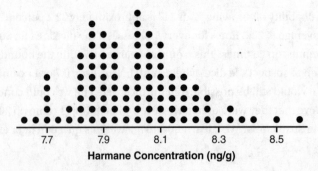

Does the simulation support the nutritionist or the chain?

✓ Solution

The simulation shows that although a sample with a mean 8.48 ng/g concentration is possible if the true mean is 8 ng/g, it is highly unlikely. The simulation supports the nutritionist.

Quiz 8

Multiple-Choice Questions

> **DIRECTIONS:** The questions that follow are each followed by five suggested answers. Choose the response that best answers the question.

1. Suppose you wish to compare the average class size of mathematics classes to the average class size of English classes in your high school. Which is the most appropriate technique for gathering the needed data?

   (A) Census
   (B) Sample survey
   (C) Experiment
   (D) Observational study
   (E) None of these methods is appropriate.

2. Two studies are run to compare the experiences of families living in high-rise public housing to those of families living in townhouse subsidized rentals. The first study interviews 25 families who have been in each government program for at least 1 year, while the second randomly assigns 25 families to each program and interviews them after 1 year. Which of the following is a true statement?

   (A) Both studies are observational studies because of the time period involved.
   (B) Both studies are observational studies because there are no control groups.
   (C) The first study is an observational study, while the second is an experiment.
   (D) The first study is an experiment, while the second is an observational study.
   (E) Both studies are experiments.

3. A consumer product agency tests miles per gallon for a sample of automobiles using each of four different octanes of gasoline. Which of the following is true?

   (A) There are four explanatory variables and one response variable.

   (B) There is one explanatory variable with four levels of response.
   (C) Miles per gallon is the only explanatory variable, but there are four response variables corresponding to the different octanes.
   (D) There are four levels of a single explanatory variable.
   (E) Each explanatory level has an associated level of response.

4. A study is made to determine whether taking AP Statistics in high school helps students achieve higher GPAs when they go to college. In comparing records of 200 college students, half of whom took AP Statistics in high school, it is noted that the average college GPA is higher for those 100 students who took AP Statistics than for those who did not. Based on this study, guidance counselors begin recommending AP Statistics for college-bound students. Which of the following is *incorrect*?

   (A) Although this study indicates a relation, it does not prove causation.
   (B) There could well be a confounding variable responsible for the seeming relationship.
   (C) Self-selection here makes drawing the counselors' conclusion difficult.
   (D) A more meaningful study would be to compare a simple random sample (SRS) from each of the two groups of 100 students.
   (E) This is an observational study, not an experiment.

5. A cardiologist surveyed a random sample of adults age 21 to 75, looking at whether or not they reported exercising regularly and then used an echocardiogram to measure heart strength of those surveyed. She found a strong positive association. Which of the following is a proper conclusion?

(A) This was a randomized experiment, which proved that regular exercise causes greater heart strength.

(B) This was a randomized experiment, which gave strong evidence that regular exercise causes greater heart strength.

(C) This was an observational study, which proved that regular exercise causes greater heart strength.

(D) This was an observational study, which gave strong evidence that regular exercise causes greater heart strength.

(E) It is likely that there is a confounding variable.

6. In designing an experiment, blocking is used

(A) to reduce bias.
(B) to reduce variation.
(C) as a substitute for a control group.
(D) as a first step in randomization.
(E) to control the level of the experiment.

Free-Response Questions

DIRECTIONS: You must show all work and indicate the methods you use. You will be graded on the correctness of your methods and on the accuracy of your final answers.

1. Rock and pop fame is associated with risk-taking and premature mortality. A study following the lives of 100 rock and pop stars periodically asked them about their frequency of substance abuse and followed their lives until death. Researchers noted the age of death and noticed that a greater amount of substance abuse was associated with an earlier age at death.

(a) What are the explanatory and response variables?

(b) Is this a prospective observational study, a retrospective observational study, or an experiment? Explain.

(c) Does this study show that greater frequency of substance abuse leads to early death?

2. Much of the profit in movie theaters comes from sales at the concession stands. A theater manager wants to run an experiment to evaluate whether subliminal advertising will help to increase sales. He has software that flashes "BUY POPCORN" on the screen for a fraction of a second.

(a) Identify the explanatory and response variables.

(b) Explain how the manager could incorporate randomization, blinding, and comparison.

3. For each of the following, would you recommend using random sampling, random assignment, both, or neither? Explain.

(a) A movie critic wants to investigate whether a greater proportion of teenagers in public or private schools have seen all three of the *Back to the Future* films.

(b) A company CEO wants to investigate whether giving hourly pay raises or end-of-the-year bonuses leads to greater productivity on an assembly line.

(c) A sportscaster wants to investigate whether the Chicago Cubs or the Chicago White Sox won a greater number of interleague games during the shortened 2020 season.

4. A six-month study looked at 800 adult volunteers who were randomly divided into four groups.

- The first group wore a clip-on fitness tracker and received a cash incentive if they walked more than 50,000 steps a week.
- The next group also wore an activity tracker and received a cash incentive for walking more than 50,000 steps a week. However, they were required to donate the cash to charity.
- A third group wore the activity tracker but did not receive a cash incentive.
- The last group did not wear a fitness tracker or get a cash incentive.

The measured outcome was "change in blood pressure over a six-month period."

(a) Was this an observational study or an experiment? Explain.

(b) Identify the treatments, experimental units, and response variable.

(c) What is the fourth group called, and what is the purpose of including it?

(d) Describe an appropriate method for assigning the subjects to the four groups so that each group will have an equal number of subjects.

(e) Is it reasonable to generalize the results of the study beyond the 800 participants? Explain why or why not.

The answers for this quiz can be found beginning on page 486.

USE CORRECT TERMINOLOGY!

Bias: when a study systematically favors certain outcomes

Blinding: refers to subjects not knowing which treatment they are receiving

Block design: an experimental design where subjects are divided into representative groups to bring certain differences into the picture and reduce variation (for example, blocking by gender, age, or race) and randomization then takes place within each block

Census: an attempt to include the entire population

Cluster sample: involves dividing the population into heterogeneous groups called clusters and then picking everyone in a random sample of the clusters

Completely randomized design: an experimental design in which everyone has an equal chance of receiving any treatment

Confounding: when there is uncertainty as to which variable is causing an effect

Control group: a group given no treatment or a sham treatment

Double-blinding: refers to subjects and those evaluating their responses not knowing who received which treatments

Experimental study: involves applying a treatment to one or more groups and observing the responses

Nonresponse bias: when a large fraction of those sampled do not respond

Observational study: researchers merely observe (no treatments are applied)

Parameter: a numerical measurement describing some characteristic of the population

Placebo: a dummy or sham treatment such as a sugar pill

Population: the entire set of items, events, people, objects, and so on that are of interest

Random assignment: in experiments, when subjects are randomly assigned to treatments (evens out effects over which we have no control)

Random sampling: use of chance in selecting a sample from a population (allows for generalization of conclusions)

Response bias: when the question itself leads to misleading results (for example, people don't want to be perceived as having unpopular, unsavory, or illegal views)

Sample: the part of the population actually examined

Simple random sample (SRS): a sample selected in such a way that every possible sample of the desired size has an equal chance of being selected (each element of the population will also have an equal chance of being selected)

Statistic: a numerical measurement describing some characteristic of the sample

Statistical significance: a resulting difference among treatment groups too large to be attributed to chance

Stratified random sample: involves dividing the population into homogeneous groups called strata and then picking random samples from each of the strata

Systematic sample: involves ordering the population, choosing a random point to start, and then picking every nth person for some n

Undercoverage bias: when part of the population is ignored (for example, some telephone surveys miss all those who only have cell phones)

Voluntary response bias: when individuals choose whether or not to respond (for example, radio call-in shows and Internet surveys)

SUMMARY

- Retrospective studies look backward, examining existing data, while prospective studies watch for outcomes, tracking individuals into the future.
- Bias is the tendency to favor the selection of certain members of a population.
- Voluntary response bias occurs when individuals choose whether to respond (for example, Internet surveys) and typically give too much emphasis to people with strong opinions (particularly strong negative opinions).
- Convenience surveys, like interviews at shopping malls, are based on choosing individuals who are easy to reach. These surveys tend to produce data highly unrepresentative of the entire population.
- Use of randomization is absolutely critical in selecting an unbiased sample.
- Use of randomization in selecting a sample is crucial in being able to generalize from a sample to the population.
- A simple random sample (SRS) is one in which every possible sample of the desired size has an equal chance of being selected.
- Stratified sampling involves dividing the population into homogeneous groups called strata and then picking random samples from each of the strata.
- Cluster sampling involves dividing the population into heterogeneous groups called clusters, randomly selecting one or more of the clusters, and then using everyone in the selected clusters.
- Sampling error (sampling variability) is not an avoidable mistake but, rather, is the natural variability among samples.
- While observational studies involve observing without causing changes, experiments involve actively imposing treatments and observing responses.
- Differences in observed responses are statistically significant when it is unlikely that they can be explained by chance variation.

- Random assignment of subjects to treatment groups is extremely important in handling unknown and uncontrollable differences.
- Random assignment refers to what is done with subjects after they've been picked for a study, whereas random sampling refers to how subjects are selected for a study.
- Variables are said to be confounded when there is uncertainty as to which variable is causing an effect.
- The placebo effect refers to the fact that many people respond to any kind of perceived treatment.
- Blinding refers to subjects not knowing which treatment they are receiving.
- Double-blinding refers to subjects and those evaluating their responses not knowing who received which treatments.
- Completely randomized designs refer to experiments in which everyone has an equal chance of receiving any treatment.
- Blocking is the process of dividing the subjects into representative groups to bring certain differences into the picture (for example, blocking by gender, age, or race/ethnicity).
- Randomized block designs refer to experiments in which the randomization occurs only within blocks.
- Randomized paired comparison designs refer to experiments in which subjects are paired and randomization is used to decide who in each pair receives what treatment.

Probability, Random Variables, and Probability Distributions

(4–8 multiple-choice questions on the AP exam)

Learning Objectives

In this unit, you will learn how

→ To understand how the law of large numbers relates to relative frequencies.

→ To be able to make basic probability calculations involving the complement, union, and intersection of events, including the use of the general addition rule.

→ To understand when events are mutually exclusive and when they are independent and how that applies to probability calculations.

→ To be able to calculate conditional probabilities.

→ To be able to calculate reverse conditional probabilities, using a tree diagram if appropriate (multistage probability calculations).

→ To be able to calculate and interpret the mean (expected value) and standard deviation of a random variable.

→ To be able to calculate and interpret the mean of the sum or difference of random variables.

→ To be able to calculate and interpret the standard deviation of the sum or difference of independent random variables.

→ To be able to calculate the probability of exactly x successes in n trials in a binomial scenario.

→ To be able to calculate the probability of at least or at most x successes in n trials in a binomial scenario.

→ To be able to calculate the mean (expected value) and standard deviation of a binomial probability distribution.

→ To be able to calculate the probabilities of the possible values of a geometric random variable.

→ To be able to calculate the mean (expected value) and standard deviation of a geometric probability distribution.

In this unit, you will learn basic probability calculations and how to calculate and interpret means and standard deviations of random variables and of binomial and geometric probability distributions. You will further study patterns and uncertainty and see how an event can be both random and still have predictability.

In the world around us, unlikely events sometimes take place. At other times, events that seem inevitable do not occur. Although we are aware of the possible outcomes, we have no idea what an individual outcome will be. Chance is everywhere! The cards you are dealt in a poker game, the particular genes you inherit from your parents, and the coin toss at the beginning of a tennis match to determine who serves first are examples of chance behavior that mathematics can help us understand. Even though we may not be able to foretell a specific result, we can sometimes assign what is called a *probability* to indicate the likelihood that a particular event will occur.

Probabilities are always between 0 and 1, with a probability close to 0 meaning that an event is unlikely to occur and a probability close to 1 meaning that the event is likely to occur. The sum of the probabilities of all the separate outcomes of an experiment is always 1.

TIP

Calculators express very small probabilities in scientific notation such as 3.4211073E−6. Know what this means, and remember that probabilities are never greater than 1.

The Law of Large Numbers

The *relative frequency* of an event is the proportion of times the event happened, that is, the number of times the event happened divided by the total number of trials. Relative frequencies may change every time an experiment is performed. The *law of large numbers* states that when an experiment is performed a large number of times, the relative frequency of an event tends to become closer to the true probability of the event; that is, *probability is long-run relative frequency*. There is a sense of predictability about the long run.

NOTE

An *event* is a collection of one or more outcomes.

The law of large numbers says nothing at all about short-run behavior. There is no such thing as a law of small numbers or a law of averages. Gamblers might say that "red" is due on a roulette table, a basketball player is due to make a shot, a player has a hot hand at the craps table, or a certain number is due to come up in a lottery. However, if events are independent, the probability of the outcome of the next trial has nothing to do with what happened in previous trials. Even though casinos and life insurance companies might lose money in the short run, they make long-term profits because of their understanding of the law of large numbers.

The idea that a few outcomes will immediately be balanced by other outcomes is also called the *gambler's fallacy*.

The law of large numbers has two conditions. First, the chance event under consideration does not change from trial to trial. Second, any conclusion must be based on a large (a very large!) number of observations.

> ### Example 4.1

There are two games involving flipping a fair coin. In the first game, you win a prize if you can throw between 45% and 55% heads. In the second game, you win if you can throw more than 60% heads. For each game, would you rather flip 20 times or 200 times?

> ### ✓ Solution

The probability of throwing heads is 0.5. By the law of large numbers, the more times you throw the coin, the more the relative frequency tends to become closer to this probability. With fewer tosses, there is greater chance of wide swings in the relative frequency. Thus, in the first game, you would rather have 200 flips, whereas in the second game, you would rather have only 20 flips.

Basic Probability Rules

> ### Example 4.2

A standard literacy test consists of 100 multiple-choice questions, each with five possible answers. There is no penalty for guessing. A score of 60 is considered passing, and a score of 80 is considered superior. When an answer

(continued)

is completely unknown, test takers employ one of three strategies: guess, choose answer (c), or choose the longest answer. The table below summarizes the results of 1,000 test takers.

		Score			
		0–59	**60–79**	**80–100**	*Total*
	Guess	40	160	100	*300*
Strategy	**Choose answer (c)**	35	170	115	*320*
	Choose longest answer	45	200	135	*380*
	Total	*120*	*530*	*350*	*1,000*

Note that in analyzing tables such as the one above, it is usually helpful to sum rows and columns.

(a)   What is the probability that someone in this group uses the "guess" strategy?

(b)   What is the probability that someone in this group scores 60–79?

(c)   What is the probability that someone in this group does not score 60–79?

(d)   What is the probability that someone in this group chooses strategy "answer (c)" *and* scores 80–100 (the intersection of the two events)?

(e)   What is the probability that someone in this group chooses strategy "longest answer" *or* scores 0–59 (the union of the two events: "or" here means one event or the other event or both events)?

> **TIP**
>
> Conditional probabilities shrink the population of interest.

(f)   What is the probability that someone in this group chooses the strategy "guess" *given that* his or her score was 0–59?

(g)   What is the probability that someone in this group scored 80–100 *given that* the person chose strategy "longest answer"?

(h)   Are the strategy "guess" and scoring 0–59 *independent events*? That is, is whether a test taker used the strategy "guess" unaffected by whether the test taker scored 0–59?

(i)   Are the strategy "longest answer" and scoring 80–100 *mutually exclusive events*? That is, are these two events disjoint and cannot simultaneously occur?

✓ Solutions

(a) $P(\text{guess}) = \frac{300}{1,000} = 0.3$

(b) $P(\text{score } 60\text{–}79) = \frac{530}{1,000} = 0.53$

(c) $P(\text{does not score } 60\text{–}79) = 1 - P(\text{score } 60\text{–}79) = 1 - 0.53 = 0.47$

(d) $P(\text{answer (c)} \cap \text{score } 80\text{–}100) = \frac{115}{1,000} = 0.115$

(e) $P(\text{longest answer} \cup 0\text{–}59) = \frac{45 + 200 + 135 + 40 + 35}{1,000} = \frac{455}{1,000} = 0.455$

Note that

$$P(\text{longest answer} \cup \text{score } 0\text{–}59) =$$
$$P(\text{longest answer}) + P(\text{score } 0\text{–}59) - P(\text{longest answer} \cap \text{score } 0\text{–}59) =$$
$$0.380 + 0.120 - 0.045 = 0.455$$

> **NOTE**
>
> The subtraction of the intersection probability is so that it is not counted twice.

(f) $P(\text{guess} \mid \text{score } 0\text{–}59) = \frac{40}{120} \approx 0.333$

(continued)

Note that we narrowed our attention to the 120 test takers who scored 0–59 to calculate this *conditional probability*.

(g) $P(\text{score } 80\text{–}100 \mid \text{longest answer}) = \frac{135}{380} \approx 0.355$

Note that we narrowed our attention to the 380 test takers who chose "longest answer" to calculate this *conditional probability*.

(h)   We must check if $P(\text{guess} \mid \text{score } 0\text{–}59) = P(\text{guess})$. From (f) and (a), we see that these probabilities are not equal ($0.333 \neq 0.3$), so the strategy "guess" and scoring 0–59 are *not* independent events.

(i)   longest answer $\cap$ score 80–100 $\neq \varnothing$ and $P(\text{longest answer} \cap \text{score } 80\text{–}100) = \frac{135}{1,000} \neq 0$, so the strategy "longest answer" and scoring 80–100 are *not* mutually exclusive events.

Recap

- The probability that an event will not occur, that is, the probability of its complement, is equal to 1 minus the probability that the event will occur. Two notations for the complement of A are A^C and A'. Then $P(A^C) = P(A') = 1 - P(A)$.
- **The general addition rule: $P(A \cup B) = P(A) + P(B) - P(A \cap B)$.**
- A and B are mutually exclusive if $A \cap B = \varnothing$ or, equivalently, $P(A \cap B) = 0$. In this case, $P(A \cup B) = P(A) + P(B)$. Don't add probabilities unless the events are mutually exclusive!
- **The general multiplication rule: $P(A \cap B) = P(A)P(B|A)$.** This comes immediately from the conditional probability formula $P(B|A) = \dfrac{P(A \cap B)}{P(A)}$.
- A and B are independent if $P(A|B) = P(A)$ or, equivalently, $P(B|A) = P(B)$. It is also true that A and B are independent if and only if $P(A \cap B) = P(A)P(B)$, that is, if and only if the probability of both events happening is the product of their probabilities. Don't multiply probabilities unless the events are independent!
- Don't confuse *mutually exclusive* with *independent*!
  - If A and B are mutually exclusive, $P(A \cap B) = 0$.
  - If A and B are independent, $P(A \cap B) = P(A)P(B)$.

❯ Example 4.3

What probability rule or formula is appropriate for solving each of the following verbal scenarios?

(a)   If 44% of Chicago residents are Cubs fans, what is the probability that a Chicago resident is not a Cubs fan?

(b)   A survey of all students at a large high school revealed that 58% pick "Fifteen" as their favorite single, 34% pick Taylor Swift as their favorite artist, and 15% pick both of these choices. If a student is selected at random, what is the probability that the student picks "Fifteen" or Taylor Swift or both?

(c)   Both blue eyes and gray eyes have a dark epithelium at the back of the iris and a relatively clear stroma at the front. In the population, 10% have blue eyes and only 1% have gray eyes. If a person is selected at random, what is the probability the person has either blue or gray eyes?

(d)   Among the students at a large state college, 72% use Pandora, 43% use Spotify, and 29% use both. If a randomly selected student uses Spotify, what is the probability the student also uses Pandora?

(e)   According to one survey, 55% of Americans "love pizza," and 68% of those who love pizza choose pepperoni as their favorite topping. What is the probability that a randomly selected American loves pizza and picks pepperoni as their favorite topping?

(f)   In 1948, Kellogg's introduced 18 prizes in their cereal boxes. Suppose 5% of boxes had a Popeye metal button, while 4% had a Superman metal button. What is the probability that for two randomly picked boxes, the first had a Popeye button and the second had a Superman button?

(continued)

✓ Solutions

(a) Complement rule: $P(A^C) = 1 - P(A)$ so $P(\text{not Cubs fan}) = 1 - P(\text{Cubs fan}) = 1 - 0.44 = 0.56$.

(b) General addition rule: $P(A \cup B) = P(A) + P(B) - P(A \cap B)$ so $P(\text{"Fifteen" or Swift}) = P(\text{"Fifteen"}) + P(\text{Swift}) - P(\text{"Fifteen" and Swift}) = 0.58 + 0.34 - 0.15 = 0.77$.

(c) Addition rule for mutually exclusive events: $P(A \cup B) = P(A) + P(B)$ so $P(\text{blue or gray}) = P(\text{blue}) + P(\text{gray}) = 0.10 + 0.01 = 0.11$.

(d) Conditional probability formula: $P(A|B) = \dfrac{P(A \cap B)}{P(B)}$ so $P(\text{Pandora|Spotify}) = \dfrac{P(\text{Pandora} \cap \text{Spotify})}{P(\text{Spotify})} = \dfrac{0.29}{0.43} = 0.674$.

(e) General multiplication rule: $P(A \cap B) = P(A) \cdot P(B|A)$ so $P(\text{love pizza} \cap \text{pepperoni}) = P(\text{love pizza}) \cdot P(\text{pepperoni | love pizza}) = (0.55)(0.68) = 0.374$.

(f) Multiplication rule for independent events: $P(A \cap B) = P(A) \cdot P(B)$ so $P(\text{Popeye} \cap \text{Superman}) = P(\text{Popeye}) \cdot P(\text{Superman}) = (0.05)(0.04) = 0.002$.

Multistage Probability Calculations

❭ Example 4.4

On a university campus, 60%, 30%, and 10% of the computers use Windows, Apple, and Linux operating systems, respectively. A new virus affects 3% of the Windows, 2% of the Apple, and 1% of the Linux operating systems.

(a) What is the probability a computer on this campus has the virus?

(b) If a randomly chosen computer on this campus has the virus, what is the probability it is a Windows machine? An Apple machine? A Linux machine?

✓ Solutions

(a) Tree diagrams can be very useful in working with conditional probabilities.

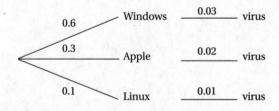

We then have:

$$P(\text{Windows} \cap \text{virus}) = (0.6)(0.03) = 0.018$$
$$P(\text{Apple} \cap \text{virus}) = (0.3)(0.02) = 0.006$$
$$P(\text{Linux} \cap \text{virus}) = (0.1)(0.01) = 0.001$$

$$P(\text{virus}) = P(\text{Windows} \cap \text{virus}) + P(\text{Apple} \cap \text{virus}) + P(\text{Linux} \cap \text{virus})$$
$$= 0.018 + 0.006 + 0.001 = 0.025$$

(b) $P(\text{Windows | virus}) = \dfrac{P(\text{Windows} \cap \text{virus})}{P(\text{virus})} = \dfrac{0.018}{0.025} = 0.72$

$P(\text{Apple | virus}) = \dfrac{P(\text{Apple} \cap \text{virus})}{P(\text{virus})} = \dfrac{0.006}{0.025} = 0.24$

$P(\text{Linux | virus}) = \dfrac{P(\text{Linux} \cap \text{virus})}{P(\text{virus})} = \dfrac{0.001}{0.025} = 0.04$

NOTE

The general multiplication rule is $P(A \cap B) = P(A)\,P(B\,|\,A)$.

Important

Naked or *bald* answers will receive little or *no* credit. You must show how you arrived at your answer.

> **Example 4.5**

A second way of analyzing Example 4.4 is to construct a table of counts for a hypothetical population. In this case, let's assume that the original given percentages hold exactly for a population of size 1,000.

	Virus	No Virus	Total
Windows			
Apple			
Linux			
Total			1,000

Numbers in parentheses indicate the order in which the cells can be calculated.

	Virus	No Virus	Total
Windows	(4) 18	(5) 582	(1) 600
Apple	(6) 6	(7) 294	(2) 300
Linux	(8) 1	(9) 99	(3) 100
Total	(10) 25	(11) 975	1,000

Then the conditional probabilities in which we are interested can be calculated straight from the first column of the table:

$$P(\text{Windows} \mid \text{virus}) = \frac{18}{25} = 0.72$$

$$P(\text{Apple} \mid \text{virus}) = \frac{6}{25} = 0.24$$

$$P(\text{Linux} \mid \text{virus}) = \frac{1}{25} = 0.04$$

Quiz 9

Multiple-Choice Questions

> **DIRECTIONS:** The questions that follow are each followed by five suggested answers. Choose the response that best answers the question.

Questions 1–4 refer to the following study: Five hundred people used a home test for strep throat, and then all underwent more conclusive testing in a medical office. The accuracy of the home test was evidenced in the following table.

	Strep Throat	Healthy	
Positive test	35	25	60
Negative test	5	435	440
	40	460	

1. What is the *predictive value* of the test? That is, what is the probability that a person has strep throat and tests positive?

    (A) $\frac{35}{500}$

    (B) $\frac{35}{60}$

    (C) $\frac{35}{40}$

    (D) $\frac{35}{35 + 25 + 5}$

    (E) $\frac{35 + 25 + 5}{500}$

2. What is the *false-positive* rate? That is, what is the probability of testing positive given that the person does not have strep throat?

    (A) $\frac{25}{460}$

    (B) $\frac{25}{60}$

    (C) $\frac{35}{40}$

    (D) $\frac{25}{500}$

    (E) $\frac{35 + 25 + 5}{500}$

3. What is the *sensitivity* of the test? That is, what is the probability of testing positive given that the person has strep throat?

    (A) $\frac{35 + 25 + 5}{500}$

    (B) $\frac{35}{35 + 25 + 5}$

    (C) $\frac{35}{500}$

    (D) $\frac{35}{60}$

    (E) $\frac{35}{40}$

4. What is the *specificity* of the test? That is, what is the probability of testing negative given that the person does not have strep throat?

    (A) $\frac{5}{40}$

    (B) $\frac{35}{60}$

    (C) $\frac{35 + 5 + 435}{500}$

    (D) $\frac{435}{500}$

    (E) $\frac{435}{460}$

5. In a 1974 "Dear Abby" letter, a woman lamented that she had just given birth to her eighth child and all were girls! Her doctor had assured her that the chance of the eighth child being a girl was less than 1 in 100. What was the real probability that the eighth child would be a girl?

   (A) 0.0039
   (B) 0.5
   (C) $(0.5)^7$
   (D) $(0.5)^8$
   (E) $\dfrac{(0.5)^7 + (0.5)^8}{2}$

6. Hospital A has about 20 births per day, and hospital B has about 40 births per day. About 50% of all babies are boys, but the percentage who are boys varies at each hospital from day to day. Over the course of a year, which hospital will have more days on which 60% or more of the births are boys?

   (A) Hospital A
   (B) Hospital B
   (C) The difference will be negligible.
   (D) By the law of small numbers, this cannot be predicted.
   (E) By the law of large numbers, this cannot be predicted.

7. The probability that a college's mathematics department will fill an opening with a female professor is 0.55, while the probability that the college's physics department will fill an opening with a female professor is 0.3. Assuming these decisions are independent, what is the probability that exactly one of these two departments will fill their position with a female professor?

   (A) $(0.55)(0.3)$
   (B) $0.55 + 0.3 - (0.55)(0.3)$
   (C) $0.55 - (0.55)(0.3)$
   (D) $(0.55)(0.7) + (0.3)(0.45)$
   (E) $(0.55)(0.7) + (0.3)(0.45) - (0.55)(0.3)$

8. Suppose that 2% of a clinic's patients are known to have cancer. A blood test is developed that is positive in 98% of patients with cancer but is also positive in 3% of patients who do not have cancer. If a person who is chosen at random from the clinic's patients is given the test and it comes out positive, what is the probability that the person actually has cancer?

   (A) 0.02
   (B) $0.02 + 0.03$
   (C) $(0.02)(0.98)$
   (D) $(0.02)(0.98) + (0.98)(0.03)$
   (E) $\dfrac{(0.02)(0.98)}{(0.02)(0.98) + (0.98)(0.03)}$

9. A teacher has 90 students. Of those students, 21 have at least two analog clocks in their homes, and 62 have at least three television sets in their homes. There are 15 students who do not have at least two analog clocks and do not have at least three television sets in their homes. Let event C represent the event of selecting a student with at least two analog clocks at home, and let event T represent the event of selecting a student with at least three television sets at home.

   Are C and T mutually exclusive events?

   (A) No, because $P(C \cap T) = \dfrac{8}{90}$
   (B) No, because $P(C \cap T) = \dfrac{75}{90}$
   (C) No, because $P(C \cap T) = \dfrac{83}{90}$
   (D) Yes, because $P(C \cap T) = \dfrac{8}{90}$
   (E) Yes, because $P(C \cap T) = \dfrac{83}{90}$

10. A random sample of adults aged 25 and up were surveyed as to whether or not they had tattoos and whether or not they had ever taken a college course. The mosaic plot below displays the data.

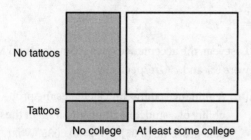

Which is greater, the probability that someone in this sample has taken a college course given that they have a tattoo or the probability that someone in this sample has a tattoo, given that they have taken a college course?

(A) The probability that someone in this sample has taken a college course given that the person has a tattoo is the greater of the two probabilities.

(B) The probability that someone in this sample has a tattoo given that the person has taken a college course is the greater of the two probabilities.

(C) The two probabilities are equal.

(D) This cannot be answered without knowing the probability that someone in this sample has taken a college course and has a tattoo.

(E) This cannot be answered without knowing (1) the probability that someone in this sample has taken a college course, (2) the probability that someone in this sample has a tattoo, and (3) the probability that someone in this sample has taken a college course and has a tattoo.

11. For which of the following probability assignments are events E and F independent?

(A) $P(E \cap F^c) = 0.2, P(E \cap F) = 0.2, P(E^c \cap F) = 0.2$

(B) $P(E \cap F^c) = 0.3, P(E \cap F) = 0.0, P(E^c \cap F) = 0.3$

(C) $P(E \cap F^c) = 0.3, P(E \cap F) = 0.2, P(E^c \cap F) = 0.1$

(D) $P(E \cap F^c) = 0.4, P(E \cap F) = 0.1, P(E^c \cap F) = 0.1$

(E) $P(E \cap F^c) = 0.2, P(E \cap F) = 0.4, P(E^c \cap F) = 0.2$

12. What is the proper order of size for the probabilities of the following three events?

(I) A randomly selected high school senior eats breakfast.

(II) A randomly selected teenager is a high school senior who eats breakfast.

(III) A randomly selected teenager who eats breakfast is a high school senior.

(A) $P(\text{I}) < P(\text{II}) < P(\text{III})$

(B) $P(\text{I}) > P(\text{II}) > P(\text{III})$

(C) $P(\text{I}) < P(\text{III}) < P(\text{II})$

(D) $P(\text{I}) > P(\text{III}) > P(\text{II})$

(E) $P(\text{II}) < P(\text{I}) < P(\text{III})$

Free-Response Questions

> **DIRECTIONS:** You must show all work and indicate the methods you use. You will be graded on the correctness of your methods and on the accuracy of your final answers.

1. A sample of applicants for a management position yields the following numbers with regard to age and experience:

	0–5	6–10	>10
Less than 50 years old	80	125	20
More than 50 years old	10	75	50

Years of experience

   (a) What is the probability that a randomly picked applicant from this sample is less than 50 years old? Has more than 10 years of experience? Is more than 50 years old and has five or fewer years of experience?

   (b) What is the probability that a randomly picked applicant from this sample is less than 50 years old, given that she has between 6 and 10 years of experience?

   (c) Are the two events "less than 50 years old" and "more than 10 years of experience" independent events? How about the two events "more than 50 years old and between 6 and 10 years of experience"? Explain.

2. Last year, the acceptance rates at Stanford and MIT were 5% and 7%, respectively.

   (a) A guidance counselor calculates the probability of a student getting into both of these colleges to be $(0.05)(0.07) = 0.0035$. What assumption is she making, and does her assumption seem reasonable? Explain.

   (b) She also calculates the probability of a student getting into at least one of these colleges to be $0.05 + 0.07 = 0.12$. What assumption is she making here, and does it seem reasonable?

3. In one community, 26% of adults identify as Generation X, 38% as Generation Y (Millennials), and 36% as Generation Z. Nine percent of Generation X, 12% of Generation Y, and 15% of Generation Z keep their smartphones at least 3 years. An adult from this community is picked at random.

   (a) What is the probability the person identifies as Generation X and keeps their smartphone at least 3 years?

   (b) What is the probability the person keeps their smartphone at least 3 years?

   (c) If the person keeps their smartphone at least 3 years, what is the probability the person identifies as Generation X?

The answers for this quiz can be found beginning on page 487.

Random Variables, Means (Expected Values), and Standard Deviations

Often each outcome of an experiment has not only an associated probability but also an associated *real number*. For example, the probability may be 0.5 that a student is taking 0 AP classes, 0.3 that she is taking 1 AP class, and 0.2 that she is taking 2 AP classes. If X represents the different numbers associated with the potential outcomes of some chance situation, we call X a random variable.

While the mean of the set $\{2, 7, 12, 15\}$ is $\dfrac{2+7+12+15}{4} = 9$, or $2(\frac{1}{4}) + 7(\frac{1}{4}) + 12(\frac{1}{4}) + 15(\frac{1}{4}) = 9$, the *expected value* or mean of a discrete random variable takes into account that the various outcomes may not be equally likely. The expected value or mean of a discrete random variable X is given by $\mu_X = E(X) = \sum xP(x)$, where $P(x)$ is the probability of outcome x. [We also write $\mu_X = \sum x_i p_i$, where p_i is the probability of outcome x_i.]

> *Discrete* random variables are discussed here, while *continuous* random variables will be looked at in Unit 5 on sampling distributions.

❯ Example 4.6

A charity holds a lottery in which 10,000 tickets are sold at $1 each and with a prize of $7,500 for one winner. What is the average result for each ticket holder?

> **NOTE**
>
> The distribution of a variable describes the possible values the variable could assume and how often each of those values occurs.

✓ Solution

The actual winning payoff is $7,499 because the winner paid $1 for a ticket.

Letting X be the payoff random variable, we have:

Outcome	Random Variable X	Probability $P(x)$
Win	7,499	$\dfrac{1}{10,000}$
Lose	-1	$\dfrac{9,999}{10,000}$

> **NOTE**
>
> The expected value is sometimes referred to as a *weighted average*.

$$\text{Expected value} = \sum xP(x) = 7{,}499\left(\frac{1}{10{,}000}\right) + (-1)\left(\frac{9{,}999}{10{,}000}\right) = -0.25$$

Thus, the *average* result for each ticket holder is a $0.25 loss. (Alternatively, we can say that the expected payoff to the charity is $0.25 for each ticket sold.)

Note that the expected value ($-$$0.25$) is not the most likely outcome ($-$$1$). In fact, in this example, the expected value ($-$$0.25$) is not expected at all as it will never occur. The expected value is simply a long-run average.

> **Important**
>
> The expected value of a random variable does not, *in any way*, imply that the value is expected to occur.

We have seen that the mean of a discrete random variable is $\sum x_i p_i$. In addition to calculating the mean, we would like to measure the variability for the values taken on by a discrete random variable. Since we are dealing with chance events, the proper tool is variance (along with standard deviation). In Unit 1, *variance* was defined to be the mean of the squared deviations $(x - \mu)^2$. If we regard the $(x - \mu)^2$ terms as the values of some discrete random variable (whose probability is the same as the probability of x), the mean of this new discrete random variable is $\sum (x_i - \mu_x)^2 p_i$, which is precisely how we define the variance, σ^2, of a discrete random variable:

$$\text{var}(X) = \sigma_x^2 = \sum (x_i - \mu_x)^2 p_i$$

> **NOTE**
>
> It is also true that var(x) = $[\sum x_i^2 p_i] - \mu_x^2$.

As before, the standard deviation σ is the square root of the variance.

> **Example 4.7**

A highway engineer knows that the workers can lay 5 miles of highway on a clear day, 2 miles on a rainy day, and only 1 mile on a snowy day. Suppose the probabilities are as follows:

Outcome	Random Variable X (Miles of highway)	Probability $P(x)$
Clear	5	0.6
Rain	2	0.3
Snow	1	0.1

(a) What are the mean (expected value), variance, and standard deviation of the random variable?

(b) Would it be surprising if the workers laid 10 miles of highway one day?

TIP

In a problem like this, there is no shortcut. You must show your work. However, the calculator shortcut of using two lists and running `1-Variable` statistics can be useful for checking your answers.

✓ **Solutions**

(a) $\mu_x = \sum x_i p_i = 5(0.6) + 2(0.3) + 1(0.1) = 3.7$

$\sigma_x^2 = \sum (x_i - \mu_x)^2 p_i = (5 - 3.7)^2(0.6) + (2 - 3.7)^2(0.3) + (1 - 3.7)^2(0.1) = 2.61$

$\sigma_x = \sqrt{2.61} = 1.62$

In the long run, the workers will lay an average of 3.7 miles of highway per day. The number of miles the workers lay on a randomly selected day will generally vary about 1.62 from the mean of 3.7 miles.

(b) 10 is $\frac{10 - 3.7}{1.62} = 3.9$ standard deviations above the mean, so this would be very surprising!

NOTE

Alternatively, you can calculate:
$\sigma_X^2 = 5^2(0.6)$
$+ 2^2(0.3)$
$+ 1^2(0.1)$
$- 3.7^2 = 2.61$

> **Example 4.8**

The World Series for Major League Baseball is decided by the first team to win four games. Letting the random variable X be the number of games needed to determine a winner in a series where the two teams are evenly matched, we have the following probabilities with accompanying histogram:

Number of Games x	Probability $P(x)$
4	0.125
5	0.25
6	0.3125
7	0.3125

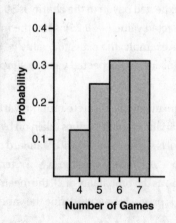

Describe the above distribution.

(continued)

✓ Solution

Shape: The distribution for the number of games needed to determine a World Series winner is skewed left.
Center: The median is 6, and the mean is $\sum xP(x) = 4(0.125) + 5(0.25) + 6(0.3125) + 7(0.3125) = 5.8125$.
Spread: The range is $7 - 4 = 3$, and the standard deviation is $\sqrt{\sum (x - \mu)^2 P(x)} = $
$\sqrt{(4 - 5.8125)^2 (0.125) + \cdots + (7 - 5.8125)^2 (0.3125)} = 1.0136$.

In the long run, the number of games needed to determine a winner in a World Series where the two teams are evenly matched will average 5.8125 games, with the length of a randomly selected individual series generally varying about 1.0136 games from the average of 5.8125 games.

Means and Variances for Sums and Differences of Random Variables

> **Important**
>
> Variances of independent random variables can be added together, but standard deviations cannot!

Many statistics problems involve combining two or more random variables. For any two random variables X and Y, the following hold true about means (expected values), variances, and standard deviations:

- $\mu_{X+Y} = \mu_X + \mu_Y$, also written as $E(X + Y) = E(X) + E(Y)$
- $\mu_{X-Y} = \mu_X - \mu_Y$, also written as $E(X - Y) = E(X) - E(Y)$
- If X and Y are independent, $(\sigma_{X+Y})^2 = (\sigma_{X-Y})^2 = (\sigma_X)^2 + (\sigma_Y)^2$
- If X and Y are independent, $\sigma_{X+Y} = \sigma_{X-Y} = \sqrt{(\sigma_X)^2 + (\sigma_Y)^2}$

❯ Example 4.9

An insurance salesperson estimates the numbers of new auto and home insurance policies she sells per day as follows:

# of auto policies	0	1	2	3
Probability	0.2	0.4	0.3	0.1

# of home policies	0	1	2
Probability	0.5	0.3	0.2

(a) What is the expected value or mean for the overall number of policies sold per day?

(b) Assuming the selling of new auto policies is independent of the selling of new home policies (which may not be true if some new customers are interested in both), what would be the standard deviation in the number of new policies sold per day?

✓ Solutions

(a) $\mu_{auto} = (0)(0.2) + (1)(0.4) + (2)(0.3) + (3)(0.1) = 1.3$,
$\quad\quad \mu_{home} = (0)(0.5) + (1)(0.3) + (2)(0.2) = 0.7$, and so
$\quad\quad \mu_{total} = \mu_{auto} + \mu_{home} = 1.3 + 0.7 = 2.0$

(b) $\sigma^2_{auto} = (0 - 1.3)^2 (0.2) + (1 - 1.3)^2 (0.4) + (2 - 1.3)^2 (0.3) + (3 - 1.3)^2 (0.1) = 0.81$,
$\quad\quad \sigma^2_{home} = (0 - 0.7)^2 (0.5) + (1 - 0.7)^2 (0.3) + (2 - 0.7)^2 (0.2) = 0.61$, and so, **assuming independence**,
$\quad\quad \sigma^2_{total} = \sigma^2_{auto} + \sigma^2_{home} = 0.81 + 0.61 = 1.42$ and $\sigma_{total} = \sqrt{1.42} = 1.192$

> Example 4.10

The attendance at tennis matches varies by who is playing. Suppose the attendance at matches held in the Arthur Ashe Stadium (AA) and the Louis Armstrong Stadium (LA) can be modeled by the following random variables:

Attendance (AA)	15,000	19,500	23,700
Probability	0.1	0.2	0.7

Attendance (LA)	7,500	11,200	14,000
Probability	0.1	0.3	0.6

(a) What is the expected value (mean) for the difference (AA – LA) in attendance on any given day?

(b) Assuming independence, what would be the standard deviation for the difference (AA – LA) in attendance on any given day?

✓ Solutions

(a) $\mu_{AA} = (15,000)(0.1) + (19,500)(0.2) + (23,700)(0.7) = 21,990$
$\mu_{LA} = (7,500)(0.1) + (11,200)(0.3) + (14,000)(0.6) = 12,510$
$\mu_{AA-LA} = \mu_{AA} - \mu_{LA} = 21,990 - 12,510 = 9,480$

(b) $\sigma^2_{AA} = (15,000 - 21,990)^2(0.1) + (19,500 - 21,990)^2 (0.2) + (23,700 - 21,990)^2(0.7) = 8,172,900$
$\sigma^2_{LA} = (7,500 - 12,510)^2(0.1) + (11,200 - 12,510)^2(0.3) + (14,000 - 12,510)^2(0.6) = 4,356,900$
Assuming independence, $\sigma^2_{AA-LA} = \sigma^2_{AA} + \sigma^2_{LA} = 12,529,800$ and $\sigma_{AA-LA} = \sqrt{12,529,800} = 3,539.75$

> Example 4.11

A company's sports shoes sell for an average of $100 with a standard deviation of $9. The different costs associated with determining a selling price for their sports shoes can be considered a set of costs that we will call S, B, A, and R. Let:

S = retail store costs and profits: average of $45 with a standard deviation of $5.
B = brand company costs and profits: average of $24 with a standard deviation of $4.
A = advertising and publicity costs: average of $11 with a standard deviation of $3.
R = remaining costs, such as materials, transportation, factory worker wages, etc.

Assuming these various costs are independent of one another, what are the mean and standard deviation for the remaining costs, R?

✓ Solution

$45 + 24 + 11 + \mu_R = 100$ gives $\mu_R = \$20$; and $5^2 + 4^2 + 3^2 + (\sigma_R)^2 = 9^2$ gives $(\sigma_R)^2 = 31$ and $\sigma_R = \sqrt{31} = \$5.57$

> Example 4.12

Suppose the mean SAT Math and Evidence-Based Reading + Writing scores for students at a particular high school are 515 and 501, respectively, with standard deviations of 65 and 40, respectively. What can be said about the mean and standard deviation of the two combined scores for these students?

(continued)

✓ Solution

The mean combined score is $515 + 501 = 1016$. Can we add the two variances, $65^2 + 40^2$, and then take the square root to find the standard deviation? No, because we don't have independence—students with high scores on one of the two sections will tend to have high scores on the other section. The standard deviation of the combined scores cannot be calculated from the given information.

Transforming Random Variables

Adding a constant to every value of a random variable X will increase the mean by that constant. However, differences between values remain the same, so measures of variability like the standard deviation will remain unchanged. Multiplying every value of a random variable X by a constant will change the mean by the same multiple. In this case, differences are also changed and the standard deviation will change by the absolute value of the same multiple.

NOTE

The results of adding and multiplying by constants is the same as what we observed in Unit 1 about sets of data.

$$E(X \pm a) = E(X) \pm a \qquad \text{var}(X \pm a) = \text{var}(X) \qquad SD(X \pm a) = SD(X)$$
$$E(bX) = bE(X) \qquad \text{var}(bX) = b^2\,\text{var}(X) \qquad SD(bX) = |b|\,SD(X)$$

▶ Example 4.13

A carnival game of chance has payoffs of \$2 with probability 0.5, of \$5 with probability 0.4, and of \$10 with probability 0.1.

(a) What are the mean and standard deviation for the winnings if you play this game? What should you be willing to pay to play the game?
(b) How do the mean and standard deviation shown in (a) change if \$4 is added to each payoff?
(c) How do the mean and standard deviation in (a) change if each payoff is tripled?

✓ Solutions

(a) $$\mu_X = E(X) = \sum xP(x) = 2(0.5) + 5(0.4) + 10(0.1) = 4$$
$$\sigma_X = \sqrt{\sum(x-\mu)^2 P(x)} = \sqrt{(2-4)^2(0.5) + (5-4)^2(0.4) + (10-4)^2(0.1)} = \sqrt{6}$$

You should be willing to pay any amount less than or equal to \$4.

(b) $$\mu_{X+4} = 6(0.5) + 9(0.4) + 14(0.1) = 8$$
$$\sigma_{X+4} = \sqrt{(6-8)^2(0.5) + (9-8)^2(0.4) + (14-8)^2(0.1)} = \sqrt{6}$$

Note that $\mu_{X+4} = \mu_X + 4$ and $\sigma_{X+4} = \sigma_X$.

(c) $$\mu_{3X} = 6(0.5) + 15(0.4) + 30(0.1) = 12$$
$$\sigma_{3X} = \sqrt{(6-12)^2(0.5) + (15-12)^2(0.4) + (30-12)^2(0.1)} = \sqrt{54} = 3\sqrt{6}$$

Note that $\mu_{3X} = 3\mu_X$ and $\sigma_{3X} = 3\sigma_X$.

The rules from algebra do not apply when combining random variables. For example, with random variables, $2X \neq X + X$. From the formulas, we have $E(2X) = 2E(X)$ and $SD(2X) = 2SD(X)$; however, while $E(X+X) = E(X) + E(X) = 2E(X)$:

$$SD(X+X) = \sqrt{[SD(X)]^2 + [SD(X)]^2} = \sqrt{2[SD(X)]^2} = \sqrt{2}\,SD(X)$$

> **Example 4.14**

Let X be what shows when rolling a single die. What are the mean and the standard deviation of X, $2X$, and $X + X$?

X takes the values $\{1, 2, 3, 4, 5, 6\}$, each with probability $\frac{1}{6}$, $\mu_X = \sum x P(x) = 3.5$, and $\sigma_X = \sqrt{\sum (x - \mu)^2 P(x)} = 1.7078$.

$2X$ takes the values $\{2, 4, 6, 8, 10, 12\}$, each with probability $\frac{1}{6}$, $\mu_{2X} = 7$, and $\sigma_{2X} = 3.4156$. (Note that $\mu_{2X} = 2\mu_X$ and $\sigma_{2X} = 2\sigma_X$.)

But $X + X$ is throwing a die twice and takes the values $\{2, 3, 4, 5, 6, 7, 8, 9, 10, 11, 12\}$ with probabilities $\left\{ \frac{1}{36}, \frac{2}{36}, \frac{3}{36}, \frac{4}{36}, \frac{5}{36}, \frac{6}{36}, \frac{5}{36}, \frac{4}{36}, \frac{3}{36}, \frac{2}{36}, \frac{1}{36} \right\}$, respectively.

Then $\mu_{X+X} = 7$ and $\sigma_{X+X} = 2.4152$. (Note that $\mu_{X+X} = 2\mu_X$ and $\sigma_{X+X} = \sqrt{2}\sigma_X$.)

$2X$ and $X + X$ are not the same thing, and although they do have the same mean, their standard deviations are different.

More generally, with n independent copies of X, we have:

$$E(X + X + \cdots + X) = E(X) + E(X) + \cdots + E(X) = nE(X)$$

$$\mathrm{var}(X + X + \cdots + X) = \mathrm{var}(X) + \mathrm{var}(X) + \cdots + \mathrm{var}(X) = n\,\mathrm{var}(X)$$

Note the following summaries.

For expected values (means):

- $E(nX) = nE(X)$
- $E(X + X + \cdots + X) = nE(X)$

For variances:

- $\mathrm{var}(nX) = n^2 \mathrm{var}(X)$
- $\mathrm{var}(X + X + \cdots + X) = n \cdot \mathrm{var}(X)$

For standard deviations:

- $\mathrm{SD}(nX) = n\mathrm{SD}(X)$
- $\mathrm{SD}(X + X + \cdots + X) = \sqrt{n}\,\mathrm{SD}(X)$

Quiz 10

Multiple-Choice Questions

> **DIRECTIONS:** The questions that follow are each followed by five suggested answers. Choose the response that best answers the question.

1. An insurance company charges $800 annually for car insurance. The policy specifies that the company will pay $1,000 for a minor accident and $5,000 for a major accident. If the probability of a motorist having a minor accident during the year is 0.2 and of having a major accident is 0.05, how much can the insurance company expect to make on a policy?

   (A) $200
   (B) $250
   (C) $300
   (D) $350
   (E) $450

2. Suppose you are one of 7.5 million people who send in their name for a drawing with 1 top prize of $1 million, 5 second-place prizes of $10,000, and 20 third-place prizes of $100. Is it worth the $0.66 postage it costs you to send in your name?

   (A) Yes, because $\dfrac{1,000,000}{0.66} = 1,515,152$, which is less than 7,500,000.

   (B) No, because your expected winnings are only $0.14.

   (C) Yes, because $\dfrac{7,500,000}{1 + 5 + 20} = 288,642$.

   (D) No, because $1,052,000 < 7,500,000$.

   (E) Yes, because $\dfrac{1,052,000}{26} = 40,462$.

3. The random variable X has mean 15 and standard deviation 4. The random variable Y is defined by $Y = 3 + 2X$. What are the mean and standard deviation of Y?

   (A) The mean is 30, and the standard deviation is 8.
   (B) The mean is 30, and the standard deviation is 11.
   (C) The mean is 33, and the standard deviation is 4.
   (D) The mean is 33, and the standard deviation is 8.
   (E) The mean is 33, and the standard deviation is 11.

4. Each hospital Internet security breach results in the theft of an average of 361 patient records with a standard deviation of 74 records. If there are 45 independent breaches during a one-year period, what is the expected value and standard deviation for the number of patient records stolen?

   (A) $E(\text{thefts}) = \sqrt{45} \times 361$ \quad $SD(\text{thefts}) = \sqrt{45} \times 74$
   (B) $E(\text{thefts}) = \sqrt{45} \times 361$ \quad $SD(\text{thefts}) = 45 \times 74$
   (C) $E(\text{thefts}) = 45 \times 361$ \quad $SD(\text{thefts}) = \sqrt{45} \times 74$
   (D) $E(\text{thefts}) = 45 \times 361$ \quad $SD(\text{thefts}) = 45 \times 74$
   (E) $E(\text{thefts}) = 45 \times 361$ \quad $SD(\text{thefts}) = 45 \times 74^2$

5. The auditor working for a veterinary clinic calculates that the mean annual cost of medical care for dogs is $98 with a standard deviation of $25 and that the mean annual cost of medical care for pets is $212 with an average deviation of $39 for owners who have one dog and one cat. Assuming expenses for dogs and cats are independent for those owning both a dog and a cat, what is the mean annual cost of medical care for cats, and what is the standard deviation?

   (A) $E(\text{Cats}) = \$114$ \quad $SD(\text{Cats}) = \$14$
   (B) $E(\text{Cats}) = \$114$ \quad $SD(\text{Cats}) = \$46$
   (C) $E(\text{Cats}) = \$114$ \quad $SD(\text{Cats}) = \$30$
   (D) $E(\text{Cats}) = \$155$ \quad $SD(\text{Cats}) = \$32$
   (E) $E(\text{Cats}) = \$188$ \quad $SD(\text{Cats}) = \$30$

6. Boxes of 50 donut holes weigh an average of 16.0 ounces with a standard deviation of 0.245 ounces. If the empty boxes alone weigh an average of 1.0 ounce with a standard deviation of 0.2 ounces, what are the mean and standard deviation of donut hole weights?

    (A) E(Donut hole) = 0.3 oz   SD(Donut hole) = 0.0063 oz
    (B) E(Donut hole) = 0.3 oz   SD(Donut hole) = 0.02 oz
    (C) E(Donut hole) = 0.3 oz   SD(Donut hole) = 0.142 oz
    (D) E(Donut hole) = 15 oz   SD(Donut hole) = 0.142 oz
    (E) E(Donut hole) = 15 oz   SD(Donut hole) = 0.445 oz

Free-Response Questions

> **DIRECTIONS:** You must show all work and indicate the methods you use. You will be graded on the correctness of your methods and on the accuracy of your final answers.

1. The two most popular women's spring semester sports at a particular school are softball and lacrosse. The softball team plays 0, 1, or 2 days a week with probabilities 0.2, 0.5, and 0.3, respectively.

    (a) What is the mean number of games played by the softball team per week?
    (b) What is the standard deviation for number of games played by the softball team per week?

    The lacrosse team plays a mean of 1.4 days per week with a standard deviation of 0.6.

    (c) What is the mean total number of softball and lacrosse games played per week?
    (d) Assuming independence, what is the standard deviation for the total number of softball and lacrosse games played per week?

2. Weights of coconuts have a mean of 3.2 pounds with a standard deviation of 0.7 pounds.

    (a) If each pound of coconut produces 5 ounces of fruit, calculate the mean and standard deviation for the amount of fruit per coconut.
    (b) If coconuts are packed five per box for shipment, calculate the mean and standard deviation for the total weight of coconuts per box.

The answers for this quiz can be found beginning on page 490.

Binomial Distribution

A probability distribution is a list or formula that gives the probability of each outcome.

In many applications, such as coin tossing, there are only *two* possible outcomes. For applications in which a two-outcome situation is repeated a certain number of times, where each repetition is independent of the other repetitions, and in which the probability of each of the two outcomes remains the same for each repetition, the resulting calculations involve what are known as *binomial probabilities*.

> **Example 4.15**

Suppose the probability that a lightbulb is defective is 0.1 (so probability of being good is 0.9).
(a)   What is the probability that four lightbulbs are all defective?
(b)   What is the probability that exactly two out of three lightbulbs are defective?
(c)   What is the probability that exactly three out of eight lightbulbs are defective?

✓ **Solutions**

(a)   Because of independence (i.e., whether one lightbulb is defective is not influenced by whether any other lightbulb is defective), we can multiply individual probabilities of being defective to find the probability that all the bulbs are defective:

$$(0.1)(0.1)(0.1)(0.1) = (0.1)^4 = 0.0001$$

(b)   The probability that the first two bulbs are defective and the third is good is $(0.1)(0.1)(0.9) = 0.009$. The probability that the first bulb is good and the other two are defective is $(0.9)(0.1)(0.1) = 0.009$. Finally, the probability that the second bulb is good and the other two are defective is $(0.1)(0.9)(0.1) = 0.009$. Summing, we find that the probability that exactly two out of three bulbs are defective is $0.009 + 0.009 + 0.009 = 3(0.009) = 0.027$.

(c)   The probability of any particular arrangement of three defective and five good bulbs is $(0.1)^3(0.9)^5 = 0.00059049$. We need to know the number of such arrangements.

The answer is given by combinations: $\binom{8}{3} = \frac{8!}{3!5!} = 56$. Thus, the probability that exactly three out of eight light bulbs are defective is $56 \times 0.00059049 = 0.03306744$. [On the TI-84, `binompdf(8,0.1,3)` = 0.03306744.]

More generally, if an experiment has two possible outcomes, called *success* and *failure*, with the probability of success equal to p and the probability of failure equal to q (of course, $p + q = 1$) and if the outcome at any particular time has no influence over the outcome at any other time, then when the experiment is repeated n times, the probability of exactly k successes (and thus $n - k$ failures) is

$$\binom{n}{k}p^k q^{n-k} = \frac{n!}{k!(n-k)!}p^k q^{n-k}$$

Software can be used to calculate binomial probabilities:

binompdf (n, p, x) gives the probability of exactly x successes in n trials, where p is the probability of success on a single trial.

binomcdf (n, p, x) gives the cumulative probability of x or fewer successes in n trials, where p is the probability of success on a single trial.

NOTE

A question on conditions for a binomial setting could appear either as a multiple-choice question or as part of a free-response question.

NOTE

The *pdf* (such as in binompdf) stands for "probability density function." That is, it's a function that tells you the probability of certain events occurring.

> **Example 4.16**

In cereal boxes that advertised an enclosed miniature toy, one-third of the toys were purple dragons. If six boxes of cereal are purchased, what is the probability of exactly two purple dragons?

✓ **Solution**

If the probability of a purple dragon is $\frac{1}{3}$, the probability of no purple dragon is $1 - \frac{1}{3} = \frac{2}{3}$. If two out of six boxes have a purple dragon, $6 - 2 = 4$ do not.

Thus, the probability of getting two purple dragons when six boxes of cereal are purchased is `binompdf(6,1/3,2) = 0.329`. [Or we could calculate $\binom{6}{2}\left(\frac{1}{3}\right)^2\left(\frac{2}{3}\right)^4 = 15\left(\frac{1}{3}\right)^2\left(\frac{2}{3}\right)^4 = \frac{80}{243} = 0.329$.]

For full credit on the exam, write either of the following two lines:

$$\text{Binomial, } n = 6, p = \frac{1}{3}, P(X = 2) = 0.329$$

$$P(X = 2) = \text{binompdf}(n = 6, p = 1/3, X = 2) = 0.329$$

In some situations it is easier to calculate the probability of the complementary event and subtract this value from 1.

> **Example 4.17**

The baseball player Joe DiMaggio had a career batting average of 0.325. What was the probability that he would get at least one hit in five official times at bat?

✓ **Solution**

This is a binomial distribution with $n = 5$ and $p = 0.325$.

$$P(X \geq 1) = 1 - P(X = 0) = 1 - \binom{5}{0}(0.325)^0(0.675)^5 = 1 - 0.140 = 0.860$$

[Or we could calculate: $1 - \text{binompdf}(n = 5, p = 0.325, X = 0) = 0.860$.]

To receive full credit for probability calculations using the probability distributions from the AP Statistics course, you need to show:

1. Name of the distribution ("binomial" in the previous example)
2. Parameters ("$n = 5$ and $p = 0.325$" in the previous example)
3. Boundary ("1" in the previous example)
4. Direction ("$\geq$" in the previous example)
5. Correct probability ("0.860" in the previous example)

Many applications of probability involve such phrases as *at least*, *at most*, *less than*, and *more than*. In these cases, solutions involve summing two or more cases. For such calculations, `binomcdf` is very useful, as it gives the probability of x or fewer successes.

> **Example 4.18**

One day a grocery store manager notes seven customers buying a particular product and is interested in the number who will make use of a store coupon to receive a discount. Suppose 35% of customers in general make use of a store coupon to receive a discount when buying that product.

(a)   Define the random variable of interest for the manager and state how the random variable is distributed.
(b)   Determine the probability that fewer than four of the customers will use a store coupon.

✓ **Solutions**

(a)   The random variable of interest is the number of customers using a store coupon to receive a discount when buying this product among seven customers. The distribution is binomial with parameters $n = 7$ and $p = 0.35$.

(b)   In this situation, "fewer than four" means zero, one, two, or three. We have a binomial with $n = 7$ and $p = 0.35$. Then $P(X \leq 3) = \texttt{binomcdf(7,0.35,3)} = 0.800$. [Or we could calculate:

$$\binom{7}{0}(0.35)^0(0.65)^7 + \binom{7}{1}(0.35)^1(0.65)^6 + \binom{7}{2}(0.35)^2(0.65)^5 + \binom{7}{3}(0.35)^3(0.65)^4$$

$$= (0.65)^7 + 7(0.35)(0.65)^6 + 21(0.35)^2(0.65)^5 + 35(0.35)^3(0.65)^4$$

$$= 0.800]$$

TIP

While using binomcdf is the quick and recommended solution if part of a free-response question, the answer choices in a multiple-choice question here could be the expanded four-term binomial formula form.

> **Example 4.19**

A manufacturer has the following quality control check at the end of a production line. If at least 8 of 10 randomly picked articles meet all specifications, the whole shipment is approved. Suppose 85% of an entire shipment meets all specifications.

(a)   Define the random variable of interest for the manufacturer and state how the random variable is distributed.
(b)   Determine the probability that a particular shipment will pass the quality control check.

✓ **Solutions**

(a)   The random variable of interest is the number of articles meeting all specifications among 10 randomly selected articles. The distribution is binomial with parameters $n = 10$ and $p = 0.85$.

(b)   We have a binomial distribution with $n = 10$ and $p = 0.85$.

$$P(X \geq 8) = 1 - P(X \leq 7) = 1 - \text{binomcdf}(10, 0.85, 7) = 0.820$$

[Or we could calculate: $P(X = 8) + P(X = 9) + P(X = 10) =$

$$\binom{10}{8}(0.85)^8(0.15)^2 + \binom{10}{9}(0.85)^9(0.15)^1 + \binom{10}{10}(0.85)^{10}(0.15)^0$$

$$= \text{binompdf}(10, 0.85, 8) + \text{binompdf}(10, 0.85, 9) + \text{binompdf}(10, 0.85, 10) = 0.820.]$$

> **Example 4.20**

Of the automobiles produced at a particular plant, 40% had a certain defect. Suppose a company purchases five of these cars and is concerned about the number with that defect.

(a)   Define the random variable of interest for the company and state how the random variable is distributed.

(b)   What is the expected value for the number of cars with defects?

(c)   What are the variance and standard deviation?

✓ **Solutions**

(a)   The random variable of interest is the number of cars with a defect among five cars. The distribution is binomial with parameters $n = 5$ and $p = 0.40$.

(b)   We might guess that the mean, or expected value, is 40% of $5 = 0.4 \times 5 = 2$, but let's calculate from the definition. Letting X represent the number of cars with the defect, we have:

$$P(0) = \binom{5}{0}(0.4)^0(0.6)^5 = (0.6)^5 = 0.07776$$

$$P(1) = \binom{5}{1}(0.4)^1(0.6)^4 = 5(0.4)(0.6)^4 = 0.25920$$

$$P(2) = \binom{5}{2}(0.4)^2(0.6)^3 = 10(0.4)^2(0.6)^3 = 0.34560$$

$$P(3) = \binom{5}{3}(0.4)^3(0.6)^2 = 10(0.4)^3(0.6)^2 = 0.23040$$

$$P(4) = \binom{5}{4}(0.4)^4(0.6)^1 = 5(0.4)^4(0.6) = 0.07680$$

$$P(5) = \binom{5}{5}(0.4)^5(0.6)^0 = (0.4)^5 = 0.01024$$

TIP

binompdf(5, 0.4) quickly gives all six probabilities.

Outcome (no. of cars):	0	1	2	3	4	5
Probability:	0.07776	0.25920	0.34560	0.23040	0.07680	0.01024
Random variable:	0	1	2	3	4	5

$$E(X) = 0(0.07776) + 1(0.25920) + 2(0.34560) + 3(0.23040)$$
$$+ 4(0.07680) + 5(0.01024)$$
$$= 2$$

Thus, the answer turns out to be the same as would be obtained by simply multiplying the probability of a "success" by the number of cases.

The following is true: **For a binomial probability distribution with the probability of a success equal to p and the number of trials equal to n, the expected value or mean number of successes for the n trials is np.**

(c) $\sigma^2 = (0 - 2)^2(0.07776) + (1 - 2)^2(0.2592) + (2 - 2)^2(0.3456) + (3 - 2)^2(0.2304)$
$+ (4 - 2)^2(0.0768) + (5 - 2)^2(0.01024) = 1.2$ and $\sigma = \sqrt{1.2} = 1.095$

[Note that with a binomial random variable, **the variance can simply be calculated as $npq = np(1 - p)$ and the standard deviation as $\sqrt{np(1 - p)}$:**

$$\sigma^2 = np(1 - p) = 5(0.4)(0.6) = 1.2 \text{ and } \sigma = \sqrt{np(1 - p)} = \sqrt{1.2} = 1.095.]$$

In random samples of five cars, the mean number of cars with the defect is 2, and in any random sample of five cars, the number of cars with defects will typically vary by about 1.095 from the mean of 2 cars.

(continued)

Thus, for a random variable X:

Mean or expected value:

$$\mu_X = \sum x_i\, p_i$$

Variance:

$$\sigma_X^2 = \sum (x_i - \mu_X)^2 p_i$$

Standard deviation:

$$\sigma_X = \sqrt{\sum (x_i - \mu_X)^2 p_i}$$

In the case of a binomial probability distribution with probability of a success equal to p and number of trials equal to n, if we let X be the number of successes in the n trials, the above equations simplify to:

Mean or expected value:

$$\mu_X = np$$

Variance:

$$\sigma_X^2 = npq = np(1 - p)$$

Standard deviation:

$$\sigma_X = \sqrt{npq} = \sqrt{np(1 - p)}$$

The shape of a binomial distribution is skewed right when $p < 0.5$, symmetric when $p = 0.5$, and skewed left when $p > 0.5$.

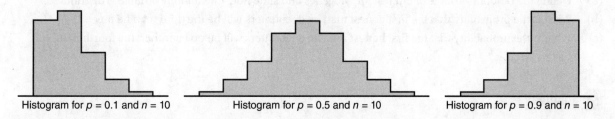

Histogram for $p = 0.1$ and $n = 10$ Histogram for $p = 0.5$ and $n = 10$ Histogram for $p = 0.9$ and $n = 10$

However, for all p, as n becomes larger without bound, the resulting histograms approach a smooth bell-shaped curve.

> Example 4.21

(a)   Ninety-five percent of all buyers of new cars choose automatic transmissions. For a group of 50 buyers of new cars, calculate and interpret the mean and standard deviation for the number of buyers choosing automatic transmissions.

(b)   Would it be surprising if only 40 buyers chose automatic transmissions in a random sample of 50 new car buyers?

✓ Solutions

(a) $\mu_X = np = 50(0.95) = 47.5$
$\sigma_X = \sqrt{np(1-p)} = \sqrt{50(0.95)(0.05)} = 1.54$

In many random samples of size 50 buyers of new cars, we can expect an average of 47.5 buyers who will choose automatic transmissions, and for any particular random sample of 50 buyers, the number of buyers choosing automatic transmissions will typically vary by about 1.54 from the mean of 47.5 buyers.

(b)   Yes, that would be surprising because 40 is $\dfrac{7.5}{1.54} = 4.87$ standard deviations below the mean of 47.5.

Geometric Distribution

Suppose an experiment has two possible outcomes, called *success* and *failure*, with the probability of success equal to p and the probability of failure equal to $q = 1 - p$ and the trials are independent. Then the probability that the first success is on trial number $X = k$ is $q^{k-1}p$ and the distribution of the random variable X is called a *geometric distribution*.

Software can be used to calculate geometric probabilities:

geometpdf(p, x) gives the probability that the first success occurs on the xth trial, where p is the probability of success on a single trial.

geometcdf(p, x) gives the cumulative probability that the first success occurs on or before the xth trial, where p is the probability of success on a single trial.

> **Example 4.22**

Suppose only 12% of men in ancient Greece were honest, and Diogenes is searching until he can find an honest man.
(a)   Define the random variable of interest for Diogenes and state how the random variable is distributed.
(b)   What is the probability that the first honest man he encounters will be the third man he meets?
(c)   What is the probability that the first honest man he encounters will be no later than the fourth man he meets?

✓ Solutions

(a)   The random variable of interest, X, is the number of men needed to be met in order to encounter an honest man. The distribution is geometric with parameter $p = 0.12$.
(b)   We have a geometric distribution with $p = 0.12$. $P(X = 3) = (0.88)^2(0.12) = 0.092928$. [Or we could calculate: geometpdf$(0.12, 3) = 0.092928$.]
(c)   This is a geometric distribution with $p = 0.12$. Then $P(X \leq 4) = $ geometcdf$(0.12, 4) = 0.40030464$. [Or we could calculate $(0.12) + (0.88)(0.12) + (0.88)^2(0.12) + (0.88)^3(0.12) = 0.40030464$.]

To receive full credit for geometric distribution probability calculations, you need to state:

1. Name of the distribution ("geometric" in the previous example)
2. Parameter ("$p = 0.12$" in the previous example)
3. The trial on which the first success occurs ("$X = 3$" in the second question of the previous example)
4. Correct probability ("0.092928" in the second question of the previous example)

In a geometric distribution, the probability of each successive value decreases by a factor of $q = 1 - p$. Thus, the most likely value of a geometric random variable is 1.

For a geometric probability distribution with probability of success equal to p, if we let X be the number of trials needed to get one success, the mean (or expected value) of X is $\frac{1}{p}$, the variance is $\frac{1-p}{p^2}$, and the standard deviation is $\sqrt{\frac{1-p}{p^2}} = \frac{\sqrt{1-p}}{p}$.

The shape of a geometric distribution is always unimodal and skewed right.

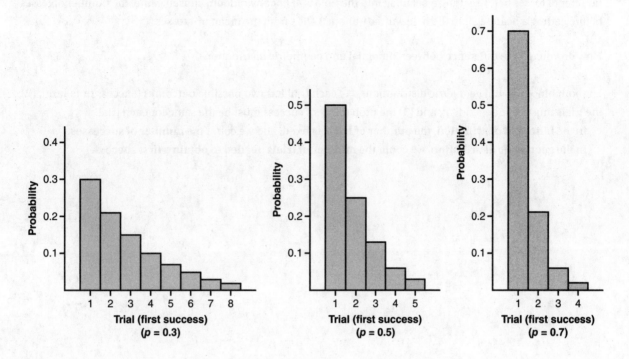

Example 4.23

Suppose in a game of cornhole, you have a 25 percent chance of tossing a beanbag into the hole, and you are interested in the number of tosses before getting a beanbag into the hole.

(a)   Define the random variable of interest and state how the random variable is distributed.
(b)   What is the probability that the first bag you get into the hole is on your fourth toss?
(c)   What is the probability that it takes more than two tosses before you get the bag into the hole?
(d)   What is the mean number of tosses you should expect before getting a bag into the hole?
(e)   What is the standard deviation for the number of tosses before getting a bag into the hole?

✓ Solutions

(a)   The random variable of interest, X, is the number of tosses in order to get a beanbag into the hole. The distribution is geometric with parameter $p = 0.25$.

(b)   This is a geometric distribution with $p = 0.25$. Then $P(X = 4) = (0.75)^3(0.25) = 0.1055$.
      [Or geometpdf(0.25, 4) = 0.1055.]

(c)   This is a geometric distribution with $p = 0.25$. Then $P(X > 2) = P(\text{miss first two tosses}) = (0.75)^2 = 0.5625$.
      [Or $1 - [0.25 + (0.25)(0.75)] = 1 - 0.4375 = 0.5625$. Or $1 - $ geometcdf(0.25, 2) = 0.5625.]

(d)   This is a geometric distribution with $p = 0.25$. $\mu = \dfrac{1}{p} = \dfrac{1}{0.25} = 4$

(e)   This is a geometric distribution with $p = 0.25$. $\sigma = \sqrt{\dfrac{1 - p}{p^2}} = \sqrt{\dfrac{1 - 0.25}{(0.25)^2}} = 3.4641$

In many random games of cornhole where there is a 25% chance of tossing a beanbag into the hole, the mean number of tosses before getting a beanbag into the hole is 4. For any random chosen game, the number of tosses before getting a beanbag into the hole will vary about 3.4641 from the mean of 4 tosses.

How do you tell the difference between binomial and geometric distributions?

For both binomial and geometric distributions, (1) each trial has two possible outcomes (success or failure), (2) the trials must be independent, and (3) the probability of success must be the same for each trial.

In the **binomial distribution**, the **number of trials is fixed**, and we count the **number of successes.**

In the **geometric distribution**, we count the **number of trials needed to obtain a first success**.

Quiz 11

Multiple-Choice Questions

DIRECTIONS: The questions that follow are each followed by five suggested answers. Choose the response that best answers the question.

1. An inspection procedure at a manufacturing plant involves picking three items at random and then accepting the whole lot if at least two of the three items are in perfect condition. If in reality 90% of the whole lot are perfect, what is the probability that the lot will be accepted?

   (A) $3(0.1)^2(0.9)$
   (B) $3(0.9)^2(0.1)$
   (C) $(0.1)^3 + (0.1)^2(0.9)$
   (D) $(0.1)^3 + 3(0.1)^2(0.9)$
   (E) $(0.9)^3 + 3(0.9)^2(0.1)$

2. Suppose we have a random variable X where, for the values $k = 0, \ldots, 10$, the associated probabilities are $\binom{10}{k}(0.37)^k(0.63)^{10-k}$. What is the mean of X?

   (A) 0.37
   (B) 0.63
   (C) 3.7
   (D) 6.3
   (E) None of the above

3. One in six adults in the workplace has experienced cyberbullying. In a random sample of five adults in the workplace, what is the distribution for the number who have experienced cyberbullying?

   (A) Binomial with $n = 6$ and $p = \frac{1}{6}$
   (B) Binomial with $n = 5$ and $p = \frac{1}{6}$
   (C) Binomial with $n = 5$ and $p = \frac{1}{5}$
   (D) Geometric with $p = \frac{1}{6}$
   (E) Geometric with $p = \frac{1}{5}$

Questions 4–5 refer to the following.

It is estimated that two out of five high school students would fall victim to a phishing email (an online scam asking for sensitive information) if it appears to originate from their high school main office.

4. In a random sample of five high school students, what is the probability that exactly two fall victim to a phishing email that appears to originate from their high school main office?

   (A) 0.4
   (B) 1.0
   (C) $\binom{5}{2}(0.4)^2(0.6)^3$
   (D) $\binom{5}{2}(0.4)^3(0.6)^2$
   (E) $(0.4)^2(0.6)^3$

5. What is the probability that the first student to fall victim will be the third student who is sent a phishing email that appears to originate from their high school main office?

   (A) $(0.4)^3$
   (B) $(0.6)^3$
   (C) $(0.6)(0.4)^2$
   (D) $(0.6)^2(0.4)$
   (E) $10(0.6)^2(0.4)$

6. Suppose that one out of 20 apples from a particular orchard is wormy. What are the mean and standard deviation for the number of apples to be sampled from this orchard before finding a wormy apple?

   (A) Mean = 5, standard deviation = 1
   (B) Mean = 5, standard deviation = 1 − 0.05
   (C) Mean = 10, standard deviation = $\sqrt{\dfrac{(0.05)^2}{1-0.05}}$
   (D) Mean = 20, standard deviation = $\sqrt{380}$
   (E) Mean = 20, standard deviation = $\sqrt{\dfrac{1-0.95}{(0.95)^2}}$

7. A fair coin is flipped 5 times. Consider the following 3 outcomes where H refers to a toss resulting in "heads" and T refers to a toss resulting in "tails":

   (I)   The result in order is HHHTT.
   (II) The result in order is HTTHH.
   (III) The result includes 3 Hs and 2 Ts.

   What is a comparison of the three probabilities?

   (A) $P(I) < P(II) < P(III)$
   (B) $P(II) < P(I) < P(III)$
   (C) $P(III) < P(I) < P(II)$
   (D) $P(I) = P(II) < P(III)$
   (E) $P(III) < P(I) = P(II)$

8. A statistician working for an online dating service determines that the binomial random variable D, representing the number of people in a sample who are happy with the service, has a mean of 5 with a standard deviation of 2. What is the probability that someone is happy with the service?

   (A) 0.1
   (B) 0.2
   (C) 0.4
   (D) 0.6
   (E) 0.8

9. A college lecture class has 500 students, of which 350 are female. Let the random variable F represent the number of females in a random sample of 10 students from the class. Random variable F follows a binomial distribution. A simulation of distribution F is performed by randomly picking 10 students from this class a large number of times. Which of the following best represents this simulation?

(A)
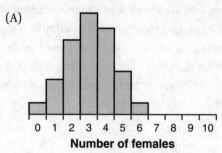

(B)
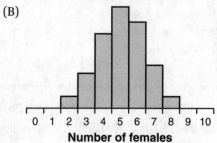

(C)
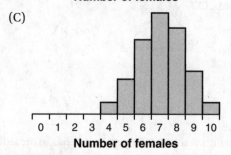

(D)
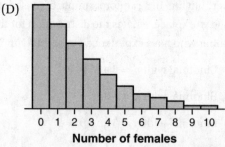

(E)
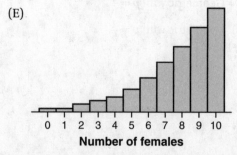

10. If it was possible, would you want to live forever? When asked, 34% of people answer yes. Let the random variable L represent the number of people who would answer yes in a random sample of 50 people. Random variable L follows a binomial distribution with a mean of 17 and a standard deviation of 3.35. Which of the following is the best interpretation of the mean?

(A) Every random sample of 50 people will have 34% answering yes.

(B) Every random sample of 50 people will have 34% answering no.

(C) On average, 17 people will have to be surveyed before finding one who answers yes.

(D) In repeated random sampling of 50 people, the average number of people answering yes is 17.

(E) In repeated random sampling of 50 people, the average number of people answering yes will vary from 17 by an average of 3.35.

11. Seventy percent of American households own an outdoor grill. Let the random variable G represent the number of randomly selected households surveyed to find one with an outdoor grill. Which of the following is the best interpretation of the mean of G?

(A) A value randomly selected from the distribution of G is expected to be 1.43.

(B) In repeated sampling from the distribution of G, the average of the values will approach 0.7.

(C) In repeated sampling from the distribution of G, the average of the values will approach 1.43.

(D) The probability is 0.7 that a value randomly selected from the distribution of G will be close to 1.43.

(E) For a sample of values randomly selected from the distribution of G, the average of the sample will vary from 1.43 by an average of 0.782.

Free-Response Questions

> **DIRECTIONS:** You must show all work and indicate the methods you use. You will be graded on the correctness of your methods and on the accuracy of your final answers.

1. Could hands-free, automatic faucets actually be housing more bacteria than the manual kind? The concern is that decreased water flow may increase the chance that bacteria grow because the automatic faucets are not being thoroughly flushed through. It is known that 15% of water cultures from manual faucets in hospital patient care areas test positive for *Legionella* bacteria. A recent study at Johns Hopkins Hospital found *Legionella* bacteria growing in 10 of cultured water samples from 20 automatic faucets.

    (a) If the probability of *Legionella* bacteria growing in a faucet is 0.15, what is the probability that in a sample of 20 faucets, 10 or more have the bacteria growing?

    (b) Does the Johns Hopkins study provide sufficient evidence that the probability of *Legionella* bacteria growing in automatic faucets is greater than 15%? Explain.

2. Consider the following probability distribution for the number of tattoos college students have:

Tattoos	0	1	2	3	4
Probability	0.85	0.11	0.02	0.014	0.006

    (a) What is the probability a student has two tattoos given that he has at least one tattoo?

    (b) In a random sample of three students, what is the probability that one of the students has at least two tattoos while the other two students have no tattoos?

    (c) What are the mean and standard deviation for number of tattoos? Interpret in context.

3. Suppose 35% of teenagers say that *The Hunger Games* is the best book series they ever read.

    (a) What is the probability that exactly two in a random sample of four teenagers say that *The Hunger Games* is the best series they ever read?

    (b) What is the probability that at least two in a random sample of four teens say that *The Hunger Games* is the best series they ever read?

    (c) What is the probability that the first teen who says that *The Hunger Games* is the best series he or she ever read is the second teen interviewed?

    (d) What are the mean and standard deviation for the number of teens interviewed before finding one who says that *The Hunger Games* is the best series he or she ever read? Interpret in context.

    (e) Would it be unusual if in a random sample of teenagers, the first to say that *The Hunger Games* is the best book series they have ever read is not until the fifth teen questioned?

The answers for this quiz can be found beginning on page 491.

Cumulative Probability Distribution

A probability distribution is a function, table, or graph that links outcomes of a statistical experiment with its probability of occurrence. A *cumulative* probability distribution is a function, table, or graph that links outcomes with the probability of less than or equal to that outcome occurring.

> **Example 4.24**

The scores received on a recent AP Statistics exam are illustrated in the following table:

Score	Probability	Cumulative Probability
1	0.231	0.231
2	0.165	0.396
3	0.234	0.630
4	0.222	0.852
5	0.148	1.000

What is the probability that a student does not receive college credit (assuming you need a 3 or higher to receive college credit)?

✓ **Solution**

The probability is $0.231 + 0.165 = 0.396$ that a student receives a 1 or 2 and thus does not receive college credit.

> **Example 4.25**

The number of minutes that teenagers spend on social media per day is illustrated in the following graphical displays: a probability distribution and the corresponding cumulative probability distribution.

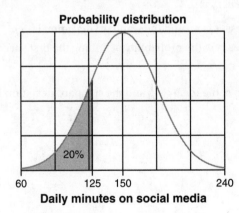

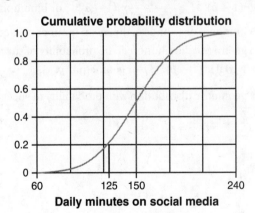

Note how, for example, 20% of the area under the probability distribution graph is to the left of 125 minutes and the point $(125, 0.2)$ is on the cumulative probability distribution graph. Both indicate that the probability is 0.2 that a teenager spends 125 minutes or less per day on social media.

SUMMARY

- The law of large numbers states that in the long run, a cumulative relative frequency tends closer and closer to what is called the probability of an event.
- The probability of the complement of an event is equal to 1 minus the probability of the event.
- Two events are mutually exclusive if they cannot occur simultaneously, and so $E \cap F = \varnothing$ and $P(E \cap F) = 0$.
- If two events are mutually exclusive, the probability that at least one will occur is the sum of their respective probabilities.
- More generally, $P(A \cup B) = P(A) + P(B) - P(A \cap B)$.
- The conditional probability of A given B is given by: $P(A|B) = \dfrac{P(A \cap B)}{P(B)}$; that is, the population of interest is shrunk.
- Two events A and B are independent if $P(A|B) = P(A)$; that is, the probability of one event occurring does not influence whether or not the other occurs.
- If two events A and B are independent, $P(A \cap B) = P(A) \times P(B)$.
- A random variable takes various numeric values, each with a given probability.
- The mean or expected value of a discrete random variable is calculated by $\mu = \sum x P(x)$, while the variance is $\sigma^2 = \sum (x - \mu)^2 P(x)$, and the standard deviation is $\sqrt{\sum (x - \mu)^2 P(x)}$.
- When transforming a random variable, $Y = a + bX$, the mean of Y is $\mu_Y = a + b\mu_X$ and the standard deviation of Y is $\sigma_Y = |b| \sigma_X$.
- When combining two random variables, $E(X + Y) = E(X) + E(Y)$ and $E(X - Y) = E(X) - E(Y)$.
- When combining two *independent* random variables, the variances always add:
  $\mathrm{var}(X \pm Y) = \mathrm{var}(X) + \mathrm{var}(Y)$.
- It can further be stated that for independent random variables, X and Y, the mean of $aX + bY$ is $a\mu_X + b\mu_Y$ and the variance is $a^2 \sigma_X^2 + b^2 \sigma_Y^2$.
- The binomial and geometric distributions arise when there are only two possible outcomes, the probability of success is constant, and the trials are independent.
- For a binomial distribution, the probability of exactly x successes in n trials is
  $\binom{n}{x} p^x (1-p)^{n-x} = \dfrac{n!}{x!(n-x)!} p^x (1-p)^{n-x}$ or binompdf(n, p, x).
- For a binomial distribution, the mean number of successes is np and the standard deviation is $\sqrt{np(1-p)}$.
- For a geometric distribution, if the probability of success is p, the probability of finding the first success on the xth trial is $(1-p)^{x-1} p$ or geometpdf(p, x).
- For a geometric distribution with probability of success p, the mean is $\dfrac{1}{p}$ and the standard deviation is $\sqrt{\dfrac{1-p}{p^2}} = \dfrac{\sqrt{1-p}}{p}$.

5

Sampling Distributions

(3–5 multiple-choice questions on the AP exam)

Learning Objectives

In this unit, you will learn how

→ To be able to find probabilities and percentiles using the normal approximation.

→ To be able to find the value that corresponds to a given percentile when the distribution is approximately normal.

→ To understand the concept of a sampling distribution.

→ To be able to describe the center, spread, and shape of the sampling distribution of a sample proportion.

→ To be able to verify that the conditions are met for normal approximation of the sampling distribution of a sample proportion.

→ To be able to describe and check the conditions for the sampling distribution for the difference of sample proportions.

→ To be able to describe the center, spread, and shape of the sampling distribution of a sample mean.

→ To understand the content and importance of the central limit theorem.

→ To be able to describe and check the conditions for the sampling distribution for the difference of sample means.

→ To understand how simulations can be used to estimate sampling distributions.

In this unit, you will learn how sample statistics, such as sample means and sample proportions, vary across samples. The population mean μ, the population standard deviation σ, the population proportion p, and the population slope β are all examples of *population parameters*. A sample mean $\overline{x}$, a sample standard deviation s, a sample proportion $\hat{p}$, and a sample slope b are all examples of *sample statistics*. While a population parameter is a fixed quantity, statistics vary depending on the particular sample chosen. The probability distribution showing how a statistic varies is called a *sampling distribution*. Although different samples from the same population produce different statistics, the distribution of statistics from many different random samples of the same size is often predictable. For large enough samples, the normal distribution can be used to approximate many of these sampling distributions. These concepts will be critical for understanding statistical inference, the key topic that runs through the rest of the curriculum.

> **Important**
>
> A *sampling distribution* is not the same thing as the distribution of a sample.

Normal Distribution Calculations

The normal distribution is valuable in describing various natural phenomena. However, the real importance of the normal distribution in statistics is that it can be used to describe sampling distributions, the topic of this unit. The normal distribution provides a valuable model for how many sample *statistics* vary, under repeated random sampling from a population. Calculations involving normal distributions are often made through *z*-scores, which measure standard deviations from the mean.

Both normalcdf and invNorm can work with *z*-scores or raw scores:

- normalcdf(lowerbd, upperbd) = probability between two *z*-scores for the standard normal distribution (mean = 0 and SD = 1)
- normalcdf(lowerbd, upperbd, mean, SD) = probability between two raw scores for a normal distribution with the given mean and SD
- invNorm(probability) = *z*-score with the given probability to the left for the standard normal distribution (mean = 0 and SD = 1)
- invNorm(probability, mean, SD) = raw score with the given probability to the left for a normal distribution with the given mean and SD

> Because it is quicker and gives more accurate calculations than the normal table, a calculator is preferred when working with the normal distribution. Many experienced AP Statistics teachers do not teach or encourage use of the normal table for this reason.

> **Example 5.1**

The life expectancy of a particular brand of lightbulb is roughly normally distributed with a mean of 1500 hours and a standard deviation of 75 hours.

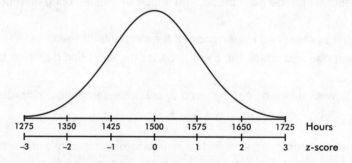

(a)   What is the probability that a lightbulb will last less than 1410 hours?
(b)   What is the probability that a lightbulb will last between 1563 and 1648 hours?
(c)   What is the probability that a lightbulb will last between 1416 and 1677 hours?

> **✓ Solutions**

(a)   Calculate normalcdf(0, 1410, 1500, 75) = 0.1151. [Or the *z*-score of 1410 is $\frac{1410 - 1500}{75} = -1.2$, and normalcdf($-100, -1.2$) = 0.1151.]

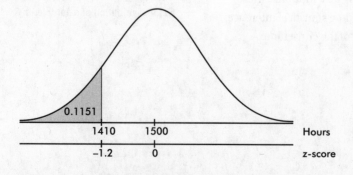

> Draw a picture! Errors are much less likely if normal probability calculations are supported with a sketch.

(continued)

Either of the following two lines will receive full credit:

Normal distribution, $\mu = 1500$, $\sigma = 75$, $P(X < 1410) = 0.1151$
normalcdf(left bd = 0, right bd = 1410, $\mu = 1500$, $\sigma = 75$) = 0.1151

(b)   For a normal distribution with $\mu = 1500$ and $\sigma = 75$,

$$P(1563 < X < 1648 = P\left(\frac{1563 - 1500}{75} < z < \frac{1648 - 1500}{75}\right) = P(0.84 < z < 1.97) = 0.176.$$

[normalcdf(1563, 1648, 1500, 75) = 0.176 and normalcdf(0.84, 1.97) = 0.176.]

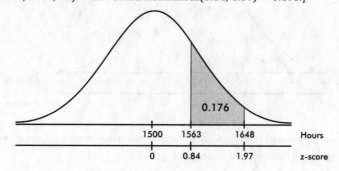

(c)   For a normal distribution with $\mu = 1500$ and $\sigma = 75$,

$$P(1416 < X < 1677) = P\left(\frac{1416 - 1500}{75} < z < \frac{1677 - 1500}{75}\right) = P(-1.12 < z < 2.36 = 0.8595.$$

[normalcdf(1416, 1677, 1500, 75) = 0.8595 and normalcdf(−1.12, 2.36) = 0.8595.]

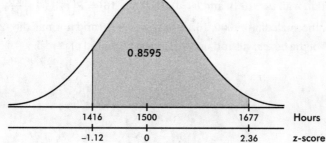

To receive full credit for probability calculations using the probability distributions, you need to show:

1. Name of the distribution ("normal" in the example above)
2. Parameters ("$\mu = 1500$, $\sigma = 75$" in the example above)
3. Boundary ("1410" in (a) of the example above)
4. Values of interest ("<" in (a) of the example above)
5. Correct probability ("0.1151" in (a) of the example above)

> **Example 5.2**

A packing machine is set to fill a cardboard box with a mean of 16.1 ounces of cereal. Suppose the amounts per box form an approximately normal distribution with a standard deviation equal to 0.04 ounce.

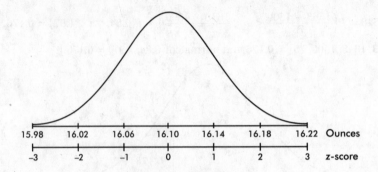

(a)   What percentage of the boxes will end up with at least 1 pound of cereal? (Note: 1 pound equals 16 ounces.)

(b)   Ten percent of the boxes will contain less than what number of ounces?

(c)   Eighty percent of the boxes will contain more than what number of ounces?

(d)   The middle 90% of the boxes will be between what two weights?

✓ **Solutions**

(a)   For a normal distribution with $\mu = 16.10$ and $\sigma = 0.04$, $P(X > 16) = P\left(z > \dfrac{16 - 16.10}{0.04}\right) =$
$P(z > -2.5) = 0.9938$. [normalcdf(16, 1000, 16.10, 0.04) = 0.9938 and normalcdf(−2.5, 10) = 0.9938.].
Conclude that 99.38% of the boxes will end up with at least 1 pound of cereal.

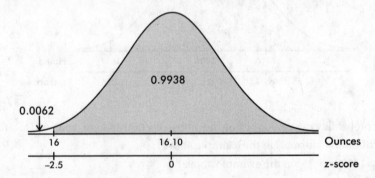

(b)   For a normal distribution with $\mu = 16.10$ and $\sigma = 0.04$, invNorm(0.1) = −1.282, and then
16.1 − 1.282(0.04) = 16.049. [invNorm(0.1, 16.10, 0.04) = 16.049.]
Conclude that 10% of the boxes will contain less than 16.049 ounces.

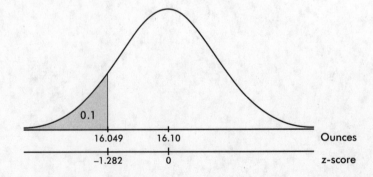

(continued)

(c)   Eighty percent to the right corresponds to a probability of 0.2 to the left. For a normal distribution with $\mu = 16.10$ and $\sigma = 0.04$, invNorm(0.2) = -0.8416, and then $16.1 - 0.8416(0.04) = 16.066$. [invNorm(0.2, 16.10, 0.04) = 16.066.]

Conclude that 80% of the boxes will contain more than 16.066 ounces.

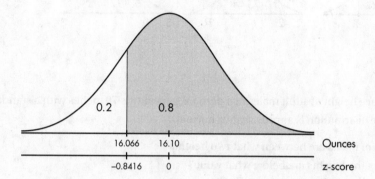

(d)   Ninety percent in the middle leaves 5% in each tail. For a normal distribution with $\mu = 16.10$ and $\sigma = 0.04$, invNorm(0.05) = -1.645, invNorm(0.95) = 1.645, $16.1 - 1.645(0.04) = 16.034$, and $16.1 + 1.645(0.04) = 16.166$. [invNorm(0.05, 16.10, 0.04) = 16.034 and invNorm(0.95, 16.10, 0.04) = 16.166.]

Conclude that the middle 90% of the boxes are between 16.034 and 16.166 ounces.

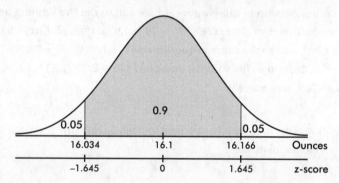

As can be seen above, there is often an interest in the limits enclosing some specified middle percentage of the data. For future reference, the limits most frequently asked for are noted below in terms of z-scores.

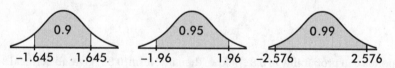

Sometimes the interest is in values with particular percentile rankings. For example,

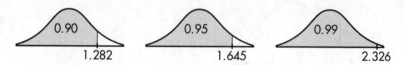

There are corresponding conclusions for negative z-scores:

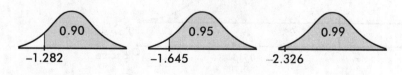

NOTE

The 2σ corresponding to 95% in the 68–95–99.7 rule is an approximation to the more exact 1.96σ.

NOTE

These critical z-scores can be found at the bottom of Table B in the Appendix.

It is also useful to note the percentages corresponding to values falling between integer z-scores. For example,

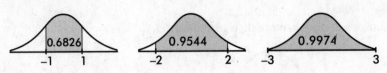

NOTE

The empirical rule's 68%, 95%, and 99.7% are approximations of 68.26%, 95.44%, and 99.74%.

> Example 5.3

Suppose that the average height of adult males in a particular locality is 70 inches with a standard deviation of 2.5 inches. Assume the distribution is approximately normal.

(a)   The middle 95% of males are between what two heights?
(b)   Ninety percent of the heights are below what value?
(c)   Ninety-nine percent of the heights are above what value?
(d)   Approximately what percentage of the heights are between z-scores of ± 1? Of ± 2? Of ± 3?

✓ Solutions

(a)   As noted above, the critical z-scores in this case are ± 1.96, and so the two limiting heights are 1.96 standard deviations from the mean. Therefore, $70 \pm 1.96(2.5) = 70 \pm 4.9$, or from 65.1 to 74.9 inches.
(b)   The critical z-score is 1.282, and so the value in question is $70 + 1.282(2.5) = 73.205$ inches.
(c)   The critical z-score is -2.326, and so the value in question is $70 - 2.326(2.5) = 64.185$ inches.
(d)   68.26%, 95.44%, and 99.74%, respectively.

If we know that a distribution is normal, we can calculate the mean μ and the standard deviation σ using percentage information from the population.

> Example 5.4

Given a normal distribution with a mean of 25, what is the standard deviation if 18% of the values are above 29?

✓ Solution

The 18% of values above corresponds to a probability of 0.82 below. Then invNorm(0.82) gives the critical z-score of 0.9154. Thus, $29 - 25 = 4$ is equal to 0.9154 standard deviations; that is, $0.9154\sigma = 4$, and $\sigma = \dfrac{4}{0.9154} = 4.37$.

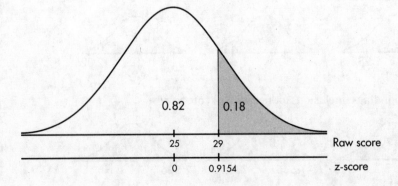

> **Example 5.5**

Given a normal distribution with a standard deviation of 10, what is the mean if 21% of the values are below 50?

✓ **Solution**

Using invNorm(0.21) gives the critical z-score of -0.8064. Thus, 50 is -0.8064 standard deviations from the mean, and so $\mu = 50 + 0.8064(10) = 58.06$.

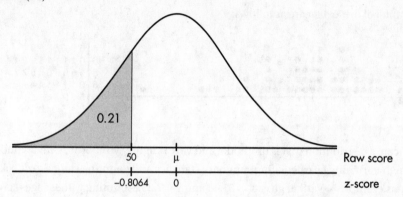

The binomial takes values only at integers, while the normal is continuous with probabilities corresponding to areas over intervals. However, for large enough n, binomial distributions can be approximated by normal distributions.

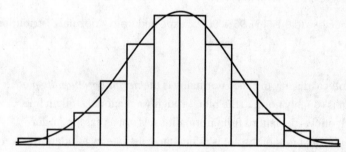

Is the normal a good approximation? The answer, of course, depends on the error tolerances in particular situations. A general rule of thumb is that **the normal is a good approximation to the binomial whenever both np and $n(1 - p)$ are at least 10.**

Proportions involve counting the presence or absence of some attribute, which suggests a binomial distribution. Inference procedures for proportions will rely on the normal distribution, and this makes sense only because the binomial can be approximated by the normal. In the above examples, we have assumed the population has a normal or approximately normal distribution. When you collect your own data, you must decide whether it is reasonable to assume the data come from a normal population before you can apply these procedures. In later units, we will have to check for normality before applying certain inference procedures.

The initial check should be to draw a picture. Dotplots, stemplots, and histograms are all useful graphical displays to show that data are unimodal and roughly symmetric.

> **Example 5.6**

The ages at inauguration of U.S. presidents from Washington to Biden were {57, 61, 57, 57, 58, 57, 61, 54, 68, 51, 49, 64, 50, 48, 65, 52, 56, 46, 54, 49, 51, 47, 55, 55, 54, 42, 51, 56, 55, 51, 54, 51, 60, 61, 43, 55, 56, 61, 52, 69, 64, 46, 54, 47, 70, 78}. Can we conclude that the distribution is roughly normal?

✓ **Solution**

A dotplot and a stemplot of these data are as follows:

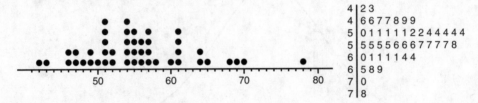

```
4 | 2 3
4 | 6 6 7 7 8 9 9
5 | 0 1 1 1 1 1 2 2 4 4 4 4 4
5 | 5 5 5 5 6 6 6 7 7 7 7 8
6 | 0 1 1 1 1 4 4
6 | 5 8 9
7 | 0
7 | 8
```

These two plots indicate a distribution that is unimodal and very roughly symmetric, with an outlier at 78 (Biden).

We can check rough normality by calculating summary statistics and counting values to check against the 68-95-99.7 rule. In this example, 1-VarStat gives $\bar{x} = 55.48$ and $s = 7.33$. Counting values (ages) gives the following:

Mean ± 1 SD	48.15 to 62.81	32 out of 46 = 69.6%
Mean ± 2 SD	40.82 to 70.14	45 out of 46 = 97.8%
Mean ± 3 SD	33.49 to 77.47	45 out of 46 = 97.8%

These results are somewhat close to the 68%, 95%, 99.7% expected from a normal distribution and so do suggest a roughly normal distribution.

A more specialized graphical display to check normality is the *normal probability plot*. When the data distribution is roughly normal, the plot is roughly a diagonal straight line. While this plot more clearly shows deviations from normality, it is not as easy to understand as a histogram. The normal probability plot is difficult to calculate by hand; however, technology such as the TI-84 readily plots the graph. (On the TI-84, in STATPLOT the sixth choice in TYPE gives the normal probability plot.)

The normal probability plot for the data in Example 5.6 is

> **NOTE**
>
> Although the normal probability plot is not part of the AP Statistics curriculum, it can be used on the exam.

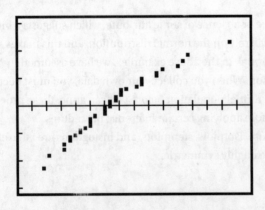

A graph close to a diagonal straight line would indicate that the data have a roughly normal distribution. This plot is roughly linear, suggesting an approximately normal distribution.

Quiz 12

Multiple-Choice Questions

DIRECTIONS: The questions that follow are each followed by five suggested answers. Choose the response that best answers the question.

1. Which of the following is a true statement?

   (A) The area under a normal curve is always equal to 1, no matter what the mean and standard deviation are.

   (B) All bell-shaped curves are normal distributions for some choice of μ and σ.

   (C) The smaller the standard deviation of a normal curve, the lower and more spread out the graph.

   (D) Depending upon the value of the standard deviation, normal curves with different means may be centered around the same number.

   (E) Depending upon the value of the standard deviation, the mean and median of a particular normal distribution may be different.

2. The continuous random variable X has a normal distribution with a mean of 20 and a standard deviation of 4. For which of the following is the probability equal to 0?

   (A) $P(X = 22)$
   (B) $P(X < 22)$
   (C) $P(X > 22)$
   (D) $P(21 < X < 22)$
   (E) $P(X < 21)$ or $P(X > 22)$

3. Populations P1 and P2 are normally distributed and have identical means. However, the standard deviation of P1 is twice the standard deviation of P2. What can be said about the percentage of observations falling within 2 standard deviations of the mean for each population?

   (A) The percentage for P1 is twice the percentage for P2.

   (B) The percentage for P1 is greater than but not twice as great as the percentage for P2.

   (C) The percentage for P2 is twice the percentage for P1.

   (D) The percentage for P2 is greater than but not twice as great as the percentage for P1.

   (E) The percentages are identical.

In Questions 4–9, assume the given distributions are approximately normal.

4. A trucking firm determines that its fleet of trucks averages a mean of 12.4 miles per gallon with a standard deviation of 1.2 miles per gallon on cross-country hauls. What is the probability that one of the trucks averages fewer than 10 miles per gallon?

   (A) $P(z < -2.4)$
   (B) $P(z < -2)$
   (C) $P(z < 10)$
   (D) $P\left(z < \frac{10}{1.2}\right)$
   (E) $P\left(z < \frac{12.4}{1.2}\right)$

5. An electronic product takes an average of 3.4 hours to move through an assembly line. If the standard deviation is 0.5 hours, what is the probability that an item will take between 3 and 4 hours to move through the assembly line?

   (A) $P(3 < z < 4)$

   (B) $P\left(\dfrac{3}{0.5} < z < \dfrac{4}{0.5}\right)$

   (C) $P(3 - 3.4 < z < 4 - 3.4)$

   (D) $P((3 - 3.4)(0.5) < z < (4 - 3.4)(0.5))$

   (E) $P\left(\dfrac{3 - 3.4}{0.5} < z < \dfrac{4 - 3.4}{0.5}\right)$

6. The mean noise level in a restaurant is 70 decibels with a standard deviation of 4 decibels. Ninety-nine percent of the time the noise level is below what value?

   (A) $70 + 1.96(4)$

   (B) $70 - 2.576(4)$

   (C) $70 + 2.576(4)$

   (D) $70 - 2.326(4)$

   (E) $70 + 2.326(4)$

7. The mean income per household in a certain state is \$9,500 with a standard deviation of \$1,750. The middle 95% of incomes are between what two values?

   (A) $9,500 \pm 1.645(1,750)$

   (B) $9,500 \pm 1.96(1,750)$

   (C) $9,500 \pm 1.645\left(\dfrac{1,750}{2}\right)$

   (D) $9,500 \pm \dfrac{1,750}{1.645}$

   (E) $9,500 \pm \dfrac{1,750}{1.96}$

8. One company produces movie trailers whose mean length is 150 seconds with a standard deviation of 40 seconds, while a second company produces movie trailers whose mean length is 120 seconds with a standard deviation of 30 seconds. What is the probability that the combined length of two randomly selected trailers, one produced by each company, will be less than 3 minutes?

   (A) 0.000

   (B) $P\left(z < \dfrac{180 - 270}{50}\right)$

   (C) $P\left(z < \dfrac{180 - 270}{70}\right)$

   (D) $P\left(z < \dfrac{180 - 270}{\sqrt{40 + 30}}\right)$

   (E) $P\left(z < \dfrac{180 - 270}{\sqrt{\dfrac{1}{40} + \dfrac{1}{30}}}\right)$

9. Cucumbers grown on a certain farm have weights with a standard deviation of 2 ounces. What is the mean weight if 85% of the cucumbers weigh less than 16 ounces?

   (A) $16 - 0.518$

   (B) $16 - 0.85(2)$

   (C) $16 - 1.036(2)$

   (D) $16 + 0.85(2)$

   (E) $16 + 1.036(2)$

10. The mean and standard deviation of the sample data collected on a continuous variable x are 10 and 5, respectively. The following table shows the relative frequencies of the data in the given intervals.

Interval	Relative Frequency
$-5 \leq x < 0$	0.03
$0 \leq x < 5$	0.12
$5 \leq x < 10$	0.35
$10 \leq x < 15$	0.34
$15 \leq x < 20$	0.14
$20 \leq x < 25$	0.02

Based on the table, do the data support using a normal model to approximate population characteristics?

(A) No, because some values are negative.

(B) No, because the relative frequency distribution doesn't follow the empirical rule.

(C) Yes, because the sum of the relative frequencies is 1.00.

(D) Yes, because the relative frequency distribution roughly follows the empirical rule.

(E) This cannot be answered without knowing the sample size.

Free-Response Questions

DIRECTIONS: You must show all work and indicate the methods you use. You will be graded on the correctness of your methods and on the accuracy of your final answers.

1. Scores on a university entrance exam are approximately normally distributed with a mean of 150 and a standard deviation of 25.

   (a) The university will accept 90% of applicants (based on entrance exam score). What is the cutoff score for acceptance (rounded to the nearest integer)?

   (b) Given that a student scores over 100, what is the probability of acceptance?

   (c) Given that three random students all score over 100, what is the probability that at least one is not accepted?

2. The following histogram shows the relative frequencies of the times of medium usage between necessary charging, recorded to the nearest 0.25 hour, of a sample of smartphones of a particular model. The mean is 8.03 hours, and the standard deviation is 0.49 hours.

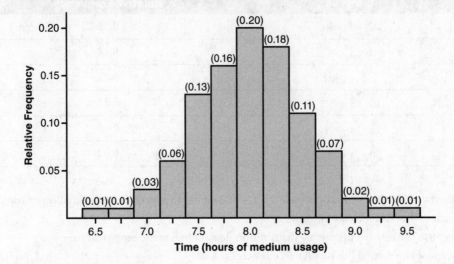

One smartphone is selected at random.

(a) Based on the histogram, what is the probability that the selected smartphone will last at least 8.5 hours of medium usage before charging is necessary?

(b) The histogram displays a discrete probability model for time. However, time is often considered a continuous variable. Assuming a roughly normal model using the given mean and standard deviation, what is the probability that the selected smartphone will last at least 8.5 hours of medium usage before charging is necessary?

(c) Explain why your answers to (a) and (b) are or are not the same.

(d) To estimate the probability that this model smartphone will last exactly 7.25 hours before charging is necessary (assuming medium usage), should the above discrete or above continuous model for time be used? Explain.

(e) To estimate the probability that this model smartphone will last at least 7.1 hours before charging is necessary (assuming medium usage), should the above discrete or above continuous model for time be used? Explain.

3. The time it takes Steve to walk to school follows an approximately normal distribution with a mean of 30 minutes and standard deviation of 5 minutes, while the time it takes Jan to walk to school follows an approximately normal distribution with a mean of 25 minutes and standard deviation of 4 minutes. Assume their walking times are independent of each other.

(a) If they leave at the same time, what is the probability that Steve arrives before Jan?

(b) How much earlier than Jan should Steve leave so that he has a 90% chance of arriving before Jan?

The answers for this quiz can be found beginning on page 492.

Central Limit Theorem

Real-life distributions are seldom, if ever, exactly normal. However, it can be shown mathematically that no matter how the original population is distributed, if n is large enough, the set of sample means is approximately normally distributed. For example, there is no reason to suppose that the amounts of money that different people spend in grocery stores are normally distributed. However, if each day we survey 50 people leaving a store and determine the average grocery bill, these daily averages will have an approximately normal distribution.

The following principle, called the *central limit theorem* of statistics (abbreviated CLT), forms the basis of much of what we discuss in this and the following units.

Start with a population with any shape distribution whatsoever. Pick n sufficiently large (at least 30), and take all samples of size n. Compute the mean of each of these samples. Then the set of all sample means is approximately normally distributed.

We say that for sufficiently large n, the *sampling distribution* of the sample mean $\bar{x}$ is approximately normal. Note that the CLT is only about the shape of the sampling distribution of the sample mean, not about its center or variability. However, we will see that the mean $\mu_{\bar{x}}$ of the sampling distribution of $\bar{x}$ equals the population mean μ, and the standard deviation $\sigma_{\bar{x}}$ of the sampling distribution of $\bar{x}$ approximately equals $\frac{\sigma}{\sqrt{n}}$, the population standard deviation divided by the square root of the sample size.

While we mention $n \geq 30$ as a rough rule of thumb (and is sufficient for the AP exam), $n \geq 40$ is often used, and n should be chosen even larger if more accuracy is required or if the original population is far from normal. However, **if the population itself has a normal distribution, then the sampling distribution of the sample mean is normal no matter what the sample size.** When working with a specific sample, we will also have the assumptions of a simple random sample and of sample size n no larger than 10% of the population.

There are six key ideas to keep in mind:

- Averages vary less than individual values.
- Averages based on larger samples vary less than averages based on smaller samples.
- The central limit theorem (CLT) states that when the sample size is sufficiently large, the sampling distribution of the mean will be approximately normal.
- The larger the sample size n, the closer the *sample distribution* is to the population distribution.
- The larger the sample size n, the closer the *sampling distribution* of $\bar{x}$ is to a normal distribution.
- If the original population has a normal distribution, then the sampling distribution of $\bar{x}$ has a normal distribution, no matter what the sample size n.

> Example 5.7

Following are simulations for sampling distributions for samples of sizes $n = 2$, $n = 10$, and $n = 30$ from three different populations. Note that as the sample size increases, the sampling distribution of $\bar{x}$ looks less like the population distribution and more like a normal distribution.

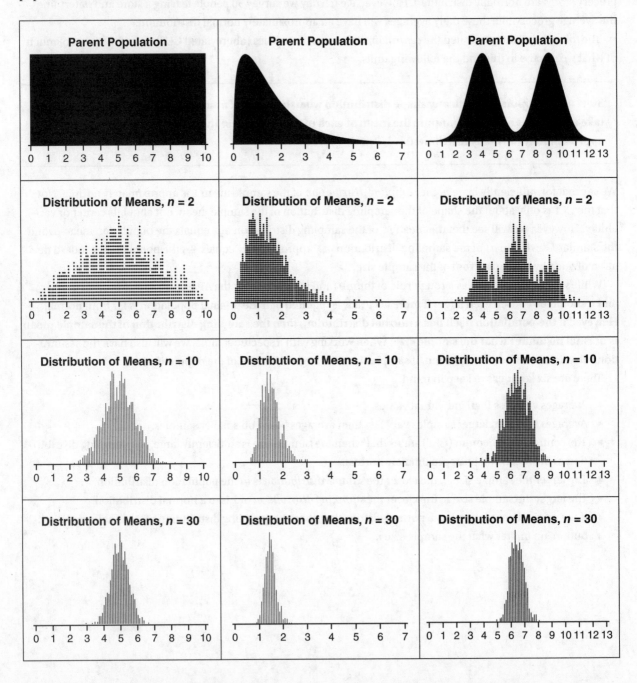

One example of a sampling distribution is a *randomization distribution*, a collection of statistics generated by simulation. For a sample survey, we can repeatedly randomly reassign individual responses to the different survey groups to see if our observed difference is significant. For an experiment, we can repeatedly randomly reassign the individual response values to treatment groups to see if our observed difference is significant.

> **Example 5.8**

In a random sample of 15 male Information Technology professionals, the average salary was $59,325, while in an independent random sample of 15 female Information Technology professionals the average salary was $51,850. The question of interest is whether or not the observed difference of $59,325 − $51,850 = $7,475 is significant. The randomization distribution method proceeds as follows. All 30 salaries are put together, and 15 are randomly reassigned to the "Male" subgroup and the remaining 15 to the "Female" subgroup. The mean difference in salary between the two groups is calculated, and the procedure is repeated over and over. Suppose the resulting histogram of differences, a simulated sampling distribution of the differences, is below.

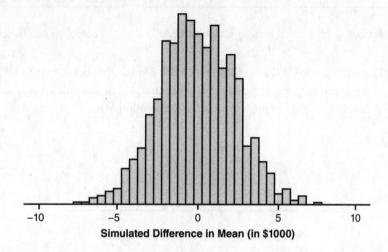

Simulated Difference in Mean (in $1000)

As can be seen, the observed difference of $7,475, while possible, is very rare among the simulated differences. We conclude that the salary gap between male and female IT professionals is statistically significant.

Don't be confused by the several different distributions being discussed! First, there's the distribution of the original population, whose shape may be uniform, bell-shaped, strongly skewed, or any other possible shape. Second, there's the distribution of the data in the sample, and the larger the sample size, the more this will look like the population distribution. Third, there's the distribution of the means of many samples of a given size. The amazing fact is that this *sampling distribution* can be described by a normal model, regardless of the shape of the original population.

Does the central limit theorem say anything about proportions? If the population values are comprised of just 0s and 1s (numerical codes for failures and successes), we have a distribution that is bimodal in the extreme. Means of samples drawn from that population are sample proportions, so the sampling distribution of sample proportions is a special case of the CLT. The CLT doesn't kick in until the sample size gets fairly large, especially if the population is far from symmetric (p far from 0.5). Requiring $np \geq 10$ and $nq \geq 10$ is a rule of thumb suggesting how large n must be.

Biased and Unbiased Estimators

In Unit 3, we discussed *bias* in the context of surveys as a tendency to favor the selection of certain members of a population. In the context of sampling distributions, *bias* means that the sampling distribution is not *centered* on the population parameter. The sampling distributions of proportions, means, and slopes are *unbiased*. That is, for a given sample size, the set of all sample proportions, $\hat{p}$, is centered on the population proportion, p; the set of all sample means, $\bar{x}$, is centered on the population mean, μ; and the set of all sample slopes, b, is centered on the population slope, β. Here are some illustrative simulations:

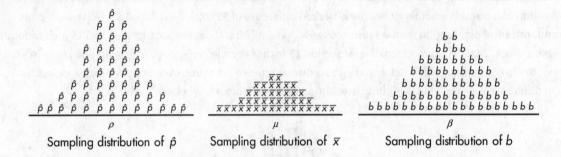

Sampling distribution of $\hat{p}$ Sampling distribution of $\bar{x}$ Sampling distribution of b

However, the sampling distribution for the maximum is clearly *biased*. That is, for a given sample size, the set of all sample maxima, $\tilde{v}$, is not centered on the population maximum, V. For example, here is one simulation of sample maxima. Note that V falls far right of the center of the distribution of $\tilde{v}$.

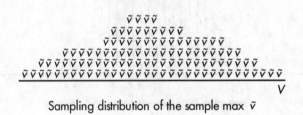

Sampling distribution of the sample max $\tilde{v}$

> **Example 5.9**

Note the following simulations for sampling distributions of $\dfrac{\Sigma(x-\bar{x})^2}{n}$ and $\dfrac{\Sigma(x-\bar{x})^2}{n-1}$, both from a population with variance $\sigma^2 = 25$.

NOTE

This shows why we divide by $n-1$ when calculating a sample variance and standard deviation.

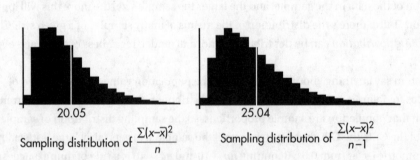

Sampling distribution of $\dfrac{\Sigma(x-\bar{x})^2}{n}$ Sampling distribution of $\dfrac{\Sigma(x-\bar{x})^2}{n-1}$

We see that $\dfrac{\Sigma(x-\bar{x})^2}{n}$ is a *biased* estimator for σ^2 (that is, the distribution, with mean 20.05, *is not* approximately centered at σ^2), while $\dfrac{\Sigma(x-\bar{x})^2}{n-1}$ is an *unbiased* estimator for σ^2 (that is, the distribution, with mean 25.04, *is* approximately centered at σ^2).

> **Example 5.10**

Five new estimators are being evaluated with regard to quality control in manufacturing professional baseballs of a given weight. Each estimator is tested every day for a month on samples of sizes $n = 10$, $n = 20$, and $n = 40$. The baseballs actually produced that month had a consistent mean weight of 146 grams. The distributions given by each estimator are as follows:

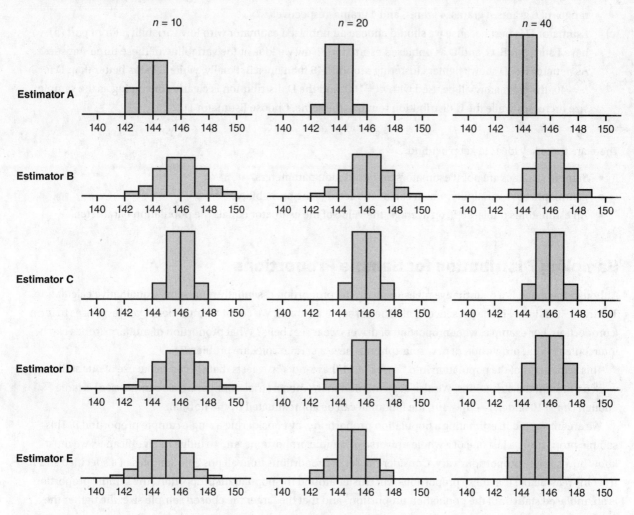

(a) Which of the above appear to be unbiased estimators of the population parameter?
(b) Which of the above exhibits the lowest variability for $n = 40$?
(c) Which of the above is the best estimator if the selected estimator will eventually be used with a sample of size $n = 100$?

(continued)

✓ **Solutions**

(a)   Estimators B, C, and D are unbiased estimators because they appear to have means approximately equal to the population mean of 146. A statistic used to estimate whether a population parameter is unbiased is if the mean of the sampling distribution of the statistic is equal to the true value of the parameter being estimated.

(b)   For $n = 40$, estimator A exhibits the lowest variability, with a range of only 2 grams compared with the other ranges of 6 grams, 4 grams, 4 grams, and 4 grams, respectively.

(c)   Estimator D is best because we should choose an unbiased estimator with low variability. From part (a), we have Estimator B, C, and D as unbiased estimators. Now we look at the variability of these three statistics. As n increases, D shows tighter clustering around 146 than does B. Finally, while C looks better than D for $n = 40$, the estimator will be used with $n = 100$, and the D distribution is clearly converging as the sample size increases while the C distribution remains the same. Choose Estimator D.

There are three key ideas to keep in mind:

- Sample statistics are point estimators for population parameters.
- Estimators have variability, which can be modeled using probability.
- An estimator is unbiased if, on average, the value of the estimator equals the population parameter.

Sampling Distribution for Sample Proportions

Although the mean is a quantitative measurement, the proportion essentially represents a qualitative calculation. The interest is simply in the presence or absence of some attribute. We count the number of yes responses to form a proportion. For example, what proportion of drivers wear seat belts? What proportion of military drones can be intercepted? What proportion of new smartphones have a certain software problem?

This separation of the population into "haves" and "have-nots" suggests that we can make use of our earlier work on binomial distributions. We also keep in mind that, when n (trials, or in this case sample size) is large enough ($np \geq 10$ and $n(1 - p) \geq 10$), the binomial can be approximated by the normal.

We are interested in estimating a population proportion p by considering a single sample proportion $\hat{p}$. This sample proportion is just one of a whole universe of sample proportions, and to judge its significance, we must know how sample proportions vary. Consider the set of proportions from all possible samples of a specified size n. It seems reasonable that these proportions will cluster around the population proportion (the sample proportion is an unbiased statistic of the population proportion) and that the larger the chosen sample size, the tighter the clustering.

How do we calculate the mean and standard deviation of the set of sample proportions? Suppose the sample size is n and the actual population proportion is p. From our work on binomial distributions, we remember that the mean and standard deviation for the number of successes in a given sample are np and $\sqrt{np(1 - p)}$, respectively, and for large values of n the complete distribution begins to look "normal."

Here, however, we are interested in the proportion rather than in the number of successes. From Unit 1, remember that when we multiply or divide every element by a constant, we multiply or divide both the mean and the standard deviation by the same constant. In this case, to change number of successes to proportion of successes, we divide by n:

$$\mu_{\hat{p}} = \frac{np}{n} = p \quad \text{and} \quad \sigma_{\hat{p}} = \sqrt{\frac{np(1-p)}{n}} = \sqrt{\frac{np(1-p)}{n^2}} = \sqrt{\frac{p(1-p)}{n}}$$

> The standard deviation of the sampling distribution of $\hat{p}$ when sampling without replacement is actually $\sigma_{\hat{p}} = \sqrt{\frac{p(1-p)}{n}}\sqrt{1 - \frac{n}{N}}$. As long as $n < 10\%(N)$, the finite population factor, $\sqrt{1 - \frac{n}{N}}$, is close to 1.

Furthermore, if each element in an approximately normal distribution is divided by the same constant, it is reasonable that the result will still be an approximately normal distribution.

Thus, note the principle forming the basis of the following discussion.

Start with a population with a given proportion p. Take all samples of size n. Compute the proportion in each of these samples:

1. **the set of all sample proportions is approximately normally distributed.**

2. **the mean $\mu_{\hat{p}}$ of the set of sample proportions equals p, the population proportion.**

3. **the standard deviation $\sigma_{\hat{p}}$ of the set of sample proportions is approximately equal to $\sqrt{\frac{p(1-p)}{n}}$.**

We say that the sampling distribution of $\hat{p}$ is approximately normal with mean p and standard deviation $\sqrt{\frac{p(1-p)}{n}}$.

Since we are using the normal approximation to the binomial, both np and $n(1-p)$ should be at least 10. Furthermore, in making calculations and drawing conclusions from a specific sample, ideally the sample should be a *simple random sample*.

Finally, when sampling is done without replacement, the sample cannot be too large; the sample size n should be no larger than 10% of the population. (We're actually worried about *independence*, but randomly selecting a sample with n less than 0.10 N allows us to assume independence. Of course, it's always better to have larger samples—it's just that if the sample is large relative to the population, the proper inference techniques are different from those taught in introductory statistics classes.)

> Example 5.11

It is estimated that 80% of people with high math anxiety experience brain activity similar to that experienced under physical pain when anticipating doing a math problem. In a simple random sample of 110 people with high math anxiety, what is the probability that less than 75% experience the physical pain brain activity?

(continued)

✓ Solution

The sample is given to be random, both $np = (110)(0.80) = 88 \geq 10$ and $n(1-p) = (110)(0.20) = 22 \geq 10$, and our sample is clearly less than 10% of all people with math anxiety. The sampling distribution of $\hat{p}$ is approximately normal with mean 0.80 and standard deviation

$\sigma_{\hat{p}} = \sqrt{\dfrac{(0.80)(0.20)}{110}} = 0.0381.$ $P(\hat{p} < 0.75) = P\left(z < \dfrac{0.75 - 0.80}{0.0381}\right) =$

$P(z < -1.312) = 0.0948$

`[normalcdf(-100, -1.312) = 0.0948 and`
`normalcdf(-100, 0.75, 0.80, 0.0381) = 0.0947.]`

> We say: "For all random samples of size $n = 110$ from the population of people with high math anxiety, the sample proportions of people who experience physical pain brain activity when anticipating doing a math problem will have a mean of 0.80 and will typically vary by about 0.0381 from the population proportion of 0.80."

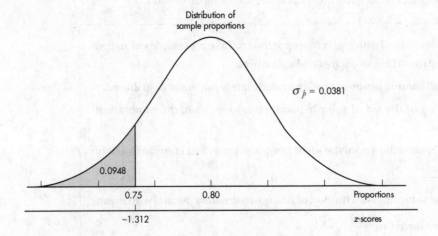

Getting a sample proportion of 0.75 or less happens in about 9.48% of all possible samples of size 110 from this population.

Sampling Distribution for Differences in Sample Proportions

Numerous important and interesting applications of statistics involve the comparison of two population proportions. For example, is the proportion of satisfied purchasers of American automobiles greater than that of buyers of Japanese cars? How does the percentage of surgeons recommending a new cancer treatment compare with the corresponding percentage of oncologists?

Our procedure involves comparing two sample proportions. When is a difference between two such sample proportions significant? Note that we are dealing with one difference from the set of all possible differences obtained by subtracting sample proportions of one population from sample proportions of a second population. To judge the significance of one particular difference, we must first determine how the differences vary among themselves. Remember that the mean of a set of differences is the difference of the means, and the variance of a set of differences is the sum of the variances of the individual sets.

$$\mu_d = \mu_1 - \mu_2 \text{ and } \sigma_d^2 = \sigma_1^2 + \sigma_2^2 \text{ with } \sigma_d = \sqrt{\sigma_1^2 + \sigma_2^2}$$

With our proportions we have $\sigma_1 = \sqrt{\dfrac{p_1(1-p_1)}{n_1}}$ and $\sigma_2 = \sqrt{\dfrac{p_2(1-p_2)}{n_2}}$ and can calculate:

$$\mu_{\hat{p}_1 - \hat{p}_2} = p_1 - p_2$$

$$\sigma_{\hat{p}_1 - \hat{p}_2} = \sqrt{\dfrac{p_1(1-p_1)}{n_1} + \dfrac{p_2(1-p_2)}{n_2}}$$

We have the following about the sampling distribution of $\hat{p}_1 - \hat{p}_2$:

Start with two populations with given proportions p_1 and p_2. Take all samples of sizes n_1 and n_2, respectively. Compute the difference $\hat{p}_1 - \hat{p}_2$ of the two proportions in each pair of samples:

1. the set of all differences of sample proportions is approximately normally distributed.

2. the mean of the set of differences of sample proportions equals $p_1 - p_2$, the difference of population proportions.

3. the standard deviation $\sigma_{\hat{p}_1 - \hat{p}_2}$ of the set of differences of sample proportions is approximately equal to:

$$\sqrt{\frac{p_1(1 - p_1)}{n_1} + \frac{p_2(1 - p_2)}{n_2}}$$

Since we are using the normal approximation to the binomial, $n_1 p_1$, $n_1(1 - p_1)$, $n_2 p_2$, and $n_2(1 - p_2)$ should all be at least 10. Furthermore, in making calculations and drawing conclusions from specific samples, ideally the samples should be *simple random samples* and be taken *independently* of each other. Finally, if sampling without replacement, the sample sizes should be no larger than 10% of the populations.

> ### Example 5.12

In a study of how environment affects our eating habits, scientists revamped one of two nearby fast-food restaurants with tablecloths, candlelight, and soft music. They then noted that at the revamped restaurant, customers ate more slowly and 25% left at least 100 calories of food on their plates. At the unrevamped restaurant, customers tended to quickly eat their food and only 19% left at least 100 calories of food on their plates. In a random sample of 110 customers at the revamped restaurant and an independent random sample of 120 customers at the unrevamped restaurant, what is the probability that the difference in the percentages of customers leaving at least 100 calories of food on their plates in the revamped setting and the unrevamped setting is more than 10% (where the difference is the revamped restaurant percent minus the unrevamped restaurant percent)?

✓ **Solution**

We have independent random samples, each less than 10% of all fast-food customers, and we note that $n_1 p_1 = 110(0.25) = 27.5$, $n_1(1 - p_1) = 110(0.75) = 82.5$, $n_2 p_2 = 120(0.19) = 22.8$, and $n_2(1 - p_2) = 120(0.81) = 97.2$ are all ≥ 10. Thus, the sampling distribution of $\hat{p}_1 - \hat{p}_2$ is roughly normal with mean $\mu_{\hat{p}_1 - \hat{p}_2} = 0.25 - 0.19 = 0.06$ and standard deviation

$$\sigma_{\hat{p}_1 - \hat{p}_2} = \sqrt{\frac{(0.25)(0.75)}{110} + \frac{(0.19)(0.81)}{120}} = 0.0547.$$

We say: "For all random samples of 110 customers at the revamped restaurant and 120 customers at the unrevamped restaurant, the differences (revamped − unrevamped) in sample proportions of customers who left at least 100 calories of food on their plates has a mean of 0.06 and will typically vary by about 0.0547 from the true difference of 0.06."

$$P(\hat{p}_1 - \hat{p}_2 > 0.10) = P\left(z > \frac{0.10 - 0.06}{0.0547}\right) = P(z > 0.731) = 0.232$$

[normalcdf(0.731, 100) = 0.232 and normalcdf(0.10, 100, 0.06, 0.0547) = 0.232.]

(continued)

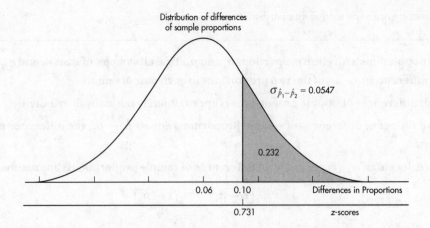

Distribution of differences of sample proportions

$\sigma_{\hat{p}_1 - \hat{p}_2} = 0.0547$

0.232

0.06   0.10   Differences in Proportions

0.731   z-scores

Getting a difference (revamped – unrevamped) in sample proportions of 0.10 or greater happens in about 23.2% of all possible samples of sizes 110 and 120, respectively, from these populations.

Sampling Distribution for Sample Means

Suppose we are interested in estimating the mean μ of a population. For our estimate we could simply randomly pick a single member of the population, but then we would have little confidence in our answer. Suppose instead that we pick 100 members and calculate their average. It is intuitively clear that the resulting sample mean has a greater chance of being closer to the mean of the whole population than the value for any individual member of the population does.

When we pick a sample and measure its mean $\bar{x}$, we are finding exactly one sample mean out of a whole universe of sample means. To judge the significance of a single sample mean, we must know how sample means vary. Consider the set of means from all possible samples of a specified size. It is both apparent and reasonable that the sample means are clustered around the mean of the whole population; furthermore, these sample means have a tighter clustering than the members of the original population. In fact, we might guess that the larger the chosen sample size, the tighter the clustering.

How do we calculate the standard deviation $\sigma_{\bar{x}}$ of the set of sample means? Suppose the variance of the population is σ^2 and we are interested in samples of size n. Sample means are obtained by first summing together n elements and then dividing by n. A set of sums has a variance equal to the sum of the variances associated with the original sets. In our case, $\sigma^2_{\text{sums}} = \sigma^2 + \cdots + \sigma^2 = n\sigma^2$. When each element of a set is divided by some constant, the new variance is the old one divided by the square of the constant. Since the sample means are obtained by dividing the sums by n, the variance of the sample means is obtained by dividing the variance of the sums by n^2. Thus, if $\sigma_{\bar{x}}$ symbolizes the standard deviation of the sample means, we find that:

$$\sigma_{\bar{x}}^2 = \frac{\sigma^2_{\text{sums}}}{n^2} = \frac{n\sigma^2}{n^2} = \frac{\sigma^2}{n}$$

In terms of standard deviations, we have $\sigma_{\bar{x}} = \frac{\sigma}{\sqrt{n}}$.

Thus, we have the following.

Start with a population with a given mean μ and standard deviation σ. Compute the mean of all samples of size n:

1. the mean $\mu_{\bar{x}}$ of the set of sample means equals μ, the mean of the population.

2. the standard deviation $\sigma_{\bar{x}}$ of the set of sample means approximately equals $\frac{\sigma}{\sqrt{n}}$, that is, the standard deviation of the whole population divided by the square root of the sample size.

We say that the sampling distribution of $\bar{x}$ has mean μ and standard deviation $\frac{\sigma}{\sqrt{n}}$.

3. if the population is approximately normal or if the sample size is large enough (at least 30) for the central limit theorem to apply, we also have that the set of sample means is approximately normally distributed.

TIP

Never talk about "the sampling distribution" without adding "of $\hat{p}$" or "of $\bar{x}$" or "of" whatever is appropriate.

Note that the variance of the set of sample means varies directly with the variance of the original population and inversely with the size of the samples; the standard deviation of the set of sample means varies directly with the standard deviation of the original population and inversely with the square root of the size of the samples. Because sampling is usually done without replacement, and independence must be considered, we cannot allow the sample size to be too large when calculating variance and standard deviation for sample means. Therefore, the sample size n should not be larger than 10% of the population.

› Example 5.13

The number of emergency room visits after drinking energy drinks is skyrocketing. One particular energy drink has an average of 200 mg of caffeine with a standard deviation of 10 mg and an approximately normal distribution. A store sells boxes of six bottles each. Describe the distribution of the average milligrams of caffeine consumers should expect from the six bottles in each box.

TIP

Now that we are dealing with quantitative variables, do not forget to include *units* when interpreting answers in context.

✓ Solution

The sampling distribution of the sample means is approximately normal because the original population is approximately normal. The mean of these sample means will equal the population mean of 200 mg. The standard deviation of these sample means will approximately equal $\sigma_{\bar{x}} = \frac{10}{\sqrt{6}} = 4.08$ mg.

› Example 5.14

It is estimated that domesticated cats who mainly live outdoors kill an average of 10 birds a year with a standard deviation of 3 birds. What is the probability an SRS of 35 such cats will kill an average of between 9 and 12 birds in a year?

✓ **Solution**

The sampling distribution of the sample mean is approximately normal (because the sample size $35 > 30$ so the CLT applies) with mean $\mu_{\bar{x}} = 10$ and standard deviation $\sigma_{\bar{x}} = \frac{3}{\sqrt{35}} = 0.507$.

$$P(9 < \bar{x} < 12) = P\left(\frac{9-10}{0.507} < z < \frac{12-10}{0.507}\right) = P(-1.972 < z < 3.945) = 0.976$$

[normalcdf($-1.972, 3.945$) = 0.976 and normalcdf(9, 12, 10, 0.507) = 0.976.]

> We say, "For all random samples of size $n = 35$ from the population of domesticated cats who mainly live outdoors, the sample mean numbers of birds killed per year will have a mean of 10 birds and will typically vary by about 0.6 birds from the population mean of 10 birds."

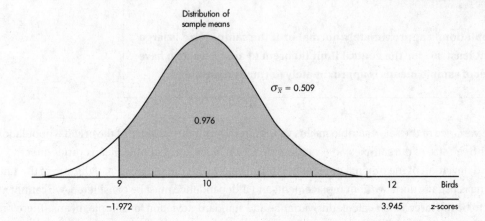

Getting a sample mean between 9 and 12 birds killed in a year happens in about 97.5% of all possible samples of size 35 from the population of domesticated cats who mainly live outdoors.

Sampling Distribution for Differences in Sample Means

Many real-life applications of statistics involve comparisons of two population means. For example, is the average weight of laboratory rabbits receiving a special diet greater than that of rabbits on a standard diet? Which of two accounting firms pays a higher mean starting salary? Is the life expectancy of a coal miner less than that of a school teacher?

> **TIP**
>
> Don't forget to check that the samples are independent!

First we consider how to compare the means of samples, one from each population. When is a difference between two such sample means significant? The answer is more apparent when we realize that what we are looking at is one difference from a set of differences. That is, there is the set of all possible differences obtained by subtracting sample means from one set from sample means from a second set. To judge the significance of one particular difference, we must first determine how the differences vary among themselves. The necessary key is the fact that the variance of a set of differences is equal to the sum of the variances of the individual sets. Thus:

$$\sigma^2_{\bar{x}_1 - \bar{x}_2} = \sigma^2_{\bar{x}_1} + \sigma^2_{\bar{x}_2}$$

Now if $\sigma_{\bar{x}_1} = \frac{\sigma_1}{\sqrt{n_1}}$ and $\sigma_{\bar{x}_2} = \frac{\sigma_2}{\sqrt{n_2}}$, then $\sigma^2_{\bar{x}_1 - \bar{x}_2} = \frac{\sigma^2_1}{n_1} + \frac{\sigma^2_2}{n_2}$ and $\sigma_{\bar{x}_1 - \bar{x}_2} = \sqrt{\frac{\sigma^2_1}{n_1} + \frac{\sigma^2_2}{n_2}}$.

Then we have the following about the sampling distribution of $\bar{x}_1 - \bar{x}_2$.

Start with two populations with means μ_1 and μ_2 and standard deviations σ_1 and σ_2. Take all samples of sizes n_1 and n_2, respectively. Compute the difference $\bar{x}_1 - \bar{x}_2$ of the two means in each pair of these samples:

1. the mean $\mu_{\bar{x}_1 - \bar{x}_2}$ of the set of differences of sample means equals $\mu_1 - \mu_2$, the difference of population means.

2. the standard deviation $\sigma_{\bar{x}_1 - \bar{x}_2}$ of the set of differences of sample means is approximately equal to
$$\sqrt{\frac{\sigma_1^2}{n_1} + \frac{\sigma_2^2}{n_2}}.$$

3. if the populations are normal or if the sample sizes are large enough (at least 30) for the central limit theorem to apply, we also have that the set of all differences of sample means is approximately normally distributed.

When analyzing two samples, one from each population, we have the assumption of independent simple random samples and, if sampling without replacement, the assumption of sample sizes no larger than 10% of their respective populations.

❯ Example 5.15

It is estimated that 40-year-old men contribute an average of 65 genetic mutations to their new children, whereas 20-year-old men contribute an average of only 25. Assuming standard deviations of 15 and 5 mutations, respectively, for the 40- and 20-year-olds, what is the probability that the mean number of mutations in a random sample of thirty-five 40-year-old new fathers is between 35 and 45 more than the mean number in a random sample of forty 20-year-old new fathers?

✓ Solution

We have independent random samples, each less than 10% of their age groups, and both sample sizes are over 30. So the sampling distribution of $\bar{x}_1 - \bar{x}_2$ is roughly normal with mean $\mu_{\bar{x}_1 - \bar{x}_2} = 65 - 25 = 40$ and standard deviation

$$\sigma_{\bar{x}_1 - \bar{x}_2} = \sqrt{\frac{15^2}{35} + \frac{5^2}{40}} = 2.656.$$

$$P(35 < \bar{x}_1 - \bar{x}_2 < 45) = P\left(\frac{35 - 40}{2.656} < z < \frac{45 - 40}{2.656}\right)$$
$$= P(-1.883 < z < 1.883) = 0.940$$

[normalcdf($-1.883, 1.883$) = 0.940 and normalcdf(35, 45, 40, 2.656) = 0.940.]

> We say, "For all random samples of thirty-five 40-year-old new fathers and forty 20-year-old new fathers, the differences (40-year-olds − 20-year-olds) in sample mean genetic mutations passed to their new children will have a mean of 40 mutations and will typically vary by about 2.656 mutations from the true difference of 40 mutations."

(continued)

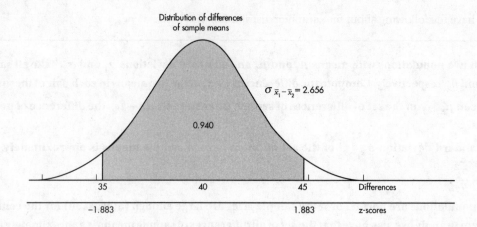

A difference (40-year-olds − 20-year-olds) in sample mean genetic mutations passed to their new children of between 35 and 45 mutations happens in about 94.0% of all possible samples of thirty-five 40-year-old new fathers and forty 20-year-old new fathers.

Simulation of a Sampling Distribution

As we've seen, the normal distribution can handle sampling distributions of the statistics we are most interested in, namely the sample proportion and sample mean. If other statistics arise, we can use simulation to obtain a rough idea of the corresponding sampling distributions.

> **Example 5.16**

A study is made of the number of dreams high school students remember having every night. The median number is 3.41 with a variance of 1.46 and a minimum of 0. Now taking a large number of random samples of 15 students, we calculate the median for each sample and graph the resulting simulated sampling distribution.

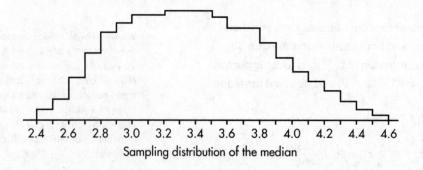

Sampling distribution of the median

The simulated sampling distribution for the sample median is roughly bell-shaped and centered around 3.4 with a range of approximately $4.6 − 2.4 = 2.2$.

Quiz 13

Multiple-Choice Questions

DIRECTIONS: The questions that follow are each followed by five suggested answers. Choose the response that best answers the question.

1. A sample of size n is to be selected from a population with population proportion p. For the sampling distribution of the sample proportion to be approximately normal, which of the following should be checked?

   (A) The sample size n should be at least 30.
   (B) The population proportion p should be close to 0.5.
   (C) The mean of the population should be np.
   (D) The standard deviation of the population should be $\sqrt{np(1-p)}$.
   (E) Both np and $n(1-p)$ should be at least 10.

2. A statistic $\hat{w}$ is an estimator for a population parameter W. The sampling distribution of $\hat{w}$ is not centered at W. What property does $\hat{w}$ exhibit?

   (A) The sampling distribution of $\hat{w}$ is not uniform.
   (B) The sampling distribution of $\hat{w}$ is normal.
   (C) The sampling distribution of $\hat{w}$ is not normal.
   (D) $\hat{w}$ is an unbiased estimator.
   (E) $\hat{w}$ is a biased estimator.

3. The distribution for the number of lies college students tell to their mothers during phone conversations is approximately normal with mean 0.81 and standard deviation 0.24. Consider two different random samples taken from the population, one of size 10 and one of size 100. Which of the following is true about the sampling distributions of the sample mean for the two sample sizes?

   (A) Both distributions are approximately normal with mean 0.81 and standard deviation 0.24.
   (B) Both distributions are approximately normal. Both the mean and the standard deviation for the $n = 10$ sampling distribution are greater than for the $n = 100$ distribution.
   (C) Both distributions are approximately normal with the same mean. The standard deviation for the $n = 10$ sampling distribution is greater than for the $n = 100$ distribution.
   (D) Only the $n = 100$ sampling distribution is approximately normal. Both distributions have mean 0.81 and standard deviation 0.24.
   (E) Only the $n = 100$ sampling distribution is approximately normal. Both distributions have the same mean. The standard deviation for the $n = 10$ sampling distribution is greater than for the $n = 100$ distribution.

4. A study is to be performed to estimate the proportion of voters who believe the economy is "heading in the right direction." Which of the following pairs of sample size and population proportion p will result in the smallest variance for the sampling distribution of $\hat{p}$?

(A) $n = 100$ and $p = 0.1$
(B) $n = 100$ and $p = 0.5$
(C) $n = 100$ and $p = 0.99$
(D) $n = 1,000$ and $p = 0.1$
(E) $n = 1,000$ and $p = 0.5$

5. Thirty-four percent of high school administrators say that math is the most important subject in school. In a random sample of 400 high school administrators, what is the probability that between 30% and 35% will say that math is the most important subject?

(A) $P(0.30 < z < 0.35)$
(B) $P(0.30 - 0.34 < z < 0.35 - 0.34)$
(C) $P\left(\dfrac{0.30 - 0.34}{\sqrt{400(0.34)(0.66)}} < z < \dfrac{0.35 - 0.34}{\sqrt{400(0.34)(0.66)}}\right)$
(D) $P\left(\dfrac{0.30 - 0.34}{\sqrt{\dfrac{(0.34)(0.66)}{400}}} < z < \dfrac{0.35 - 0.34}{\sqrt{\dfrac{(0.34)(0.66)}{400}}}\right)$
(E) $P\left(\dfrac{0.30 - 0.34}{\sqrt{\dfrac{(0.30)(0.70)}{400}}} < z < \dfrac{0.35 - 0.34}{\sqrt{\dfrac{(0.35)(0.65)}{400}}}\right)$

6. Suppose that, using accelerometers in helmets, researchers determine that boys playing high school football absorb an average of 355 hits to the head with a standard deviation of 80 hits during a season (including both practices and games). What is the probability on a randomly selected team of 48 players that the average number of head hits per player is between 340 and 360?

(A) $P(340 < z < 360)$
(B) $P(340 - 355 < z < 360 - 355)$
(C) $P\left(\dfrac{340 - 355}{80} < z < \dfrac{360 - 355}{80}\right)$
(D) $P\left(\dfrac{340 - 355}{\sqrt{48}} < z < \dfrac{360 - 355}{\sqrt{48}}\right)$
(E) $P\left(\dfrac{340 - 355}{\dfrac{80}{\sqrt{48}}} < z < \dfrac{360 - 355}{\dfrac{80}{\sqrt{48}}}\right)$

7. A 2021 survey of private, not-for-profit, four-year colleges indicated the mean spent on tuition and fees per year is $39,070. A simulation was conducted to create a sampling distribution of the sample mean for a population with a mean of $39,070. The following histogram shows the result of the simulation.

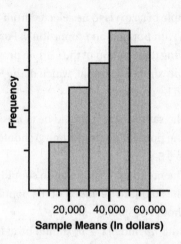

Among the following, which would be the best reason why the simulation of the sampling distribution does not appear roughly normal?

(A) The population distribution is roughly normal.
(B) The population distribution has large variability.
(C) The samples were not randomly selected.
(D) The samples were selected without replacement.
(E) The sample size was small.

8. A human resources specialist for a large inter-
    national company wanted to estimate the mean
    amount of wasted time during 8-hour workdays
    for all employees of the company. The specialist
    took repeated random samples of size 25 employ-
    ees and estimated the mean and standard devia-
    tion of the sampling distribution to be 95 minutes
    and 4 minutes, respectively. Based on the esti-
    mates for the sampling distribution, which of the
    following provides the best estimates of the popu-
    lation parameters?

    (A) Mean of 95 minutes and standard deviation
        of 0.16 minutes
    (B) Mean of 95 minutes and standard deviation
        of 0.8 minutes
    (C) Mean of 95 minutes and standard deviation
        of 4 minutes
    (D) Mean of 95 minutes and standard deviation
        of 20 minutes
    (E) Mean of 95 minutes and standard deviation
        of 100 minutes

9. In a study of a large number of patients arriving
    at a city hospital emergency room, the propor-
    tion of initial misdiagnoses was 0.14. The hospital
    typically averages between 100 and 200 emer-
    gency arrivals per day. Two simulations will be
    conducted for the sampling distribution of a
    sample proportion with a true proportion of 0.14.
    Simulation A will consist of 4,000 trials with a
    sample size of 100, while simulation B will consist
    of 2,000 trials with a sample size of 200. Which of
    the following statements correctly compares the
    center and variability of the two simulations?

    (A) The centers will be roughly equal, and the
        variabilities will be roughly equal.
    (B) The centers will be roughly equal; however,
        the variability of simulation A will be less
        than the variability of simulation B.

    (C) The centers will be roughly equal; however,
        the variability of simulation A will be greater
        than the variability of simulation B.
    (D) The center of simulation A will be less than
        the center of simulation B; however, the vari-
        ability of simulation A will be greater than the
        variability of simulation B.
    (E) The center of simulation A will be greater
        than the center of simulation B; however, the
        variability of simulation A will be less than
        the variability of simulation B.

10. In coffee shops throughout a large city, research-
    ers surveyed a random sample of 60 men and
    50 women from populations of over 1,000 men
    and 1,000 women. The sampled individuals were
    asked how many seconds wait time they experi-
    enced before their orders were taken. Let $\bar{x}_M$ be
    the calculated mean waiting time of the 60 men,
    and let $\bar{x}_W$ be the calculated mean waiting time
    of the 50 women. Which of the following is the
    best explanation for why the sampling distribu-
    tion of $\bar{x}_M - \bar{x}_W$ can be modeled with a normal
    distribution?

    (A) There are at least 30 people in each
        population.
    (B) The sample sizes are sufficiently large.
    (C) The population distributions are assumed to
        be roughly normal.
    (D) The sample distributions are assumed to be
        unimodal and roughly symmetric.
    (E) The two sample standard deviations are
        assumed to be roughly equal.

11. Three populations all have the same mean of 50 and same standard deviation of 10. They are represented by the following very different probability distributions:

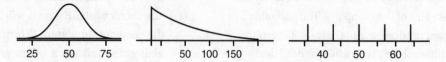

Using software, a large number of samples, each of size 100, are selected from each population. Which of the following gives the resulting three most likely distributions of sample means?

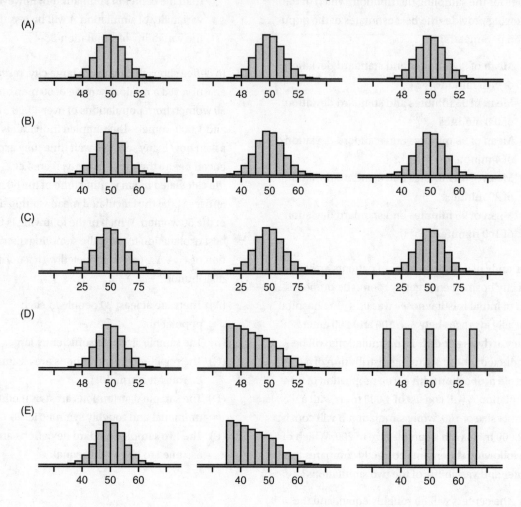

Free-Response Questions

DIRECTIONS: You must show all work and indicate the methods you use. You will be graded on the correctness of your methods and on the accuracy of your final answers.

1. Suppose that 16% of teenagers who play video games would enjoy playing a new Super Mario game. The manufacturer doesn't know this actual proportion and plans on sampling a random sample of 80 teenagers who play video games.

   (a) Calculate $\sigma_{\hat{p}}$, the standard deviation of the sampling distribution of $\hat{p}$.
   (b) Are the conditions for inference met? Explain.
   (c) If a different sample size would result in $\sigma_{\hat{p}} = 0.037$, is the sample size smaller or larger than 80? Explain.

2. It is estimated that 58% of all Americans sleep on their sides.

   (a) What is the probability that a randomly chosen American sleeps on his or her side?
   (b) In a random sample of five Americans, what is the probability that at least three sleep on their sides?
   (c) In a random sample of 350 Americans, what is the probability that at least 50% sleep on their sides?

3. The amount of fuel used by jumbo jets to take off is approximately normally distributed with a mean of 4,000 gallons and a standard deviation of 125 gallons.

   (a) What is the probability that a randomly selected jumbo jet will need more than 4,250 gallons of fuel to take off?
   (b) What are the shape, mean, and standard deviation of the sampling distribution of the mean gallons of fuel to take off of 40 randomly selected jumbo jets?
   (c) In a random sample of 40 jumbo jets, what is the probability that the mean number of gallons of fuel needed to take off is less than 3,950?
   (d) Would the answer to (a), (b), or (c) be affected if the original population of gallons of fuel used by jumbo jets to take off were skewed instead of normal? Explain.

4. College coaches often strongly stress study hours out of concern that their players remain academically eligible to play. In one study at a Division III college, 25 students participating and 25 students not participating in a varsity sport were surveyed as to the number of hours per week spent with academic studies outside the classroom. The survey results are summarized in the following parallel boxplots.

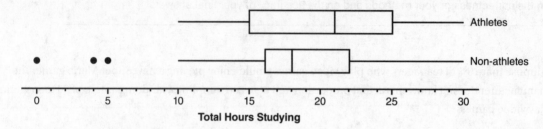

(a) Compare the two distributions.

To test whether the observed difference in medians is significant, a simulation is conducted. For each trial of the simulation, the 25 values from each survey were combined into one group of size 50 and then randomly split into two groups of 25. The difference in medians (athletes minus nonathletes) of the groups were calculated for each trial. The following dotplot shows the difference in the medians for 110 trials of the simulation.

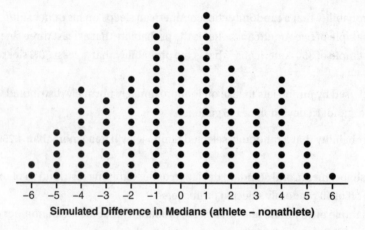

(b) In how many of the trials was the athletes' median 3 or more than the nonathletes' median?

(c) Using the observed difference (athletes – nonathletes) in median study hours and using the results of the simulation, what is the estimated probability of the observed difference occurring if there were really no difference in study hours (the assumption of the simulation methodology)?

(d) Based on the answer to (b), is there convincing statistical evidence that athletes participating in varsity sports at this college spend more hours per week with academic studies outside the classroom than do nonathletes?

The answers for this quiz can be found beginning on page 494.

SUMMARY

- Drawing pictures of normal curves with horizontal lines showing raw scores and z-scores is usually helpful.
- Areas (probabilities) under a normal curve and critical values corresponding to given areas (probabilities) can be found using calculator software, such as `normalcdf` and `invNorm`.
- If the probability distribution of each random variable is approximately normal, then the distribution of the sum (or the difference) of two independent random variables is also approximately normal.
- To check whether a given distribution is approximately normal, first check if it is unimodal, roughly symmetric, and roughly bell-shaped. A further check is whether the empirical rule (the 68-95-99.7 rule) seems to apply.
- Although not required, a normal probability plot is very useful in gauging whether a given distribution of data is approximately normal. If the plot is nearly straight, the data are nearly normal.
- A sampling distribution of a statistic is the distribution of values for the statistic for all possible samples of the same size from a given population.
- The central limit theorem (CLT) states that when the sample size is sufficiently large, the sampling distribution of the mean will be approximately normal.
- An estimator is *unbiased* if, on average, the value of the estimator is equal to the population parameter.
- Sampling distribution conditions

  1. **INDEPENDENCE:** Satisfied by random sampling, independent samples for differences, and the sample size less than 10% of the population when sampling without replacement.

  2. **NORMALITY:** Satisfied by $np \geq 10$ and $n(1 - p) \geq 10$ for proportions and by either approximately normal populations or large sample sizes ($n \geq 30$) for means.

- Provided that conditions are met, the sampling distribution of sample proportions is approximately normal with mean p and standard deviation $\sqrt{\dfrac{p(1 - p)}{n}}$.
- Provided that conditions are met, the sampling distribution of sample means is approximately normal with mean μ and standard deviation $\dfrac{\sigma}{\sqrt{n}}$.
- Provided that conditions are met, the sampling distribution of the difference of sample proportions is approximately normal with mean $p_1 - p_2$ and standard deviation $\sqrt{\dfrac{p_1(1 - p_1)}{n_1} + \dfrac{p_2(1 - p_2)}{n_2}}$.
- Provided that conditions are met, the sampling distribution of the difference of sample means is approximately normal with mean $\mu_1 - \mu_2$ and standard deviation $\sqrt{\dfrac{\sigma_1^2}{n_1} + \dfrac{\sigma_2^2}{n_2}}$.

6

Inference for Categorical Data: Proportions

(5–6 multiple-choice questions on the AP exam)

Learning Objectives

In this unit, you will learn how

→ To be able to check conditions to perform inference involving population proportions.

→ To be able to construct and interpret a confidence interval for a population proportion and for the difference of population proportions.

→ To understand the relationship among confidence level, sample size, and confidence interval width.

→ To be able to interpret confidence levels.

→ To be able to perform a significance test to evaluate a claim about a population proportion and about a difference between population proportions.

→ To be able to interpret Type I and Type II errors and their possible consequences in context.

→ To understand the relationship between significance level, probability of Type II error, and power.

We want to know some important truth about a population, but in practical terms this truth is unknowable. What's the average adult human body weight? What proportion of people have high cholesterol? What we can do is carefully collect data from as large and representative a group of individuals as practically possible and then use this information to estimate the value of the population parameter. How close are we to the truth? We know that different samples will give different estimates, and so sampling error (sampling variability) is unavoidable. What is wonderful, and what we will learn in this unit, is that we can quantify this sampling error! We can make statements like "the average weight must almost surely be within 5 pounds of 176 pounds" or "the proportion of people with high cholesterol must almost surely be within ±3% of 37%."

When using a measurement from a sample, we are never able to say *exactly* what a population proportion or mean or slope is; rather, we always say we have a certain *confidence* that the population proportion or mean or slope lies in a particular *interval*. The particular interval is centered around a sample statistic and can be expressed as the sample statistic plus or minus an associated *margin of error*.

In this unit, you will use what you learned from the previous unit on sampling distributions in analyzing data to make inferences about a population proportion and about a difference between two population proportions. You will also understand the importance of checking conditions before applying inference procedures and, as always, make conclusions in context. In addition, you will be able to interpret confidence intervals and confidence levels. You will deepen your understanding of statistical significance. Finally, you will learn to interpret Type I and Type II errors and their possible consequences.

> **NOTE**
>
> The phrase "margin of error" does not mean that a mistake has been made. It derives from the variability resulting from taking a random sample from a population. It does not account for any data collection mistake.

The Meaning of a Confidence Interval

By using what we know about sampling distributions, we are able to establish a certain confidence that a sample proportion or mean lies within a specified interval around the population proportion or mean. However, we then have the same confidence that the population proportion or mean lies within a specified interval around the sample proportion or mean (e.g., the distance from Missoula to Whitefish is the same as the distance from Whitefish to Missoula).

Typically, we consider 90%, 95%, and 99% confidence interval estimates, but any percentage is possible. The percentage is the percentage of samples that would pinpoint the unknown p or μ within plus or minus respective margins of error. We do *not* say there is a 0.90, 0.95, or 0.99 probability that p or μ is within a certain margin of error of a given sample proportion or mean. For a given sample proportion or mean, p or μ either is or isn't within the specified interval, and so the probability is either 1 or 0.

As will be seen, there are two aspects to this concept. First, there is the *confidence interval,* which gives an interval of plausible values, and is expressed in the form:

$$\text{estimate} \pm \text{margin of error}$$

Second, there is the *success rate for the method,* called the *confidence level,* that is, the proportion of times repeated applications of this method would capture the true population parameter.

Don't confuse margin of error with standard error. The standard error is a measure of how far the sample statistic typically varies from the population parameter. The margin of error is a multiple of the standard error, with that multiple determined by how confident we wish to be of our procedure.

All of the above assume that certain conditions are met. For inference on population proportions, means, and slopes, we must check for independence in data collection methods and for selection of the appropriate sampling distribution.

> **NOTE**
> The margin of error will be determined by two factors: how much the statistic typically varies from the parameter, that is, $\sigma_{\hat{p}}$ or $\sigma_{\bar{x}}$, and how confident we want to be in our answer, determined by a critical z^* or t^*.

Conditions for Inference

The following are the two standard assumptions for our inference procedures and the "ideal" way they are met. Of course, we seldom, if ever, meet ideals!

1. **INDEPENDENCE ASSUMPTION:** Individuals in a sample or an experiment must be independent of each other, and this is obtained through random sampling or a randomized experiment. Ideally, for samples, they should be large, simple, random samples. For experiments, the subjects should be randomly assigned to treatments. Independence across samples is obtained by selecting two (or more) separate random samples. Always examine how the data were collected to check if the assumption of independence is reasonable. Sample size can also affect independence. Because sampling is usually done without replacement, if the sample is too large, lack of independence becomes a concern. So, we typically require that the sample size n be no larger than 10% of the population (the 10% Rule).

2. **NORMALITY ASSUMPTION:** Inference for proportions is based on a normal model for the sampling distribution of $\hat{p}$, but actually we have a binomial distribution. Fortunately, the binomial is approximately normal if both np and $nq \geq 10$. Inference for means is based on a normal model for the sampling distribution of $\bar{x}$; this is true if the population is normal and is approximately true (thanks to the central limit theorem (CLT)) if the sample size is large enough (typically we accept $n \geq 30$). With regard to means, this is referred to as the *Normal/Large Sample* condition.

> **NOTE**
> Of course, to obtain better estimates, it's always better to have larger samples—it's just that if the sample size is large relative to the population, the proper inference techniques are different from those taught in the AP Statistics curriculum.

> **Example 6.1**

If we pick a simple random sample of size 80 from a large population, which of the following values of the population proportion p would allow use of the normal model for the sampling distribution of $\hat{p}$?

(A)   0.10
(B)   0.15
(C)   0.90
(D)   0.95
(E)   0.99

✓ **Solution**

(B) The relevant condition is that both np and $nq \geq 10$. In (A), $np = (80)(0.10) = 8$; in (C), $nq = (80)(0.10) = 8$; in (D), $nq = (80)(0.05) = 4$; and in (E), $nq = (80)(0.01) = 0.8$. However, in (B), $np = (80)(0.15) = 12$ and $nq = (0.85)(80) = 68$ are both ≥ 10.

> **Example 6.2**

It is hypothesized that the ratio $\dfrac{\text{outer edges of lips}}{\text{upper ridges of lips}}$ is most pleasing when equal to the golden ratio, $\varphi \approx 1.618$.

A student would like to calculate the mean ratio for students in her high school and plans to randomly sample 50 out of the 780 students at the school. Does this satisfy the 10% Rule, and what is the purpose of the rule?

✓ **Solution**

10% of 780 = 78, and 50 < 78, so yes, the 10% Rule is satisfied. In picking the sample of 50 students, the sampling will obviously be done without replacement, because the same student will not be used twice. If the sample were too large compared with the size of the population, lack of independence would have become a concern.

> **Example 6.3**

A human resources director is interested in the proportion of employees in an 85-member department who believe they are saving enough for retirement. She randomly selects 12 employees to interview. Explain whether conditions for inference are satisfied.

✓ **Solution**

Even though it may or may not have been the ideal method of sampling, it is stated that a random sample was taken, satisfying the randomization condition. The 10% condition (related to independence) is violated because 10% of 85 is 8.5, and 12 is greater than 8.5. The normality assumption is also violated because np and nq total 12, so both cannot be ≥ 10.

> **Example 6.4**

An English teacher is interested in the mean length of words in a Hemingway novel. She picks a random sample of 25 words. Explain whether conditions for inference are satisfied.

(continued)

✓ **Solution**

Even though it may or may not have been the ideal method of sampling, it is stated that a *random* sample was taken, satisfying the randomization condition. The 10% condition (related to independence) is also satisfied because clearly 25 is less than 10% of the number of words in an entire novel. However, the Normal/Large Sample condition is not satisfied because it is not given that the distribution of the lengths of all words in the novel is approximately normal and the sample size is not large; that is, $n = 25 < 30$.

Confidence Interval for a Proportion

We are interested in estimating a population proportion p by considering a single sample proportion $\hat{p}$. This sample proportion is just one of a whole universe of sample proportions, and from Unit 5 we remember the following:

1. The set of all sample proportions is approximately normally distributed.
2. The mean $\mu_{\hat{p}}$ of the set of sample proportions equals p, the population proportion.
3. The standard deviation $\sigma_{\hat{p}}$ of the set of sample proportions is approximately equal to $\sqrt{\dfrac{p(1-p)}{n}}$.

Remember that we are really using a normal approximation to the binomial, so $n\hat{p}$ and $n(1 - \hat{p})$ should both be at least 10. Furthermore, in making calculations and drawing conclusions from a specific sample, it is important that the sample be randomly drawn. Ideally, it should be a *simple random sample*. Finally, the population should be large, typically checked by the assumption that the sample is less than 10% of the population. (If the population is small and the sample exceeds 10% of the population, models other than the normal are more appropriate.)

In finding confidence interval estimates of the population proportion p, how do we find $\sigma_{\hat{p}} = \sqrt{\dfrac{p(1-p)}{n}}$ since p is unknown? The reasonable procedure is to use the sample proportion $\hat{p}$:

$$\sigma_{\hat{p}} \approx \sqrt{\frac{\hat{p}(1-\hat{p})}{n}}$$

When the standard deviation is estimated in this way (using the sample), we use the term *standard error*:

$$SE(\hat{p}) = SE_{\hat{p}} = \sqrt{\frac{\hat{p}(1-\hat{p})}{n}}$$

Later in this unit we will directly assess the evidence for a claim about a proportion using a significance test. However, a confidence interval can provide us with some evidence about a claim. For example, if the claim is that $p > 0.75$ and all values in the confidence interval are >0.75, then there is convincing evidence in support of the claim. However, if a single value in the confidence interval of plausible values is not >0.75, then there is not convincing evidence of the claim.

> **Example 6.5**

(a) If 42% of a simple random sample of 550 young adults say that whoever asks for the date should pay for the first date, determine a 99% confidence interval estimate for the true proportion of all young adults who would say that whoever asks for the date should pay for the first date.

(b) Does this confidence interval give convincing evidence in support of the claim that fewer than 50% of young adults would say that whoever asks for the date should pay for the first date?

(c) Give an interpretation of the confidence level.

(continued)

✓ Solutions

(a) The parameter is p, which represents the proportion of the population of young adults who would say that whoever asks for the date should pay for the first date. We check both *independence* and *normality*. For independence, we are given that the sample is an SRS, and 550 is clearly less than 10% of all young adults. For normality, $n\hat{p} = 550(0.42) = 231 \geq 10$ and $n(1 - \hat{p}) = 550(0.58) = 319 \geq 10$. Since $\hat{p} = 0.42$, the standard error of the set of sample proportions is

$$SE(\hat{p}) = \sqrt{\frac{(0.42)(0.58)}{550}} = 0.021$$

From Unit 5, we know that 99% of the sample proportions should be within 2.576 standard deviations of the population proportion. Equivalently, we are 99% certain that the population proportion is within 2.576 standard deviations of any sample proportion. Thus, the 99% confidence interval estimate for the population proportion is $0.42 \pm 2.576(0.021) = 0.42 \pm 0.054$. We say that the *margin of error* is 0.054. We are 99% confident that the true proportion of young adults who would say that whoever asks for the date should pay for the first date is between 0.366 and 0.474. [Alternatively, we can say: "We are 99% confident that the interval from 0.366 to 0.474 captures the true proportion of young adults who would say that whoever asks for the date should pay for the first date."]

> **Important**
>
> Verifying assumptions and conditions means more than simply listing them with check marks. You must show work or give some reason to confirm verification.

> **Important**
>
> After finding a confidence interval, you must always interpret the interval in *context*, including mentioning the *confidence level* and *parameter* and referring to the *population*, not the sample.

> **TIP**
>
> The words "true" and "all" in the conclusion satisfy the requirement of referring to the population.

(b) Yes, the confidence interval (0.366 to 0.474) gives an interval of plausible values. Because all these values are less than 0.50, this confidence interval gives convincing evidence in support of the claim that fewer than 50% of young adults would say that whoever asks for the date should pay for the first date.

(c) If all possible samples of 550 young adults were interviewed, about 99% of the resulting confidence intervals would capture the true proportion of young adults who would say that whoever asks for the date should pay for the first date.

❯ Example 6.6

(a) In a simple random sample of couples kissing in public places, such as airports, beaches, and parks, 80 out of 124 leaned their heads to the right. Establish a 95% confidence interval for the true proportion of public kissing couples who lean their heads to the right.

(b) Suppose a sociologist claimed that over 60% of all public kissing couples lean their heads to the right. Does this confidence interval give convincing evidence in support of the sociologist's claim?

(c) Give an interpretation of the confidence level.

✓ Solutions

(a) *Parameter:* Let p represent the proportion of the population of public kissing couples who lean their heads to the right.

Procedure: One sample z-interval for a population proportion.

Checks: We check *independence* and *normality*. For independence, we are given that the sample is an SRS, and we assume that 124 is less than 10% of all public kissing couples. For normality, $n\hat{p} = 80 \geq 10$ and $n(1 - \hat{p}) = 124 - 80 = 44 \geq 10$.

> **NOTE**
>
> $n\hat{p} = (124)\left(\frac{80}{124}\right) = 80$ is the number of "successes."

(continued)

Mechanics: Calculator software (such as 1-PropZInt) gives (0.56095, 0.72938).

[For instructional purposes, we note that $\hat{p} = \frac{80}{124} = 0.64516$,

$SE(\hat{p}) = \sqrt{\frac{(0.64516)(0.35484)}{124}} = 0.04297$, and $0.64516 \pm 1.96(0.04297) =$

0.64516 ± 0.08422.]

Conclusion in context: We are 95% confident that the true proportion of all public kissing couples who lean their heads to the right is between 0.561 and 0.729. Alternatively, we can say the following: "We are 95% confident that the interval from 0.561 to 0.729 captures the true proportion of all public kissing couples who lean their heads to the right."

(b)   No, the confidence interval (0.561 to 0.729) gives an interval of plausible values. Because all the values in the interval are not greater than 0.60, this confidence interval does not give convincing evidence in support of the sociologist's claim that over 60% of all public kissing couples lean their heads to the right.

(c)   If all possible samples of 124 couples kissing in public places were observed, about 95% of the resulting confidence intervals would capture the true proportion of couples who lean their heads to the right.

> **TIP**
>
> On the TI-84, for a confidence interval of a proportion, use `1-PropZInt`, *not* `Zinterval`; and *x* must be an integer.

> **NOTE**
>
> The fact that $\hat{p} = 0.64516 > 0.60$ is not enough evidence to support the sociologist's claim.

〉 Example 6.7

In a telephone survey of a random sample of 1,000 high school seniors, 832 answered yes to the question: "Do you want to get married someday?"

(a)   What is the parameter of interest?

(b)   Are the conditions for inference satisfied?

(c)   With what confidence can it be asserted that 83.2% ± 3% of the high school population want to get married someday?

✓ Solutions

(a)   The parameter is *p*, which represents the proportion of the population of high school seniors who would answer yes to the question: "Do you want to get married someday?"

(b)   We check *independence* and *normality*. For independence, we are given that there was a random sample, and clearly 1,000 is less than 10% of all high school seniors. For normality, $n\hat{p} = 832 \geq 10$ and $n(1 - \hat{p}) = 1{,}000 - 832 = 168 \geq 10$.

(c) $\hat{p} = \frac{832}{1{,}000} = 0.832$ and so $SE(\hat{p}) = \sqrt{\frac{(0.832)(0.168)}{1{,}000}} = 0.0118$

The relevant *z*-scores are $\pm \frac{0.03}{0.0118} = \pm 2.54$. `Normalcdf(-2.54,2.54) = 0.9889`.

In other words, 83.2% ± 3% is a 98.89% confidence interval estimate for the proportion of all high school seniors who would answer yes to the question: "Do you want to get married someday?"

As we have seen, there are two types of statements that come out of confidence intervals. First, we can interpret the **confidence interval** and say we are 90% confident that, for example, between 60% and 66% of all voters favor a bond issue. Second, we can interpret the **confidence level** and say that if this survey were conducted many times, about 90% of the resulting confidence intervals would contain the true proportion of voters who favor the bond issue.

If (0.60, 0.66) is a 90% confidence interval for the proportion of all voters who favor a bond issue, following are three common misconceptions:

MISCONCEPTION #1: The true proportion of all voters who support the bond issue is between 0.60 and 0.66.
    FACT: The true proportion may or may not be in the specified interval; we are only 90% confident that it is in the interval.

MISCONCEPTION #2: There is a 0.90 probability that the true proportion of all voters who favor the bond issue is between 0.60 and 0.66.
    FACT: For a given confidence interval, the true proportion either is or is not within the specified interval, and so the probability is either 1 or 0.

MISCONCEPTION #3: If this survey were conducted many times, about 90% of the sample proportions would be in the interval (0.60, 0.66).
    FACT: (0.60, 0.66) is one specific interval derived from one sample proportion and has nothing to do with other samples and other sample proportions.

> **NOTE**
>
> **Finding confidence intervals *using formulas* is not necessary on free-response questions, but answer choices using formulas may well appear on multiple-choice questions.**

One important consideration in setting up a survey is the choice of sample size. To obtain a smaller, more precise interval estimate of the population proportion, we must either decrease the degree of confidence or increase the sample size. Similarly, if we want to increase the degree of confidence, we can either accept a wider interval estimate or increase the sample size. Again, while choosing a larger sample size may seem desirable, in the real world this decision involves time and cost considerations.

> **NOTE**
>
> **Interpreting confidence level never involves a specific interval.**

In setting up a survey to obtain a confidence interval estimate of the population proportion, what should we use for $\sigma_{\hat{p}}$? Algebraically it can be shown that $\sqrt{p(1-p)}$ is greatest when $p = 0.5$, and in this case $\sqrt{p(1-p)} = \sqrt{0.5(1-0.5)} = 0.5$. Thus, $\sigma_{\hat{p}} = \sqrt{\frac{p(1-p)}{n}}$ is at most $\frac{0.5}{\sqrt{n}}$. We make use of this fact to determine sample sizes in questions such as those in Examples 6.8 and 6.9.

❯ Example 6.8

A CDC investigator wants to know the proportion of all adults who are fully vaccinated and boostered against Covid-19. If the answer must be within ±0.03 at the 96% confidence level, how many adults should be in the selected sample?

> For 96% confidence, there is 0.02 in each tail and `invNorm(0.02) = -2.05` or `invNorm(0.98)=2.05`.

✓ Solution

The parameter is p, which represents the proportion of all adults who are fully vaccinated and boostered against Covid-19. We want $2.05\sigma_{\hat{p}} \leq 0.03$. From the above remark, $\sigma_{\hat{p}}$ is at most $\frac{0.5}{\sqrt{n}}$, and so it is sufficient to consider $2.05\left(\frac{0.5}{\sqrt{n}}\right) \leq 0.03$. Algebraically, we have $\sqrt{n} \geq \frac{2.05(0.5)}{0.03} = 34.17$ and $n \geq 1{,}167.4$. Therefore, choosing a sample of 1,168 adults gives the true vaccinated and boostered proportion to within ±0.03 at the 96% level.

> **TIP**
>
> **Noting the inequality, you must round a result "up" if it is a decimal.**

Note that the accuracy of the estimate does *not* depend on what fraction of the whole population we have sampled. What is critical is the *absolute size* of the sample. Is some minimal value of n necessary for the procedures we are using to be meaningful? Since we are using the normal approximation to the binomial, both np and $n(1-p)$ should be at least 10.

> **Example 6.9**

A study is undertaken to determine the proportion of industry executives who believe that workers' pay should be based on individual performance. How many executives should be interviewed if an estimate is desired at the 99% confidence level to within ± 0.06? To within ± 0.03? To within ± 0.02?

✓ **Solution**

The parameter is p, which represents the proportion of the population of industry executives who believe that workers' pay should be based on individual performance. Algebraically, $2.576\left(\frac{0.5}{\sqrt{n}}\right) \leq 0.06$ gives $\sqrt{n} \geq \frac{2.576(0.5)}{0.06} = 21.5$, so $n \geq 462.25$. Similarly, $2.576\left(\frac{0.5}{\sqrt{n}}\right) \leq 0.03$ gives $\sqrt{n} \geq \frac{2.576(0.5)}{0.03} = 42.9$, so $n \geq 1{,}840.4$. Finally, $2.576\left(\frac{0.5}{\sqrt{n}}\right) \leq 0.02$ gives $\sqrt{n} \geq \frac{2.576(0.5)}{0.02} = 64.4$, so $n \geq 4{,}147.4$. Thus, at the 99% confidence level, at least 463 executives should be interviewed for a margin of error of ± 0.06, at least 1,841 should be interviewed for a margin of error of ± 0.03, and at least 4,148 should be interviewed for a margin of error of ± 0.02.

Note that to cut the interval estimate in half (from ± 0.06 to ± 0.03), we would have to increase the sample size fourfold. To cut the interval estimate to a third (from ± 0.06 to ± 0.02), a ninefold increase in the sample size would be required (answers are not exact because of round-off error).

More generally, to divide the interval estimate by d without affecting the confidence level, we must increase the sample size by a multiple of d^2.

Quiz 14

Multiple-Choice Questions

> **DIRECTIONS:** The questions that follow are each followed by five suggested answers. Choose the response that best answers the question.

1. For a class project, an AP Statistics student wishes to determine a 95% confidence interval for the proportion of students in the school building who have social media profile pages. The total number of students in the building is 3,520. Which of the following combinations of n and $\hat{p}$ will satisfy the conditions for inference?

   (A) $n = 15$ and $\hat{p} = 0.67$
   (B) $n = 25$ and $\hat{p} = 0.35$
   (C) $n = 50$ and $\hat{p} = 0.17$
   (D) $n = 75$ and $\hat{p} = 0.15$
   (E) $n = 100$ and $\hat{p} = 0.91$

2. In a simple random sample of 60 teenagers, two-thirds said they would rather text a friend than call. What is the 98% confidence interval for the proportion of all teenagers who would rather text than call a friend?

   (A) $\frac{2}{3} \pm 1.96\sqrt{\dfrac{(2/3)(1/3)}{60}}$

   (B) $\frac{2}{3} \pm 2.326\sqrt{\dfrac{(2/3)(1/3)}{60}}$

   (C) $\frac{2}{3} \pm 2.405\sqrt{\dfrac{(2/3)(1/3)}{60}}$

   (D) $\frac{2}{3} \pm 2.96\sqrt{\dfrac{(2/3)(1/3)}{60}}$

   (E) $\frac{2}{3} \pm 3.326\sqrt{\dfrac{(2/3)(1/3)}{60}}$

3. A school district has 250 teachers. A member of the Board of Education wanted to estimate the percentage of teachers who would be willing to accept a lower pay raise in return for a substantial increase in health benefits. Thirty randomly selected teachers were surveyed. Of the 30 selected teachers, 14 expressed a willingness to accept a lower pay raise in return for a substantial increase in health benefits. Have the conditions been met for inference with a one-sample z-interval?

   (A) Yes, the conditions for inference have been met.
   (B) No, because the sample likely has undercoverage bias.
   (C) No, because the sample size is not large enough to ensure that the sampling distribution is approximately normal.
   (D) No, because the sample size violates the independence condition.
   (E) No, because both $n\hat{p}$ and $n(1 - \hat{p})$ have to be ≥ 10.

4. In a random sample of 150 adults, 60 admitted to feeling lonely at least once during the previous week. A 95 percent confidence interval for the proportion of all adults who would admit to feeling lonely at least once during the previous week is given by (0.3216, 0.4784). Which of the following is a correct interpretation of the confidence level of 95 percent?

   (A) There is a 0.95 probability that the true proportion of all adults who would admit to feeling lonely at least once during the previous week is between 0.3216 and 0.4784.

   (B) We are 95 percent confident that the true proportion of all adults who would admit to feeling lonely at least once during the previous week is between 0.3216 and 0.4784.

   (C) If the proportion of adults who would admit to feeling lonely at least once during the previous week is calculated for repeated samples of 150 adults, 95 percent of these sample proportions will fall between 0.3216 and 0.4784.

   (D) In repeated samplings of 150 adults, approximately 95 percent of the calculated confidence intervals for the proportion of adults who would admit to feeling lonely at least once during the previous week will capture the associated sample proportion.

   (E) If a confidence interval for the proportion of adults who would admit to feeling lonely at least once during the previous week is calculated for repeated samples of 150 adults, 95 percent of these intervals will contain the true population proportion.

5. A 2020 random survey of 1,470 adult Americans found that 29% of the American adult population would be willing to pay higher taxes if the government offered "Medicare for All." With what degree of confidence can it be concluded that 29% ± 3% of the American adult population would be willing to pay higher taxes if the government offered "Medicare for All"?

   (A) 49.5%
   (B) 90%
   (C) 95%
   (D) 98%
   (E) 99%

6. The population of the United States is about 330,000,000. During a time that it is strongly suggested to regularly wear face masks, suppose that a public health official wants to estimate, with 95 percent confidence, the proportion of Americans who do so within a margin of error of 2.5 percentage points. Approximately how many people would she need to randomly sample?

   (A) 150
   (B) 1,500
   (C) 15,000
   (D) 150,000
   (E) 1,500,000

7. From a random sample of dog owners, 43% indicated that they allow their dogs to sleep in bed with them. A 95 percent confidence interval was constructed from the sample, and the margin of error for the estimate was 3%. Which of the following is the best interpretation of the interval?

   (A) There is a 0.95 probability that between 40% and 46% of the sample dog owners allow their dogs to sleep in bed with them.

   (B) There is a 0.95 probability that between 40% and 46% of the population of dog owners allow their dogs to sleep in bed with them.

   (C) We are 95 percent confident that between 40% and 46% of the sample dog owners allow their dogs to sleep in bed with them.

   (D) We are 95 percent confident that between 40% and 46% of the population of dog owners allow their dogs to sleep in bed with them.

   (E) If 95 percent confidence intervals are calculated from all possible samples of the given size, 95 percent of these intervals will capture the true percentage of the population of dog owners who allow their dogs to sleep in bed with them.

Free-Response Questions

DIRECTIONS: You must show all work and indicate the methods you use. You will be graded on the correctness of your methods and on the accuracy of your final answers.

1. Net neutrality is the idea that all Internet traffic should be treated equally—that is, no Internet service provider (ISP) should have the power to favor one source over another by blocking or paid prioritization. An online survey, responded to by 155,612 Internet users, reported that 93% favor net neutrality with a margin of error of 0.1%.

   (a) Is "93%" a statistic or a parameter? Explain.
   (b) Without doing an actual calculation, explain why the margin of error is so small.
   (c) How confident should you be that between 92.9% and 93.1% of all Internet users favor net neutrality?

2. In a random sample of 500 American women between the ages of 45 and 64, 305 said that breast cancer was the medical condition they feared most.

   (a) Calculate a 99% confidence interval for the proportion of all American women between the ages of 45 and 64 who would say that breast cancer was the medical condition they fear most.
   (b) Explain the meaning of "99% confidence level" in this example.
   (c) Currently, about 3% of female deaths are due to breast cancer. Is the value 3% consistent with the confidence interval obtained in (a)? Explain.

The answers to this quiz can be found beginning on page 496.

Quiz 15

Multiple-Choice Questions

> **DIRECTIONS:** The questions that follow are each followed by five suggested answers. Choose the response that best answers the question.

1. An environmental activist, interested in advocating for the implementation of sustainable energy utilization, interviews a random sample of 30 white-collar workers. Only eight of them said they would be willing to accept cuts in their standard of living to protect the environment. Letting p be the proportion of all white-collar workers who would be willing to accept cuts in their standard of living to protect the environment, is it appropriate to assume that the corresponding sampling distribution of $\hat{p}$ is approximately normal?

   (A) Yes, because the sample was chosen randomly.
   (B) Yes, because sampling distributions of the sample proportion are modeled with normal distributions.
   (C) Yes, because the sample size is ≥ 30.
   (D) No, because the population size is unknown.
   (E) No, because the sample size is not large enough for the normality condition.

2. In a random sample of 500 adults, 480 said that they found it unacceptable that the federal government would hack into their emails. A 90% confidence interval for the proportion of all adults who would find it unacceptable that the federal government would hack into their emails is (0.946, 0.974). Which of the following is a correct interpretation of the confidence level 90 percent?

   (A) We are 90% confident that the true proportion of all adults who would find it unacceptable that the federal government would hack into their emails is between 0.946 and 0.974.
   (B) There is a 0.90 probability that the true proportion of all adults who would find it unacceptable that the federal government would hack into their emails is between 0.946 and 0.974.

   (C) If the proportion of adults who would find it unacceptable that the federal government would hack into their emails is calculated for repeated samples of 500 adults, 90% of these sample proportions will fall between 0.946 and 0.974.
   (D) If a confidence interval for the proportion of adults who would find it unacceptable that the federal government would hack into their emails is calculated for repeated samples of 500 adults, 90% of these intervals will contain the true population proportion.
   (E) In repeated samplings of 500 adults, approximately 90% of the calculated confidence intervals for the proportion of adults who would find it unacceptable that the federal government would hack into their emails will capture the associated sample proportion.

3. Consider a sample with sample proportion $\hat{p}$. Which of the following best describes the purpose of a one-sample z-interval?

   (A) To estimate $\hat{p}$
   (B) To establish whether or not the sampling distribution of $\hat{p}$ is approximately normal
   (C) To estimate a margin of error for $\hat{p}$
   (D) To estimate the probability of observing a value as extreme or more extreme than $\hat{p}$
   (E) To ascertain whether or not there is convincing evidence that $\hat{p}$ is the true population proportion

4. One month, the actual unemployment rate in Spain was 13.4%. If during that month you took a simple random sample of 100 Spaniards of working age and constructed a confidence interval estimate of the unemployment rate, which of the following would be true?

   (A) The center of the interval is 13.4.
   (B) The interval contains 13.4.
   (C) A 99% confidence interval estimate contains 13.4.
   (D) The z-score of 13.4 is between ± 2.576.
   (E) None of the above are true statements.

5. In a confidence interval using z-scores, the margin of error covers which of the following?

   (A) Sampling error (sampling variability)
   (B) Errors due to undercoverage and non-response in obtaining sample surveys
   (C) Errors due to using sample standard deviations as estimates for population standard deviations
   (D) Type I errors
   (E) Type II errors

6. In a random sample of voters, 87% indicated that they do not have confidence in what the White House physician says about the president's health. A 95% confidence interval was constructed from the sample, and the margin of error for the estimate was 2%. Which of the following is the best interpretation of the interval?

   (A) There is a 0.95 probability that between 85% and 89% of the sample voters do not have confidence in what the White House physician says about the president's health.
   (B) There is a 0.95 probability that between 85% and 89% of the population of voters do not have confidence in what the White House physician says about the president's health.
   (C) We are 95% confident that between 85% and 89% of the sample voters do not have confidence in what the White House physician says about the president's health.
   (D) We are 95% confident that between 85% and 89% of the population of voters do not have confidence in what the White House physician says about the president's health.
   (E) If 95% confidence intervals are calculated from all possible samples of the given size, 95% of these intervals will capture the true percentage of the population of voters who do not have confidence in what the White House physician says about the president's health.

7. A politician wants to know what percentage of all voters support her position on the issue of forced busing for the racial integration of public schools. What size voter sample should be obtained to determine with 90% confidence the support level to within 4%?

   (A) 21
   (B) 25
   (C) 423
   (D) 600
   (E) 1,691

Free-Response Questions

> **DIRECTIONS:** You must show all work and indicate the methods you use. You will be graded on the correctness of your methods and on the accuracy of your final answers.

1. During the H1N1 pandemic, one published study concluded that if someone in your family had H1N1, you had a 1 in 8 chance of also contracting the disease. A state health officer tracked a random sample of new H1N1 cases in her state and noted that 129 out of 876 family members later contracted the disease.

    (a) Calculate a 90% confidence interval for the proportion of family members who contract H1N1 after an initial family member does in this state.
    (b) Based on this confidence interval, is there evidence that the proportion of family members who contract H1N1 after an initial family member does in this state is different from the 1 in 8 chance concluded in the published study? Explain.
    (c) Would the conclusion in (b) be any different with a 99% confidence interval? Explain.

2. In a July 2020 study of 1,050 randomly selected smokers, 74% said they would like to give up smoking.

    (a) At the 95% confidence level, what is the margin of error?
    (b) Calculate and explain the meaning of the "95% confidence interval" in this example.
    (c) Explain the meaning of "95% confidence level" in this example.
    (d) Give an example of possible response bias in this example.
    (e) If we want instead to be 99% confident, would our confidence interval need to be wider or narrower?
    (f) If the sample size were greater, would the margin of error be smaller or greater?

The answers to this quiz can be found beginning on page 497.

Logic of Significance Testing

Closely related to the problem of estimating a population proportion or mean is the problem of testing a hypothesis about a population proportion or mean. For example, a travel agency might determine an interval estimate for the proportion of sunny days in the Virgin Islands or, alternatively, might test a tourist bureau's claim about the proportion of sunny days. A consumer protection agency might determine an interval estimate for the mean nicotine content of a particular brand of cigarettes or, alternatively, might test a manufacturer's claim about the mean nicotine content of its cigarettes. In each of these cases, the experimenter must decide whether the interest lies in estimating a population proportion or mean or in testing a claimed proportion or mean.

The general testing procedure is to choose a specific hypothesis to be tested, called the *null hypothesis*, pick an appropriate random sample, and then use measurements from the sample to determine the likelihood of the null hypothesis. If the sample statistic is far enough away from the claimed population parameter, we say that there is sufficient evidence to reject the null hypothesis. We attempt to show that the null hypothesis is unacceptable by showing that it is improbable.

Consider the context of the population proportion. The null hypothesis H_0 is stated in the form of an equality statement about the *population* proportion (for example, $H_0: p = 0.37$). There is an *alternative hypothesis*, stated in the form of a strict inequality (for example, $H_a: p < 0.37$ or $H_a: p > 0.37$ or $H_a: p \neq 0.37$). The strength of the sample statistic $\hat{p}$ can be gauged through its associated *P-value*, which is the probability of obtaining a sample statistic as extreme (or more extreme) as the one obtained if the null hypothesis is assumed to be true. The smaller the *P*-value, the more significant the difference between the null hypothesis and the sample results.

Several points worth noting:

- The equality reference in the null hypothesis can be stated using $\geq$ or $\leq$ when the alternative is one-sided in the other direction.
- Never base your hypotheses on what you find in the sample data.
- Note that only population parameter symbols like μ, p, and β appear in H_0 and H_a. Sample statistics like $\bar{x}$, $\hat{p}$, and b do NOT appear in H_0 or H_a.
- For a one-sided test (the alternative is $<$ or $>$), the *P*-value is the proportion of values at or more extreme than the test statistic (in the direction of the alternative). For a two-sided test (the alternative is $\neq$), the *P*-value is the proportion of values at or more extreme than the test statistic in either direction; that is, it is twice the calculated tail probability.
- The *P*-value is NOT the probability of the null hypothesis being true.

There are two types of possible errors: the error of mistakenly rejecting a true null hypothesis and the error of mistakenly failing to reject a false null hypothesis. The α-risk, also called the *significance level* of the test, is the probability of committing a *Type I error* and mistakenly rejecting a true null hypothesis. A *Type II error*, a mistaken failure to reject a false null hypothesis, has associated probability β. There is a different value of β for each possible correct value for the population parameter p. For each β, $1 - \beta$ is called the "power" of the test against the associated correct value. The *power* of a hypothesis test is the probability that a Type II error is not committed. That is, given a true alternative, the power is the probability of rejecting the false null hypothesis. Increasing the sample size and

> **TIP**
>
> **Understand that the sample statistic almost never equals the claimed population parameter. Either the sample statistic differs by random chance or the claimed parameter is wrong. If you can eliminate random chance as a plausible explanation for the sample statistic, then you can say that there is convincing evidence that the claimed parameter is wrong.**

> **NOTE**
>
> **The *P*-value is a *conditional probability*; it assumes the null hypothesis is true. Assuming that H_0 is true, the *P*-value measures how likely it is to get evidence for H_a as strong as or stronger than the observed evidence by chance alone.**

> **The significance level, α, is fixed at the beginning of the hypothesis test. The conclusion to the test will be based upon whether or not the calculated *P*-value is less than or greater than α. If the *P*-value $\leq \alpha$, reject H_0; there is convincing evidence for H_a. If the *P*-value $> \alpha$, fail to reject H_0; there is not convincing evidence for H_a.**

increasing the significance level are both ways of increasing the power. Also note that a true null that is further away from the hypothesized null is more likely to be detected, thus offering a more powerful test.

Population truth

		H_0 true	H_0 false
Decision based on sample	Reject H_0	Type I error	Correct decision
	Fail to reject H_0	Correct decision	Type II error

A simple illustration of the difference between a Type I and a Type II error is as follows. Suppose the null hypothesis is that all systems are operating satisfactorily with regard to a NASA launch. A Type I error would be to delay the launch mistakenly thinking that something was malfunctioning when everything was actually OK. A Type II error would be to fail to delay the launch mistakenly thinking everything was OK when something was actually malfunctioning. The power is the probability of recognizing a particular malfunction. (Note the complementary aspect of power, a "good" thing, with Type II error, a "bad" thing.)

TIP
Be able to identify Type I and Type II errors in context and give possible consequences of each in context.

The U.S. justice system provides another often-quoted illustration. If the null hypothesis is that a person is innocent, a Type I error results when an innocent person is found guilty while a Type II error results when a guilty person is not convicted. We try to minimize Type I errors in criminal trials by demanding unanimous jury guilty verdicts. In civil suits, however, many states try to minimize Type II errors by accepting simple majority verdicts.

It should be emphasized that with regard to calculations, questions like "What is the *power* of this test?" and "What is the probability of a *Type II error* in this test?" cannot be answered without reference to a specific alternative hypothesis. Furthermore, AP students are not required to know how to calculate these probabilities. However, you are required to understand the concepts and interactions among the concepts.

Significance Test for a Proportion

When conducting a hypothesis test for a proportion, we must check *independence* and *normality*. For independence, we have a random sample, ideally a simple random sample, and if sampling without replacement, we have that the sample size is less than 10% of the population. For normality, both np_0 and $n(1 - p_0)$ are at least 10, where p_0 is the claimed proportion.

TIP
For a hypothesis test, we check if np_0 and $n(1 - p_0)$ are $\geq$ 10. For a confidence interval, we check if $n\hat{p}$ and $n(1 - \hat{p})$ are $\geq$ 10.

It is important to understand that because the P-value is a conditional probability, calculated based on the assumption that the null hypothesis, H_0: $p = p_0$, is true, we use the claimed proportion p_0 both in checking the $np_0 \geq 10$ and $n(1 - p_0) \geq 10$ conditions and in calculating the standard deviation $\sigma_{\hat{p}} = \sqrt{\dfrac{(p_0)(1 - p_0)}{n}}$.

> **Example 6.10**

(a)   A union spokesperson claims that 75% of union members will support a strike if their basic demands are not met. A company negotiator believes the true percentage is lower and runs a hypothesis test. What is the conclusion if 87 out of a simple random sample of 125 union members say they will strike? (Give conclusions both at the 0.05 and the 0.10 significance levels.)

(b)   For each of the two possible answers to part (a), what error might have been committed, Type I or Type II, and what would be a possible consequence?

(continued)

✓ Solutions

(a) *Parameter:* Let p represent the proportion of all union members who will support a strike if their basic demands are not met.

Hypotheses: H_0: $p = 0.75$ and H_a: $p < 0.75$.

Procedure: One-sample z-test for a population proportion.

Checks: We check *independence* and *normality*. For independence, it is given that we have an SRS, and we must assume that 125 is less than 10% of the total union membership. For normality, $np_0 = 125(0.75) = 93.75 \geq 10$ and $n\left(1 - p_0\right) = 125(0.25) = 31.25 \geq 10$).

Mechanics: Calculator software (`1-PropZTest`) gives $z = -1.394$ and $P = 0.0816$.

[For instructional purposes, we note that the observed sample proportion is $\hat{p} = \frac{87}{125} = 0.696$, and using the claimed proportion, 0.75, we have

$$\sigma_{\hat{p}} = \sqrt{\frac{(0.75)(0.25)}{125}} = 0.03873 \text{ and } z = \frac{0.696 - 0.75}{0.03873} = -1.394, \text{ with a}$$

resulting P-value of 0.0817.]

Conclusion in context with linkage to the P-value: There are two possible answers depending on choice of significance level:

1. With this large of a P-value, $0.0816 > 0.05$, there is not sufficient evidence to reject H_0; that is, there is not convincing evidence at the 5% significance level that the true percentage of union members who support a strike is less than 75%.

2. With this small of a P-value, $0.0816 < 0.10$, there is sufficient evidence to reject H_0; that is, there is convincing evidence at the 10% significance level that the true percentage of union members who support a strike is less than 75%.

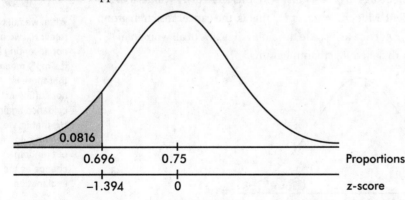

(b) If the P-value is considered large, $0.0816 > 0.05$, so that there is not sufficient evidence to reject the null hypothesis, there is the possibility that a false null hypothesis would mistakenly not be rejected and thus a Type II error would be committed. In this case, the union might call a strike thinking they have greater support than they actually do. If the P-value is considered small, $0.0816 < 0.10$, so that there is sufficient evidence to reject the null hypothesis, there is the possibility that a true null hypothesis would mistakenly be rejected, and thus a Type I error would be committed. In this case, the union might not call for a strike thinking they don't have sufficient support when they actually do have support.

TIP

Hypotheses are always about the population, never about a sample; note that H_0 is $p = 0.75$, not $\hat{p} = 0.75$.

Interpretation of P-value: Assuming that 75% of union members will support a strike, there is a probability of 0.0817 of getting a sample proportion of 0.696 or lower by chance alone in a random sample of 125 union members.

Important

We never "accept" a null hypothesis; we either do or do not have sufficient evidence to reject it.

TIP

Always give the conclusion in context, and, in general, the conclusion should be about the alternative hypothesis, H_a.

NOTE

Absence of evidence does not constitute evidence of absence. Just because there is not convincing evidence to reject H_0 does not mean that there is evidence that H_0 is true or that H_a is false.

> **Example 6.11**

(a)   A cancer research group surveys a random sample of 500 women more than 40 years old to test the hypothesis that 28% of women in this age group have regularly scheduled mammograms. Should the hypothesis be rejected at the 5% significance level if 151 of the women respond affirmatively?

(b)   In the answer to (a) above, what error might have been committed, Type I or Type II, and what would be a possible consequence?

✓ **Solutions**

(a)   *Parameter:* Let p represent the proportion of the population of women over 40 years old who have regularly scheduled mammograms.

Hypotheses: H_0: $p = 0.28$ and H_a: $p \neq 0.28$. (Since no suspicion is voiced that the 28% claim is low or high, we run a two-sided z-test for proportions.)

Procedure: One-sample z-test for a population proportion.

Checks: We check *independence* and *normality*. For independence, we are given a random sample, and clearly 500 < 10% of all women over 40 years old. For normality, $np_0 = 500(0.28) = 140 \geq 10$ and $n(1 - p_0) = 500(0.72) = 360 \geq 10$.

Mechanics: Calculator software gives $z = 1.0956$ and $P = 0.2732$.

[For instructional purposes, we note that the observed $\hat{p} = \dfrac{151}{500} = 0.302$,

$\sigma_{\hat{p}} = \sqrt{\dfrac{(0.28)(0.72)}{500}} = 0.02008$ and $z = \dfrac{0.302 - 0.28}{0.02008} = 1.0956$, which

corresponds to a tail probability of 0.1366. Doubling this value (because the test is two-sided) gives a *P*-value of 0.2732.]

Conclusion in context with linkage to the P-value: With this large of a *P*-value, 0.2732 > 0.05, there is not sufficient evidence to reject H_0; that is, the cancer research group does not have convincing evidence to dispute the claim that 28% of all women in this age group have regularly scheduled mammograms.

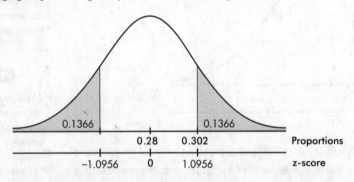

(b)   There was not sufficient evidence to reject the null hypothesis. So a Type II error might be committed, that is, failing to reject a false null hypothesis. If so, the cancer research group would be doing their research assuming that 28% of women over 40 years old have regularly scheduled mammograms whereas the true figure is different.

If $H_0 : p = 0.28$ and $H_a : p \neq 0.28$, where p = the proportion of women over 40 years old who have regularly sched-uled mammograms, with $\hat{p} = 0.302$ and a P-value of 0.2732, and with a significance level of $\alpha = 0.05$, following are three common misconceptions:

> **MISCONCEPTION #1:** H_0 and H_a are alternative *conclusions*.
>
> **FACT:** A significance test is a decision-making process in order to reject or fail to reject a null hypothesis, H_0. You are not deciding that one of H_0 or H_a is true. Your conclusion should be either that there is sufficient evidence to reject H_0 (that is, there is convincing evidence in support of H_a) or that there is not sufficient evidence to reject H_0 (that is, there is not convincing evidence in support of H_a).

> **MISCONCEPTION #2:** The significance level $\alpha = 0.05$ is the probability that the alternative hypothesis is true. That is, the probability is 0.05 that the proportion of women over 40 years old who have regularly scheduled mammograms is different from 0.28.
>
> **FACT:** The significance level is a fixed value we use to decide how small a P-value is required to be in order to reject the null hypothesis.

> **MISCONCEPTION #3:** The P-value of 0.2732 is the probability of obtaining by chance alone a statistic as extreme or more extreme than the observed event of $\hat{p} = 0.302$.
>
> **FACT:** This misses that the P-value is a conditional probability. The P-value is the probability of obtaining a statistic as extreme or more extreme than the one actually observed, by chance alone, when assuming that the null hypothesis is true.

Confidence Interval for the Difference of Two Proportions

From Unit 5, we have the following information about the sampling distribution of $\hat{p}_1 - \hat{p}_2$:

1. The set of all differences of sample proportions is approximately normally distributed.
2. The mean of the set of differences of sample proportions equals $p_1 - p_2$, the difference of the population proportions.
3. The standard deviation $\sigma_d = \sigma_{\hat{p}_1 - \hat{p}_2}$ of the set of differences of sample proportions is approximately equal to:

$$\sqrt{\frac{p_1(1 - p_1)}{n_1} + \frac{p_2(1 - p_2)}{n_2}}$$

Remember that we are using the normal approximation to the binomial, so $n_1\hat{p}_1$, $n_1(1 - \hat{p}_1)$, $n_2\hat{p}_2$, and $n_2(1 - \hat{p}_2)$ should all be at least 10. In making calculations and drawing conclusions from specific samples, it is important both that the samples be random samples, ideally *simple random samples*, and that they be taken *independently* of each other. Finally, if sampling without replacement, the original populations should be large compared with the sample sizes; that is, check that $n_1 \leq 10\% N_1$ and $n_2 \leq 10\% N_2$.

> **NOTE**
>
> That is, check that the number of "successes" in each sample and the number of "failures" in each sample are all at least 10.

> **NOTE**
>
> In randomized experiments, the 10% condition does not apply, but you must check that the two groups have been randomly assigned.

> **Example 6.12**

Is high-speed Internet access critical to educational opportunities? A researcher believes that a greater proportion of adults living in urban areas than rural areas are in favor of spending more government funds to expand high-speed Internet access. To investigate this belief, independent simple random samples of 1,017 adults living in urban settings and 801 adults living in rural areas were selected. Of those surveyed, 612 of the urban residents and 137 of the rural residents said they were in favor of spending more government funds to expand high-speed Internet access.

(a)   Use a 99% confidence interval to estimate the difference in the proportions of urban residents and rural residents who are in favor of spending more government funds to expand high-speed Internet access.

(b)   Based on only this confidence interval, is there convincing evidence that a greater proportion of urban residents than rural residents are in favor of spending more government funds to expand high-speed Internet access?

✓ **Solutions**

(a)   *Parameters:* Let p_U represent the proportion of the population of urban residents who are in favor of spending more government funds to expand high-speed Internet access. Let p_R represent the proportion of the population of rural residents who are in favor of spending more government funds to expand high-speed Internet access.

Procedure: Two-sample z-interval for the difference between population proportions, $p_U - p_R$.

Checks: We check *independence* and *normality*. For independence, we are given independent SRSs. The sample sizes are clearly less than 10% of the population of all urban residents and of all rural residents, respectively. For normality, $n_U\hat{p}_U = 612 \geq 10$, $n_U(1 - \hat{p}_U) = 1,017 - 612 = 405 \geq 10$, $n_R\hat{p}_R = 137 \geq 10$, and $n_R(1 - \hat{p}_R) = 801 - 137 = 664 \geq 10$.

Mechanics: Computer software (`2-PropZInt`) gives (0.378, 0.483).

[For instructional purposes, we have $\hat{p}_U = \frac{612}{1,017} = 0.602$,

$\hat{p}_R = \frac{137}{801} = 0.171$, and $SE(\hat{p}_U - \hat{p}_R) =$

$\sqrt{\frac{(0.602)(0.398)}{1,017} + \frac{(0.171)(0.829)}{801}} = 0.0203.$

The observed difference is $0.602 - 0.171 = 0.431$, and the critical z-scores are ± 2.576. The confidence interval is $0.431 \pm 2.576(0.0203) = 0.431 \pm 0.052$ or (0.379, 0.483).]

> **Interpretation of the confidence level:**
> If all possible random samples of 1,017 urban residents and 801 rural residents are surveyed and if a confidence interval for the difference in proportions of those who are in favor of spending more government funds to expand high-speed Internet access (urban residents minus rural residents) is constructed from each pair of samples, then 99% of these intervals would capture the true difference in proportions between urban residents and rural residents who are in favor of spending more government funds to expand high-speed Internet access.

Conclusion in context: We are 99% confident that the true proportion of urban residents who are in favor of spending more government funds to expand high-speed Internet access is between 0.378 and 0.483 higher than the true proportion of rural residents who are in favor of spending more government funds to expand high-speed Internet access. [Or we are 99% confident that the interval from 0.378 and 0.483 captures the true difference (urban residents minus rural residents) in the proportions of urban residents and rural residents who are in favor of spending more government funds to expand high-speed Internet access.]

─── **NOTE** ───

If the value in question was ANYWHERE in the interval, even at the very edge, then there is not convincing evidence of a significant difference from that value.

(b)   Yes, because the entire confidence interval from 0.378 to 0.483 is positive, there is convincing evidence that a greater proportion of urban residents than rural residents are in favor of spending more government funds to expand high-speed Internet access.

⟩ Example 6.13

A dermatologist asked a random sample of 950 teenagers suffering from moderate cases of acne to choose from among the following three statements:

- **Glycemic diet statement**: A low-glycemic diet is the most important action one can take to control acne outbreaks.
- **Dairy statement**: Avoiding dairy products is the most important action one can take to control acne outbreaks.
- **No right diet**: Diets are of minor importance in controlling acne outbreaks.

The number of teenagers choosing each statement is given in the following table.

	Glycemic Diet	Dairy	No Right Diet
Number of teenagers	285	323	342

(a) Assuming all conditions for inference are met, find a 95% confidence interval for the proportion of teenagers suffering from moderate cases of acne who would choose the glycemic diet statement.

(b) One of the conditions that was met is that the number of teenagers who chose the glycemic diet statement and the number who did not choose the glycemic diet statement are both at least 10. Explain why it is necessary to satisfy this condition.

(c) The dermatologist would like to use a two-sample z-interval to find a confidence interval for the difference in proportions of teenagers who would choose the glycemic diet statement and the proportion who would choose the dairy statement. Is this an appropriate procedure using this data? Explain.

✓ Solutions

(a) *Parameter:* Let p represent the proportion of the population of teenagers suffering from moderate cases of acne who would choose the glycemic diet statement.

Procedure: One-sample z-interval for a population proportion.

Checks: It is given that all conditions for inference are met.

Mechanics: Calculator software gives $(0.271, 0.329)$.

Conclusion in context: We are 95% confident that the true proportion of teenagers suffering from moderate cases of acne who would choose the glycemic diet statement is between 0.271 and 0.329.

(b) The formula for the confidence interval relies on the fact that the binomial distribution can be approximated by a normal distribution so that the sampling distribution of $\hat{p}$ is approximately normal. This works if both $n\hat{p}$ and $n(1 - \hat{p})$ are at least 10.

(c) No, it would not be appropriate, because to use a two-sample z-interval to find a confidence interval for the difference in proportions, there has to be two independent random samples. The data here are from only a single sample.

Significance Test for the Difference of Two Proportions

The null hypothesis for a difference between two proportions is $H_0: p_1 = p_2$, and so the normality condition becomes that $n_1\hat{p}_c$, $n_1(1 - \hat{p}_c)$, $n_2\hat{p}_c$, and $n_2(1 - \hat{p}_c)$ should all be at least 10, where $\hat{p}_c$ is the combined (or pooled) proportion, $\hat{p}_c = \dfrac{x_1 + x_2}{n_1 + n_2}$. The other important conditions to be checked are both that the samples be random samples, ideally *simple random samples*, and that they be taken *independently* of each other. If sampling without replacement, the original populations should also be large compared with the sample sizes; that is, check that $n_1 \leq 10\% N_1$ and $n_2 \leq 10\% N_2$.

NOTE

We use a pooled proportion here because we hypothesize that the proportions are equal.

The fact that a sample proportion from one population is greater than a sample proportion from a second population does not automatically justify a similar conclusion about the population proportions themselves. Remember that sample proportions from the same population can vary from each other.

For many problems, the null hypothesis states that the population proportions are equal or, equivalently, that their difference is 0:

$$H_0: p_1 - p_2 = 0$$

The alternative hypothesis is then:

$$H_a: p_1 - p_2 < 0, \quad H_a: p_1 - p_2 > 0, \quad \text{or} \quad H_a: p_1 - p_2 \neq 0$$

where the first two possibilities lead to one-sided tests and the third possibility leads to a two-sided test.

Since the null hypothesis is that $p_1 = p_2$, we call this common value p_c and use this pooled value in calculating σ_d:

> **TIP**
>
> When using a calculator for this hypothesis test, the pooled p_c is automatically calculated.

$$\sigma_d = \sqrt{\frac{p_c(1 - p_c)}{n_1} + \frac{p_c(1 - p_c)}{n_2}} = \sqrt{p_c(1 - p_c)\left(\frac{1}{n_1} + \frac{1}{n_2}\right)}$$

In practice, if $\hat{p}_1 = \frac{x_1}{n_1}$ and $\hat{p}_2 = \frac{x_2}{n_2}$, we use $\hat{p}_c = \frac{x_1 + x_2}{n_1 + n_2}$ as an estimate of p_c in calculating σ_d.

> Example 6.14

FEMA aid denial was a big problem after Hurricane Harvey hit the Gulf Coast in 2018. In a random sample of 775 District X homeowners, 403 said they were able to get the help they needed. In an independent random sample of 825 District Y homeowners, 264 said they were able to get the help they needed.

(a)   Do the above data give convincing evidence that a greater proportion of District X homeowners than District Y homeowners were able to get the help they needed?

(b)   The 95% confidence interval for the difference in proportions is (0.153, 0.257). Is this result consistent with the above conclusion?

(c)   Based on your conclusion in part (a), which type of error, Type I or Type II, could have been made? What is one potential consequence of this error?

✓ Solutions

(a)   *Parameters:* Let p_X represent the proportion of the population of District X homeowners who were able to get the help they needed. Let p_Y represent the proportion of the population of District Y homeowners who were able to get the help they needed.

    Hypotheses: $H_0: p_X - p_Y = 0$ or $H_0: p_X = p_Y$ and $H_a: p_X - p_Y > 0$ or $H_a: p_X > p_Y$

    Procedure: Two-sample z-test for a difference of two population proportions.

    Checks: We check *independence* and *normality*. For independence, we are given independent random samples and must assume that the sample sizes are less than 10% of the respective populations. For normality and with $\hat{p}_c = \frac{403 + 264}{775 + 825} = 0.417$, we have $n_X \hat{p}_c = 775(0.417) = 323.2$, $n_X(1 - \hat{p}_c) = 775(0.583) = 451.8$, $n_Y \hat{p}_c = 825(0.417) = 344.0$, and $n_Y(1 - \hat{p}_c) = 825(0.583) = 481.0$ are all ≥ 10.

(continued)

Mechanics: Calculator software (`2-PropZTest`) gives $z = 8.11$ and $P = 0.000$. [For instructional purposes, we note that $\hat{p}_X = \frac{403}{775} = 0.52$ and $\hat{p}_Y = \frac{264}{825} = 0.32$. The observed difference is $0.52 - 0.32 = 0.20$, $SE(\hat{p}_X - \hat{p}_Y) = \sqrt{(0.417)(0.583)\left(\frac{1}{775} + \frac{1}{825}\right)} = 0.02467$, $z = \frac{0.20 - 0}{0.02467} = 8.11$, and the tail probability gives a P-value of `normalcdf(8.11, 1000) = 0.000`.]
Conclusion in context with linkage to the P-value: With this small a P-value, $0.000 < 0.05$, there is sufficient evidence to reject H_0. In other words, there is convincing evidence that after Hurricane Harvey hit the Gulf Coast, a greater proportion of all District X homeowners than all District Y homeowners were able to get the help they needed.

(b) We are 95% confident that the true difference in proportions (true proportion of all District X homeowners who received the help they needed minus the true proportion of District Y homeowners who received the help they needed) is between 0.153 and 0.247. Since this interval is entirely positive, it is consistent with the conclusion from the hypothesis test.

(c) Since we rejected H_0, there is the possibility that a true null hypothesis was mistakenly rejected, which would have been a Type I error. A possible consequence would have been claims of district inequities in FEMA aid distribution when none actually existed.

> **Interpretation of P-value:** Assuming that there is no difference in the proportions of District X and District Y homeowners who were able to get the help they needed, there is a probability of 0.000 of getting an observed proportion difference of 0.20 or greater by chance alone in random samples of 775 District X homeowners and 825 District Y homeowners.

Quiz 16

Multiple-Choice Questions

DIRECTIONS: The questions that follow are each followed by five suggested answers. Choose the response that best answers the question.

---

1. Consider the null and alternative hypotheses in a significance test. Are these statements about a parameter or a statistic?

   (A) Null hypothesis: parameter
   Alternative hypothesis: parameter

   (B) Null hypothesis: parameter
   Alternative hypothesis: statistic

   (C) Null hypothesis: statistic
   Alternative hypothesis: parameter

   (D) Null hypothesis: statistic
   Alternative hypothesis: statistic

   (E) This depends on the context.

2. Past studies indicate that 70 percent of teens in a large city pass their driving test on the first attempt. A city administrator believes that because of an economic slowdown resulting in fewer cars on the road, the proportion might be greater than 0.70. A random sample of teens taking their driving test for the first time was surveyed. With $H_0 : p = 0.70$ and $H_a : p > 0.70$, the P-value of the test is 0.035. Which of the following is a correct interpretation of the P-value?

   (A) If it is true that 70 percent of teens in the city pass their driving test on the first attempt, 0.035 is the probability of obtaining a population proportion greater than 0.70.

   (B) If it is true that 70 percent of teens in the city pass their driving test on the first attempt, 0.035 is the probability of obtaining a sample proportion as small as or smaller than the one obtained by the administrator.

   (C) If it is true that 70 percent of teens in the city pass their driving test on the first attempt, 0.035 is the probability of obtaining a sample proportion as large as or larger than the one obtained by the administrator.

   (D) If it is *not* true that 70 percent of teens in the city pass their driving test on the first attempt,

   0.035 is the probability of obtaining a sample proportion as large as or larger than the one obtained by the administrator.

   (E) If it is *not* true that 70 percent of teens in the city pass their driving test on the first attempt, 0.035 is the probability of obtaining a population proportion greater than 0.70.

3. One extrasensory perception (ESP) test asks the subject to view the backs of cards and identify whether a circle, square, star, or cross is on the front of each card. If p is the proportion of correct answers, this may be viewed as a hypothesis test with $H_0 : p = 0.25$ and $H_a : p > 0.25$. The subject is recognized to have ESP when the null hypothesis is rejected. What would a Type II error result in?

   (A) Correctly recognizing someone has ESP

   (B) Mistakenly thinking someone has ESP

   (C) Not recognizing that someone really has ESP

   (D) Correctly realizing that someone doesn't have ESP

   (E) Failing to understand the nature of ESP

4. A research dermatologist believes that skin cancer presenting on the head or neck will occur most often on the left side, the side next to a window when a person is driving. In a review of 565 cases of skin cancer on the head or neck, 305 occurred on the left side. What is the resulting P-value?

   (A) $P\left(z > \dfrac{0.54 - 0.50}{\sqrt{(0.5)(0.5)/565}}\right)$

   (B) $2P\left(z > \dfrac{0.54 - 0.50}{\sqrt{(0.5)(0.5)/565}}\right)$

   (C) $P\left(z > \dfrac{0.54 - 0.50}{\sqrt{(0.54)(0.46)/565}}\right)$

   (D) $P\left(z \geq \dfrac{0.54 - 0.50}{\sqrt{(0.54)(0.46)/565}}\right)$

   (E) $2P\left(z > \dfrac{0.54 - 0.50}{\sqrt{(0.54)(0.46)/565}}\right)$

5. Is Internet usage different in the Middle East from Internet usage in Latin America? In a random sample of 500 adults in the Middle East, 151 claimed to be regular Internet users, while in a random sample of 1,000 adults in Latin America, 345 claimed to be regular users. What is the P-value for the appropriate hypothesis test?

(A) $P\left(z < \dfrac{0.302 - 0.345}{\sqrt{\dfrac{(0.302)(0.698)}{500} + \dfrac{(0.345)(0.655)}{1,000}}}\right)$

(B) $2P\left(z < \dfrac{0.302 - 0.345}{\sqrt{\dfrac{(0.302)(0.698)}{500} + \dfrac{(0.345)(0.655)}{1,000}}}\right)$

(C) $P\left(z < \dfrac{0.302 - 0.345}{\sqrt{(0.331)(0.669)\left(\dfrac{1}{500} + \dfrac{1}{1,000}\right)}}\right)$

(D) $2P\left(z < \dfrac{0.302 - 0.345}{\sqrt{(0.331)(0.669)\left(\dfrac{1}{500} + \dfrac{1}{1,000}\right)}}\right)$

(E) $2(0.095167)$

6. The greater the difference between the null hypothesis claim and the true value of the population parameter,

(A) the smaller the risk of a Type II error and the smaller the power.
(B) the smaller the risk of a Type II error and the greater the power.
(C) the greater the risk of a Type II error and the smaller the power.
(D) the greater the risk of a Type II error and the greater the power.
(E) the greater the probability of no change in Type II error or in power.

7. Suppose that a two-sided significance test of whether two population proportions differ results in a P-value of 0.069. What can be said about a 95% confidence interval for the difference in population proportions, calculated from the same sample data?

(A) It includes only negative values.
(B) It includes both positive and negative values.

(C) It includes only positive values.
(D) This cannot be answered without knowing which proportion is subtracted from which.
(E) This cannot be answered without knowing the sample sizes.

8. Consider a claimed population proportion p_0, and a sample with sample proportion $\hat{p}$. Which of the following best describes the purpose of a one-sample z-test?

(A) To estimate $\hat{p}$
(B) To estimate the population proportion
(C) To estimate a margin of error for $\hat{p}$
(D) To estimate the probability of observing a value as extreme or more extreme than $\hat{p}$ given that p_0 is true
(E) To prove that p_0 either is or is not true

9. A company is considering investing in an expensive new computer security system, and their engineers are running a test to gauge whether the new software is superior to what they have. The null hypothesis is that the new software is not any better than what they have. Their major worry is that the new software is not superior, but the test indicates it is, and they end up with an inadvisable costly investment. Which of the following should the testers do to help avoid making this costly mistake?

(A) Increase the significance level to increase the probability of a Type I error.
(B) Increase the significance level to decrease the probability of a Type I error.
(C) Decrease the significance level to increase the probability of a Type I error.
(D) Decrease the significance level to decrease the probability of a Type I error.
(E) Decrease the significance level to decrease the probability of a Type II error.

Free-Response Questions

> **DIRECTIONS:** You must show all work and indicate the methods you use. You will be graded on the correctness of your methods and on the accuracy of your final answers.

1. In a random sample of 500 new births in the United States, 41.2% were to unmarried women, while in a random sample of 400 new births in the United Kingdom, 46.5% were to unmarried women.

   (a) Calculate a 95% confidence interval for the difference in the proportions of new births to unmarried women in the United States and the United Kingdom.
   (b) Does the confidence interval support the belief by a UN health care statistician that the proportions of new births to unmarried women is different in the United States and the United Kingdom? Explain.

2. It is estimated that 17.4% of all U.S. households own a Roth IRA, a retirement savings account. The American Association of University Professors (AAUP) believes this figure is higher among their members and commissions a study. In the study, 150 out of a random sample of 750 AAUP members own Roth IRAs.

   (a) Is this convincing evidence to support the AAUP belief?
   (b) What error might have been committed, Type I or Type II, and what would be a possible consequence?

3. A national survey before the death of Ruth Bader Ginsburg studied whether or not a random sample of American voters could name a single Supreme Court justice. The study compared survey responses between voters under the age of 35 and voters over the age of 65. The following table displays some of the results.

	All Voters	Voters under 35	Voters over 65	Difference (percentage points)
Number surveyed	2,050	1,500	550	
% who could not name a justice	54%	57%	46%	11%

   (a) Is random sampling, random assignment, both, or neither involved in this study? Explain.
   (b) What are the appropriate hypotheses for testing whether the two populations of voters differ with regard to the proportion who cannot name a single Supreme Court justice?
   (c) The test statistic is $z = 4.43$. What does this value mean in context?
   (d) Is there convincing evidence to reject the null hypothesis at the 0.01 significance level? Explain.
   (e) A 99% confidence interval based on this data is (0.04612, 0.17388). Interpret this interval in context.
   (f) Is this confidence interval consistent with the test decision in part (d)? Explain.

4. A randomized, double-blind, placebo-controlled trial across the United States and parts of Canada tested hydroxychloroquine as a postexposure prophylaxis for asymptomatic participants reporting a high-risk exposure to a confirmed Covid-19 contact. The incidence of Covid-19 among participants receiving hydroxychloroquine was 49 out of 414, while the incidence among participants receiving a placebo was 58 out of 407.

(a) Explain the purpose of random assignment in this experiment.

(b) State the hypotheses the researchers tested about hydroxychloroquine use as a postexposure prophylaxis for Covid-19.

(c) Identify the significance test the researchers used to analyze the results of this experiment and show that the conditions for this test were met.

(d) The P-value of this test is 0.152. What conclusion should be made at the $\alpha = 0.05$ significance level?

(e) What error might have been committed, Type I or Type II, and what would be a possible consequence?

The answers for this quiz can be found beginning on page 498.

Quiz 17

Multiple-Choice Questions

> **DIRECTIONS:** The questions that follow are each followed by five suggested answers. Choose the response that best answers the question.

1. In a simple random sample of 300 elderly men, 65% were married, while in an independent simple random sample of 400 elderly women, 48% were married. Determine a 99% confidence interval estimate for the difference between the proportions of all elderly men and women who are married.

    (A) $(0.65 - 0.48) \pm 2.326 \sqrt{\dfrac{(0.65)(0.35)}{300} + \dfrac{(0.48)(0.52)}{400}}$

    (B) $(0.65 - 0.48) \pm 2.576 \sqrt{\dfrac{(0.65)(0.35)}{300} + \dfrac{(0.48)(0.52)}{400}}$

    (C) $(0.65 - 0.48) \pm 2.576 \left(\dfrac{(0.65)(0.35)}{\sqrt{300}} + \dfrac{(0.48)(0.52)}{\sqrt{400}} \right)$

    (D) $\left(\dfrac{0.65 + 0.48}{2} \right) \pm 2.576 \sqrt{\dfrac{(0.65)(0.35)}{300} + \dfrac{(0.48)(0.52)}{400}}$

    (E) $\left(\dfrac{0.65 + 0.48}{2} \right) \pm 2.807 \sqrt{(0.565)(0.435)\left(\dfrac{1}{300} + \dfrac{1}{400} \right)}$

2. A college guidance counselor suspects that less than 15 percent of students at her college are willing to report cheating by other students. She surveys a random sample of students at her college and conducts a test of $H_0 : p = 0.15$ and $H_a : p < 0.15$. The P-value of the test is 0.078. Which of the following is a correct interpretation of the P-value?

    (A) If it is true that 15 percent of students at her college are willing to report cheating by other students, 0.078 is the probability of obtaining a population proportion greater than 0.15.

    (B) If it is true that 15 percent of students at her college are willing to report cheating by other students, 0.078 is the probability of obtaining a sample proportion as small as or smaller than the one obtained by the guidance counselor.

    (C) If it is true that 15 percent of students at her college are willing to report cheating by other students, 0.078 is the probability of obtaining a sample proportion as large as or larger than the one obtained by the guidance counselor.

    (D) If it is *not* true that 15 percent of students at her college are willing to report cheating by other students, 0.078 is the probability of obtaining a sample proportion as small as or smaller than the one obtained by the guidance counselor.

    (E) If it is *not* true that 15 percent of students at her college are willing to report cheating by other students, 0.078 is the probability of obtaining a population proportion smaller than 0.15.

3. Laura has a penny and a nickel and wonders if they have the same likelihood of showing heads when they are spun. She spins each coin 200 times to test if there is a significant difference in the proportion of spins that they each land showing heads. Suppose the test with $H_0 : p_{penny} = p_{nickel}$ and $H_a : p_{penny} \neq p_{nickel}$ results in a P-value of 0.032. Which of the following is a proper conclusion?

(A) The probability that the null hypothesis is true is 0.032.

(B) The probability that the alternative hypothesis is true is 0.032.

(C) The difference in sample proportions is 0.032.

(D) The difference in population proportions is 0.032.

(E) None of the above are proper conclusions.

4. A teacher is concerned that less than 50% of her students are getting at least eight hours of sleep on school nights. She plans to test at a significance level of $\alpha = 0.05$. She is considering consequences of increasing the sample size to $n = 100$ from her original plan to use a sample of size $n = 50$. What is the effect on the probability of committing a Type I error if the sample size is increased with everything else unchanged?

(A) The probability of committing a Type I error decreases.

(B) The probability of committing a Type I error is unchanged.

(C) The probability of committing a Type I error increases.

(D) The effect cannot be determined without knowing the relevant standard deviation.

(E) The effect cannot be determined without knowing if a Type II error is committed.

5. A sociologist claims that over two-thirds of White adults have entirely White social networks without any minority presence. In a random sample of 200 White adults, 150 of them have entirely White social networks. Is this convincing evidence in support of the sociologist's claim?

(A) Yes, because the P-value is 0.0062.

(B) Yes, because the P-value is 0.0124.

(C) No, because the P-value is 0.0062.

(D) No, because the P-value is 0.0124.

(E) No, because the P-value is 0.9938.

6. Choosing a smaller level of significance, that is, a smaller α-risk, results in

(A) a lower risk of Type II error and lower power.

(B) a lower risk of Type II error and higher power.

(C) a higher risk of Type II error and lower power.

(D) a higher risk of Type II error and higher power.

(E) no change in risk of Type II error or in power.

7. Suppose that you want to compare the percentage of male high school students who hope someday to run for political office to the percentage of female high school students who hope someday to run for political office. A random sample of male high school students and a random sample of female high school students are selected. Which of the following results provides the strongest evidence that the percentages differ between male and female high school students who hope someday to run for political office?

(A) 10 of 100 males hope someday to run for political office compared with 30 out of 100 females.

(B) 15 of 100 males hope someday to run for political office compared with 25 out of 100 females.

(C) 15 of 150 males hope someday to run for political office compared with 45 out of 150 females.

(D) 20 of 200 males hope someday to run for political office compared with 60 out of 200 females.

(E) 25 of 200 males hope someday to run for political office compared with 55 out of 200 females.

8. The risk ratio (or relative risk) is the ratio of proportions from two groups. Suppose that a 90% confidence interval for the risk ratio is calculated to be (1.58, 2.41). Based on this interval, can we conclude that the sample data provide convincing evidence that the proportions in the two groups differ?

    (A) Yes, because the confidence interval is entirely greater than zero.
    (B) No, because the confidence interval is entirely greater than zero.
    (C) Yes, because the confidence interval is entirely greater than one.
    (D) No, because the confidence interval is entirely greater than one.
    (E) Yes, because a 90% confidence interval corresponds to a significance test at the 0.10 significance level.

9. Medical researchers are working around the clock to develop a vaccine for a newly discovered virus. A large sample study is being run with the null hypothesis that a tested vaccine is not effective. The researchers are concerned that the new vaccine really is effective but that the stringent test will not detect the effectiveness of the vaccine. Which of the following should the researchers do to help minimize the probability of this concern?

    (A) Increase the significance level to increase the probability of a Type I error.
    (B) Increase the significance level to decrease the probability of a Type I error.
    (C) Decrease the significance level to increase the probability of a Type I error.
    (D) Decrease the significance level to decrease the probability of a Type I error.
    (E) Decrease the significance level to decrease the probability of a Type II error.

Free-Response Questions

> **DIRECTIONS:** You must show all work and indicate the methods you use. You will be graded on the correctness of your methods and on the accuracy of your final answers.

1. No vaccination is 100% risk free, and the theoretical risk of rare complications always has to be balanced against the severity of the disease. Suppose the CDC (Centers for Disease Control) decides that a risk of one in a million is the maximum acceptable risk of Guillain-Barré syndrome (GBS) complications for a new vaccine for a particularly serious strain of influenza. A large sample study of the new vaccine is conducted with the following hypotheses:

    H_0: The proportion of GBS complications is 0.000001 (one in a million).
    H_a: The proportion of GBS complications is greater than 0.000001 (one in a million).

    The P-value of the test is 0.138.

    (a) Interpret the P-value in the context of this study.
    (b) What conclusion should be drawn at the $\alpha = 0.10$ significance level?
    (c) Using your conclusion from part (b), what possible error, Type I or Type II, might be committed? Give a possible consequence of committing this error.

2. A behavior study of high school students looked at whether a higher proportion of boys than girls meets a recommended level of physical activity (increased heart rate for 60 minutes per day for at least 5 days during the 7 days before the survey). What is the proper conclusion if 370 out of a random sample of 850 boys and 218 out of an independent random sample of 580 girls met the recommended level of activity?

3. A national survey studied whether or not a random sample of American voters believed a particular false political claim. The study compared survey responses between voters 18 to 25 years old and voters over the age of 65. The following table displays some of the results.

	All Voters	Voters 18 to 25	Voters over 65	Difference (percentage points)
Number surveyed	1,500	900	600	
% who believed the false claim	10.8%	12%	9%	3%

(a) Is random sampling, random assignment, both, or neither involved in this study? Explain.

(b) What are the appropriate hypotheses for testing whether the two populations of voters differ with regard to the proportion who believe the false political claim?

(c) The test statistic is $z = 1.834$. What does this value mean in context?

(d) Is there convincing evidence to reject the null hypothesis at the 0.05 significance level? Explain.

(e) A 95% confidence interval based on this data is $(-0.0012, 0.06123)$. Interpret this interval in context.

(f) Is this confidence interval consistent with the test decision in part (d)? Explain.

4. Exercise addiction, an uncontrollable desire to exercise, is one example of behavior addiction. In one study of 1,000 randomly selected adults from Boulder, Colorado, psychologists investigated the relationship between exercise addiction and negative body image. The following two-way table summarizes the data:

Exercise addiction diagnosis

		None	Mild	Severe	Total
Negative	Yes	301	68	26	395
body image	No	500	83	22	605
	Total	801	151	48	1,000

(a) In the sample, 4.8% of the adults were diagnosed with severe exercise addiction. To what population can this result be generalized? Explain.

(b) If one person from the sample is selected at random, what is the probability the person is diagnosed to have some form of exercise addiction?

(c) If one person from the sample is selected at random, what is the probability the person is diagnosed to have some form of exercise addiction given that the person has negative body image?

Let p_{neg} = the proportion of people with severe exercise addiction among those in the population who have negative body image, and let p_{pos} = the proportion of people with severe exercise addiction among those in the population who do not have negative body image. A 95% confidence interval for $p_{neg} - p_{pos}$ is $(0.000815, 0.0581)$.

(d) One condition for constructing this confidence interval is that the number of people with severe exercise addiction and the number of people who do not have severe exercise addiction in each of the two groups (with and without negative body image) is at least 10. Explain why it is necessary for this condition to be satisfied.

(e) Interpret the confidence interval.

(f) Even though all of the values in the confidence interval are positive, a cause-and-effect relationship between negative body image and severe exercise addiction cannot be established from this study due to possible confounding. Identify a potential confounding variable in this context and explain your answer.

The answers for this quiz can be found beginning on page 501.

SUMMARY

- **Statistical inference** draws a conclusion (i.e., *infers* something) about a population parameter based on a sample statistic.

  1. A **confidence interval** is used to estimate the value of a parameter with a range of plausible values. We estimate a population proportion p with a z-interval.
  2. A **significance test** is used to assess the plausibility of a particular claim about a parameter. We test a claim about a population proportion p with a z-test.

- Important assumptions and conditions always must be checked before calculating a confidence interval or proceeding with a significance test. For proportions, these include:

  1. *Independence*, which involves a random sample, and if sampling without replacement, $n < 0.10N$.
  2. *Normality*, where for confidence intervals, both $n\hat{p}$ and $n(1 - \hat{p})$ are ≥ 10, and for hypothesis tests, both np_0 and $n(1 - p_0)$ are ≥ 10.

- Conditions when constructing a confidence interval or for performing a significance test about a difference in proportions:

  1. *Independence* for two independent random samples, and if sampling without replacement, $n_1 < 0.10N_1$ and $n_2 < 0.10N_2$.
  2. *Normality*, where for confidence intervals, $n_1\hat{p}_1$, $n_1(1 - \hat{p}_1)$, $n_2\hat{p}_2$, and $n_2(1 - \hat{p}_2)$ are all ≥ 10, and for hypothesis tests, $n_1\hat{p}_c$, $n_1(1 - \hat{p}_c)$, $n_2\hat{p}_c$, and $n_2(1 - \hat{p}_c)$ are all ≥ 10 where $\hat{p}_c = \dfrac{x_1 + x_2}{n_1 + n_2}$.

- We are never able to say exactly what a population parameter is equal to; rather, we say that we have a certain confidence that it lies within a certain interval.

- The width of a confidence interval is proportional to $\dfrac{1}{\sqrt{n}}$.

- The margin of error is exactly one-half the width of the confidence interval.

- If we want a narrower interval, we must either decrease the confidence level or increase the sample size.

- If we want a higher level of confidence, we must either accept a wider interval or increase the sample size.

- The confidence level refers to the percentage of samples that produce intervals that capture the true population parameter.

- The margin of error is determined by two factors:

  1. How much the statistic typically varies from the parameter; that is, $\sigma_{\hat{p}} \approx SE(\hat{p})$.
  2. How confident we want to be in our answer, determined by a critical z^*.

- The margin of error formula, $z^*\sqrt{\dfrac{\hat{p}(1 - \hat{p})}{n}}$, can be solved for the minimum sample size n needed to achieve a given margin of error (use a supposition or previous value for $\hat{p}$ or use $\hat{p} = 0.5$).

- When scoring a confidence interval free-response question on the exam, readers will look for whether you:

  1. Identify the procedure (such as one-sample z-interval for a population proportion) and check the conditions.
  2. Compute a proper interval. (It is sufficient to simply copy the interval from your calculator calculation.)
  3. Interpret the interval in context. (The conclusion must mention the confidence level and the parameter and must refer to the population, not to the sample.)

- We can justify a claim based on a confidence interval only if *all* the values in the confidence interval are consistent with the claim.

- The null hypothesis is stated in the form of an equality statement about a population parameter, while the alternative hypothesis is in the form of a strict inequality statement about the same population parameter.
- We attempt to show that a null hypothesis is unacceptable by showing it is improbable.
- The test statistic for a population proportion is $z = \dfrac{\hat{p} - p_0}{\sqrt{\dfrac{p_0(1 - p_0)}{n}}}$.

- The significance level α is the conditional probability of rejecting the null hypothesis given that it is true.
- Lack of sufficient evidence for the alternative hypothesis is not the same as evidence for the null hypothesis. That is, *absence of evidence does not constitute evidence of absence.*
- The P-value is the probability of obtaining a sample statistic as extreme as (or more extreme than) the one obtained if the null hypothesis is assumed to be true.
- When the P-value is small (typically, less than 0.05 or 0.10), we say we have sufficient evidence to reject the null hypothesis.
- A Type I error is the probability of mistakenly rejecting a true null hypothesis.
- A Type II error is the probability of mistakenly failing to reject a false null hypothesis.
- The power of a significance test is the probability that a Type II error is not committed.
- When scoring a significance test free-response question on the exam, readers will look for whether you:

  1. State the hypotheses (both H_0 and H_a) and define the parameter of interest. (The hypotheses must refer to the population, not to the sample.)
  2. Identify the procedure (such as one-sample z-test for a population proportion) and check the conditions.
  3. Compute the test statistic and the P-value. (It is sufficient to simply copy these from your calculator calculation.)
  4. Give a conclusion in context with linkage to the P-value. (The conclusion must refer to the parameter and to the population, not to the sample.)

- A recommended way of formatting conclusions to significance tests is as follows:

  1. Because the P-value is small (less than alpha), there is sufficient evidence to reject H_0; that is, there is convincing evidence that (alternative in context).
  2. Because the P-value is large (greater than alpha), there is not sufficient evidence to reject H_0; that is, there is not convincing evidence that (alternative in context).
  3. Note that this also makes it easier to talk about Type I errors, Type II errors, and power, as the language closely matches these conclusions:

     - Type I error: Finding convincing evidence that (alternative in context) is true when it really isn't
     - Type II error: Not finding convincing evidence that (alternative in context) is true when it really is
     - Power: The probability that we do find convincing evidence that the (alternative in context) is true when it really is

- Assuming everything else remains the same, the *power* of a test will be greater (and the probability of a Type II error will be less) when:

  1. The sample size n increases.
  2. The significance level α increases.
  3. The standard error decreases.
  4. The true parameter is further from the null.

- One suggested acronym for confidence intervals is PANIC:

  - **P**arameter (state it)
  - **A**ssumptions (state and justify)
  - **N**ame the interval
  - **I**nterval (find it)
  - **C**onclusion (in context)

- One suggested acronym for significance tests is PHANTOMS:

  - **P**arameter (state it)
  - **H**ypotheses (about the population)
  - **A**ssumptions (state and justify)
  - **N**ame the test
  - **T**est statistic (calculate z or t or χ^2)
  - **O**btain a P-value
  - **M**ake a decision (reject or fail to reject the null hypothesis)
  - **S**tate a conclusion (in context with linkage to the P-value)

7

Inference for Quantitative Data: Means

(4–7 multiple-choice questions on the AP exam)

Learning Objectives

In this unit, you will learn how

→ To be able to check conditions to perform inference involving population means.

→ To understand that when the population standard deviation σ is unknown and provided that either the original population is normally distributed or the sample size n is large enough, the statistic $\dfrac{\bar{x} - \mu}{s/\sqrt{n}}$ follows a t-distribution with $n-1$ degrees of freedom.

→ To be able to construct and interpret a confidence interval for a population mean and for the difference of population means.

→ To understand the relationship between confidence level, sample size, sample standard deviation, and confidence interval width.

→ To be able to interpret confidence levels.

→ To be able to perform a significance test to evaluate a claim about a population mean and about a difference between population means.

→ To be able to interpret Type I and Type II errors and their possible consequences in context.

→ To understand the relationships among significance level, probability of Type II error, and power.

→ To understand how simulation can be used to estimate a P-value.

In this unit, you will learn inference procedures about population means. You will understand when to use t- versus z-procedures. Finally, you will see the similarities and differences between the conditions for inference with proportions and means. The terminology, procedures, and much of what you learned in Unit 6 will carry over here with some modification.

The *t*-Distribution

When the population standard deviation σ is unknown, we use the sample standard deviation s as an estimate for σ. But then $\dfrac{\bar{x} - \mu}{s/\sqrt{n}}$ does not follow a normal distribution. There is a distribution that can be used when working with the $s/\sqrt{n}$ ratios. This *Student's t-distribution* was introduced in 1908 by W. S. Gosset, a British mathematician employed by the Guinness Breweries.

For a sample from a *normally distributed population*, we work with the variable

$$t = \frac{\bar{x} - \mu}{s/\sqrt{n}}$$

> **NOTE**
>
> **When we are working with *small* samples from a population that is *not* nearly normal, we must use very different "nonparametric" techniques not discussed in this review book.**

with a resulting *t*-distribution that is bell-shaped and symmetric but lower at the mean, higher at the tails, and so more spread out than the normal distribution.

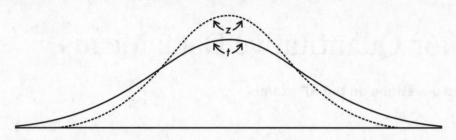

NOTE

In the real world, there is no such thing as a truly normal distribution, and instead we assume a *roughly* normally distributed population.

Like the binomial distribution, *t*-distributions are different for different values of *n*. In the tables, these distinct *t*-distributions are associated with the values for degrees of freedom (*df*). For this discussion, the *df* value is equal to the sample size minus 1. The smaller the *df* value, the larger the dispersion in the distribution. The larger the *df* value, that is, the larger the sample size, the closer the distribution to the normal distribution.

Since there is a separate *t*-distribution for each *df* value, fairly complete tables would involve many pages; therefore, in Table B of the Appendix, we list areas and *t*-values for only the more commonly used percentages or probabilities. The last row of Table B is the normal distribution, which is a special case of the *t*-distribution taken when *n* is infinite. Technology (such as `tcdf` and `invT`) easily handle calculations involving the *t*-distribution.

Thus, the *t*-distribution is the proper choice whenever the population standard deviation σ is unknown. In the real world, σ is almost always unknown, and so we should almost always use the *t*-distribution. The check that must be satisfied before using the *t*-distribution is that either the original population is roughly normal or the sample size is large enough ($n \geq 30$) for the central limit theorem to apply. What happens if the sample size is less than 30 and we are not told that the parent population is roughly normal? Then we must check that the sample is very roughly unimodal and symmetric, showing no outliers and no strong skewness, so that we can say it is not unreasonable to assume the parent population is roughly normal. (This check can quickly be made with a hand-drawn dotplot or stemplot of the sample data.)

> **Important**
>
> For $\frac{\overline{x} - \mu}{s/\sqrt{n}}$ to have a *t*-distribution, we actually need that the sampling distribution of $\overline{x}$ is normal (that is, that $\frac{\overline{x} - \mu}{\sigma/\sqrt{n}}$ has a *z*-distribution).
>
> This follows either if the original population has a normal distribution or if the sample size is large enough (from the central limit theorem).

Confidence Interval for a Mean

We are interested in estimating a population mean μ by considering a single sample mean $\overline{x}$. This sample mean is just one of a whole universe of sample means, and from Unit 5 we remember that if *n* is sufficiently large:

1. the set of all sample means is approximately normally distributed.
2. the mean of the set of sample means equals μ, the mean of the population.
3. the standard deviation $\sigma_{\overline{x}}$ of the set of sample means is approximately equal to $\frac{\sigma}{\sqrt{n}}$, that is, equal to the standard deviation of the whole population divided by the square root of the sample size.

Typically, we do not know σ, the population standard deviation. In such cases, we must use *s*, the *standard deviation of the sample*, as an estimate of σ. In this case $\frac{s}{\sqrt{n}}$ is called the *standard error*, $SE(\overline{x})$, and is used as an estimate for $\sigma_{\overline{x}} = \frac{\sigma}{\sqrt{n}}$. (We will use *t*-distributions instead of the standard normal curve whenever σ is unknown, no matter what the sample size.)

Remember that in making calculations and drawing conclusions from a specific sample, it is important that the sample be a random sample, ideally a *simple random sample*, and be no more than 10% of the population. This satisfies the independence condition.

NOTE

You cannot conclude "nearly normal" or even "roughly symmetric" from a boxplot.

The normality condition to be checked is that either the parent population is approximately normal ("given" or "stated") or the sample size is large enough ($n \geq 30$) for the central limit theorem to apply. If we are not given that the parent population is approximately normal, for small samples ($n < 30$) the sample data should be free from strong skewness and outliers (which we can check using a dotplot, stemplot, or histogram).

> **TIP**
>
> Understand that you are not trying to prove that the sample data are normal but, rather, trying to infer something about the parent population.

⟩ Example 7.1

When a random sample of 10 cars of a new model was tested for gas mileage, the results showed a mean of 27.2 miles per gallon with a standard deviation of 1.8 miles per gallon. (Assume that the population of mpg results for all the new model cars is approximately normally distributed.)

(a)   What is a 95% confidence interval estimate for the mean gas mileage achieved by this model?

(b)   Based on this confidence interval, do you think that the true mean mileage is significantly different from 25 mpg?

(c)   Determine a 99% confidence interval. Note what effect going from 95% to 99% confidence has on the interval size.

(d)   What would the 95% confidence interval be if the sample mean of 27.2 and standard deviation of 1.8 had come from a sample of 20 cars? Note what effect going from a sample size of 10 to one of 20 has on the interval size.

(e)   With the original data ($\bar{x} = 27.2$, $s = 1.8$, $n = 10$), with what confidence can we assert that the true mean gas mileage is 27.2 ± 1.04?

> **NOTE**
>
> The population standard deviation is unknown, so we must use sample standard deviations together with a t-distribution. The confidence interval takes the form $\bar{x} \pm t^{*}SE(\bar{x})$ $= \bar{x} \pm t^{*}\frac{s}{\sqrt{n}}$ where $t^{*}SE(\bar{x}) = t^{*}\frac{s}{\sqrt{n}}$ is called the margin of error.

✓ Solutions

(a)   *Parameter:* Let μ represent the mean gas mileage (in miles per gallon) in the population of cars of a new model.
Procedure: A one-sample t-interval for a population mean (population SD is unknown).
Checks: We check *independence* and *normality*. For independence, the sample is given as random, and 10 cars are assumed to be less than 10% of all cars of the new model. For normality, the population is stated to be approximately normal. So, $\frac{\bar{x} - \mu}{s/\sqrt{n}}$ follows a t-distribution.
Mechanics: Calculator software (`TInterval`) gives (25.912, 28.488).

[For instructional purposes, we note that the standard error of the sample means is $SE(\bar{x}) = \frac{1.8}{\sqrt{10}} = 0.569$. With $n - 1 = 10 - 1 = 9$ degrees of freedom and 2.5% in each tail, the appropriate t-scores are $\pm \text{invT}(0.975, 9) = \pm 2.262$. Then $27.2 \pm 2.262(0.569) = 27.2 \pm 1.29$.]

Conclusion in context: We are 95% confident that the true mean gas mileage of all cars of the new model is between 25.91 and 28.49 miles per gallon.

> Interpretation of the confidence level: If all possible random samples of 10 cars of this model are tested for gas mileage, and a confidence interval for the mean is constructed from each sample, then 95% of all these intervals would capture the true mean gas mileage.

(continued)

[Alternatively we can say: "We are 95% confident that the interval from 25.91 to 28.49 mpg captures the true mean gas mileage of all cars of the new model."]

(b)   Yes, because 25 is not in the interval of plausible values from 25.9 to 28.5, there is convincing evidence that the true mean mileage is significantly different from 25 mpg.

(c)   Here, TInterval gives (25.35, 29.05). We are 99% confident that the true mean gas mileage of all cars of the new model is between 25.35 and 29.05 miles per gallon.

    Note that when we want a higher confidence (99% instead of 95%), we have to settle for a larger, less specific interval (±1.85) instead of (±1.29).

(d)   Here, TInterval gives (26.36, 28.04). We are 95% confident that the true mean gas mileage of all cars of the new model is between 26.36 and 28.04 miles per gallon.

    Note that when the sample size increased (from $n = 10$ to $n = 20$), the same sample mean and standard deviation resulted in a narrower, more specific interval (±0.84) instead of (±1.29).

(e)   Converting ±1.04 to t-scores yields $\dfrac{\pm1.04}{1.8/\sqrt{10}} = \pm1.827$ and tcdf$(-1.827, 1.827, 9) = 0.899 = 89.9\%$.

    We are 89.9% confident that the true mean gas mileage of all cars of the new model is between 26.16 and 28.24 miles per gallon.

⟩ Example 7.2

Maximum bench press weights were recorded for a random sample of ten 17-year-old males. The resulting data are displayed in the following dotplot.

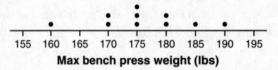

Max bench press weight (lbs)

(a)   Find a 90% confidence interval for the mean maximum bench press weight for all 17-year-old males.

(b)   Based on this interval, is there convincing evidence that the mean maximum bench press weight for all 17-year-old males is less than 180 lbs?

✓ Solutions

(a)   *Parameter:* Let μ represent the mean maximum bench press weight for all 17-year-old males.

Procedure: One-sample t-interval for a population mean (population SD is unknown)

Checks: We check *independence* and *normality*. For independence, the sample is given to be random, and the ten 17-year-old males are less than 10% of all 17-year-old males. For normality, a dotplot of the sample values is unimodal, reasonably symmetric, and with no outliers and no skewness, indicating it is not unreasonable to assume the data come from a roughly normal distribution.

Mechanics: Calculator software (such as TInterval with the data put in a List) gives (171.11, 180.89).

> Interpretation of the confidence level: If all possible random samples of ten 17-year-old males are taken and if a confidence interval for the mean maximum bench press weight is constructed from each sample, then 90% of all these intervals would capture the true mean maximum bench press weight for all 17-year-old males.

(continued)

[For instructional purposes, we note the following:

$$\bar{x} = \frac{\sum x}{n} = \frac{1{,}760}{10} = 176,$$

$$s = \sqrt{\frac{\sum(x - \bar{x})^2}{n - 1}} = \sqrt{\frac{\sum(x - 176)^2}{9}} = 8.433, \text{ and}$$

$$SE(\bar{x}) = \frac{s}{\sqrt{n}} = \frac{8.433}{\sqrt{10}} = 2.667.$$

With $df = 10 - 1 = 9$ and 5% in each tail, the critical t-scores are $\pm \text{invT}(0.95, 9) = \pm 1.833$. Then $176 \pm 1.833(2.667) = 176 \pm 4.89$.]

Conclusion in context: We are 90% confident that the true mean maximum bench press weight for all 17-year-old males is between 171.11 and 180.89 lbs.

(b)   No, because the entire interval from 171.11 to 180.89 is not less than 180, there is not convincing evidence that the mean maximum bench press weight for all 17-year-old males is less than 180 lbs. [It is not sufficient that 176 < 180, and it is not sufficient that almost the entire interval is < 180. All values in the interval are considered to be "plausible" values, and 180 is in the interval.]

Statistical principles are useful not only in analyzing data but also in setting up surveys and experiments. One important consideration is the choice of a sample size. In making interval estimates of population means, we have seen that each inference must go hand in hand with an associated confidence level statement. Generally, if we want a smaller, more precise interval estimate, we either decrease the degree of confidence or increase the sample size. Similarly, if we want to increase the degree of confidence, we either accept a wider interval estimate or increase the sample size. Thus, choosing a larger sample size always seems desirable; in the real world, however, time and cost considerations may limit the sample size.

❯ Example 7.3

Ball bearings are manufactured by a process that results in a standard deviation in diameter of 0.025 inch. What sample size should be chosen if we wish to be 99% sure of knowing the mean diameter within ±0.01 inch?

✓ Solution

We have $\sigma_{\bar{x}} = \frac{\sigma}{\sqrt{n}} = \frac{0.025}{\sqrt{n}}$ and $2.576\, \sigma_{\bar{x}} \leq 0.01$.

Thus, $2.576\left(\frac{0.025}{\sqrt{n}}\right) \leq 0.01$, and algebraically we find that $\sqrt{n} \geq \frac{2.576(0.025)}{0.01} = 6.44$, and so $n \geq 41.5$. We choose a sample size of 42.

Quiz 18

Multiple-Choice Questions

> **DIRECTIONS:** The questions that follow are each followed by five suggested answers. Choose the response that best answers the question.

1. What happens to a *t*-distribution as *df*, the degrees of freedom, increase?

   (A) The center remains the same, and the area in the tails decreases.

   (B) The center remains the same, and the area in the tails remains the same.

   (C) The center remains the same, and the area in the tails increases.

   (D) The center increases, and the area in the tails decreases.

   (E) The center increases, and the area in the tails increases.

2. The mayor of a small city would like to know how many weeks women think should be a standard for paid maternity leave. He asked the 38 women working in his office to anonymously answer the query. He plans to construct a one-sample *t*-interval for a population mean. Have the conditions for inference been met?

   (A) Yes, all conditions have been met.

   (B) No, the sample size is too small.

   (C) No, the normality condition is violated.

   (D) No, the independence condition cannot be verified.

   (E) No, the 10% condition cannot be verified.

3. To estimate the average number of books high school students read during their four years in high school, a random sample of 95 high school graduates are interviewed, and a one-sample *t*-interval is found. However, it is noted that the sample data are skewed right. Have all conditions for inference been met?

   (A) Yes, all conditions have been met.

   (B) No, while a random sample was selected, there is no mention of random assignment.

   (C) No, the normality condition is violated.

   (D) No, independence is violated.

   (E) No, it is not clear whether the 10% condition is satisfied.

4. Most recent tests and calculations estimate at the 95% confidence level that mitochondrial Eve, the maternal ancestor to all living humans, lived $138,000 \pm 18,000$ years ago. What is meant by "95% confidence" in this context?

   (A) A confidence interval of the true age of mitochondrial Eve has been calculated using *z*-scores of ± 1.96.

   (B) A confidence interval of the true age of mitochondrial Eve has been calculated using *t*-scores consistent with $df = n - 1$ and tail probabilities of ± 0.025.

   (C) There is a 0.95 probability that mitochondrial Eve lived between 120,000 and 156,000 years ago.

   (D) If 20 random samples of data are obtained by this method and a 95% confidence interval is calculated from each, the true age of mitochondrial Eve will be in 19 of these intervals.

   (E) Of all random samples of data obtained by this method, 95% will yield intervals that capture the true age of mitochondrial Eve.

5. A confidence interval estimate is determined from the GPAs of a simple random sample of *n* students. All other things being equal, which of the following will result in a smaller margin of error?

   (A) A smaller confidence level

   (B) A larger sample standard deviation

   (C) A smaller sample size

   (D) A larger population size

   (E) A smaller sample mean

6. One gallon of gasoline is put into each of an SRS of 30 test autos, and the resulting mileage figures are tabulated with $\bar{x} = 28.5$ and $s = 1.2$. Determine a 95% confidence interval estimate of the mean mileage of all comparable autos. (Assume that all conditions for inference are met.)

 (A) $28.5 \pm 2.045(1.2)$

 (B) $28.5 \pm 2.045\left(\dfrac{1.2}{\sqrt{29}}\right)$

 (C) $28.5 \pm 2.045\left(\dfrac{1.2}{\sqrt{30}}\right)$

 (D) $28.5 \pm 1.96\left(\dfrac{1.2}{\sqrt{29}}\right)$

 (E) $28.5 \pm 1.96\left(\dfrac{1.2}{\sqrt{30}}\right)$

7. Studies have found an association between drinking sugary beverages such as soda and the likelihood of developing liver cancer. A high school has 3,000 students, of whom 300 drink sugary beverages. In order to estimate the total quantity of sugary beverages consumed daily by students at this high school, two methods are proposed.

Method I: Sample 30 students at random. Calculate a confidence interval for the mean quantity of sugary beverages consumed daily by those 30 students. Multiply both ends of the interval by 3,000 to get a confidence interval estimate of the total quantity of sugary beverages consumed daily by students at this high school.

Method II: Identify the 300 students who drink sugary beverages. Sample 30 of these 300 students at random. Calculate a confidence interval for the mean quantity of sugary beverages consumed daily by those 30 students. Multiply both ends of the interval by 300 to get a confidence interval estimate of the total quantity of sugary beverages consumed daily by students at this high school.

Which is the better method for estimating the total quantity of sugary beverages consumed daily by students at this high school?

 (A) Method I is better.
 (B) Method II is better.
 (C) Both methods are good and will produce equivalent results.
 (D) Neither method will estimate the total quantity of sugary beverages consumed.
 (E) There is insufficient information given to compare the two methods.

8. To assess the effectiveness of a new over-the-counter sleep aid, 30 adults from a random sample were each assessed as to the time in minutes it took to fall asleep before beginning the medication regimen and also as to the time in minutes it took to fall asleep after being on the regimen for one week. For each adult, the difference in time was calculated (before minus after) and used to create the 95 percent confidence interval (4.6, 7.3). Assuming all conditions for inference are met, which of the following is a correct interpretation of the interval?

 (A) There is a 0.95 probability that the mean difference in times for all adults in the sample is between 4.6 and 7.3 minutes.
 (B) There is a 0.95 probability that the mean difference in times for all adults taking this sleep aid is between 4.6 and 7.3 minutes.
 (C) For all adults taking this sleep aid, 95 percent will have a mean difference in times of between 4.6 and 7.3 minutes.
 (D) We are 95 percent confident that the mean difference in times for all adults in this sample is between 4.6 and 7.3 minutes.
 (E) We are 95 percent confident that the mean difference in times for all adults taking this sleep aid is between 4.6 and 7.3 minutes.

9. Hospital administrators wish to learn the average length of stay of all surgical patients. A statistician determines that, for a 95% confidence level estimate of the average length of stay to within ± 0.5 days, 50 surgical patients' records will have to be examined. How many records should be looked at to obtain a 95% confidence level estimate to within ± 0.25 days?

 (A) 25
 (B) 50
 (C) 100
 (D) 150
 (E) 200

10. Researchers collected data from two different random samples, X and Y, on weight gain (in pounds) by teenagers with anorexia taking part in family therapy. The confidence interval 8.3 ± 0.9 was constructed from sample X, and the confidence interval 8.3 ± 1.4 was constructed from sample Y. Both samples had roughly the same standard deviation. Which of the following statements could explain why the width of the confidence interval from X is less than the width of the confidence interval from Y?

(A) The sample sizes of X and Y are the same, the confidence level is the same for both intervals, and the calculated difference in widths is due to sampling variability.

(B) The sample size of X is less than the sample size of Y, and the confidence level is the same for both intervals.

(C) The sample size of X is less than the sample size of Y, and the confidence level used for X is greater than that used for Y.

(D) The sample sizes of X and Y are the same, and the confidence level used for X is less than that used for Y.

(E) The sample sizes of X and Y are the same, and the confidence level used for X is greater than that used for Y.

Free-Response Questions

> **DIRECTIONS:** You must show all work and indicate the methods you use. You will be graded on the correctness of your methods and on the accuracy of your final answers.

1. Students gather a simple random sample of "11 oz" bags of tortilla chips and weigh the contents. Software output of the sample data follow:

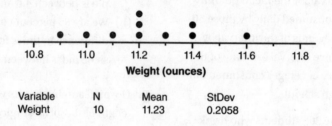

Variable	N	Mean	StDev
Weight	10	11.23	0.2058

(a) Calculate a 95% confidence interval for the population mean.
(b) What percentage of the bags in the sample have weights that fall within this interval?
(c) Is the percentage in (b) close to 95%? Explain what went wrong or explain that nothing went wrong.

2. In a random sample of 30 white-collar crimes (such as embezzlement, securities fraud, and money laundering), the average monetary amount involved was $2,900,000 with a standard deviation of $850,000.

(a) What is the point estimate of the population mean?
(b) Find the standard error of the sample mean.
(c) Are conditions for inference met? Explain.
(d) The 95% confidence interval is (2,580,000, 3,222,000). Interpret this in context.
(e) What does this confidence interval say, if anything, about the plausibility of a white-collar crime with a monetary amount of $2,000,000?
(f) What does this confidence interval say, if anything, about the plausibility that the mean of all white-collar crime monetary amounts is $2,000,000?

The answers for this quiz can be found beginning on page 504.

Significance Test for a Mean

To conduct a significance test for a mean, we must check that we have a simple random sample, that the sample size is less than 10% of the population, and that either the sample size is large enough ($n \geq 30$) for the CLT to apply or the population has an approximately normal distribution (either stated or if we are given the sample data, a plot should be very roughly unimodal and symmetric, showing no outliers and no strong skewness).

❯ Example 7.4

(a)   A manufacturer claims that a new brand of air-conditioning units uses only 6.5 kilowatts of electricity per day. A consumer agency believes the true figure is higher and runs a test on a random sample of size 50. If the sample mean is 7.0 kilowatts with a standard deviation of 1.4, should the manufacturer's claim be rejected at a significance level of 5%?

(b)   Given the above conclusion, what type of error, Type I or Type II, might have been committed, and what would be a possible consequence?

✓ Solutions

(a)   *Parameter:* Let μ represent the mean electricity usage (in kilowatts per day) of the population of this new brand of air-conditioning unit.
Hypotheses: $H_0: \mu = 6.5$ and $H_a: \mu > 6.5$.
Procedure: A one-sample t-test for a population mean (population SD is unknown).
Checks: We check *independence* and *normality*. For independence, we are given a random sample, and we assume that 50 is less than 10% of all the new AC units. For normality, $n = 50$, so the CLT applies.
Mechanics: Calculator software (such as `T-Test` on the TI-84) gives $t = 2.525$ and $P = 0.0074$.

[For instructional purposes, we note that $SE(\overline{x}) = \dfrac{1.4}{\sqrt{50}} = 0.198$ and $t = \dfrac{7.0 - 6.5}{0.198} = 2.525$ with a resulting P-value of `tcdf(2.525,1000,49)` $= 0.0074$.]

Conclusion in context with linkage to the P-value: With this small of a P-value, $0.0074 < 0.05$, there is sufficient evidence to reject H_0; that is, there is convincing evidence for the consumer agency to reject the manufacturer's claim that the new unit uses a mean of only 6.5 kilowatts of electricity per day.

> **NOTE**
>
> **There are two possible explanations for the sample mean of 7.0 being greater than the claimed mean of 6.5. Either the manufacturer's claim is wrong, or by random chance (sample variability), the sample picked by the agency happened to have a mean greater than the claimed mean.**

> Interpretation of the *P*-value: Assuming that the true mean electricity usage of the new brand of this AC unit is 6.5 kilowatts per day, there is a probability of 0.0074 of getting a sample mean of 7.0 or greater by chance alone in a random sample of 50 units.

```
EDIT CALC TESTS      T-Test               T-Test
1:Z-Test…            Inpt:Data Stats      µ>6.5
2:T-Test…            µ0:6.5               t=2.525381361
3:2-SampZTest…       x̄:7                  p=0.0074210807
4:2-SampTTest…       Sx:1.4               x̄=7
5:1-PropZTest…       n:50                 Sx=1.4
6:2-PropZTest…       µ:≠µ0 <µ0 >µ0        n=50
7:ZInterval…         Calculate Draw
8:TInterval…
9↓2-SampZInt…
```

(continued)

(b) There was sufficient evidence to reject the null hypothesis. If the null hypothesis were true, we would be committing a Type I error, that is, mistakenly rejecting a true null hypothesis. A possible consequence here is that the consumer agency would discourage customers from purchasing a new brand of air-conditioning unit that really was saving on electricity consumption as advertised.

> Example 7.5

An ice cream parlor manager believes that the mean annual ice cream consumption for high school students is 44 pints. For a class project to test that claim, a random sample of 25 high school students kept careful track of the quantity of ice cream each consumed for a full year. Following are the data and summary statistics.

```
0 |
1 | 5
2 | 5 6                    x̄ = 50.04
3 | 1 7 9                  s = 15.78
4 | 2 2 5 7 8
5 | 0 0 1 3 6 6 9
6 | 2 4 4 7
7 | 1 5 6
```

(1 I 5 means 15 pints)

(a) At a significance level of $\alpha = 0.05$, is there convincing evidence to reject the manager's belief?
(b) In context of this study, what is the meaning of the P-value calculated in part (a)?
(c) Given the conclusion in part (a), which kind of error, Type I or Type II, might have been committed, and what would be a possible consequence?

✓ Solutions

(a) *Parameter:* Let μ represent the true mean annual ice cream consumption for the population of all high school students.
Hypotheses: H_0: $\mu = 44$ and H_a: $\mu \neq 44$.
Procedure: A one-sample t-test for a population mean (population SD is unknown)
Checks: We check *independence* and *normality*. For independence, the sample is given to be random, and the 25 high school students are less than 10% of all high school students. For normality, a stemplot of the sample values is unimodal, reasonably symmetric, and with no strong skewness or outliers, indicating it is not unreasonable to assume the data come from a roughly normal distribution.

> **NOTE**
> The question does not suggest a direction for the alternative so a two-sided test is called for.

Mechanics: Calculator software (T-Test) gives $t = 1.914$ and $P = 0.0675$.
[For instructional purposes, we note that $SE(\bar{x}) = \frac{15.78}{\sqrt{25}} = 3.156$ and $t = \frac{50.04 - 44}{3.156} = 1.914$ with a resulting P-value of $2 \times \texttt{tcdf(1.914,1000,24)} = 0.0676$.]

Conclusion in context with linkage to the P-value: With this large a P-value, $0.0676 > 0.05$, there is not sufficient evidence to reject H_0. In other words, there is not convincing evidence to reject the manager's belief that the true mean annual ice cream consumption for high school students is 44 pints.

(b) Assuming that the true mean annual ice cream consumption for high school students is 44 pints, there is a probability of 0.0675 of getting a sample mean as extreme or more extreme than 50.04 pints in either direction by chance alone in a random sample of 25 high school students.

(c) There was not sufficient evidence to reject H_0. If the null hypothesis were false, we would be committing a Type II error, that is, mistakenly failing to reject a false null hypothesis. A possible consequence here is that the manager will incorrectly estimate future sales, either underordering or overordering for future demand.

Confidence Interval for the Difference of Two Means

We have the following information about the sampling distribution of $\bar{x}_1 - \bar{x}_2$:

1. The set of all differences of sample means is approximately normally distributed.
2. The mean of the set of differences of sample means equals $\mu_1 - \mu_2$, the difference of population means.
3. The standard deviation $\sigma_{\bar{x}_1 - \bar{x}_2}$ of the set of differences of sample means is approximately equal to $\sqrt{\dfrac{\sigma_1^2}{n_1} + \dfrac{\sigma_2^2}{n_2}}$.

In making calculations and drawing conclusions from specific samples, it is important that the samples be random samples, ideally *simple random samples,* that they be taken *independently* of each other, and that if sampling without replacement, each is less than 10% of their respective population.

When the population standard deviations are unknown, use a *t*-distribution with $SE(\bar{x}_1 - \bar{x}_2) = \sqrt{\dfrac{s_1^2}{n_1} + \dfrac{s_2^2}{n_2}}$. There is the additional condition that either the original populations are roughly normal or the sample sizes are large enough ($n_1 \geq 30$ and $n_2 \geq 30$) for the CLT to apply.

> ## Example 7.6

College professors' lives are often so immersed in their careers that they find it difficult to accept retirement. Although police officers identify with their careers, their stressful discipline takes its toll. A sociologist surveys 30 randomly chosen college professors and 30 randomly chosen police officers as to their planned retirement ages. In the sample of college professors, the mean planned retirement age was 66 with a standard deviation of 3.5, while in the sample of police officers, the mean planned retirement age was 55 with a standard deviation of 5.1.

(a) Determine a 95% confidence interval for the difference in mean planned retirement ages (professors minus police).

(b) Based on this confidence interval, is there convincing evidence that the mean retirement age for college professors is more than 5 years greater than for police officers?

> The confidence interval takes the form $(\bar{x}_1 - \bar{x}_2) \pm t^* SE(\bar{x}_1 - \bar{x}_2)$
> $$= (\bar{x}_1 - \bar{x}_2) \pm t^* \sqrt{\dfrac{s_1^2}{n_1} + \dfrac{s_2^2}{n_2}}$$
> where $t^* SE(\bar{x}_1 - \bar{x}_2)$
> $$= t^* \sqrt{\dfrac{s_1^2}{n_1} + \dfrac{s_2^2}{n_2}}$$
> is called the margin of error.

✓ Solutions

(a) *Parameters:* Let μ_c represent the true mean planned retirement age for college professors. Let μ_p represent the true mean planned retirement age for police officers.
Procedure: A two-sample *t*-interval for the difference of two population means, $\mu_c - \mu_p$
Checks: We check *independence* and *normality.* For independence, we are given independent random samples, and the sample sizes, $n_c = n_p = 30$, are less than 10% of the respective populations of professors and police. For normality, the sample sizes, $n_c = n_p = 30 \geq 30$, are large enough for the CLT to apply.
Mechanics: Calculator software (`2-SampTInt`) gives (8.73, 13.27) with $df = 51.36$.
Conclusion in context: We are 95% confident that the true difference in mean planned retirement ages (professors minus police) between all college professors and police officers is between 8.73 and 13.27.

(b) Yes, there is convincing evidence that the mean retirement age for college professors is at least 5 years greater than for police officers because the entire interval (8.73, 13.27) is greater than 5.

> **NOTE**
>
> Interpretation of the confidence level: If all possible random samples of 30 college professors and 30 police officers are analyzed as to planned retirement ages and if a confidence interval for the difference in means is constructed for each pair of samples, then 95% of these intervals would capture the true difference (professors minus police) in mean planned retirement ages.

Significance Test for the Difference of Two Means

A significance test for the difference of two means requires us to check that we have two independent random samples, ideally two independent *simple random samples*, that either the original two populations are roughly normal or the sample sizes are each large enough ($n_1 \geq 30$ and $n_2 \geq 30$) for the CLT to apply, and if sampling without replacement, the samples should be less than 10% of the respective populations.

In this situation, the null hypothesis is usually that the means of the populations are the same or, equivalently, that their difference is 0:

> As in the previous unit about proportions, the null hypothesis is usually a claim about "no difference" or "no change" or "no treatment effect" and is stated as an equality. The alternative hypothesis is a claim that we hope to support with evidence from the collected data and is stated as a strict inequality. Both hypotheses are about population parameters.

$$H_0: \mu_1 - \mu_2 = 0$$

The alternative hypothesis is then:

$$H_a: \mu_1 - \mu_2 < 0, \quad H_a: \mu_1 - \mu_2 > 0, \quad \text{or} \quad H_a: \mu_1 - \mu_2 \neq 0$$

The first two possibilities lead to one-sided tests, and the third possibility leads to two-sided tests.

› Example 7.7

A study is conducted to determine whether or not there is convincing evidence that professional tennis players and professional basketball players put on different mean numbers of miles during a game. Thirty-five players are randomly selected from each sport and are outfitted with pedometers to measure distances run during their next game. Summary statistics are shown in the table.

Sport	Sample Size	Mean (miles)	Standard Deviation (miles)
Tennis	35	2.8	0.41
Basketball	35	2.5	0.32

(a) Do the data provide convincing evidence at the $\alpha = 0.01$ significance level of a difference in mean miles put on during a game between professional tennis players and professional basketball players?

(b) Interpret in context the P-value from part (a).

✓ Solutions

(a) *Parameters:* Let p_T represent the true mean miles put on during games by the population of professional tennis players. Let p_B represent the true mean miles put on during games by the population of professional basketball players.
Hypotheses: $H_0: p_T = p_B$ and $H_a: p_T \neq p_B$.
Procedure: A two-sample t-test for the difference of two population means
Checks: We check *independence* and *normality*. For independence, we are given independent random samples and assume that the sample sizes are less than 10% of the respective populations of professional players. For normality, the sample sizes, $n_T = 35 \geq 30$ and $n_B = 35 \geq 30$, are large enough for the CLT to apply.

> **NOTE**
> Always remember to define each parameter so as to identify which parameter refers to which subscript.

(continued)

Mechanics: Calculator software gives $t = 3.41$ and $P = 0.0011$ with $df = 64.21$.
Conclusion in context with linkage to the P-value: With this small a *P-value*,
$0.0011 < 0.01$, there is sufficient evidence to reject H_0. In other words, there is
convincing evidence that there is a difference in mean miles put on during a game
between professional tennis players and professional basketball players.

(b)   Assuming that there is no difference in the mean miles put on during a game
between professional tennis players and professional basketball players, there is a
probability of 0.0011 of getting an observed mean difference of 0.3 or greater in either direction by chance
alone in random samples of 35 professional tennis players and 35 professional basketball players.

> **NOTE**
>
> You should *not* pool in a two-sample *t*-test like we did in a two-proportion *z*-test.

❯ Example 7.8

A city council member claims that male and female officers wait equal times for promotion in the police department. A women's spokesperson, however, argues that women wait longer than men. She obtains a random sample
of five promoted men officers with wait times of 8, 7, 10, 5, and 7 years for promotion and an independent random
sample of promoted four women officers with wait times of 9, 5, 12, and 8 years.

(a)   What are the parameters and hypotheses of interest for the women's spokesperson?
(b)   Are the conditions for a two-sample *t*-test for the difference of population means satisfied? Explain.

The *t*-test yields $t = -0.6641$ and $P = 0.2685$.

(c)   Interpret the *P*-value in context.
(d)   Give a conclusion in context.
(e)   Given the above conclusion, what type of error, Type I or Type II, might have been committed, and what
would be a possible consequence?

✓ Solutions

(a)   *Parameters:* Let μ_M represent the mean wait time for promotion in the population of male officers. Let μ_F represent the mean wait time for promotion in the population of female officers.
Hypotheses: $H_0: \mu_M = \mu_F$ and $H_a: \mu_M < \mu_F$
(b)   We check *independence* and *normality*. For independence, we are given random samples and must assume
that they are independent and less than 10% of total numbers of promoted male and female officers. For
normality, dotplots of the sample data show no strong skewness or outliers, so it is not unreasonable to
assume the data come from roughly normal population distributions.

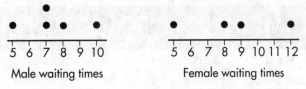

Male waiting times   Female waiting times

(c)   Assuming that the mean wait times for promotion for male and female officers are the same, there is a
0.2685 probability in getting a difference (male minus female) in sample means of $7.4 - 8.5 = -1.1$ or less
(more negative) by chance alone in random samples of five promoted male officers and four promoted
female officers.
(d)   With this large a *P*-value, $0.2685 > 0.05$, there is not sufficient evidence to reject H_0. In other words, there is
not convincing evidence of the women's spokesperson's claim that the mean wait time for promotion is longer for women than for men.

(continued)

(e)   There was not sufficient evidence to reject the null hypothesis. If the null hypothesis were false, we would be committing a Type II error, that is, mistakenly failing to reject a false null hypothesis. A possible consequence would be that female officers really do have to wait longer times for promotions but nothing is done to remedy the situation.

The above example could have been approached using a simulation. Could the above observed difference of -1.1 have occurred by chance? We can randomly assign the observed nine waiting times, (5, 7, 7, 8, 10, 5, 8, 9, 12), into two groups, five to the male group and the remaining four to the female group. For each such regrouping, the difference in mean waiting times (male $-$ female) is calculated. One such simulation is illustrated in the dotplot below.

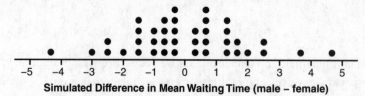

Simulated Difference in Mean Waiting Time (male − female)

As can be seen from this simulation, a difference of -1.1 or less is not unusual. We conclude that the observed difference could reasonably have occurred by chance, and is thus not significant. (Note that there are 9 out of 40 dots which are -1.1 or less. This gives $\frac{9}{40} = 0.225$ as a simulated estimate for the P-value.)

The analysis and procedure described above require that the two samples being compared be independent of each other. However, many experiments and tests involve comparing pairs of data. These pairs either occur in a single population or in two non independent populations. In this case, the proper procedure is to run a one-sample test on a single variable consisting of the differences from the paired data.

Paired Data

We've looked at confidence intervals and significance tests about the difference between two means when data come from two independent random samples. However, when we have a quantitative variable measured twice for the same individual or for two very similar individuals, inference on the true mean difference involves one-sample analysis on the single variable consisting of the differences from the paired data.

❯ Example 7.9

An SAT preparation class of 30 randomly selected students produces the following total score summary:

	First Score	Second Score	Improvement (2nd Score–1st Score)
Mean	1093.33	1135.58	42.25
Standard Deviation	87.76	85.73	27.92

Find a 90% confidence interval of the mean improvement in test scores.

✓ Solution

It would be wrong to calculate a confidence interval for a difference between two means using the means and standard deviations of the first and second scores. The independence condition between the two samples is

(continued)

violated! The proper procedure is a one-sample t-interval on the set of differences (improvement) between the scores for each of the 30 students.

Parameter: Let μ represent the mean improvement (2nd score minus 1st score) in the SAT scores of the population of students who take this SAT preparation class.

Procedure: A one-sample t-interval for the mean of a population of differences in paired data

Checks: We check *independence* and *normality*. For independence, we are given a random sample and $n = 30$ is less than 10% of all students. For normality, $n = 30$ is large enough for the CLT to apply.

Mechanics: With an unknown population SD, it's necessary to find a t-interval.

Using $\bar{x} = 42.25$ and $s = 27.92$, calculator software (`TInterval`) gives (33.59, 50.91).

Conclusion in context: We are 90% confident that the true mean improvement in test scores is between 33.59 and 50.91.

> **NOTE**
> Alternatively we can say: "We are 90% confident that the interval from 33.59 to 50.91 captures the true mean improvement in test scores."

The choice between doing a two-sample analysis and doing a matched pair analysis can be confusing. The determining factor is whether or not the two groups are independent. The difference has to do with design. What is the average reaction time of people who are sober compared with that of people who have had two beers? If our experimental design calls for randomly assigning half of a group of volunteers to drink two beers and then comparing reaction times with the remaining sober volunteers, a two-sample analysis is called for. However, if our experimental design calls for testing reaction times of all the volunteers and then giving all the volunteers two beers and retesting, a matched pair analysis is appropriate. Matching may come about as above because you have made two measurements on the same person, or measurements might be made on sets of identical twins, or between salaries of president and provost at a number of universities, etc. The key is whether the two sets of measurements are independent or related in some way relevant to the question under consideration.

> Example 7.10

Does a particular drug slow reaction times? If so, the government might require a warning label concerning driving a car while taking the medication. An SRS of 30 people who might benefit from the drug are tested before and after taking the drug, and their reaction times (in seconds) to a standard testing procedure are noted. The resulting histograms and summary statistics are as follows:

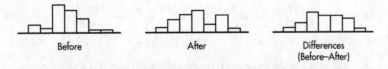

Before   After   Differences (Before–After)

	n	Mean	SD
Before	30	1.46433	0.26267
After	30	1.50200	0.20942
Differences	30	−0.037667	0.11755

(a) Perform the appropriate inference test.

(b) Given your conclusion, what type of error, Type I or Type II, might have been committed, and what would be a possible consequence?

(continued)

✓ **Solutions**

(a) If we performed a two-sample test, we would calculate the P-value to be 0.2708 and would conclude that with such a large P, the observed rise is *not* significant. However, this would not be the proper test or conclusion! The two-sample test works for *independent* sets. Yet, in this case, there is a clear relationship between the data, in pairs, and this relationship is completely lost in the procedure for the two-sample test. The proper procedure is to form the set of 30 differences and to perform a single sample significance test as follows.

Parameter: Let μ represent the mean difference in reaction time (before taking drug minus after taking drug) in the population of people taking a particular drug.

Hypotheses:

H_0: The reaction times of individuals to a standard testing procedure are the same before and after they take a particular drug; the mean difference is zero: $\mu_d = 0$.

H_a: The reaction time is greater after they take the drug; the mean difference (before minus after) is less than zero: $\mu_d < 0$.

Procedure: A paired t-test, that is, a one-sample t-test of a population mean (the mean difference of reaction times, before minus after)

Checks:

1. The *data are paired* because they are measurements on the same individuals before and after taking the drug.

2. The reaction times of any individual are independent of the reaction times of the others, so the *differences are independent*.

3. A *random sample* of people are tested.

4. The histogram of the differences looks *nearly normal* (roughly unimodal and symmetric), or we could use that the sample size, $n = 30$, is large enough for the CLT to apply.

5. The sample size, $n = 30$, is assumed to be less than 10% of the population of all people who may take the drug.

> **TIP**
>
> **Use proper terminology! In a two-sample t-test, the hypotheses are about the difference of means. In a paired t-test, the hypotheses are about the mean difference.**

Mechanics: `T-Test` gives $t = -1.755$ with $P = 0.0449$.

[Or we can calculate $SE(\bar{x}) = \frac{s}{\sqrt{n}} = \frac{0.11755}{\sqrt{30}} = 0.021462$ and $t = \frac{\bar{x} - 0}{\sigma_{\bar{x}}} = \frac{-0.037667}{0.021462} = -1.755$.

With $df = n - 1 = 29$, the P-value is $P(t < -1.755) = $ `tcdf(-1000,-1.755,29)` $= 0.0449$.]

Conclusion in context with linkage to the P-value: With this small of a P-value, $0.0449 < 0.05$, there is sufficient evidence to reject H_0; that is, there is convincing evidence that the true mean of the observed rise in reaction times after taking the drug is significant.

(b) There was sufficient evidence to reject the null hypothesis. If the null hypothesis were true, we would be committing a Type I error, that is, mistakenly rejecting a true null hypothesis. A possible consequence here is that the drug really doesn't slow reaction times, but the government requires the company to put an unnecessary, and possibly costly, label on the drug. Furthermore, people might choose not to take necessary medication.

Quiz 19

Multiple-Choice Questions

> **DIRECTIONS:** The questions that follow are each followed by five suggested answers. Choose the response that best answers the question.

1. It is hypothesized that grandmothers spend more hours per week on FaceTime with their grandchildren than do grandfathers. A random sample of 100 couples who have at least one grandchild are interviewed, and a two-sample t-test is performed. Have all conditions for inference been met?

   (A) Yes, all conditions have been met.
   (B) No, while a random sample was selected, there is no mention of random assignment.
   (C) No, there is no indication of whether the sample distribution is approximately normal.
   (D) No, independence is violated.
   (E) No, it is not clear whether the 10% condition is satisfied.

2. A consumer group believes that a wine-dispensing vending machine is pouring less than the advertised 5-ounces. They randomly sample 22 glasses of wine from the machine. In the sample, most of the wine weights are between 4.95 and 5.03 ounces with a few of the weights being substantially larger. Is it appropriate for the consumer group to perform a one-sample t-test for the mean wine weight poured by the machine?

   (A) Yes, all conditions have been met.
   (B) No, np is not given to be ≥ 10.
   (C) No, the independence condition is violated.
   (D) No, the normality condition is violated.
   (E) No, with no blinding, any conclusion is suspect.

3. A pharmaceutical company claims that a medicine will produce a desired effect for a mean time of 58.4 minutes. A government researcher runs a significance test of 40 patients and calculates a mean of $\bar{x} = 59.5$ with a standard deviation of $s = 8.3$. What is the P-value? (Assume that all conditions for inference are met.)

   (A) $P\left(t > \dfrac{59.5 - 58.4}{8.3/\sqrt{40}}\right)$ with $df = 39$

   (B) $P\left(t > \dfrac{59.5 - 58.4}{8.3/\sqrt{40}}\right)$ with $df = 40$

   (C) $2P\left(t > \dfrac{59.5 - 58.4}{8.3/\sqrt{40}}\right)$ with $df = 39$

   (D) $2P\left(t > \dfrac{59.5 - 58.4}{8.3/\sqrt{40}}\right)$ with $df = 40$

   (E) $2P\left(z > \dfrac{59.5 - 58.4}{8.3/\sqrt{40}}\right)$

4. In a study aimed at reducing developmental problems in low birth weight babies (under 2,500 grams), 347 infants were exposed to a special educational curriculum while 561 did not receive any special help. After 3 years, the children exposed to the special curriculum showed a mean IQ of 93.5 with a standard deviation of 19.1; the other children had a mean IQ of 84.5 with a standard deviation of 19.9. Find a 95% confidence interval estimate for the difference in mean IQs of all low birth weight babies who receive special intervention and those who do not. (Assume that all conditions for inference are met.)

   (A) $(93.5 - 84.5) \pm 1.97\sqrt{\dfrac{(19.1)^2}{347} + \dfrac{(19.9)^2}{561}}$

   (B) $(93.5 - 84.5) \pm 1.97\left(\dfrac{19.1}{\sqrt{347}} + \dfrac{19.9}{\sqrt{561}}\right)$

   (C) $(93.5 - 84.5) \pm 1.65\sqrt{\dfrac{(19.1)^2}{347} + \dfrac{(19.9)^2}{561}}$

   (D) $(93.5 - 84.5) \pm 1.65\left(\dfrac{19.1}{\sqrt{347}} + \dfrac{19.9}{\sqrt{561}}\right)$

   (E) $(93.5 - 84.5) \pm 1.65\sqrt{\dfrac{(19.1)^2 + (19.9)^2}{347 + 561}}$

5. A Fish and Game Department inspector is testing mercury content of fish in an area and will close the area to fishing if fish have a mean mercury content over 5 ppm.

   Suppose that a significance test results in a *t*-statistic of 0.035. Which of the following is the best interpretation of this value?

   (A) The probability is 0.035 that the null hypothesis is true.
   (B) The probability is 0.035 that the null hypothesis is false.
   (C) If the null hypothesis were true, the probability would be 0.035 of obtaining a sample mean as great as or greater than the observed sample mean.
   (D) The sample mean is 3.5% of the standard error.
   (E) The sample mean is 0.035 standard errors greater than the hypothesized mean of 5.

6. A published study states that U.S. beer consumption per adult age 21 and over is 28.2 gallons per person, per year. A brewer believes that because of difficult economic times during the past several years, this number has gone up. He surveys 50 randomly selected adults over age 21 as to yearly beer consumption and obtains a sample mean of 29.4 gallons. A one-sample *t*-test results in a *P*-value of 0.051. Which of the following is a correct interpretation of the *P*-value?

   (A) The probability is 0.051 that the mean beer consumption per adult age 21 and over is greater than 28.2 gallons.
   (B) The probability is 0.051 that the mean beer consumption per adult age 21 and over is greater than 29.4 gallons.
   (C) The probability is 0.051 that the mean beer consumption per adult age 21 and over is less than 29.4 gallons.
   (D) If the mean beer consumption per adult age 21 and over is 28.2 gallons, the probability is 0.051 that a randomly selected group of adults age 21 and over consumes a mean of 29.4 gallons or more.
   (E) If the mean beer consumption per adult age 21 and over is greater than 28.2 gallons, the

probability is 0.051 that a randomly selected group of adults age 21 and over consumes a mean of 29.4 gallons or more.

7. A researcher believes a new diet should improve weight gain in laboratory mice. If ten control mice on the old diet gain an average of 4 ounces with a standard deviation of 0.3 ounces, while the average gain for ten mice on the new diet is 4.8 ounces with a standard deviation of 0.2 ounces, what is the test statistic? (Assume that all conditions for inference are met.)

   (A) $t = \dfrac{4 - 4.8}{\sqrt{\dfrac{(0.3)^2}{10} + \dfrac{(0.2)^2}{10}}}$
   (B) $z = \dfrac{4 - 4.8}{\sqrt{\dfrac{(0.3)^2}{10} + \dfrac{(0.2)^2}{10}}}$
   (C) $t = \dfrac{4 - 4.8}{\dfrac{0.3}{\sqrt{10}} + \dfrac{0.2}{\sqrt{10}}}$
   (D) $z = \dfrac{4 - 4.8}{\dfrac{0.3}{\sqrt{10}} + \dfrac{0.2}{\sqrt{10}}}$
   (E) $t = \dfrac{4 - 4.8}{\dfrac{0.3}{\sqrt{10-1}} + \dfrac{0.2}{\sqrt{10-1}}}$

8. A company manufactures a synthetic rubber bungee cord with a braided covering of natural rubber and a minimum breaking strength of 450 kg. If the mean breaking strength of a sample drops below a specified level, the production process is halted and the machinery inspected. Which of the following would result from a Type I error?

   (A) Halting the production process when too many cords break
   (B) Halting the production process when the breaking strength is below the specified level
   (C) Halting the production process when the breaking strength is within specifications
   (D) Allowing the production process to continue when the breaking strength is below specifications
   (E) Allowing the production process to continue when the breaking strength is within specifications

9. Suppose you do five independent tests of the form $H_0: \mu = 38$ versus $H_a: \mu > 38$, each at the $\alpha = 0.01$ significance level. What is the probability of committing a Type I error and incorrectly rejecting a true null hypothesis with at least one of the five tests?

   (A) 0.01
   (B) 0.049
   (C) 0.05
   (D) 0.226
   (E) 0.951

10. Given an experiment with $H_0: \mu = 35$, $H_a: \mu < 35$, and a possible correct value of 32, you obtain a sample statistic of $\bar{x} = 33$. After doing analysis, you realize that the sample size n is actually larger than you first thought. Which of the following results from reworking with the increase in sample size?

   (A) Decrease in probability of a Type I error; decrease in probability of a Type II error; decrease in power
   (B) Increase in probability of a Type I error; increase in probability of a Type II error; decrease in power
   (C) Decrease in probability of a Type I error; decrease in probability of a Type II error; increase in power
   (D) Increase in probability of a Type I error; decrease in probability of a Type II error; decrease in power
   (E) Decrease in probability of a Type I error; increase in probability of a Type II error; increase in power

11. Do high school girls apply to more colleges than high school boys? A two-sample t-test of the hypotheses $H_0: \mu_{girls} = \mu_{boys}$ versus $H_a: \mu_{girls} > \mu_{boys}$ results in a P-value of 0.02. Which of the following statements must be true?

   I.   A 90% confidence interval for the difference in means contains 0.
   II. A 95% confidence interval for the difference in means contains 0.
   III. A 99% confidence interval for the difference in means contains 0.

   (A) I only
   (B) III only
   (C) I and II only
   (D) II and III only
   (E) I, II, and III

Free-Response Questions

DIRECTIONS: You must show all work and indicate the methods you use. You will be graded on the correctness of your methods and on the accuracy of your final answers.

1. A particular wastewater treatment system aims at reducing the most probable number per mL (MPN per mL) of E. coli to 1,000 MPN per 100 mL. A random study of 40 of these systems in current use is conducted with the data showing a mean of 1,002.4 MPN per 100 mL and a standard deviation of 7.12 MPN per 100 mL. A test of significance is conducted with:

   H_0: The mean concentration of E. coli after treatment under this system is 1,000 MPN per 100 mL.
   H_a: The mean concentration of E. coli after treatment under this system is greater than 1,000 MPN per 100 mL.

   The resulting P-value is 0.0197 with $df = 39$ and $t = 2.132$.

   (a) Interpret the P-value in the context of this study.
   (b) What conclusion should be drawn at a 5% significance level?
   (c) Given this conclusion, what possible error, Type I or Type II, might be committed? Give a possible consequence of committing this error.

2. Bisphenol A (BPA), a synthetic estrogen found in packaging materials, has been shown to leach into infant formula and beverages. More recently, BPA has been detected in high concentrations in cash register receipts. Random sample biomonitoring data to see if retail workers carry higher amounts of BPA concentration in their bodies than nonretail workers is as follows:

		BPA concentration (μg/L)	
	Sample size	Mean	Standard deviation
Nonretail workers	528	2.43	0.45
Retail workers	197	3.28	0.48

   (a) Calculate a 99% confidence interval for the difference in mean BPA body concentrations of nonretail and retail workers.

   (b) Does the confidence interval support the belief that retail workers carry higher amounts of BPA in their bodies than nonretail workers? Explain.

3. A random sample of 50 married couples were surveyed as to age at wedding. Following are data with regard to husbands, wives, and couple differences (husband's age minus wife's age).

Ages at date of wedding		
	Sample size n	Mean	Standard deviation
Husbands	50	30.6	8.792
Wives	50	28.6	7.977
Differences	50	2.0	3.742

   (a) A sociologist believes that the above table is evidence that the mean age at wedding for all husbands is greater than the mean age at wedding for all wives. He calculates $(-1.332, 5.332)$ to be a two-sample, 95% confidence interval for the true difference in mean ages of husbands and wives. He then concludes that there is not convincing evidence that the true mean age at wedding for husbands is greater than the true mean age at wedding for wives because the interval contains both positive and negative values. Is this a statistically valid argument? Explain.

   (b) Perform a hypothesis test at the 0.05 significance level with regard to the sociologist's claim.

The answers for this quiz can be found beginning on page 505.

Simulations and *P*-Values

When our learned test procedures do not apply, either because conditions are not met or because we are working with a parameter for which we have not learned a test procedure, we may be able to proceed with a simulation. We can use a simulation to determine what values of a test statistic are likely to occur by random chance alone, assuming the null hypothesis is true. Then looking at where our test statistic falls, we can estimate a *P*-value.

❭ Example 7.11

Researchers want to study whether Paul the octopus, who lives in a tank at Sea Life Centre in Oberhausen, Germany, could correctly predict the winner of soccer matches. In a random sample of 14 matches, Paul correctly predicted the winner in 12 out of 14 soccer matches (by choosing to eat mussels from the boxes labeled with national flags of the eventual winning teams).

NOTE

These numbers are too small to use in our standard proportion hypothesis tests, so we use a simulation.

(a)   What is the parameter of interest?

(b)   What are the null and alternative hypotheses?

(c)   If you conduct a simulation to investigate whether the observed result provides strong evidence that Paul can correctly predict winners, what would you use for the probability of success, for the sample size, and for the number of samples?

(d)   Which of the following graphs could reasonably have come from the simulation?

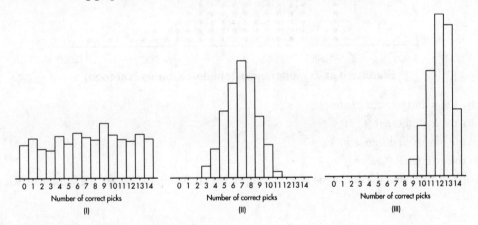

(e)   Estimate the *P*-value, and interpret it in context.

(f)   Make a conclusion based on the simulation and *P*-value.

✓ Solutions

(a)   Let *p* represent the proportion of correct predictions in the population of all soccer match winner predictions by Paul the octopus.

(b) H_0: Paul can do no better than guesswork in predicting winners, H_0: $p = 0.5$.
      H_a: Paul can correctly predict winners better than guesswork, H_a: $p > 0.5$.

(c) $p = 0.5$, $n = 14$, and a large number for number of samples.

(d)   Graph (II). The simulation should be centered at the null hypothesis expected value, not the observed value.

(continued)

(e)   Looking at the correct graph, (II), 12 or greater correct picks seems to have a probability of 0. If the null hypothesis were true, that is, if Paul can do no better than guesswork, the probability of Paul correctly predicting 12 or more out of 14 soccer matches is close to 0.00.

(f)   With this small of a P-value, $0.00 < 0.05$, there is strong evidence to reject H_0; that is, there is convincing evidence that Paul can pick the winner of soccer matches better than guesswork.

> Example 7.12

Childcare work is critically important, but it isn't well paid. Do childcare workers have greater salaries than fast-food workers? A random sample of 30 childcare workers' salaries has a median of \$23,240, while a random sample of 35 fast-food workers' salaries has a median of \$21,740. Could the observed difference have occurred by chance? We can randomly assign the 65 observed salaries into two groups, 30 to the childcare workers group and the remaining 35 to the fast-food workers group. For each regrouping, the difference in median salaries (childcare minus fast food) is calculated. One such simulation of 100 such regroupings is illustrated in the dotplot below.

NOTE

We have not learned a test procedure for medians, so a simulation is needed.

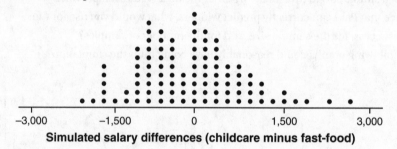

Simulated salary differences (childcare minus fast-food)

(a)   What are the parameters of interest?
(b)   What are the hypotheses?
(c)   What is the observed difference?
(d)   What is the estimated P-value?
(e)   Based on the estimated P-value, what is the appropriate conclusion in context at a significance level of 0.05?

✓ Solution

(a)   Let M_C represent the median salary in the population of childcare workers. Let M_F represent the median salary in the population of fast-food workers.

(b) $H_0: M_C = M_F$ and $H_a: M_C > M_F$.

(c)   The observed difference is $23,240 - 21,740 = \$1,500$.

(d)   There are 6 differences 1,500 or greater. So the estimated P-value is $\frac{6}{100} = 0.06$.

(e)   With this large of a P-value, $0.06 > 0.05$, there is not sufficient evidence to reject the null hypothesis. In other words, there is not convincing evidence that the median salary of all childcare workers is greater than the median salary of all fast-food workers.

Quiz 20

Multiple-Choice Questions

> **DIRECTIONS:** The questions that follow are followed by five suggested answers. Choose the response that best answers the question.

1. A company engineer creates a diagnostic measurement, $G = \dfrac{\text{Max} + \text{Min}}{2}$, which should be at least 34.60 in a sample of size 10 if certain machinery is operating correctly. To explore this diagnostic measurement, the machine is perfectly calibrated, and 100 random samples of size 10 of the product are taken from the assembly line. For each of these 100 samples, the diagnostic measurement G is calculated and shown plotted below.

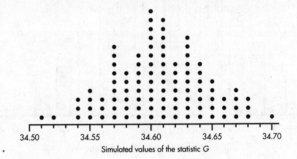

Simulated values of the statistic G

Each day, one sample of size 10 is taken from the assembly line and the diagnostic measurement G is calculated. If G drops too low, a decision to recalibrate the machinery is made.

One day the random sample is {34.89, 35.55, 33.93, 34.4, 35.51, 33.51, 35.0, 34.73, 35.19, 33.71}.

For the hypothesis test H_0: $G = 34.60$, H_a: $G < 34.60$, what is the estimated P-value?

(A) 0.01
(B) 0.02
(C) 0.03
(D) 0.04
(E) 0.05

2. A dietitian compares the effects of two diet pills. Using 25 participants taking each pill, she calculates a statistic W by computing a mean difference in weight loss expressed in pooled standard deviation units. The dietitian calculates $W = -0.7$ in her experiment. One thousand simulations assuming no real difference between the two pills gives the plot below for W.

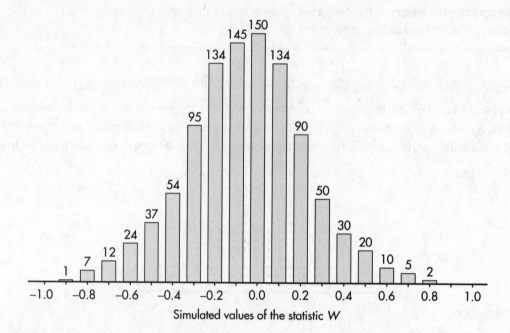

Simulated values of the statistic W

What is the estimated P-value for her observed value of $W = -0.7$ for the null hypothesis of no difference in effect versus the alternative hypothesis that there is a difference in effect?

(A) 0.007

(B) 0.008

(C) 0.01

(D) 0.02

(E) 0.027

Free-Response Questions

DIRECTIONS: You must show all work and indicate the methods you use. You will be graded on the correctness of your methods and on the accuracy of your final answers.

1. A battery manufacturer advertises that a 12-volt lawn and garden battery it sells has an average cranking power of 350 CCA (cold cranking amps). A consumer group believes that the true figure is lower. The consumer group obtains a random sample of 5 of the company's batteries and determines an average of 320 CCA. Is this significant? The company claims that perfectly good batteries do vary in CCA. To illustrate its point, the company runs a simulation by randomly picking 5 batteries 160 times from a group of new batteries manufactured to have 350 CCA strength and calculating the resulting averages to show that the consumer group's sample average of 320 CCA was possible. The company makes the following dotplot.

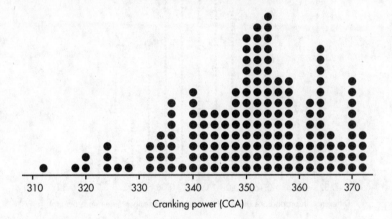

Cranking power (CCA)

    (a) What are the hypotheses for an appropriate test?
    (b) Explain why or why not the conditions for a *t*-test of a mean are met.
    (c) What is the estimated *P*-value?
    (d) Based on the estimated *P*-value, what is the appropriate conclusion in context?

2. A medical researcher believes that the antidepressant desipramine is more effective than the mood stabilizer lithium in treating cocaine addiction. In a randomized experiment, 50 cocaine addicts were randomly assigned to take either desipramine or lithium. Fourteen out of the 25 addicts taking desipramine had not relapsed after one year, while only 6 of the 25 addicts taking lithium had not relapsed. These numbers were too small to use in our standard proportion hypothesis tests. In a simulation of 500 sets of 50 addicts undergoing this experiment, assuming the null hypothesis of no difference in the treatments, the difference in proportions of addicts not relapsing (desipramine minus lithium) yields:

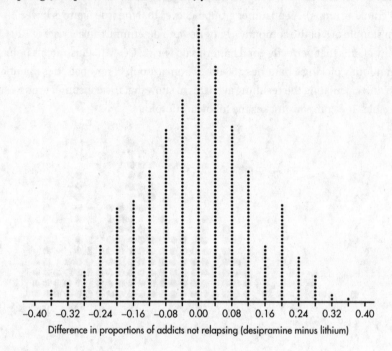

Difference in proportions of addicts not relapsing (desipramine minus lithium)

(a) In terms of appropriate proportions, what are the hypotheses?

(b) What is the observed difference in proportions?

(c) What is the estimated P-value?

(d) What is the appropriate conclusion in context?

The answers for this quiz can be found beginning on page 508.

More on Power, Type I Errors, and Type II Errors

Given a specific alternative hypothesis, a Type II error is a mistaken failure to reject the false null hypothesis, while the *power* is the probability of rejecting that false null hypothesis.

> **Example 7.13**

A candidate claims to have the support of 70% of the people, but you believe that the true figure is lower. You plan to gather an SRS and will reject the 70% claim if your sample shows 65% or less support. What if, in reality, only 63% of the people support the candidate?

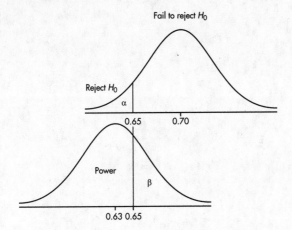

The upper graph shows the null hypothesis model with the claim that $p_0 = 0.70$ and the plan to reject H_0 if $\hat{p} < 0.65$. The lower graph shows the true model with $p = 0.63$. When will we fail to recognize that the null hypothesis is incorrect? Answer: precisely when the sample proportion is greater than 0.65. This is a Type II error with probability β. When will we rightly conclude that the null hypothesis is incorrect? Answer: when the sample proportion is less than 0.65. This is the "power" of the test and has probability $1 - \beta$.

The following points should be emphasized:

- Power gives the probability of avoiding a Type II error.
- Power has a different value for different possible correct values of the population parameter; thus, it is actually a *function* where the independent variable ranges over specific alternative hypotheses.
- Choosing a smaller α (that is, a tougher standard to reject H_0) results in a higher risk of Type II error and a lower power—observe in the above graphs how making α smaller (in this case moving the critical cutoff value to the left) makes the power less and β more!
- The greater the difference between the null hypothesis p_0 and the true value p, the smaller the risk of a Type II error and the greater the power—observe in the above picture how moving the lower graph to the left makes the power greater and β less. (The difference between p_0 and p is sometimes called the *effect*—thus the greater the effect, the greater is the power.)
- A larger sample size n will reduce the standard deviations, making both graphs narrower, and—if the rejection value is unchanged—the result is smaller α, smaller β, and larger power!

TIP

While calculations are not called for, you must understand the interactions among the errors, confidence level, power, *effect*, and sample size.

TIP

Both blocking in experiments and stratification in sampling reduce variability and so can improve power.

- A general conclusion is that if the other three values below are unchanged, the power will increase (and the probability of a Type II error will decrease) if any of the following occur:

    1. Significance level, *alpha*, increases (but this also increases the chance of a Type I error).

    2. Standard error decreases (for example, blocking on something strongly related to the response variable can do this).

    3. Sample size, *n*, increases (of course, more data also means greater expense in time and money).

    4. True parameter is farther from the null (this is called the *effect size* and is something you really have no control over).

Quiz 21

Multiple-Choice Questions

> **DIRECTIONS:** The questions that follow are each followed by five suggested answers. Choose the response that best answers the question.

1. An air quality monitor gives a warning when the air quality is unhealthy for sensitive groups. The situation can be stated in terms of hypotheses as follows.

   H_0: The air quality is good.

   H_a: The air quality is poor.

   Which of the following best describes the power of the test?

   (A) The probability of a warning when the air quality is poor.
   (B) The probability of a warning when the air quality is good.
   (C) The probability of not giving a warning when the air quality is poor.
   (D) The probability of not giving a warning when the air quality is good.
   (E) The probability that the monitor is working correctly.

2. Which of the following gives the probability of committing a Type I error?

   (A) The sample size
   (B) The standard error
   (C) The significance level
   (D) The P-value
   (E) The power

3. If the significance level is 0.1, and the probability of a Type II error is 0.4, what is the resulting power?

   (A) 0.1
   (B) 0.4
   (C) 0.5
   (D) 0.6
   (E) 0.9

4. A nutritionist conducts a study as to whether children prefer candy or toys for Halloween treats. The hypotheses are as follows.

   H_0: The probability a child chooses candy is 0.50.

   H_a: The probability a child chooses candy is greater than 0.50.

   Suppose the null hypothesis is false. What is the effect on power and the probability of a Type II error if the significance level is changed from 0.05 to 0.01?

   (A) Neither the power of the test nor the probability of a Type II error would change.
   (B) Both the power of the test and the probability of a Type II error would increase.
   (C) Both the power of the test and the probability of a Type II error would decrease.
   (D) The power of the test would increase, and the probability of a Type II error would decrease.
   (E) The power of the test would decrease, and the probability of a Type II error would increase.

5. Which of the following is the best explanation of the power of a test?

   (A) The probability of correctly detecting an effect which exists.
   (B) The probability of incorrectly detecting an effect when no effect exists.
   (C) The probability of incorrectly not detecting an effect which exists.
   (D) The probability of correctly not detecting an effect when no effect exists.
   (E) The probability of detecting an effect whether or not one exists.

6. A guidance counselor believes that students studying for an AP exam get less than 8 hours of sleep the night before the exam. The hypotheses are as follows:

$H_0: \mu = 8$

$H_a: \mu < 8$

Assume that the guidance counselor is correct that $\mu < 8$. For a fixed sample size and significance level, the power of the test will be greatest if the actual mean hours of sleep before the AP exam is which of the following?

(A) 6
(B) 7
(C) 8
(D) 9
(E) 11

7. A better business agency claims that over 20 percent of merchants favor eliminating the penny from circulation. To test this claim at a 5 percent significance level, a random sample of n merchants will be interviewed. The hypotheses are as follows.

H_0: Twenty percent of merchants favor eliminating the penny from circulation.

H_a: More than twenty percent of merchants favor eliminating the penny from circulation.

Which of the following would result in the greatest power for this test?

(A) A sample size of 300 and a true proportion of 40 percent
(B) A sample size of 500 and a true proportion of 20 percent
(C) A sample size of 500 and a true proportion of 40 percent
(D) A sample size of 700 and a true proportion of 30 percent
(E) A sample size of 700 and a true proportion of 40 percent

8. Suppose that your friend claims that he can shoot free throws at a 70 percent rate. You don't believe he is that good and decide to test him. The hypotheses are as follows.

$H_0: p = 0.70$

$H_a: p < 0.70$

Suppose his real rate is 65 percent. Which of the following would result in the highest power for this test?

(A) A significance level of 0.01 and 100 shots
(B) A significance level of 0.01 and 200 shots
(C) A significance level of 0.01 and 300 shots
(D) A significance level of 0.05 and 200 shots
(E) A significance level of 0.05 and 300 shots

Free-Response Question

> **DIRECTIONS:** You must show all work, and indicate the methods you use. You will be graded on the correctness of your methods and on the accuracy of your final answers.

1. The amount of active ingredient per tablet of an advertised medication has an approximately normal distribution with mean 81 mg and standard deviation 0.3 mg. To pass quality inspection, tablets should have between 80.5 and 81.5 mg of active ingredient.

   (a) What is the probability that a single selected tablet passes inspection?

   As each shipment arrives, an inspector randomly chooses five tablets from that shipment. She finds the mean active ingredient in the five tablets and rejects the entire shipment if the sample mean is less than 80.5 mg or greater than 81.5 mg.

   (b) Describe the sampling distribution of the sample mean for samples of size five.
   (c) What is the probability that a Type I error is committed?

   Suppose the true mean active ingredient in a particular shipment is 80.4 mg with a standard deviation of 0.3 mg.

   (d) What is the probability a Type II error is committed for this shipment?

The answers to this quiz are found beginning on page 508.

Confidence Intervals Versus Hypothesis Tests

A question sometimes arises as to whether a problem calls for calculating a confidence interval or performing a hypothesis test. Generally, a claim about a population parameter indicates a hypothesis test, while an estimate of a population parameter asks for a confidence interval. Sometimes, however, it is possible to conduct a hypothesis test by constructing a confidence interval estimate of the parameter and then checking whether or not the null parameter value is in the interval. While, when possible, this alternative approach is accepted on the AP exam, it

For example:

- If a 95% confidence interval for μ captures μ_0, then there is not sufficient evidence to reject H_0: $\mu = \mu_0$ in a two-sided test with $\alpha = 0.05$.
- If a 95% confidence interval for μ does not capture μ_0, then there is sufficient evidence to reject H_0: $\mu = \mu_0$ in a two-sided test with $\alpha = 0.05$.

does require very special care in still remembering to state hypotheses, in dealing with one-sided versus two-sided, and in how standard errors are calculated in problems involving proportions. So, the recommendation is to avoid the alternative approach and instead to conduct the test as you normally would for a hypothesis test. Carefully read the question. If it asks whether or not there is evidence, the question requires a hypothesis test; if it involves how much or how effective, the question requires a confidence interval calculation. However, it should be noted that there have been AP free-response questions where part (a) calls for a confidence interval calculation and part (b) asks if this calculation provides evidence relating to a hypothesis test.

› Example 7.14

Suppose it is reported that random samples of 3-point shots by basketball players Stephen Curry and Michael Jordan show a 43% rate for Curry and a 33% rate for Jordan.

(a)   Are these numbers parameters or statistics?
(b)   State appropriate hypotheses for testing whether the difference is statistically significant.
(c)   Suppose the sample sizes were both 100. Find the z-statistic and the P-value, and give an appropriate conclusion at the 5% significance level.
(d)   Calculate and interpret a 95% confidence interval for the difference in population proportions.
(e)   Are the test decision and confidence interval consistent with each other?
(f)   Repeat (c), (d), and (e) as if the sample size had been 200 for each.

✓ Solutions

(a)   These two numbers are statistics because they describe samples, not all 3-point shots ever taken by these two players.
(b) H_0: $p_{Curry} = p_{Jordan}$ and H_a: $p_{Curry} \neq p_{Jordan}$
(c)   Calculator software gives $z = 1.46$ and $P = 0.145$. With this large of a P-value, $0.145 > 0.05$, there is not sufficient evidence to reject H_0; that is, there is not convincing evidence of a difference in the true 3-point percentage rates of Curry and Jordan.
(d)   Calculator software gives $(-0.034, 0.234)$. We are 95% confident that the true difference in 3-point percentage rates (Curry minus Jordan) is between -3.4% and 23.4%.
(e)   Yes, they are consistent. We did not conclude that the two percentages differ, and the confidence interval (for the difference in population proportions) includes the value zero.
(f)   With a sample size of 200, calculator software gives $z = 2.06$ and $P = 0.039$. With this small of a P-value, $0.039 < 0.05$, there is sufficient evidence to reject H_0; that is,

TIP

Sample size plays a substantial role in inferential statistics!

(continued)

there is convincing evidence of a difference in the true 3-point percentage rates of Curry and Jordan. With a sample size of 200, calculator software gives a confidence interval of (0.005, 0.195). We are 95% confident that the true difference in 3-point percentage rates (Curry minus Jordan) is between 0.5% and 19.5%. This interval does not include zero, so it is convincing evidence of a difference in the true 3-point percentage rates of Curry and Jordan. This is again consistent with the hypothesis test.

SUMMARY

- Statistical inference draws a conclusion (i.e., infers something) about a population parameter based on a sample statistic.

  1. A confidence interval estimates the value of a parameter with a range of values, and we estimate a population mean μ with a t-interval.
  2. A significance test assesses the plausibility of a particular claim about a parameter, and we test a claim about a population mean μ with a t-test.

- Important assumptions and conditions always must be checked before calculating a confidence interval or proceeding with a hypothesis test for a population mean. These include:

  1. *Independence:* the data come from a random sample from the population of interest, and when sampling without replacement, $n < 0.10N$.
  2. *Normality:* either the population has a roughly normal distribution or the sample size is large enough, $n \geq 30$, for the CLT to apply; if the population distribution is unknown and $n < 30$, use a graph of the sample data to assess whether it is reasonable to assume the data come from a roughly normal population distribution.

- When the population standard deviation σ is unknown (which is almost always the case) and provided that the original population is normally distributed or the sample size n is large enough, the statistic $\frac{\bar{x} - \mu}{s/\sqrt{n}}$ follows a t-distribution with $n - 1$ degrees of freedom.

- To decrease the margin of error ($\text{ME} = t^* SE(\bar{x}) = t^* \frac{s}{\sqrt{n}}$) you can choose to either:

  1. Increase the sample size n or
  2. Decrease the confidence level (this will make t^* smaller) while everything else remains the same.

- When scoring a confidence interval free-response question on the exam, readers will look for whether you:

  1. Identify the procedure (such as one-sample t-interval for a population mean) and check the conditions.
  2. Compute a proper interval (it is sufficient to simply copy the interval from your calculator calculation).
  3. Interpret the interval in context (the conclusion must mention the confidence level and the parameter and refer to the population, not to the sample).

- When asked to justify a claim based on a confidence interval:

  1. If all the values in the confidence interval are consistent with the claim, then the confidence interval gives convincing evidence in support of the claim.
  2. If even a single value in the confidence interval is not consistent with the claim, then the confidence interval does not give convincing evidence in support of the claim.

- When scoring a hypothesis test free-response question on the exam, readers will look for whether you:

  1. State the hypotheses (both H_0 and H_a) and define the parameter of interest. (The hypotheses must refer to the population, not to the sample. The alternative hypothesis must be stated in terms of a strict inequality.)
  2. Identify the procedure (such as one-sample t-test for a population mean) and check the conditions.
  3. Compute the test statistic, P-value, and degrees of freedom if appropriate. (It is sufficient to simply copy all of these from your calculator calculation.)
  4. Give a conclusion in context with linkage to the P-value. (The conclusion must refer to the parameter and to the population, not to the sample.)

NOTE

You do not need to interpret the P-value unless specifically asked to do so.

- The conclusion to a significance test should start with a comparison between the *P*-value and a significance level.

  1. With small *P*-values, that is, when $P \leq \alpha$, the test statistic is unlikely to have occurred by random chance alone. We have sufficient evidence to reject H_0, and there is convincing evidence in support of H_a.
  2. With large *P*-values, that is, when $P > \alpha$, the test statistic is likely to have occurred by random chance alone. We do not have sufficient evidence to reject H_0, and there is not convincing evidence in support of H_a.

- Conditions when constructing a confidence interval or for performing a hypothesis test about a difference in means:

  1. *Independence:* the data come from two independent random samples, and when sampling without replacement, $n_1 < 0.10N$ and $n_2 < 0.10N_2$.
  2. *Normality:* for *each* sample, the population distribution is roughly normal or the sample size is large enough, $n \geq 30$, for the CLT to apply; for each sample, if the population distribution is unknown and $n < 30$, use a graph of the sample data to assess whether it is reasonable to assume the data come from a roughly normal population distribution.

- To distinguish between matched pair versus two-sample inference procedures:

  1. If the data come from separate (independent) random samples or are obtained from randomly assigned treatment groups in an experiment, then a two-sample inference procedure is appropriate.
  2. If the data come from two measurements taken from a pairing or from a single individual completing two tasks or if they are obtained from a matched pairs experiment, then a matched pair (one-sample) inference procedure is appropriate.

> **NOTE**
>
> **For paired data, be sure to indicate the direction of the difference.**

- A general conclusion about power is that the power will increase (and the probability of a Type II error will decrease) if any of the following occur:

  1. Significance level, *alpha*, increases.
  2. Standard error decreases.
  3. Sample size, *n*, increases.
  4. True parameter is farther from the null.

- When our learned test procedures do not apply, we may be able to proceed with a simulation.

8

Inference for Categorical Data: Chi-Square

(1–2 multiple-choice questions on the AP exam)

Learning Objectives

In this unit, you will learn how

→ To be able to conduct a chi-square test for goodness-of-fit (for a distribution of proportions of one categorical variable).

→ To be able to conduct a chi-square test for independence (for associations between two categorical variables within a single population).

→ To be able to conduct a chi-square test for homogeneity (for comparing distributions of a categorical variable across two or more populations).

In this unit, you will be introduced to the chi-square statistic, which gives a statistical measurement of the difference between observed and expected counts. You will learn how to determine if observed counts in categorical data are consistent with expected counts due to random variation. You will learn when to use each of three different chi-square tests: the goodness-of-fit test, which regards the distribution of proportions of a categorical variable; the test for independence, which looks for an association between two categorical variables in a single population; and the test for homogeneity, which compares distributions of a categorical variable across two or more populations. In Units 6 and 7, we carried out significance tests about population parameters, such as a population proportion p or a population mean μ. In this unit, we are carrying out significance tests about *distributions* of categorical variables.

Chi-Square Test for Goodness-of-Fit

A critical question is often whether or not an observed pattern of data fits some given distribution. A perfect fit cannot be expected, and so we must look at discrepancies and make judgments as to the *goodness-of-fit*.

Our approach is similar to that developed earlier. There is the null hypothesis of a good fit, that is, the hypothesis that a given theoretical distribution correctly describes the situation, problem, or activity under consideration. Our observed data consist of one possible sample from a whole universe of possible samples. We ask about the chance of obtaining a sample with the observed discrepancies if the null hypothesis is really true. Finally, if the chance is too small, we reject the null hypothesis and say that there is convincing evidence that the fit is not a good one.

How do we decide about the significance of observed discrepancies? It should come as no surprise that the best information is obtained from squaring the discrepancy values, as this has been our technique for studying variances from the beginning. Since, for example, an observed difference of 23 is more significant if the original values are 105 and 128 than if they are 10,602 and 10,625, we must appropriately *weight* each difference. Such weighting is accomplished by dividing each squared difference by the expected value. The sum of these

> **TIP**
>
> As long as the observed values do not exactly equal the expected values, then $\chi^2 \neq 0$, and the explanation is either random chance or that the claimed distribution is incorrect.

weighted differences or discrepancies is called the *chi-square statistic* and is denoted as χ^2 (χ is the lowercase Greek letter chi):

$$\chi^2 = \sum \frac{(\text{observed count} - \text{expected count})^2}{\text{expected count}}$$

The smaller the resulting χ^2-value, the better the fit. The *P*-value is the probability of obtaining a χ^2-value as extreme as (or as more extreme than) the one obtained if the null hypothesis is assumed true. If the χ^2-value is large enough, that is, if the *P*-value is small enough, we say there is sufficient evidence to reject the null hypothesis and convincing evidence that the fit is poor.

To decide how large a calculated χ^2-value must be to be significant, that is, to choose a critical value, we must understand how χ^2-values are distributed. A χ^2-distribution has only nonnegative values, is not symmetric, and is always skewed to the right. There are distinct χ^2-distributions, each with an associated number of degrees of freedom (*df*). The larger the *df* value, the less pronounced is the skew, and the closer the χ^2-distribution is to a normal distribution. Note, for example, that squaring the often-used *z*-scores 1.645, 1.96, and 2.576 results in 2.71, 3.84, and 6.63, respectively, which are entries found in the first row of the χ^2-distribution table.

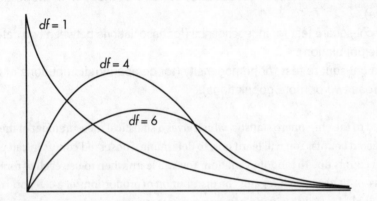

> **NOTE**
>
> While we will be using chi-square for categorical data, the chi-square distribution as seen in these graphs is a *continuous* distribution, and applying it to counting data is just an approximation.

For inference about the distribution of a single categorical variable, such as a goodness-of-fit test, we will use a chi-square distribution with degrees of freedom, *df* = *number of categories* − 1.

A large χ^2-value may or may not be significant—the answer depends on which χ^2-distribution we are using. Table C in the Appendix gives critical χ^2-values for the more commonly used percentages or probabilities, and probabilities can be calculated using χ^2cdf.

χ^2cdf(lowerbound, upperbound, *df*) gives the probability a score is between two bounds (χ^2-values) for the specified *df* (degrees of freedom).

To use the χ^2-distribution for approximations in goodness-of-fit problems, the individual expected values cannot be too small. Our rule of thumb is that all expected values should be greater than 5. Finally, as in all hypothesis tests we've looked at, the sample should be randomly chosen from the given population, and, if sampling without replacement, the sample size should be less than 10% of the population size.

❯ Example 8.1

Is there a birthday effect in college athletics? A sportscaster believes that the birthdays of male D1 tennis players are not uniformly distributed across the four quarters of the year. He selects a random sample of 90 male D1 tennis players and tabulates their birthday months in the following table.

Birthday	Jan–Mar	Apr–Jun	Jul–Sep	Oct–Dec
Number of players	34	24	13	19

(continued)

(a)   What are the hypotheses to be tested, and what is the procedure to be used?

(b)   Confirm that the conditions are satisfied.

Calculator software gives $\chi^2 = 10.53$.

(c)   What is the *P*-value? Interpret it in context.

(d)   Using a significance level of $\alpha = 0.05$, give a conclusion to the hypothesis test.

✓ Solutions

(a)   *Hypotheses:*

H_0: The birthdays of all male D1 tennis players are uniformly distributed across the four quarters of the year.

H_a: The birthdays of all male D1 tennis players are not uniformly distributed across the four quarters of the year.

Procedure: Chi-square test for goodness-of-fit.

(b)   *Checks:* We check *independence* and *large counts.* For independence, we are given a random sample, and the sample size, 90, is less than 10% of all male D1 tennis players. For large counts, if H_0 were true, all four expected counts would be $\left(\frac{1}{4}\right)(90) = 22.5$ and $22.5 > 5$.

(c)   With $df =$ categories $-1 = 4 - 1 = 3$, the *P*-value $= \chi^2\text{cdf}(10.53, 1000, 3) = 0.015$. *Interpretation of the P-value:* There is a 0.015 probability of getting a chi-square statistic of 10.53 or greater just by random chance in selecting the 90 male D1 tennis players, assuming that birthdays of all male D1 tennis players are uniformly distributed across the four quarters of the year.

(d)   *Conclusion in context with linkage to the P-value:* With this small a *P*-value, $0.015 < 0.05$, there is sufficient evidence to reject H_0. In other words, there is convincing evidence that the birthdays of all male D1 tennis players are not uniformly distributed across the four quarters of the year.

> **NOTE**
>
> In Units 6 and 7, the tests could be one-sided or two-sided, but with χ^2, H_a is simply that H_0 is incorrect.

> **NOTE**
>
> Even though the observed counts are integers, the expected counts might not be integers and should not be rounded to integers.

> Example 8.2

American roulette wheels have 38 slots: 18 black, 18 red, and 2 green. An inspector records data from a random sample of 500 spins. The following table displays the results.

Color	Black	Red	Green
Count	247	219	34

(a)   What conclusion should be drawn as to whether or not the machine is operating correctly? Use a significance level of $\alpha = 0.05$.

(b)   Given the above conclusion, what type of error, Type I or Type II, might have been committed, and what would be a possible consequence?

✓ Solution

(a)   *Hypotheses:*

H_0: The machine is operating correctly with black, red, and green appearing in the ratio 18:18:2.

H_a: The machine is operating incorrectly; black, red, and green are not appearing in the ratio 18:18:2.

(continued)

Procedure: Chi-square test for goodness-of-fit.

Checks: We check *independence* and *large counts*. For independence, we are given a random sample. For large counts, if H_0 were true, the 3 expected counts would be $\left(\frac{18}{38}\right)(500)$ = 236.8, $\left(\frac{18}{38}\right)(500) = 236.8$, and $\left(\frac{2}{38}\right)(500) = 26.3$, all three of which are > 5.

Mechanics: $\chi^2 = \frac{(247 - 236.8)^2}{236.8} + \frac{(219 - 236.8)^2}{236.8} + \frac{(34 - 26.3)^2}{26.3} = 4.03$.

With $df = 3 - 1 = 2$, we calculate the *P*-value to be $\chi^2\text{cdf}(4.03, 1000, 2) = 0.133$.

Conclusion in context with linkage to the P-value: With this large a *P*-value, $0.133 > 0.05$, there is not sufficient evidence to reject H_0. In other words, there is not convincing evidence that the machine is operating incorrectly.

(b) There was not sufficient evidence to reject the null hypothesis. If the null hypothesis were false, we would be committing a Type II error, that is, mistakenly failing to reject a false null hypothesis. A possible consequence would be that the machine is operating incorrectly but nothing is done to fix it.

NOTE

We do not check the 10% condition here because there is no sampling without replacement.

Interpretation of the *P*-value: There is a 0.133 probability of getting a chi-square statistic of 4.03 or greater just by random chance in selecting the 500 spins, assuming that the machine is operating correctly.

Chi-Square Test for Independence

In the previous goodness-of-fit problems, a set of expectations was based on an assumption about how the distribution should turn out. We then tested whether an observed sample distribution could reasonably have come from a larger set based on the assumed distribution.

There is a class of real-world problems in which we want to determine if there is a significant association between two categorical variables. The procedure is called a *test of independence*. For example, do nonsmokers, light smokers, and heavy smokers all have the same likelihood of eventually being diagnosed with cancer, heart disease, or emphysema? Is there a relationship (association) between smoking status and being diagnosed with one of these diseases?

We classify our **single sample** observations in two ways and then ask whether the two ways are independent of each other. For example, we might consider several age groups and within each group ask how many employees show various levels of job satisfaction. The null hypothesis is that age and job satisfaction are independent, that is, that the proportion of employees expressing a given level of job satisfaction is the same no matter which age group is considered.

Analysis involves calculating a table of *expected values*, assuming the null hypothesis about independence is true. We then compare these expected values with the observed values and ask whether the differences are reasonable if H_0 is true. The significance of the differences is gauged by the same χ^2-value of weighted squared differences. The smaller the resulting χ^2-value, the more reasonable the null hypothesis of independence is. If the χ^2-value is large enough, that is, if the *P*-value is small enough, we can say that the evidence is sufficient to reject the null hypothesis and to claim that there *is* convincing evidence of some relationship between the two variables or methods of classification.

As with the goodness-of-fit test, the test for independence requires that we have a simple random sample. The sample should be less than 10% of the population (if sampling without replacement), and expected values should all be > 5. (It is not enough to say that all expected cells are > 5; you must list the expected values.)

When testing for independence,

$$df = (r - 1)(c - 1)$$

where *df* is the number of degrees of freedom, *r* is the number of rows, and *c* is the number of columns.

TIP

In a chi-square test for independence, the null hypothesis is always that the two categorical variables are independent, or that there is no association between them, or that there is no relationship between them.

A point worth noting is that even if there is sufficient evidence to reject the null hypothesis of independence, we cannot necessarily claim any direct *causal* relationship. In other words, although we can make a statement about some link or relationship between two variables, we are *not* justified in claiming that one causes the other. For example, we may demonstrate a relationship between salary level and job satisfaction, but our methods would not show that higher salaries cause higher job satisfaction. Perhaps an employee's higher job satisfaction impresses his superiors and thus leads to larger increases in pay. Or perhaps there is a third variable, such as training, education, or personality, that has a direct causal relationship to both salary level and job satisfaction.

> ### Example 8.3

A growing number of states have legalized marijuana for medical or recreational purposes. In a nationwide telephone poll of 1000 randomly selected adults representing Democrats, Republicans, and Independents, respondents were asked two questions: their party affiliation and if they supported the legalization of marijuana. The answers, cross-classified by party affiliation, are given in the following two-way table (also called a *contingency table*).

Support Legalization

	Yes	**No**	**No Opinion**	*Totals*
Democrats	280	110	15	*405*
Republicans	155	190	10	*355*
Independents	180	45	15	*240*
Totals	*615*	*345*	*40*	*1,000*

Test the null hypothesis that support for legalizing marijuana is independent of party affiliation. Use a 5% significance level.

✓ Solution

Before beginning the formal solution, we note how a mosaic plot visually shows information about the joint proportions of both variables. The plot indicates that the proportion of supporters for legalization is greatest among the Independents and the least among the Republicans.

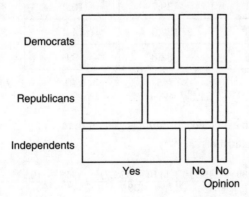

> **NOTE**
>
> Another acceptable statement of the hypotheses:
>
> H_0: There is no association between party affiliation and support for legalizing marijuana.
>
> H_a: There is an association between party affiliation and support for legalizing marijuana.

Hypotheses: H_0: Party affiliation and support for legalizing marijuana are independent.

H_a: Party affiliation and support for legalizing marijuana are not independent.

(continued)

Procedure: Chi-square test for independence.

Checks: We check *independence* and *large counts*. For independence, we are given a random sample, and $n = 1{,}000$ is less than 10% of all adults. For large counts, by putting the observed data into a calculator matrix, calculator software such as χ^2-Test gives the expected cells in a second matrix and all are >5:

NOTE

Put the data, but not the totals, into a matrix on your calculator.

249.1	139.7	16.2
218.3	122.5	14.2
147.6	82.8	9.6

Mechanics: The χ^2-Test software gives $\chi^2 = 94.5$ and $P = 0.000$.

Conclusion with linkage to the P-value: With this small of a *P*-value, $0.000 < 0.05$, there is sufficient evidence to reject H_0; that is, among all adults there is convincing evidence of a relationship between party affiliation and support for legalizing marijuana.

[For instructional purposes, we note that the expected values, χ^2, and *P* can be calculated as follows:

To calculate, for example, the expected value in the upper left cell, we can proceed in any of several equivalent ways. First, we could note that the proportion of Democrats is $\frac{405}{1{,}000} = 0.405$; and so, if independent, the expected number of Democrat *yes* responses is $0.405(615) = 249.1$. Instead, we could note that the proportion of *yes* responses is $\frac{615}{1{,}000} = 0.615$; and so, if independent, the expected number of Democrat *yes* responses is $0.615(405) = 249.1$. Finally, we could note that both these calculations simply involve $\frac{(405)(615)}{1{,}000} = 249.1$.

In other words, the expected value of any cell can be calculated by multiplying the corresponding row total by the corresponding column total and then dividing by the table total. Thus, for example, the expected value for the middle cell, which corresponds to Republican *no* responses, is $\frac{(355)(345)}{1{,}000} = 122.5$.

The formula for an expected cell count is the following: *expected count* = $\frac{(row\ total)(column\ total)}{table\ total}$.

Continuing in this manner, we fill in the table of expected values as follows:

	Support Legalization			
	Yes	**No**	**No Opinion**	*Totals*
Democrats	249.1	139.7	16.2	*405*
Republicans	218.3	122.5	14.2	*355*
Independents	147.6	82.8	9.6	*240*
Totals	*615*	*345*	*40*	*1,000*

(An appropriate check at this point is that each expected cell count is greater than 5.)
Next we calculate the value of the chi-square statistic:

$$\chi^2 = \sum \frac{(\text{observed} - \text{expected})^2}{\text{expected}} = \frac{(280 - 249.1)^2}{249.1} + \frac{(110 - 139.7)^2}{139.7} + \frac{(15 - 16.2)^2}{16.2} + \frac{(155 - 218.3)^2}{218.3}$$

$$+ \frac{(190 - 122.5)^2}{122.5} + \frac{(10 - 14.2)^2}{14.2} + \frac{(180 - 147.6)^2}{147.6} + \frac{(45 - 82.8)^2}{82.8} + \frac{(15 - 9.6)^2}{9.6} = 94.5$$

$$df = (r - 1)(c - 1) = (3 - 1)(3 - 1) = 4$$

The P-value is then calculated using calculator software such as $\chi^2\text{cdf}(94.5, 1000, 4) =$ 0.000.]

NOTE

While it is important to understand these calculations, on the exam the calculations should be made quickly by putting the observed data into a matrix and using the χ^2-Test on your calculator.

If asked, the largest contributions to the chi-square statistics were two of the Republican responses, with fewer yes responses and more no responses than would be expected with independence.

As for conditions to check for chi-square tests for independence, we should check that the sample is randomly chosen, that the expected values for all cells are greater than 5, and that the sample size is less than 10% of the population (if sampling without replacement).

❯ Example 8.4

Young adults from different generations have had entirely different religious and social experiences that have led to profound religious changes over the generations. Has this affected spirituality? In a nationwide poll of 500 randomly selected adults, respondents were asked two questions: their age and whether spirituality was important in their lives. The cross-classified answers are given in the following two-way table:

		Age (Generation)			
		Baby boomers	Generation X	Millenials	Generation Z
Is spirituality in one's life important?	Yes	73	71	79	69
	No	32	54	61	61

(a) Test the null hypothesis that importance of spirituality is independent of generational age. Use a 5% significance level.

(b) Interpret the P-value in context.

✓ Solution

(a) *Hypotheses:*

H_0: Importance of spirituality and generational age are independent.
H_a: Importance of spirituality and generational age are not independent.

Procedure: Chi-square test of independence.

Checks: We are given a random sample, and $n = 500$ is less than 10% of all adults. All expected cells (calculated below) are greater than 5.

Mechanics: Using a calculator matrix and χ^2-Test software gives $\chi^2 = 7.220$, $P = 0.0652$, $df = 3$, and the following expected cell counts:

61.3	73	81.8	75.9
43.7	52	58.2	54.1

Conclusion in context with linkage to the P-value: With this large a P-value, $0.0652 > 0.05$, there is *not* sufficient evidence to reject the null hypothesis. There is *not* convincing evidence of a relationship between importance of spirituality and generational age.

(b) Assuming that importance of spirituality and generational age are independent, there is a 0.0652 probability of getting a χ^2-value of 7.220 or greater by chance alone in a random sample of 500 adults.

Chi-Square Test for Homogeneity

In chi-square goodness-of-fit tests, we work with a single variable in comparing a single sample to a population model. In chi-square independence tests, we work with a single sample classified on two variables. Chi-square procedures can also be used with a single variable to compare samples from two or more populations. Conditions to check include that the samples ideally be *simple random samples*, that they be taken *independently* of each other, that the original populations be large compared with the sample sizes (if sampling without replacement), and that the expected values for all cells be greater than 5. The contingency table used has a row for each sample. The resulting procedure is called a *chi-square test for homogeneity*.

> ## Example 8.5

In a large city, a group of AP Statistics students work together on a project to determine which group of school employees has the greatest proportion who are satisfied with their jobs. In independent simple random samples of 100 teachers, 60 administrators, 45 custodians, and 55 secretaries, the numbers satisfied with their jobs were found to be 82, 38, 34, and 36, respectively. Is there convincing evidence that the proportion of employees satisfied with their jobs is different in different school system job categories?

✓ Solution

Hypotheses:

H_0: The proportion of employees satisfied with their jobs is the same across the various school system job categories.

H_a: At least two of the job categories differ in the proportion of employees satisfied with their jobs.

Procedure: A chi-square test for homogeneity with observed data:

	Teachers	Administrators	Custodians	Secretaries
Satisfied	82	38	34	36
Not satisfied	18	22	11	19

Checks: We check *independence* and *large counts*. For independence, we are given independent random samples, and we assume that the four samples are each less than 10% of their respective populations in the large city. For large counts, using a calculator matrix and χ^2-Test gives the expected cells in a second matrix, and these are all >5:

73.1	43.8	32.9	40.2
26.9	16.2	12.1	14.8

Mechanics: The χ^2-Test software gives $\chi^2 = 8.707$, $P = 0.0335$, and $df = 3$.

Conclusion in context with linkage to the P-value:

With this small of a *P*-value, $0.0335 < 0.05$, there is convincing evidence to reject H_0; that is, there is convincing evidence that the true proportion of employees satisfied with their jobs is not the same across all the school system job categories.

NOTE

Interpretation of the *P*-value: Assuming there is no difference in the proportion of employees satisfied with their jobs across the various school system job categories, there is a 0.0335 probability of getting a χ^2 of 8.707 or greater, by chance alone in the random samples of the designated sizes.

The difference between the test for independence and the test for homogeneity can be confusing. When we're simply given a two-way table, it's not obvious which test should be performed. The crucial difference is in the *design* of the study. Did we pick samples from each of two or more populations to compare the distribution of some variable (one question) among the different populations? If so, we are doing a test for homogeneity. Did we pick one sample from a single population and cross-categorize on two variables (two questions) to see if there is an association between the variables? If so, we are doing a test for independence.

For example, if we separately sample Democrats, Republicans, and Independents to determine whether they are for, against, or have no opinion with regard to stem cell research (several samples, one variable), we do a test for homogeneity with a null hypothesis that the distribution of opinions on stem cell research is the same among Democrats, Republicans, and Independents. However, if we sample the general population, noting the political preference and opinions on stem cell research of the respondents (one sample, two variables), we do a test for independence with a null hypothesis that political preference is independent of opinion on stem cell research.

Finally, it should be remembered that while we used χ^2 for categorical data, the χ^2-distribution is a *continuous* distribution. Applying it to counting data is just an approximation.

Quiz 22

Multiple-Choice Questions

> **DIRECTIONS:** The questions that follow are each followed by five suggested answers. Choose the response that best answers the question.

1. A random sample of 100 former student-athletes are picked from each of the eleven colleges that are members of the Big East conference. Students are surveyed about whether or not they feel they received a quality education while participating in varsity athletics. Which of the following is the most appropriate test to determine whether there is a difference among these schools as to the student-athlete perception of having received a quality education?

    (A) A chi-square goodness-of-fit test for a uniform distribution
    (B) A chi-square test of independence
    (C) A chi-square test of homogeneity
    (D) A z-test of eleven sample proportions
    (E) A z-test of eleven population proportions

2. Given a two-way table, it's not obvious whether to perform a test for independence or a test for homogeneity. What is the main difference between the tests?

    (A) How expected counts are calculated from the table
    (B) How df, the degrees of freedom, is calculated from the table
    (C) The number of rows in the table
    (D) The number of columns in the table
    (E) The number of samples from which the data in the table is obtained

3. Is there a relationship between education level and sports interest? A study cross-classified 1,500 randomly selected adults in three categories of education level (not a high school graduate, high school graduate, and college graduate) and five categories of major sports interest (baseball, basketball, football, hockey, and tennis). The χ^2-value is 13.95. Is there evidence of a relationship between education level and sports interest?

    (A) The data prove there is a relationship between education level and sports interest.
    (B) The evidence points to a cause-and-effect relationship between education and sports interest.
    (C) There is sufficient evidence at the 5% significance level of a relationship between education level and sports interest.
    (D) There is sufficient evidence at the 10% significance level, but not at the 5% significance level, of a relationship between education level and sports interest.
    (E) The P-value is greater than 0.10, so there is not sufficient evidence of a relationship between education level and sports interest.

4. A geneticist claims that four species of fruit flies should appear in the ratio 1:3:3:9. Suppose that a sample of 2,000 flies contained 110, 345, 360, and 1,185 flies of each species, respectively. For a chi-square goodness-of-fit test, what is the expected value for the second species?

    (A) 125
    (B) 250
    (C) 345
    (D) 375
    (E) 500

5. Random samples of 25 students are chosen from each high school class level, students are asked whether or not they are satisfied with the school cafeteria food, and the results are summarized in the following table:

	Freshmen	Sophomores	Juniors	Seniors
Satisfied	15	12	9	7
Dissatisfied	10	13	16	18

A chi-square test for homogeneity gives $\chi^2 = 5.998$. Is there sufficient evidence of a difference in cafeteria food satisfaction among the class levels?

(A) The data prove that there is a difference in cafeteria food satisfaction among the class levels.

(B) There is sufficient evidence of a linear relationship between food satisfaction and class level.

(C) There is sufficient evidence at the 1% significance level of a difference in cafeteria food satisfaction among the class levels.

(D) There is sufficient evidence at the 5% significance level, but not at the 1% significance level, of a difference in cafeteria food satisfaction among the class levels.

(E) With $P = 0.1117$, there is not sufficient evidence of a difference in cafeteria food satisfaction among the class levels.

6. With a chi-square test of independence and a 3 by 4 table, we have $\chi^2 = 12.7$. Assuming a significance level of 0.05, which of the following is true?

(A) While a Type II error was not committed, a Type I error may have been committed.

(B) While a Type I error was not committed, a Type II error may have been committed.

(C) Either a Type I or a Type II error may have been committed.

(D) Neither a Type I nor a Type II error may have been committed.

(E) With the information provided, it's not possible to say whether or not a Type I or a Type II error may have been committed.

Free-Response Questions

> **DIRECTIONS:** You must show all work and indicate the methods you use. You will be graded on the correctness of your methods and on the accuracy of your final answers.

1. A candy manufacturer advertises that its fruit-flavored sweets have hard sugar shells in five colors with the following distribution: 35% cherry red, 10% vibrant orange, 10% daffodil yellow, 25% emerald green, and 20% royal purple. A random sample of 300 sweets yielded the counts in the following table:

	Cherry Red	Vibrant Orange	Daffodil Yellow	Emerald Green	Royal Purple
Observed counts	94	34	22	77	73

   (a) Is there convincing evidence at the 10% significance level that the distribution is different from that which is claimed by the manufacturer?

   (b) A significance test with $H_0 : p = 0.20$ and $H_a : p \neq 0.20$, where p = proportion of the population of fruit-flavored sweets that are royal purple, results in a test statistic of $z = 1.876$. What is the conclusion at the 10 percent significance level?

   (c) Are the answers to (a) and (b) consistent with each other? Explain.

2. Optometrists are interested in various causes of eye abnormalities. In a Philadelphia eye institute study, a random sample of parents were asked whether their children slept primarily in room light, darkness, or with a night-light before the age of two. They were also asked about each child's eyesight diagnosis (nearsighted, farsighted, or normal vision) from the child's most recent examination. The resulting counts are as follows:

	Room Light	Darkness	Night-light	Total
Nearsighted	41	18	78	137
Farsighted	12	40	39	91
Normal	22	114	115	251
Total	75	172	232	479

   (a) What were the explanatory and response variables?

   (b) Was this an observational study or an experiment? Explain.

   (c) What was the probability that a child who slept in darkness was farsighted? What was the probability that a farsighted child slept in darkness?

   (d) Is there convincing evidence to say that there is an association between the bedroom lighting condition and a child's eyesight? Perform an appropriate test.

   (e) Some researchers proposed that parents' eyesight might be a confounding variable in this study. How would that explain the observed association between the bedroom lighting condition and the child's eyesight?

The answers for this quiz can be found beginning on page 509.

SUMMARY

- Chi-square analysis here is an important tool for inference on distributions of *counts.*
- Chi-square distributions take only positive values and are skewed right. With increasing degrees of freedom, the skew is less pronounced and the distribution looks more like a normal distribution.
- The chi-square statistic is found by summing the weighted squared differences between observed and expected counts, that is, $\chi^2 = \sum \dfrac{(\text{observed count} - \text{expected count})^2}{\text{expected count}}$.
- A chi-square goodness-of-fit test compares an observed distribution to some expected distribution.
- A chi-square test of independence tests for evidence of an association between two categorical variables from a single random sample from one population.
- A chi-square test of homogeneity compares independent random samples from two or more populations with regard to distributions of a single categorical variable.
- For these three chi-square tests, the observed counts must be integers; however, this is not true for the calculated expected counts.
- Expected counts for a chi-square goodness-of-fit test are (sample size) × (claimed proportions).
- Expected count for a particular cell of a two-way table is $\dfrac{(\text{row total})(\text{column total})}{\text{table total}}$.
- Conditions to check for these chi-square tests are the following:

  1. Random samples (independent random samples for homogeneity tests).
  2. Samples should be less than 10% of the populations (if sampling without replacement).
  3. All expected counts should be greater than 5.

- Just like with proportions and means, the *P*-value for a chi-square test is a conditional probability; it is the probability, given that the null hypothesis is true, of obtaining a test statistic as extreme as, or more extreme than, the observed value.
- Just like with proportions and means, the decision to either reject or fail to reject the null hypothesis for a chi-square test is based on comparison of the *P*-value to the significance level, α.
- When scoring a chi-square free-response question on the exam, readers will look for whether you:

  1. State the hypotheses (both H_0 and H_a) and define the parameter of interest. (The hypotheses must refer to the population, not to the sample.)
  2. Name the test: goodness-of-fit, independence, or homogeneity. (How the hypotheses are stated determines which test you choose.)
  3. Check the conditions. This requires showing the expected values (which can be copied from the calculator).
  4. Report the resulting chi-square value, the *P*-value, and *df* (the number of degrees of freedom).
  5. Give a conclusion in context with linkage to the *P*-value.

9

Inference for Quantitative Data: Slopes

(1–2 multiple-choice questions on the AP exam)

Learning Objectives

In this unit, you will learn how

→ To be able to describe the center, spread, and shape of the sampling distribution of a sample slope.

→ To be able to read generic computer regression output and identify the quantities needed for inference for the slope of the regression line (sample slope, *SE* of the slope, and the degrees of freedom).

→ To be able to state and verify whether or not the conditions are met for inference on the slope of the regression line (based on using the *t*-distribution).

→ To conduct a confidence interval procedure for the slope of the regression line.

→ To conduct a hypothesis test for the slope of the regression line.

In this unit, you will learn about the sampling distribution for the sample slope. You will then use this concept in finding confidence intervals and performing hypothesis tests on the slope. Just like with inference on means and proportions, you will see conditions to be checked and understand how to give proper conclusions in context.

If a scatterplot shows a linear relationship between two quantitative variables, we can calculate the least squares line from the data. As seen in Unit 2, the resulting equation, $\hat{y} = a + bx$, can be used to predict y for a given value of x. Statistical inference will help us answer two questions:

1. What is a confidence interval for the true slope, β? This will give the margin of error for predicting the change in y for a unit increase in x.

2. Is it plausible that the pattern observed in the scatterplot happened by chance, or is there convincing evidence of a linear relationship between x and y?

Inference on slopes depends on knowing the standard error of the slope, written as s_b or $SE(b)$. While this can be calculated by $s_b = \dfrac{s}{s_x \sqrt{n-1}}$ (where s is the standard deviation of the residuals, and s_x is the standard deviation of the x-values), it is almost always given in generic computer output.

> **NOTE**
>
> We see that increasing *s* (that is, increasing spread about the line) *increases* the slope's standard error, while increasing s_x (that is, increasing the spread of *x*-values) or increasing *n* (the sample size) *decreases* the slope's standard error.

Sampling Distribution for the Slope

Just as with the sampling distributions of $\bar{x}$ and $\hat{p}$, when certain conditions are met, we can model the sampling distribution of the sample slope b with a normal distribution with mean μ_b and standard deviation σ_b. Working with the standard error s_b as an estimate for σ_b leads to a *t*-distribution with $df = n - 2$.

If μ_y is the mean value of the response variable y for a given value of the explanatory variable x, then the population regression model is given by $\mu_y = \alpha + \beta x$.

Start with a bivariate population with a given slope β, a standard deviation of residuals σ, and a standard deviation of x-values σ_x. Take all samples of size n. Compute the slope in each of these samples:

1. The distribution of sample slopes is approximately normal.
2. The mean μ_b of the set of sample slopes equals β, the population slope. (That is, b is an unbiased estimator for β.)
3. The standard deviation σ_b of the set of sample slopes is approximately equal to $\dfrac{\sigma}{\sigma_x\sqrt{n}}$.

Alternatively, we say that the sampling distribution of b is approximately normal with mean β and standard deviation $\dfrac{\sigma}{\sigma_x\sqrt{n}}$.

The theoretical conditions for inference on the slope are

1. The true relationship between the response and explanatory variables is linear.
2. The standard deviation of y, σ_y, does not vary with x.
3. The responses (y-values) for each x are approximately normally distributed.

While the above are the theoretical conditions that should be met, we will be working with data from a single sample; therefore, we will be approximating the sampling distribution and need to give conditions based on the sample slope b, a standard deviation of the sample residuals s, and a standard deviation of the sample x-values s_x. Using s and s_x as estimates for σ and σ_x, respectively, leads us to estimating σ_b with s_b and a resulting t-distribution with $df = n - 2$. That is, the statistic $t = \dfrac{b - \beta}{s_b}$ has a t-distribution with $df = n - 2$.

Fortunately, $s_b = SE(b) = \dfrac{s}{s_x\sqrt{n-1}} =$ standard error of the sample slopes is typically given to you in generic computer output just to the right of the sample slope b.

NOTE

In a simple linear regression model with *one* parameter, β, (for example, if the *y*-intercept is fixed at 0), the test for the slope has *df* = *n* − 1.

Confidence Interval for the Slope of a Least Squares Regression Line

The slope b of the regression line and the standard error $s_b = SE(b)$ of the slope are listed explicitly in the computer output. A confidence interval for β can be found using t-scores with $df = n - 2$. If given raw data, a confidence interval can readily be found using the statistical software on a calculator.

Conditions for finding a confidence interval for the slope include:

1. *Independence:* a random sample and if sampling without replacement, $n \leq 0.10N$.
2. *Normality:* for each x-value, the associated y-values are normally distributed; we rarely have enough data for each x-value, so check if a histogram, dotplot, stemplot, or normal probability plot of all the residuals indicates normality, or $n \geq 30$.
3. *Linearity:* the true relationship between x and y is linear; the scatterplot is roughly linear, and the residual plot should have no apparent pattern.
4. *Equal SD:* the standard deviation of y is the same for all x-values; the residual plot should show approximately equal SDs for all x, for example, no "fanning."

> **Example 9.1**

Information concerning SAT Reading and Writing scores and SAT Math scores was collected from 15 randomly selected students. A linear regression performed on the data using a statistical software package produced the following printout:

```
Dependent variable:        Math
Variable       ,           Coef        SE Coef          T        Prob
Constant                92.5724         31.75         2.92       0.012
Reading and writing     0.763604       0.05597        13.6       0.000
S = 16.69          R-Sq = 93.5%     R-Sq(adj) = 93.0%
```

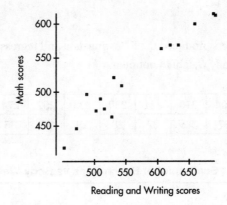

Reading and Writing scores

Assume that all conditions for regression are met.

(a) What is the regression equation?
(b) What is a 95% confidence interval estimate for the slope of the regression line? Show your work.
(c) Interpret the confidence level in context.
(d) Does the confidence interval provide convincing evidence that SAT Math scores are linearly related to SAT Reading and Writing scores?

✓ Solutions

(a) The *y*-intercept and slope of the equation are found in the Coef column of the above printout.

$$\widehat{\text{Math}} = 92.57 + 0.764(\text{Reading and Writing})$$

> Note the similarities between $b \pm t^*SE(b)$, $\hat{p} \pm z^* SE(\hat{p})$, and $\bar{x} \pm t^* SE(\bar{x})$.

(b) *Parameter:* Let β represent the slope of the true regression line for predicting SAT Math scores from SAT Reading and Writing scores.

Procedure: One-sample *t*-interval for a population slope β.

Conditions: Given that all conditions are met.

Mechanics: Standard error of the slope is $s_b = SE(b) = 0.05597$. With 15 data points, $df = 15 - 2 = 13$, and the critical *t*-values are $\pm \text{invT}(0.975, 13) = \pm 2.160$. The 95% confidence interval of the true slope is:

$$b \pm t^* s_b = 0.764 \pm 2.160(0.05597) = 0.764 \pm 0.121 \text{ or } (0.643, 0.885)$$

Conclusion in context: We are 95% confident that the interval from 0.64 to 0.89 captures the slope of the true regression line relating the SAT Math score, *y*, and SAT Reading and Writing score, *x*. (Or we are 95% confident that for every 1-point increase in Reading and Writing SAT score, the average increase in SAT Math score is between 0.64 and 0.89.)

(continued)

(c)   If all possible samples of SAT Reading and Writing and SAT Math scores were collected, each from 15 randomly selected students, and if each sample was used to construct a 95% confidence interval for the slope of the regression line relating SAT Math scores y and SAT Reading and Writing scores x, about 95% of these intervals would capture the true slope.

(d)   Note that $\beta = 0$ would indicate a line with slope 0 is the model for predicting SAT Math scores from SAT Reading and Writing scores; that is, the model would predict the same SAT Math score no matter what the SAT Reading and Writing score, and there would not be convincing evidence of a linear relationship. In this example, because the confidence interval (0.64 to 0.89) does not contain 0 as a plausible value of the slope of the population regression line, there is convincing evidence that SAT Math and Reading and Writing scores are linearly related.

❯ Example 9.2

A random sample of 12 football players produced the following data and regression analysis for number of bench press reps with 225 pounds versus body weight in pounds.

Body weight	180	204	210	221	230	233	257	274	295	298	311	324
Bench press reps	23	27	32	22	28	29	33	35	41	37	45	40

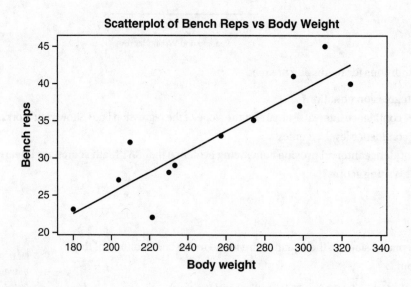

Scatterplot of Bench Reps vs Body Weight

Regression analysis is as follows.

```
Predictor           Coef        SE Coef          T            P
Constant          -2.553          5.390       -0.47        0.646
Body weight       0.13916        0.02097        6.64        0.000

S = 3.261          R-Sq = 81.5%         R-Sq(adj) = 79.6%
```

(continued)

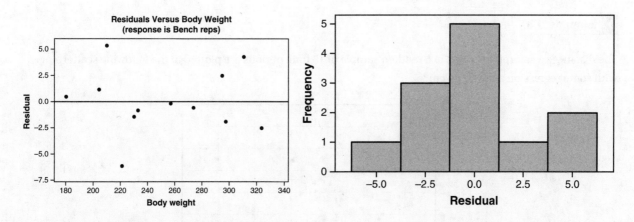

(a) What is the equation of the regression line that predicts number of bench press reps with 225 pounds as a function of body weight in pounds?

(b) Interpret the slope of the regression line.

(c) Determine a 95% confidence interval for the true slope.

(d) Interpret the confidence level.

(e) A sports trainer believes that every 10 pounds more in body weight is associated with an additional two bench press reps with 225 pounds. Does the above analysis support the trainer's belief?

✓ Solution

(a) Predicted bench reps $= -2.553 + (0.13916)(\text{Body weight})$.

(b) Each additional pound of body weight is *predicted* to increase the number of bench reps by 0.13916.

(c) *Parameter:* Let β represent the slope of the true regression line for predicting number of bench press reps with 225 pounds from body weight in pounds.

   Procedure: One-sample t-interval for a population slope β.

   Conditions: We are told the data come from a random sample, the sample size of $n = 12$ is less than 10% of all football players, the scatterplot looks roughly linear, there is no pattern in the residual plot, and a histogram of the residuals is roughly unimodal and symmetric with no strong skew or outliers.

   Mechanics: With $df = n - 2 = 12 - 2 = 10$ and a 95% confidence level, the critical t-scores are $\pm \text{invT}(0.975, 10) = \pm 2.228$. The confidence interval for the slope is $b \pm t^*SE(b) = 0.13916 \pm 2.228(0.02097) = 0.13916 \pm 0.04672$ or $(0.092, 0.186)$.

   Conclusion in context: We are 95% confident that the interval from 0.092 to 0.186 captures the slope of the true regression line relating number of bench press reps with 225 pounds and body weight in pounds. [Or we are 95% confident that each additional pound of body weight is associated with a mean increase in bench reps of between 0.092 and 0.186.]

(d) If all possible samples of bench reps and body weights are collected, each from 12 randomly selected football players, and if each sample was used to construct a 95% confidence interval for the slope of the regression line relating bench reps y and body weight x, about 95% of these intervals would capture the true slope.

(e) The interval of (0.092, 0.186) additional bench reps per additional pound of body weight gives an interval of (0.92, 1.86) additional bench reps per additional 10 pounds of body weight. Since 2 is not in this interval, the analysis does not support the trainer's belief.

> **Example 9.3**

The distances a caterpillar crawls in a random sample of 15 time periods are plotted in the following scatterplot with some regression analysis given:

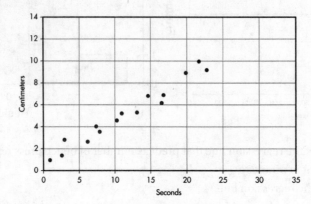

The 95 percent confidence interval for the slope β is (0.325, 0.430).

(a) Another measurement of 14 centimeters in 35 seconds is obtained. With the addition of this data point, would the confidence interval change, and, if so, would the margin of error increase or decrease?

(b) What would happen if instead the data point (12, 1) is added?

☑ **Solutions**

(a) (35, 14) appears to follow the linear trend of the rest of the data. The slope of the regression line will change little, if at all. However, with more points following the same trend, the standard error of the slope will be smaller, and thus the margin of error will decrease. [In this example, the 95% confidence interval for the slope with the additional data point is actually (0.339, 0.415).]

(b) With the x-coordinate, 12, appearing close to the mean of all the x-coordinates, the slope of the regression line will change a little, and the y-intercept will change. Both the standard error of the slope and the margin of error of the confidence interval will also increase. [In this example, the 95% confidence interval for the slope with the additional data point is actually (0.272, 0.479).]

Hypothesis Test for Slope of Least Squares Regression Line

In addition to finding a confidence interval for the true slope, we can also perform a hypothesis test for the value of the slope. Often we use the null hypothesis H_0: $\beta = 0$, that is, that there is no linear relationship between the two variables.

Conditions for a hypothesis test are the same as for a confidence interval:

1. *Independence:* a random sample and if sampling without replacement, $n \leq 0.10N$.
2. *Normality:* for each x-value, the associated y-values are normally distributed; we rarely have enough data for each x-value, so check if a histogram, dotplot, stemplot, or normal probability plot of all the residuals indicates normality, or $n \geq 30$.
3. *Linearity:* the true relationship between x and y is linear; the scatterplot is roughly linear, and the residual plot should have no apparent pattern.
4. *Equal SD:* the standard deviation of y is the same for all x-values; the residual plot should show approximately equal SDs for all x, for example, no "fanning."

Note that a low *P*-value tells us that if the two variables did not have some linear relationship, it would be highly unlikely to find such a random sample. However, **strong evidence that there is some linear association does not mean the association is strong**.

⟩ Example 9.4

A company offers a 10-lesson program of study to improve students' SAT scores. A survey is made of a random sampling of 25 students. A scatterplot of improvement in total (Math and Reading and Writing) SAT score versus number of lessons taken is as follows:

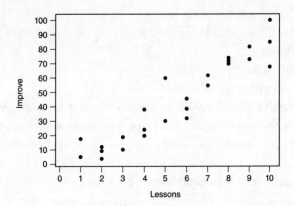

Some computer output of the regression analysis follows, along with a plot and a histogram of the residuals.

Regression Analysis: Improvement Versus Lessons

```
Dependent variable: Improvement

Predictor        Coef        SE Coef          T            P
Constant        -7.718        4.272        -1.81        0.084
Lessons          9.2854       0.6789       13.68        0.000

S = 9.744           R-Sq = 89.1%
```

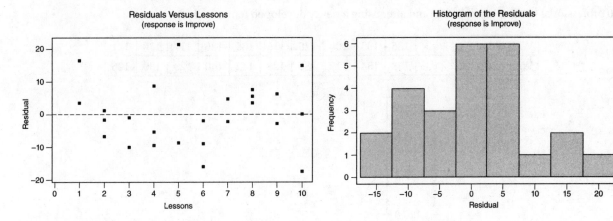

(a)   What is the equation of the regression line that predicts score improvement as a function of number of lessons?

(b)   Interpret, in context, the slope of the regression line.

(c)   What is the meaning of R-Sq in the context of this study?

(d)   Give the value of the correlation coefficient.

(e)   Perform a test of significance for the slope of the regression line.

(continued)

✓ Solutions

(a)   Predicted Improvement $= -7.718 + 9.2854$(Lessons)

(b)   The slope of the regression line is 9.2854, meaning that each additional lesson is *predicted* to improve one's total SAT score by 9.2854.

(c)   R-Sq $= 89.1\%$ means that 89.1% of the variation in total SAT score improvement can be explained by the linear relationship between total SAT score improvement and number of lessons taken.

(d) $r = +\sqrt{0.891} = 0.944$, where the sign is positive because the slope is positive.

(e)   A test of significance of the slope of the regression line with $H_0: \beta = 0$ and $H_a: \beta \neq 0$ (test of significance of correlation) is as follows:

Parameter: Let β represent the slope of the true regression line for predicting improvement in SAT scores from number of lessons taken.

Hypotheses: $H_0: \beta = 0$, $H_a: \beta \neq 0$.

Procedure: t-test for the slope of a regression model.

Checks: We are told the data came from a random sample, the scatterplot appears to be approximately linear, there is no apparent pattern in the residuals plot, the histogram of residuals is roughly unimodal with no outliers or strong skewness, and we assume the sample of 25 is less than 10% of all students taking the 10-lesson program of study.

Mechanics: Reading directly from the computer output, $t = 13.68$ and $P = 0.000$. [For instructional purposes, we note that we could have calculated $t = \dfrac{9.2854 - 0}{0.6789} = 13.68$ and with $df = 25 - 2 = 23$, $P = \text{tcdf}(13.68, 1000, 23)$ $= 0.000$.]

Conclusion in context with linkage to the P-value: With this small of a P-value, $0.000 < 0.05$, there is very strong evidence to reject H_0 and conclude that there is convincing evidence of a linear relationship between improvement in total SAT score and number of lessons taken.

❯ Example 9.5

The following table gives serving speeds in mph using a "cannonball" serve from a random sample of 10 professional tennis players before and after using a newly developed tennis racket.

With old racket	125	133	108	128	115	135	125	117	130	121
With new racket	133	134	112	139	123	142	140	129	139	126

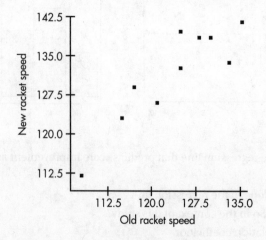

Some computer output of the regression analysis follows along with a plot and histogram of the residuals.

(continued)

Predictor	Coef	SE Coef	T	P
Constant	8.76	21.05	0.416	0.6883
Old speed	0.9938	0.1698	5.853	0.0002

S = 4.32977   R-Sq = 81.1%   R-Sq(adj) = 78.7%

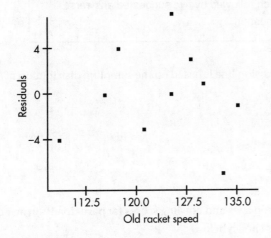

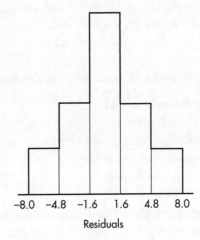

(a) Is there evidence of a linear relationship with positive slope between serving speeds of professionals using their old and the new rackets?

(b) What do the slope and intercept of the regression line reveal about the difference in serving speeds between using the old and new rackets?

(c) Interpret the *P*-value in context.

✓ Solutions

(a) *Parameter:* Let β represent the slope of the true regression line for predicting serving speed using the newly developed tennis racket from serving speed using the old rackets.

Hypotheses: $H_0: \beta = 0$, $H_a: \beta > 0$. (Note that we are testing for a *positive* slope.)

Procedure: t-test for the slope of a regression line.

Checks: We are given that the data come from a random sample, the sample size of $n = 10$ is less than 10% of all professional tennis players, the scatterplot appears to be roughly linear, there is no apparent pattern in the residuals, and the histogram of residuals appears roughly normal.

Mechanics: The computer output gives $t = 5.853$ and $P = \left(\frac{1}{2}\right)(0.0002) = 0.0001$.

Conclusion in context with linkage to the P-value: With this small a *P*-value, $0.0001 < 0.05$, there is sufficient evidence to reject H_0. In other words, there is convincing evidence of a linear relationship with positive slope between serving speeds of professional tennis players using their old and the new rackets.

> **NOTE**
> Generic computer output gives the *P*-value for a two-sided test; we divide by 2 for a one-sided test.

(b) The regression line is:

Predicted speed with new racket $= 8.76 + 0.9938$(Speed with old racket)

With a slope of approximately 1 and a *y*-intercept of 8.76, the regression line indicates that use of the new racket increases serving speed on the average by 8.76 mph regardless of the old racket speed. That is, players with lower and higher racket speeds experience on the average the same numerical (rather than percentage) increase when using the new racket.

(c) Assuming there is no linear relationship between serving speeds before and after using the new racket, there is a 0.0001 probability of getting a sample regression line with a slope of 0.9938 or greater by chance alone in a random sample of 10 professional tennis players.

Quiz 23

Multiple-Choice Questions

DIRECTIONS: The questions that follow are each followed by five suggested answers. Choose the response that best answers the question.

1. Inference about the slope of a least squares regression line is based on the sampling distribution of b being

   (A) approximately normal.
   (B) a chi-square distribution with $df = n - 1$.
   (C) a chi-square distribution with $df = n - 2$.
   (D) a t-distribution with $df = n - 1$.
   (E) a t-distribution with $df = n - 2$.

2. A statistician investigated the relationship between ages and bowling scores for participants in an adult bowling league. The regression analysis for her data is shown below.

Predictor	Coef	SE Coef	T	P
Constant	122.79	94.44	1.30	0.218
Age	3.342	1.200	2.78	0.017
S = 37.3598	R—Sq = 84.0%			

   By how much do bowling score estimates produced by this model typically differ from the actual scores of the bowlers?

   (A) 1.200
   (B) 3.342
   (C) 37.3598
   (D) 94.44
   (E) 122.79

3. A random sample of 25 geographic locations was selected to investigate the relationship between poverty rate and violent crime rate. A regression analysis of the data was conducted where poverty rate is the explanatory variable x, and violent crime rate is the response variable y. If a 90 percent confidence interval is constructed for the slope of the population regression line, which of the following is a condition that must be checked?

   (A) The confidence interval is not biased.
   (B) The correlation between poverty rate and violent crime rate is not equal to zero.
   (C) The slope is not equal to zero.
   (D) The scatterplot of violent crime rate vs poverty rate is roughly linear.
   (E) The point $(\bar{x}, \bar{y})$ is on the regression line.

4. Researchers studying the relationship between cannabis use and cocaine use surveyed a random selection of ten world regions. For each location, the prevalence of cannabis and cocaine use as a percentage of the population aged 15 to 64 was found and is shown in the following scatterplot.

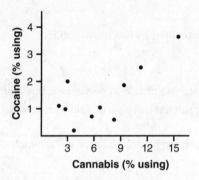

The 90 percent confidence interval for the slope is given by 0.18 ± 0.11. Assuming that the conditions for inference for the slope of the regression equation are met, which of the following is the correct interpretation of the calculated confidence interval?

(A) We are 90 percent confident that the increase in cocaine usage in world regions, when the cannabis usage increases by 1 percent, is between 0.07 and 0.29 percent.

(B) We are 90 percent confident that the increase in cocaine usage in this sample of world regions, when the cannabis usage increases by 1 percent, is between 0.07 and 0.29 percent.

(C) We are 90 percent confident that the average increase in cocaine usage in world regions, when the cannabis usage increases by 1 percent, is between 0.07 and 0.29 percent.

(D) We are 90 percent confident that the average increase in cannabis usage in this sample of world regions, when the cocaine usage increases by 1 percent, is between 0.07 and 0.29 percent.

(E) We are 90 percent confident that the average increase in cannabis usage in world regions, when the cocaine usage increases by 1 percent, is between 0.07 and 0.29 percent.

5. A nutritionist claims that the resting metabolic rate decreases by 20 kcal/day for each decrease of 1 kg in body mass. To investigate the relationship between resting metabolic rate y, in kcal/day, and body mass x, in kg, data are collected on a random sample of 85 adults. Assume the conditions for inference for the slope of a regression line are met. The resulting 95 percent confidence interval for the population slope of the regression line relating resting metabolic rate and body mass is given by (21.3, 29.7). Which of the following best describes the conclusion that can be reached about the nutritionist's claim based on the confidence interval?

(A) The claim is supported by the interval because the interval does not contain the value 0.

(B) The claim is supported by the interval because the interval does not contain the value 20.

(C) The claim is not supported by the interval because the interval does not contain the value 20.

(D) The claim is not supported by the interval because all values in the interval are positive, and the claim is about decreases.

(E) The claim is not supported by the interval because 20 is positive, and the claim is about decreases.

6. Can percentage body fat for men be predicted from waist size? In a random sample of 300 adult males, percentage body fat versus waist size (in inches) gives the following regression results:

Predictor	Coef	Coef SE	T	P
Constant	−44.23	2.83	−15.63	0.000
Waist Size	1.68	0.072	23.33	0.000

What is the correct interpretation of 1.68?

(A) 1.68 percent of the variation in percentage body fat can be explained by this linear model.

(B) 1.68 is a measure of the variability in waist size.

(C) 1.68 is a measure of the variability in the set of residuals.

(D) For every 1 inch increase in waist size, there is a predicted increase of 1.68 in percentage body fat.

(E) For every 1 unit increase in percentage body fat, there is a predicted increase of 1.68 inch in waist size.

7. A 95% confidence interval for the slope of a regression line is calculated to be (−0.783, 0.457). Which of the following must be true?

(A) The slope of the regression line is 0.

(B) The slope of the regression line is −0.326.

(C) A scatterplot of the data would show a linear pattern.

(D) A residual plot would show no pattern.

(E) The correlation is negative.

Questions 8-11 refer to the following setting. In a random sample of 25 professional baseball players, their salaries (in millions of dollars) and batting averages result in the following regression analysis:

Predictor	Coef	SE Coef	T	P
Constant	0.2336	0.005883	39.71	0.0000
Salary	0.008051	0.002825	2.850	0.0091

S = 0.01695   R-Sq = 71.6%   R-Sq(adj) = 70.3%

8. What is the equation of the least squares regression line?

(A) Batting Average = 0.008051 + 0.2336(Salary)

(B) Predicted Batting Average = 0.008051 + 0.2236(Salary)

(C) Predicted Batting Average = 0.2336 + 0.008051(Salary)

(D) Predicted Batting Average = 0.002825 + 0.008051(Salary)

(E) Predicted Salary = 0.2336 + 0.008051(Batting Average)

9. Which of the following gives a 95 percent confidence interval for the slope of the regression line?

(A) $0.008051 \pm 1.96(0.002825)$

(B) $0.008051 \pm 1.96(0.01695)$

(C) $0.008051 \pm 2.0639(0.002825)$

(D) $0.008051 \pm 2.0687(0.002825)$

(E) $0.008051 \pm 2.85(0.002825)$

10. In testing the hypothesis $H_0: \beta = 0$ versus $H_a: \beta > 0$, what is the t-statistic?

    (A) 2.850
    (B) (0.5)(2.850)
    (C) 39.71
    (D) (0.5)(39.71)
    (E) This cannot be answered without first deciding on an α-risk.

11. In testing the hypothesis $H_0: \beta = 0$ versus $H_a: \beta > 0$, what is the P-value?

    (A) 0.0045
    (B) 0.0091
    (C) 0.0250
    (D) 0.7030
    (E) 0.7160

Free-Response Questions

> **DIRECTIONS:** You must show all work and indicate the methods you use. You will be graded on the correctness of your methods and on the accuracy of your final answers.

1. Information with regard to the assessed values (in $1,000) and the selling prices (in $1,000) of a random sample of homes sold in a Northeast market yields the following computer output:

   ```
   Dependent variable is   Price

   R squared = 89.8%   R squared (adjusted) = 89.2%
   s = 9.508 with   20 - 2 = 18  degrees of freedom

   Variable      Coeff      s.e. of coeff      t-ratio      prob
   Constant      0.890087        16.16           0.055      0.956
   Assessed      1.0292         0.08192          12.6       0.000
   ```

   Assume all conditions for inference are met.

   (a) Determine the equation of the least squares regression line.
   (b) Construct a 99% confidence interval estimate for the slope of the regression line, and interpret this in context.
   (c) What conclusion can be reached given where the interval lies in relation to the value of 0?

2. A study of 100 randomly selected teenagers, ages 13–17, looked at number of texts per waking hour versus age, yielding the computer regression output below:

   ```
   Predictor      Coef        SE Coef          T           P
   Constant      -1.055        2.815         -0.37       0.709
   Age            0.4577       0.1866         2.45       0.016

   S = 2.66501        R-Sq = 5.8%        R-Sq(adj) = 4.8%
   ```

   (a) Interpret the slope in context.
   (b) What three graphs should be checked with regard to conditions for a test of significance for the slope of the regression line?
   (c) Assuming all conditions for inference are met, perform this test of significance.
   (d) Give a conclusion in context, taking into account both your answer to the hypothesis test as well as the value of R-Sq.

The answers for this quiz can be found beginning on page 511.

Quiz 24

Multiple-Choice Questions

> **DIRECTIONS:** The questions that follow are each followed by five suggested answers. Choose the response that best answers the question.

1. Researchers are investigating a linear association between humidity (percent) and attendance (1,000s) at major league baseball games for a random sample of days and stadiums. A 95 percent confidence interval for the slope is constructed for predicting attendance based on humidity, obtaining $(-1.32, 0.18)$. Which statement is correct?

   (A) The association is probably negative because most of the interval is below 0.

   (B) The association must be negative because the interval contains -1.

   (C) The association is not statistically significant because 0 is in the interval.

   (D) The relatively high margin of error, 0.75, makes any conclusion suspect.

   (E) A calculation error was made because the lower endpoint is below -1.

2. Which of the following is a necessary assumption for performing inference analysis on the slope of a least squares regression line?

   (A) There is no strong skew or outliers in the data.

   (B) A straight line can be drawn through the set of paired observations in the scatterplot.

   (C) The distribution of the residuals is approximately uniform.

   (D) The distribution of the residuals is approximately linear.

   (E) The distribution of the residuals is approximately normal.

3. Can either the correlation coefficient or the P-value ever be negative?

   (A) Correlation coefficient: never
       P-value: never

   (B) Correlation coefficient: never
       P-value: sometimes

   (C) Correlation coefficient: sometimes
       P-value: never

   (D) Correlation coefficient: sometimes
       P-value: sometimes

   (E) The answer depends upon context.

4. An engineer tests battery life (in hours) as a function of screen brightness (in cd/m^2) for a sampling of brightness settings. A 99 percent confidence interval for estimating the population slope of the linear regression line predicting battery life based on screen brightness is given by -0.015 ± 0.009. Assuming that the conditions for inference for the slope of the regression equation are met, which of the following is the correct interpretation of the calculated confidence interval?

   (A) We are 99 percent confident that the mean decrease in battery life for each one cd/m^2 increase in screen brightness is between 0.006 and 0.024 hours.

   (B) We are 99 percent confident that a one cd/m^2 increase in screen brightness will result in a decrease in battery life of between 0.006 and 0.024 hours.

   (C) We are 99 percent confident that for every one cd/m^2 increase in screen brightness in the sample, the average decrease in battery life is between 0.006 and 0.024 hours.

   (D) We are 99 percent confident that the mean increase in screen brightness for each one hour increase in battery life is between 0.006 and 0.024 cd/m^2.

   (E) We are 99 percent confident that all random samples of the same size will result in confidence intervals with a margin of error of ± 0.009 cd/m^2.

5. A sociologist investigates whether or not there is a linear relationship between how long customers stay at buffets and how large the tips they leave. The sociologist collects data and performs a hypothesis test with $H_0: \beta = 0$, $H_a: \beta \neq 0$, and $\alpha = 0.05$. With a resulting P-value of 0.18, which of the following is a correct conclusion?

(A) The null hypothesis is accepted because $0.18 > 0.05$. There is convincing evidence that there is not a linear relationship between how long customers stay at buffets and how large the tips they leave.

(B) The null hypothesis is rejected because $0.18 > 0.05$. There is convincing evidence that there is a linear relationship between how long customers stay at buffets and how large the tips they leave.

(C) The null hypothesis is not rejected because $0.18 > 0.05$. There is convincing evidence that there is a linear relationship between how long customers stay at buffets and how large the tips they leave.

(D) The null hypothesis is rejected because $0.18 > 0.05$. There is not convincing evidence that there is a linear relationship between how long customers stay at buffets and how large the tips they leave.

(E) The null hypothesis is not rejected because $0.18 > 0.05$. There is not convincing evidence that there is a linear relationship between how long customers stay at buffets and how large the tips they leave.

6. Can points per game (PPG) be predicted based on NBA players' salaries? For the New York Knicks 2019–2020 season, a linear association study yields the following computer output:

```
Regression Analysis: PPG versus Salary (in Millions of $)

Variable      N      Mean    SE Mean    StDev
Salary        14     7.34     1.45       5.43
PPG           14     8.08     1.52       5.70

The regression equation is PPG = 4.39 + 0.656 Salary

Predictor        Coef       SE Coef
Constant        3.6648       1.5246
Salary          0.6013       0.2488

S = 4.8695     R-Sq = 32.7%
```

Which of the following is an appropriate test statistic for testing the null hypothesis that the slope of the regression line is greater than 0? (Assume all conditions for inference are met.)

(A) $\dfrac{8.08}{1.52}$

(B) $\dfrac{8.08}{5.70}$

(C) $\dfrac{3.6648}{1.5246}$

(D) $\dfrac{0.6013}{0.2488}$

(E) $\dfrac{0.6013}{4.8695}$

Questions 7-11 refer to the following setting. Smoking is a known risk factor for cardiovascular disease and cancer. A study analyzing smoking levels (seven men smoking at levels from 0.5 to 3.0 packs per day) and risk of dementia (as measured by the Cox hazard ratio) resulted in the following regression analysis:

```
Predictor        Coef      SE Coef         T          P
Constant       0.7737      0.07833      9.88     0.0000
Packs          0.4341      0.07675      5.656    0.0024

S = 0.09817  R-Sq = 95.3%  R-Sq(adj) = 94.4%
```

7. What is the equation of the least squares regression line?

    (A) Predicted Cox Ratio = 0.4341 + 0.07675(Packs)
    (B) Predicted Cox Ratio = 0.07675 + 0.4341(Packs)
    (C) Predicted Cox Ratio = 0.4341 + 0.7737(Packs)
    (D) Predicted Cox Ratio = 0.7737 + 0.4341(Packs)
    (E) Predicted Cox Ratio = 0.7737 + 0.07675(Packs)

8. By how much do Cox Ratio estimates produced by this model typically differ from the actual ratios of the smokers?

    (A) 0.07675
    (B) 0.07833
    (C) 0.09817
    (D) 0.4341
    (E) 0.7737

9. Which of the following gives a 90 percent confidence interval for the slope of the regression line?

    (A) 0.4341 ± 1.645(0.07675)
    (B) 0.4341 ± 1.645(0.09817)
    (C) 0.4341 ± 1.943(0.07675)
    (D) 0.4341 ± 1.943(0.09817)
    (E) 0.4341 ± 2.015(0.07675)

10. In testing the hypothesis $H_0: \beta = 0$ versus $H_a: \beta > 0$, what is the P-value?

    (A) 0.0012
    (B) 0.0024
    (C) 0.0048
    (D) 0.07675
    (E) 0.09817

11. Which of the following is a correct interpretation of the P-value?

    (A) The probability of making a Type I error
    (B) The probability that there is a linear relationship between smoking levels and risk of dementia of the seven smokers in the sample
    (C) The probability that there is a linear relationship between smoking levels and risk of dementia
    (D) If there is a linear relationship between smoking levels and risk of dementia, the probability of getting a random sample of seven smokers that yields a least squares regression line with a slope of 0.4341 or greater
    (E) If there is no linear relationship between smoking levels and risk of dementia, the probability of getting a random sample of seven smokers that yields a least squares regression line with a slope of 0.4341 or greater

Free-Response Questions

> **DIRECTIONS:** You must show all work and indicate the methods you use. You will be graded on the correctness of your methods and on the accuracy of your final answers.

1. A sociologist is researching a possible link between household income and self-reported life satisfaction. A least squares regression on "Self-reported life satisfaction" measured on a 10-point scale versus "Household income" in $1,000 among 50 randomly selected adults yields the following computer printout:

   ```
                  Coef        SE Coef
   Intercept      1.415       0.0943
   Income         0.054       0.00358

   S = 0.977     R-Sq = 0.862
   ```

   Assume all conditions for inference are met.

   (a) Determine the equation of the least squares regression line.
   (b) Interpret R-Sq = 0.862 in context.
   (c) Interpret $S = 0.977$ in context.
   (d) What is the test statistic for testing the null hypothesis that the slope of the regression line is greater than 0?
   (e) Find the 90 percent confidence interval for the slope of the regression line, and interpret it in context.

2. Can the average women's life expectancy in a country be predicted from the average fertility rate (children per woman) in that country? The following are recent data from 11 randomly selected countries:

Country	Fertility Rate (children per woman)	Life Expectancy (women)
Afghanistan	6.25	45.5
Angola	5.33	51.4
Argentina	2.16	80.0
Australia	1.85	84.4
France	1.85	85.1
Liberia	4.69	61.5
Nepal	2.66	69.0
Netherlands	1.77	82.6
Pakistan	3.58	68.3
Poland	1.29	80.4
Singapore	1.29	83.4

   (a) Construct a scatterplot in which fertility rate is the explanatory variable and life expectancy is the response variable. Based on your graph, does a linear model seem appropriate?
   (b) Find the least squares regression line to predict life expectancy from fertility rate.
   (c) What are the hypotheses to test whether there is a linear relationship between life expectancy and fertility rate? Be sure to define the parameter.
   (d) The test statistic is $t = -12.53$. Do the data provide convincing statistical evidence of a linear relationship between life expectancy and fertility rate?

The answers for this quiz can be found beginning on page 513.

SUMMARY

- The sampling distribution for the slope of a regression line can be modeled by a t-distribution with $df = n - 2$; that is, the sampling distribution of $t = \dfrac{b - \beta}{SE(b)}$ has a t-distribution with $df = n - 2$.
- One suggested acronym for conditions for linear regression inference is LINER:
  - **Linear** (the scatterplot looks roughly linear with a residual plot showing random scatter around a horizontal line)
  - **Independent** (the observations are independent; check 10% Rule if sampling without replacement)
  - **Normal** (for each fixed value of x, the response y varies according to a normal distribution; we often don't have enough values for each x to check this, so in that case we lump all the residuals together and look for rough normality)
  - **Equal SD** (the SD of y is the same for all values of x; check if the residual plot shows a similar amount of scatter for each x; for example, a fan shaped pattern would violate this)
  - **Random** (the data come from a random sample)
- The standard error of the slope can be read directly from generic computer output of regression.
- With regard to confidence intervals, note the similarities between $b \pm t^* SE(b)$, $\hat{p} \pm z^* SE(\hat{p})$, and $\bar{x} \pm t^* SE(\bar{x})$.
- The t-statistic and the P-value (for a two-sided test) for a hypothesis test of the slope can be read directly from generic computer output of regression.
- It is important to be able to read generic computer output (where a is the y-intercept, b is the slope, r^2 is the coefficient of determination, and the P-values are for a two-sided test).

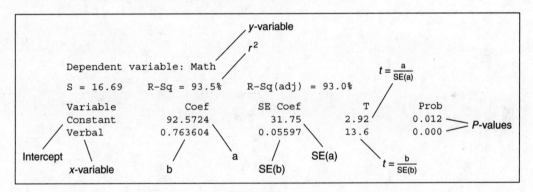

PART 3
Final Review

Final Review

After completing Units 1–9, their associated quizzes, and the following review sections, you'll be ready to try the practice tests. Wishing you good luck will not be appropriate, because you'll be so well prepared that you won't need luck!

Selecting an Appropriate Inference Procedure

Of the six free-response questions on the AP Statistics exam, at least one—and often two—involves inference. For these problems, the first thought should be to identify the appropriate inference method. While knowing how to perform the different procedures is relatively easy when the question appears on a page with the unit heading at the top, the exam provides no such help! Thus, time devoted to recognition and decision-making should be an important part of any review. This unit provides that valuable practice.

To increase success, stop and think before proceeding. Hasty decisions often lead to choosing the wrong statistical procedure needed to solve the problem. Take a few moments to answer several questions that should be at the forefront of your mind. Use the test booklet to jot down notes. The following are important questions and tips:

Are there key terms or phrases? (Underline or circle them!)

- Words such as *proportion, mean*, or *association*?
- Expressions such as *find a confidence interval* or *do a hypothesis test*?

What kind of data are presented?

- Categorical (indicating proportions or chi-square)?
- Quantitative (indicating means or slopes)?

What summary statistics are given?

- A sample proportion?
- A sample mean and standard deviation?
- Sample counts (leading to chi-square inference)?
- A sample slope?

What specifically are you asked to do?

- Estimate a quantity (indicating a confidence interval)?
- Look for evidence to test a claim (indicating a hypothesis test)?

How was the data collected?

- From one sample or two samples?
- Is there pairing of data (is information lost if the data in either set are shuffled around)?

If a hypothesis test is indicated, should you use a one-tailed or two-tailed test?

- Are we looking for a difference from some particular value or to see whether nothing has changed (indicating a two-tailed test)?
- Are we looking for a change in a specific direction (indicating a one-tailed test)?

Which distribution should be used for inference problems?

- Proportions? Use z.
- Means? Use t.
- Counts? Use χ^2.
- Slopes? Use t.

What is the context of the question?

- Underline or circle the context!
- A response should always refer to the context in the conclusion.

What conditions need to be addressed?

- Independence assumption
    1. Random sampling or random assignment
    2. $n \leq 0.10N$ if sampling without replacement
    3. Independence across samples if comparing more than one population
- Normality assumption
    1. For proportions, check that np and nq are both ≥ 10.
    2. For means, check that either the population is roughly normal or the sample size is large.
    3. For slopes, check that the distribution of residuals is roughly normal.
    4. For counts, check that all expected cells are > 5 (for increasing df, the χ^2 distribution approaches a normal distribution).

❯ Example

What procedure is called for with the data below?

$$29 \quad 32 \quad 50 \quad 41 \quad 32$$
$$34 \quad 50 \quad 48 \quad 44 \quad 40$$

Of course this is impossible to answer without knowing context!

(a) A random sample of five students signing up for a private tutoring service take a pretest and a posttest with scores on the 50-point tests given below.

Student	A	B	C	D	E
Pretest score	29	32	50	41	32
Posttest score	34	50	48	44	40

What procedure should be used to find a 95% confidence interval for the mean difference in score results (posttest score minus pretest score) for students who use the tutoring service?

(b) In a random sample of 400 attendees at a Cubs–White Sox game, favorite team is cross-classified against favorite reading genre. The following table shows the data.

	Science Fiction	Romance	Sports	Mystery	History
Cubs	29	32	50	41	32
White Sox	34	50	48	44	40

Which procedure should be used to test whether or not there is an association between favorite team and favorite reading genre?

(c) A miniature golf course offers two nine-hole courses. A random sample of five players who play both courses shows the following scores.

	A	B	C	D	E
Course 1	29	32	50	41	32
Course 2	34	50	48	44	40

The franchise owner is interested in whether a player's score on one of the courses can be used to predict his or her score on the other course. What procedure should be used?

(d) A school's AP coordinator picks a random sample of five students who claim to have stayed up all night studying for a next day's AP exam and an independent random sample of five students who claim to have had a good night's sleep the night before the AP exam. Their scores on the exam are in the table below.

Student	A	B	C	D	E
No sleep	29	32	50	41	32
Sleep	34	50	48	44	40

What procedure should the AP coordinator use to determine if there is a statistically significant difference in the mean scores obtained by students who stay up all night studying and those who have a good night's sleep before the exam?

✓ Solutions

(a) The pretest and posttest data are not independent. This is a paired data test, so we cannot use two-sample inference. No population standard deviation is given, so we will use a t-procedure rather than a z-procedure. Use a one-sample t-interval on the set of differences.

(b) There is a single sample cross-classified on two variables (rather than samples from two or more populations and a single categorical variable), so use a chi-square test of independence (rather than a chi-square test of homogeneity).

(c) Use a linear regression t-test to determine if there is a statistically significant linear relationship between how players do on the two courses.

(d) We have two independent samples, are asked to determine if there is a significant difference (rather than a confidence interval) between two means, and are not given any population standard deviation (so we must use a t-procedure rather than a z-procedure). Use a two-sample t-test for the difference of two population means.

The following quizzes, Quiz 25 and Quiz 26, evaluate your ability to name procedures, define parameters, list conditions to be checked, and state hypotheses if appropriate.

Quiz 25

> **DIRECTIONS:** For each of the following problems, name the procedure you would use, define any parameter you use, list the conditions to be checked, and state hypotheses if appropriate.

1. At the high school level, do men and women play varsity sports for the same reasons? Of all students who tried out for varsity sports, a random sample of men and a random sample of women are picked. The selected students are given questionnaires asking which of the following is their primary reason for playing sports: social, health, or status.

2. Estimate the proportion of AP Statistics students who will graduate in the top ten percent of their senior class. There were 150 AP Statistics students tracked.

3. Estimate the average number of books read by high school students during their four years in secondary school. A random sample of 125 students are followed during their high school years.

4. To test the claim that over 70 percent of racially motivated hate crimes are motivated by anti-Black bias, a criminologist obtains data from a random sample of police commissioners.

5. It is believed that in cases of identity theft, 30% of the victims used their mother's maiden name for their banking password, 25% used their pet's name, 20% used "password," and the rest used something else. A study is to be made to test this claimed distribution.

6. Do baseball pitchers throw harder if they participate in ten-minute yoga sessions before games? A pitcher who normally averages 95 mph with his fastballs participated in ten-minute yoga sessions for a random sample of his games. His fastballs for those games were clocked.

7. Does family income have an impact on school performance? In a random sample of 100 students, family income and student GPA were recorded. Is there significant statistical evidence of a linear association?

8. A study is planned to find the difference in the proportion of patients with warts who are cured with a treatment of cryotherapy and the proportion cured with a treatment of duct tape occlusion.

9. A study is planned to determine if there is significant evidence that the mean cholesterol level (measured in milligrams per deciliter) for people living in "Western" countries is higher than that of people living in "non-Western" countries.

10. What is the 95 percent confidence interval for the average increase in CO_2 emissions (in tons per person) for each unit increase in income level (in GDP per capita) among countries of the world?

11. An economist is interested in whether there is an association between race/ethnicity and whether an adult is unbanked (does not have a checking or savings account). A random sample of 500 adults gives the following table:

	White	Black	Hispanic (non-White)	Native American
Banked	203	98	77	49
Unbanked	18	25	20	10

12. For the prevention of strokes, at-risk patients are often given anticlotting drugs. A study is being designed to compare the proportion of patients receiving rivaroxaban who still have strokes to the proportion receiving warfarin who still have strokes.

13. Malaria is endemic in Liberia. Many people contract and are cured of malaria repeatedly during their lifetimes. A new prophylactic medication was developed as a pill taken weekly to prevent the contraction of malaria. A random sample of adults took the pill for one year and took a placebo pill for another year. The year each adult took the prophylactic was decided randomly for each participant. Summary statistics were as follows.

	n	Mean	Standard Deviation
Episodes on the prophylactic	1,500	0.22	0.11
Episodes on the placebo	1,500	0.84	1.04
Difference (on − off)	1,500	−0.62	0.35

14. In a random sample of 5,000 live births in the U.S., the infant mortality rate was 5.8 (infant deaths per 1,000 live births). In a random sample of 2,000 live births in Japan, the infant mortality rate was 2.0 (infant deaths per 1,000 live births). What is the 99 percent confidence interval for the difference in mortality rates in the United States and Japan?

15. How much better do students do on national standardized exams if they take them a second time? The scores of 500 students who took a national standardized exam twice are charted.

16. What is the mean waist size of male high school math teachers? The waist sizes of a random sample of 50 male high school math teachers are measured. From past studies, it is known that the standard deviation of such waist sizes is 2.15 inches.

The answers for this quiz can be found beginning on page 514.

Quiz 26

> **DIRECTIONS:** For each of the following problems, name the procedure you would use, define any parameter you use, list the conditions to be checked, and state hypotheses if appropriate.

1. College advisers suspect that first-year students study more hours per week than sophomores.

2. Are students willing to report cheating by other students? Students from a random sample are asked to fill out an anonymous questionnaire asking whether they would report cheating by other students.

3. A cardiologist is interested in whether there is a relationship between blood pressure level (low, average, or high) and personality type (high-strung or easygoing).

4. During a previous season, Division 1 college basketball players successfully made 34% of their 3-point shot attempts. A sports magazine is interested in determining if the percentage is lower this year.

5. The director of a health clinic is interested in finding out if participating in a particular month-long workout program will increase the average number of push-ups that can be completed by clinic members. Enrollees will be asked to do as many push-ups as possible in a 90-second period before the program begins and then again a month later when the program ends.

6. It is hypothesized that the average body temperature is less than 98.6°F.

7. By how much does exercise lower resting heart rates for teenagers? Ninety teenagers took part in a survey where their resting heart rates were tested before and after an extensive exercise program.

8. A study about racism in the business world involved sending resumes with stereotypically "White" names and identical resumes except with stereotypically "Black" names. Twelve out of 110 job applications with stereotypically "White" names received callbacks, while only 10 out of 155 job applications with stereotypically "Black" names received callbacks. Establish a 90 percent confidence interval for the difference in proportions of callbacks between applications with stereotypically "White" and stereotypically "Black" names.

9. Is there a significant difference in the distribution of highest school level attained between Whites, Blacks, Asians, and Hispanics (non-White)? Random samples of 200 people from each population interviewed resulted in the following summary table.

	High School or Less	Undergraduate Degree	Graduate Degree
White	130	58	12
Black	168	27	5
Asian	98	82	20
Hispanic (non-White)	172	24	4

10. A reporter for a national car magazine plans to sample used cars of a certain model in hopes of being able to predict the average decrease in selling price for each additional 1000 miles noted on the odometer.

11. After answering your phone, if there is an initial pause on the other end, it often means that the call is from a telemarketer. A study is conducted to estimate the average pause length for telemarketing calls.

12. In a random sample of 550 White students, 4.1 percent reported that, at times, they felt unsafe to go to school, while in an independent random sample of 425 non-White students, 7.8 percent reported that, at times, they felt unsafe to go to school. A sociologist believes that the difference is significant, meaning non-White students are more likely than White students to feel unsafe to go to school, at times.

13. In 2010, the U.S. population of incarcerated adults by race/ethnicity was as follows: 39 percent White (non-Hispanic), 19 percent Hispanic, 40 percent Black, 2 percent other. A random sample of 1,000 incarcerated adults from January 2020 had the following distribution: 451 White (non-Hispanic), 147 Hispanic, 400 Black, and 2 other. Is there sufficient evidence that the proportions of the U.S. incarcerated adult population by race/ethnicity has changed?

14. Conservationists plan to determine the mean difference in weights between elephants living in captivity and those in the wild.

15. Is there a positive linear relationship between the quantity of caffeine consumed in a sitting rate and pulse rate?

16. It is hypothesized that the mean hourly wage of summer work available for high school students is $10.50. A city planner believes the true mean is lower. She randomly surveys 75 high school students with summer jobs. From past studies, it is known that the standard deviation of such jobs is $1.83.

The answers for this quiz can be found beginning on page 517.

Statistical Insights into Social Issues

Quiz 27 Topics
1. **DISTRIBUTION OF WEALTH**
2. **INCARCERATION BY AGE AND RACE/ETHNICITY**
3. **FAMILY INCOME AND SAT SCORES**
4. **DEVELOPMENT AID**
5. **REPORTING CONFLICT VIOLENCE**
6. **JURY SELECTION**
7. **ESSENTIAL HISTORY**
8. **ENVIRONMENTAL RACISM**
9. **DEATH PENALTY CONSIDERATIONS**
10. **FACIAL RECOGNITION**
11. **FLAWED FORENSIC TESTIMONY**
12. **MODERN-DAY SLAVERY**
13. **HATE CRIMES**

Quiz 28 Topics
1. **BULLYING**
2. **FIREARM SUICIDES**
3. **DANGEROUS LEAD LEVELS**
4. **RACIALLY IDENTIFIED NAMES**
5. **CHILDREN AND GUN SAFETY**
6. **LAW ENFORCEMENT–ASSISTED DIVERSION**
7. **POLICE PROFILING**
8. **INEQUALITY AND HEALTH**
9. **WHY JAILS ARE FULL**
10. **VACCINES AND AUTISM**
11. **GLOBAL WARMING**

In the spirit of AP Statistics, the following two quizzes encompass comprehensive review questions that aim to give an appreciation of the power of statistics to better understand some of society's most pressing issues. As you review and apply the statistical tools you have learned, you should also reflect on the content of the examples and the empowerment that statistics gives you to be aware of, to discuss, and maybe someday to help overcome some of the ills present in the broad social world in which we live. More specifically, statistics can help you understand the relationships among power, resource inequities, and uneven opportunities among different social groups and to understand explicit discrimination based on race/ethnicity, class, and gender. In many ways, the AP Statistics curriculum is just as much a civics course as it is a math course. Of course, in addition to reflecting on these significant issues, do not forget to read carefully what the questions ask, to check assumptions and conditions when appropriate, to show your work, and to communicate your answers clearly.

These exercises explore sociopolitical contexts using statistics and mathematics. They are purposefully **nonpartisan**. Students are asked to analyze these data sets critically, and multiple perspectives are given consistent with the data analyzed. Over the years, these exercises have been carefully vetted by AP Statistics teachers who have given their feedback with regard to fairness, balance, and rigor.

There is great interest in how statistics can help better society. Statistics can be an incredibly powerful tool both in helping you understand social issues in terms of our current information age and in helping you effect social change. From disparities in income levels to disparities in the distribution of wealth, from unequal access to education to unequal access to health care, from unfair criminal justice enforcement to unfair housing market discrimination, statistics can be the key for you to become an engaged participant in our democracy. Based on your background, your interests, and your available time, you can decide how far to take the opportunities for deeper discussion raised in the following exercises.

Following are two quizzes. Quiz 27 has 13 multisection free-response questions reviewing the concepts and skills from Units 1–5 on exploratory analysis, collecting data, and probability. Quiz 28 has 11 multisection free-response questions reviewing the concepts and skills from Units 6–9 on statistical inference. All the quiz questions are set in the context of contemporary social justice issues, and they are specifically designed to evaluate your understanding of the entire curriculum and reinforce your communication skills. The answers can be found beginning on page 517.

Quiz 27

1. A 2023 Oxfam press release stated that the richest 1% take in twice as much new wealth as the rest of the world put together.[1] One way in which the distribution of wealth throughout the world can be analyzed is illustrated below (where GDP refers to Gross Domestic Product and Oceania is a region centered on the islands of the tropical Pacific Ocean).

   (a) Fill in the blank cells in the table below.

Region	Population (in millions)	% of World Population	Wealth (GDP) (in billions)	% of World GDP
Africa	1,341		2,995	
Asia	4,641		40,248	
Oceania	43		2,002	
Europe	748		24,881	
U.S./Canada	369		31,319	
Latin America	654		4,068	
World Total	7,796	100.0	105,513	100.0

   (b) Construct a segmented bar chart with two bars from the above table, and comment on its significance.

   (c) Consider the following graph:

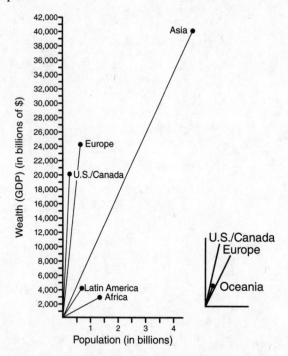

   What do the slopes of the six slanted lines represent? Use these slopes to give an appropriate ranking of the six regions.

---

[1]Oxfam International, January 16, 2023.

(d) The particular quantitative index used in a statistical study can make a marked difference. For example, an alternative to the GDP for measuring the wealth of a nation is the GPI (Genuine Progress Indicator).[2] The GPI takes into account negative factors, such as industrial pollution, depletion of natural resources, and crime rates, as well as positive factors, such as socially productive time uses like volunteerism. Give an example of how changing the quantitative index from GDP to GPI might affect the relative position of U.S./Canada and Oceania in part (c).

2. The United States has the second highest incarceration rate in the world (second only to the small African country of Seychelles).[3] One study[4] of incarceration rates and ages summarized their findings in the following graph:

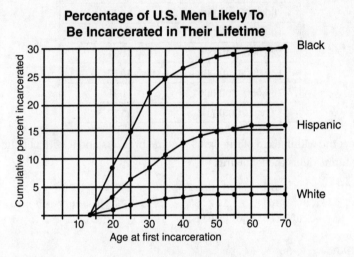

Percentage of U.S. Men Likely To Be Incarcerated in Their Lifetime

(a) What is the meaning of the point (25, 15) on the Black graph?

(b) What is the meaning of the horizontal segment on the White graph from ages 60 to 70?

(c) What is the median age at first incarceration of Hispanics who ever went to prison?

(d) Some states are considering legislation to address the disproportionate school suspensions among students of color. How might this affect the above graphs?

---

[2]Talberth, Cobb, and Slattery, "Gross Domestic Product, A Tool for Sustainable Development," February 2007.
[3]"U.S. Prison Boom: Impact on Age of State Prisoners," *Journalist's Resource,* February 2016.
[4]"The Use of Incarceration in the United States," *American Society of Criminology,* November 2000.

3. SAT used to stand for "Scholastic Aptitude Test," but some researchers suggest it also stands for "Student Affluence Test."[5] There appears to be an association between mean total SAT scores and family income level. One study is summarized in the following table:

2015 College-Bound High School Students

Income Bracket	Total SAT Score
0–20,000	876
20,000–40,000	933
40,000–60,000	972
60,000–80,000	1000
80,000–100,000	1029
100,000–120,000	1054
120,000–140,000	1061
140,000–160,000	1077
160,000–200,000	1091
Over 200,000	1147

Using the median income within each of the first 9 income brackets and a $210,000 income from the last bracket, computer output yields the following:

```
Predictor      Coef       SE Coef       T       P
Constant     903.36        112.57   71.85   0.000
Income      0.00116      0.000104   11.17   0.000

S = 20.936     R-Sq = 94.0%
```

(a) What is the equation of the regression line?
(b) Interpret the slope of the regression line in context.
(c) What is the coefficient of determination? Interpret it in context.
(d) What is the residual if a family has $63,500 income and a student has a score of 970? What does this mean in context?
(e) What is the meaning of $S = 20.936$?
(f) Why is it not reasonable to conclude that the best way to raise SAT scores is to raise family incomes? Explain, giving a possible confounding variable that could suggest an alternative approach to raising SAT scores.

[5]"SAT Scores and Income Inequality: How Wealthier Kids Rank Higher," *Wall Street Journal*, October 2014.

4. It is estimated that 700 million people worldwide live in extreme poverty. With regard to foreign policy, governments invest in hard power (military) and soft power (development aid). The cost of deploying one soldier overseas is the same as producing 100 village wells. Countries spend widely different percentages of their Gross National Income (GNI) on development aid. A random sample of 12 countries plus the United States yields the following:

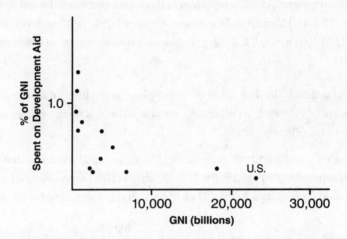

(a) Describe the relationship between GNI and percentage of GNI spent on development aid.

(b) Explain how the U.S. compares to the other countries in the sample with regard to GNI and percentage of GNI spent on development aid.

(c) How would the answer to (b) differ if the graph showed the amount of development aid (rather than percentage of GNI) versus GNI?

(d) The five lowest values for GNI in this sample come from Belgium, Denmark, Netherlands, Norway, and Sweden. What is apparent about the use of soft power by these countries?

Following is a scatterplot and some computer linear regression output showing the relationship between the natural logarithm of percentage of GNI spent on development aid and the natural logarithm of GNI, where GNI is in billions of dollars.

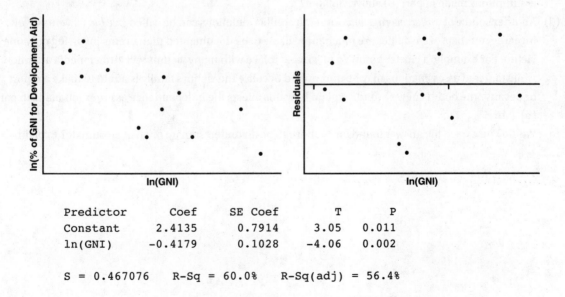

Predictor	Coef	SE Coef	T	P
Constant	2.4135	0.7914	3.05	0.011
ln(GNI)	-0.4179	0.1028	-4.06	0.002

S = 0.467076   R-Sq = 60.0%   R-Sq(adj) = 56.4%

(e) Describe the relationship between ln(GNI) and ln(percentage of GNI spent on development aid).

(f) Based on this output, explain why it would be reasonable to use this regression to describe the relationship between ln(GNI) and ln(percentage of GNI spent on development aid).

(g) What is the correlation coefficient r for this regression?

(h) Interpret the slope for this regression line in context.

(i) Using the computer output, give the equation of the least-squares regression line.

(j) Using the model from (i) above, predict the percentage of GNI spent on development aid for the U.S. with a GNI value of $23,400 (billion). Is this prediction an overestimate or an underestimate for the true value of 0.15 percent?

5. In analyzing conflict violence, the data are often incomplete, as obtaining data in such situations is very difficult if not impossible. However, accurate data are crucial for systematic studies of the origins of conflict, conflict dynamics, and conflict resolution.

(a) Suppose during one conflict month early during the Syrian civil war, the Syrian Observation for Human Rights (SOHR) reported 3,060 deaths, the United Nations (UN) reported 3,627 deaths, and 2,635 of the deaths were found on *both* lists. Letting D be the actual number of deaths, fill in the following table:

SOHR

		Reported Deaths	Unreported Deaths	*Totals*
	Reported Deaths			3,627
UN	**Unreported Deaths**			$D - 3,627$
	Totals	3,060	$D - 3,060$	D

(b) Assuming the two reports above represented independent random samples of all deaths that month, solve for D.

(c) In reality, only *convenience samples* were available—perhaps posted videos of public executions, perhaps text messages sent out by protesters, or perhaps victims' testimonies recorded by local reporters. In general, why do convenience samples often lead to misleading conclusions? In our context, how are assumptions made in part (b) above violated?

(d) Another source of *undercoverage bias* found in conflict situations can be called *size bias*. Events involving large numbers of casualties are much more likely to be documented than events involving only one victim. For example, a cluster bombing of a marketplace with many victims will attract media attention while targeted assassinations may be unreported because the victim's family is afraid to make a report or because the body is never found. Explain how this affects the independence assumption made in part (b) above.

(e) Would you expect the answer found for D above to be an underestimate or an overestimate? Explain.

6. In 1986, the Supreme Court ruled that it is unconstitutional to exclude potential jurors because of race/ethnicity.[6] However, many studies since then have shown that racial discrimination in jury selection remains widespread, particularly in murder trials. A university law department plans a study using mock trials to determine if outcomes are different with all-White versus mixed-race/ethnicity juries. The law department will first randomly select two counties from among its state's 67 counties.

   (a) What is this sampling method called, and what is an advantage of using this method over a simple random sample?

   (b) Give a procedure for randomly selecting the two counties.

   (c) Forty-five people will then be randomly selected from among the pool of all eligible White jurors in each of the two counties to obtain a sample of size 90. What is the purpose of randomization in this two-step process?

   (d) The proposed study will involve 10 mock cases involving a young Black man charged with murdering an elderly White woman. The sample of 90 White jurors will be randomly assigned to two groups—60 to make up five all-White juries (each of size 12) and 30 to be part of five mixed-race/ethnicity juries (each with 6 White jurors and 6 jurors of color). What is the purpose of random assignment here? How might allowing jurors to self-select whether or not to be part of mixed-race/ethnicity juries lead to bias?

7. Book banning is a controversial topic in school districts across the country. Suppose a questionnaire is to be designed as to parents' opinions on whether books with depictions of slavery should be banned from school libraries. *Wording bias* occurs when the wording of a question influences the responses.

   (a) Give an example of a neutrally designed question on this topic.

   (b) Give an example of a question on this topic illustrating wording bias that will lead to an underestimate of the true proportion of parents in favor of banning books with depictions of slavery.

   (c) Give an example of a question on this topic illustrating wording bias that will lead to an overestimate of the true proportion of parents in favor of banning books with depictions of slavery.

   In a recent survey, 26% of American adults described their political views as liberal, 38% as moderate, and 36% as conservative. Given a neutrally designed question, 9% percent of liberals, 12% of moderates, and 15% of conservatives answered that books with depictions of slavery should be banned from school libraries. An American adult is picked at random.

   (d) What is the probability the person is liberal and believes that books depicting slavery should be banned from school libraries?

   (e) What is the probability the person believes that books depicting slavery should be banned from school libraries?

   (f) If the person believes that books depicting slavery should be banned from school libraries, what is the probability the person is liberal?

---

[6]*Batson v. Kentucky*, 476 U.S. 79 (1986).

8. "Environmental racism" is said to occur when race/ethnicity is associated with the location of hazardous sites. A 2015 California study was designed to assess who was at the greatest risk from oil trains, which can derail and explode. Of the 5.5 million Californians who live within a blast zone (defined as within one mile of the railroad tracks), 75% are Latino and only 10% are White. California's population of 38.8 million is 39.0% Latino and 38.9% White.

    (a) What is the probability a randomly selected Californian is a Latino living within a blast zone?
    (b) What is the probability that a randomly selected Californian Latino lives within a blast zone?
    (c) In California, are being Latino and living within a blast zone independent events? Justify your answer mathematically.
    (d) In a random sample of 5 Californians, what is the probability that at least 2 of them live in a blast zone?
    (e) What is another explanatory variable that could account for the relatively high probability that someone living in a blast zone is a person of color?

9. Consider the following mosaic plot of murder cases.

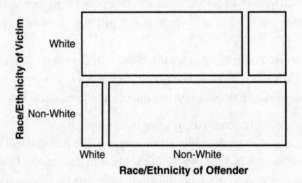

    (a) What can be concluded from the above mosaic plot?

    Studies seem to indicate that Black lives matter less than White lives with regard to the death penalty.[7] One long-term study found that while the race/ethnicity of the offender did not seem to predict the likelihood of a death sentence, the race/ethnicity of the victim did. In this study, 50.5 percent of murder victims were White, 49.5 percent were non-White. In the cases which resulted in executions, 76% of the murder victims were White and 24% were non-White. Finally, 1.2% of the murder cases resulted in the execution of the defendant. Use these findings to answer the following questions.

    (b) What is the probability that there is a White murder victim and the defendant is executed? What is the probability that there is a non-White murder victim and the defendant is executed?
    (c) Given that the murder victim is White, what is the probability that the defendant is executed? Given that the murder victim is non-White, what is the probability that the defendant is executed?
    (d) In (c), what can be concluded about the likelihood of a victim being White or non-White if the defendant is executed?
    (e) If murder cases are randomly selected, what is the mean number of cases sampled until one is found involving the execution of the defendant?

[7]"A Vast Racial Gap in Death Penalty Cases," Adam Liptak, *New York Times,* Aug. 3, 2020.

10. Law enforcement agencies across the United States are rapidly adopting facial-recognition surveillance.[8] A false positive, which incorrectly identifies an innocent person as being a criminal, generally occurs less than 10% of the time. While this percent may sound low, the question is whether it is a small enough risk when the end result may be the arrest of, or even worse, the use of force (including deadly force) against an innocent person.[9]

    Suppose facial recognition software used by a police department to identify possible suspects is 99 percent accurate. That is, the software will correctly identify a criminal as being a criminal 99 percent of the time, and it will correctly identify an innocent person as being innocent 99 percent of the time. Assume 0.88 percent of the adult population are criminals (in 2020, about 0.88% of the United States adult population were behind bars).

    (a) What is the probability that someone is identified to be a criminal by this software?
    (b) What is the probability that someone identified as being a criminal by this software really is a criminal?
    (c) What concern regarding the use of this software is raised by the answer to (b)? Explain.
    (d) Statisticians should always consider ethical issues when gathering data. Numerous studies have found that facial recognition is significantly less accurate when identifying people of color. Does this raise ethical issues? Explain.

11. Although there is no accepted research on how often hair from different people may appear the same, for over two decades the FBI used visual hair comparison matches to help obtain convictions in major criminal cases. Acknowledging possible flawed forensic testimony based on unsound science, the FBI agreed to review over 2,500 cases. Flawed forensic testimony was found in 257 of the first 268 cases reviewed.[10] In the following, assume that the sample proportion of cases involving flawed testimony is an accurate measure of the population proportion.

    (a) Was this a retrospective observational study, a prospective observational study, or an experiment? Explain.
    (b) The FBI agreed to review 2,500 cases involving possible flawed forensic testimony. Assuming this is a random sample of relevant cases, what is the expected number of these 2,500 cases with actual flawed forensic testimony? What is the standard deviation for the expected number of cases with actual flawed forensic testimony?
    (c) Among the 257 cases with flawed forensic testimony, 32 resulted in death penalty convictions. Assume this is illustrative of the overall proportion. In examining a random sequence of cases involving flawed testimony, what is the probability that the first case with a resulting death penalty doesn't occur until at least the third case examined?
    (d) According to the Innocence Project, the misapplication of forensic science is the second-leading contributor to wrongful conviction, playing a role in 50% of known wrongful convictions, while eyewitness misidentification has played a role in nearly 75% of known wrongful convictions.[11] In wrongful conviction cases, can "the misapplication of forensic science" and "eyewitness misidentification" be mutually exclusive? Can they be independent? Explain.

---

[8]Laperruque, Jake, "Unmasking the Realities of Facial Recognition," in POGO (Project on Government Oversight), December 5, 2018.
[9]Wagner, Peter, and Wanda Bertram, "What percent of the U.S. is incarcerated?", in Prison Policy Initiative, January 16, 2020.
[10]National Association of Criminal Defense Lawyers and Innocence Project Analysis of FBI and Justice Department Data as of March 2015.
[11]*http://www.innocenceproject.org*

12. We often think that slavery ended in the 19th century, but the sad fact is that there are more people enslaved today than at any other time in human history. According to the not-for-profit Force4Compassion, an average of 3,287 people are sold or kidnapped and forced into slavery every day.[12] Assume this daily total has a roughly normal distribution with a standard deviation of 250 people.

(a) What is the probability that on a randomly selected day, over 3,500 people are forced into slavery?

(b) In a random sample of 5 days, what is the probability that on a majority of the days over 3,500 people are forced into slavery?

(c) In a random sample of 5 days, what is the probability that the average number of people forced into slavery is over 3,500?

(d) Are there any statistics that, if more widely disseminated, might mobilize the world to fight this social blight?

13. When bias-motivated crimes (triggered by prejudice and hate) are committed, whole communities can feel victimized. A study[13] of profiles on America's most popular online hate site revealed that the site members are very young. Consider the following boxplot of self-reported ages of the site members:

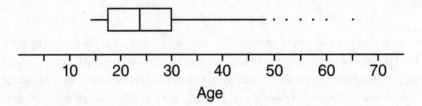

Age

(a) Describe the distribution of ages.

(b) Is the quotient $\dfrac{\text{Mean age}}{\text{Median age}}$ greater than 1, less than 1, or approximately 1? Explain.

(c) Members of online social networks tend to be younger than the general population. What does this possibly indicate about the median age of all hate group members? Explain.

(d) According to recent FBI statistics,[14] 18.6% of bias incidents are religious based. In a random sample of 3 bias incidents, what is the probability that a majority are religious based?

The answers for this quiz can be found beginning on page 519.

[12]*http://www.f-4-c.org/statistics/*
[13]The Data of Hate," *New York Times*, July 13, 2014.
[14]"Bias Breakdown," *www.fbi.gov*, November 2015.

Quiz 28

1. Schoolyard and online bullying are leading to increased levels of violence and depression among students. In a national survey[15] of 15,686 students in grades six through ten, 19% were involved in moderate to frequent bullying as perpetrators, 17% as victims, and 6% as both victims and perpetrators.

    (a) What proportion of the students were involved in moderate to frequent bullying?

    (b) Find a 95% confidence interval for the proportion of students in grades six through ten who are victims of moderate to frequent bullying.

    (c) Based on this confidence interval, is there convincing evidence that the proportion of students who are victims of moderate to frequent bullying is different from the 22% claimed in one study?[16] Explain.

    (d) School administrators tend to underestimate the rates of bullying because students rarely report it and it often happens when adults are not present. However, knowing the numbers and types of bullying at all grade levels in your school is critical before a plan for bullying prevention and intervention can be developed. How would you plan a survey at your school to assess the frequency and types of bullying present?

2. Using the Second Amendment to the U.S. Constitution, gun rights proponents have argued for nearly or completely unchecked rights of personal gun ownership. However, because the U.S. has one of the highest rates of gun violence in the world, others have sought to regulate gun ownership. In the news, deaths by firearms are most often associated with murders; however, most firearm-related deaths in the United States are suicides, and most suicides are by guns. According to the CDC,[17] in 2021 there were 26,328 suicides in the U.S. using firearms.

    (a) Assume we have a random sample of 7 years that are considered representative with regard to numbers of firearm suicides:

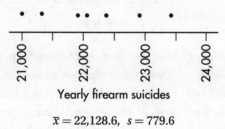

Yearly firearm suicides

$$\bar{x} = 22{,}128.6, \quad s = 779.6$$

   Calculate a 95% confidence interval estimate for the mean number of firearm suicides in the U.S. each year.

    (b) According to a major recognized survey,[18] the U.S. has the highest rate of gun ownership in the world with 120.5 guns per 100 people. The same survey reports Israel with 6.7 guns per 100 people. Given that these figures report only privately owned guns and exclude weapons that are technically owned by the government, how might the apparent wide discrepancy between the U.S. and Israel not be as wide as indicated?

[15]Nansel, T. R., Overpeck, M., Pilla, R. S., Ruan, W. J., Simons-Morton, B., & Scheidt, P. "Bullying Behaviors Among U.S. Youth: Prevalence and Association with Psychosocial Adjustment," *Journal of the American Medical Association*, 285(16), 2094–2100.

[16]National Center for Educational Statistics, 2015.

[17]*https://www.cdc.gov/nchs/fastats/suicide.htm*

[18]*https://www.cnn.com/2021/11/26/world/us-gun-culture-world-comparison-intl-cmd/index.html*

(c) When presented with any data set, we should always try to understand what and whose values are implicitly represented. Israel rejects 40% of gun applications and requires yearly renewals for continuing gun ownership. How would this affect the conversation from the statistics quoted in part (b)?

(d) The average U.S. gun owner owns 3 guns.[19] Comment on measuring rates of suicide by firearms in terms of guns per 100 people (120.5 in the U.S.) versus measuring rates of suicide by firearms in terms of gun owners per 100 people.

3. As a cost-cutting measure in April 2014, Flint, Michigan, began temporarily drawing its drinking water from the Flint River. The result was dangerous lead contamination in the water supply. (Lead poisoning especially impacts a child's brain and nervous system.)

(a) An independent research team[20] tested the Flint water supply for lead. The resulting cumulative frequency plot of the lead levels in ppb (parts per billion) in the water supply of sampled homes is as follows:

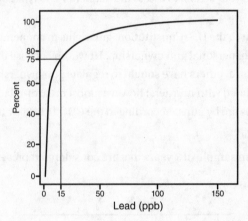

What is the shape of the histogram of the lead levels in the water supply of sampled homes? How would you expect the mean and median lead levels to compare with each other? Explain.

(b) It is recognized that if over 10% of a city's homes have lead levels in their water supply of over 5 ppb, the city has "a very serious problem with lead."[21] The research team determined that 15 ppb corresponded to the 75th percentile of Flint home water supply lead levels. Does this indicate a very serious problem?

(c) Suppose the concern is whether or not the mean lead level in the water supply of the homes is greater than 10 ppb. (The EPA regards 15 ppb in the water supply as a very dangerous level.[22]) Describe Type I and Type II errors in context. Which is of greater concern, a Type I or Type II error, and what would be a possible consequence? Explain.

[19]https://qz.com/1095899/gun-ownership-in-america-in-three-charts/
[20]2015 Virginia Tech Study.
[21]http://flintwaterstudy.org
[22]CDC 24/7: "Saving Lives, Protecting People."

4. Even in the highest-paying occupations, non-White workers face racism in the workplace.[23] In one study, racially identified names were randomly assigned to resumes, and the number of interviews offered (positive responses) for each group was tabulated.

    (a) The positive response rate for applicants with "White" names had a 95% confidence interval of (0.21882, 0.28258). What was the sample size for applicants with "White" names?

    (b) A mosaic plot of responses is shown below. What does it show?

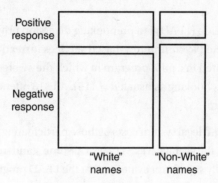

    (c) A significance test, with $H_a: p_W > p_{NW}$ where $p_W =$ the true proportion of resumes with "White" names that result in positive responses, and $p_{NW} =$ the true proportion of resumes with "Non-White" names that result in positive responses, yields $z = 2.585$ and $P = 0.00486$. Is there convincing evidence that resumes with "White" names result in a greater proportion of positive responses than resumes with "Non-White" names?

    (d) A measure of discrimination, the *discrimination ratio*, for African Americans contrasted to Whites (with positive values indicating discrimination against African Americans and negative values indicating discrimination against Whites) was calculated over a 25-year period, resulting in a confidence interval for the slope of $(-0.007, 0.015)$. Is there convincing evidence that discrimination against African Americans is trending higher?

5. Gun violence is now the leading cause of death for children in the United States, and all sides of the political spectrum are concerned. In a recent study, 226 children, ages 8 to 12, were tested in pairs of kids who knew each other, including siblings, cousins, and friends. One of each pair was randomly picked to watch a short video on car safety, while the other of each pair was shown a short video on gun safety. A week later, all 226 children came to a large playroom filled with toys and with two disabled 9 mm handguns. Of those who watched the gun safety video, 34% told an adult that there were guns in the room, while of those who watched the car safety video, 11% told an adult there were guns in the room.[24]

    (a) Is random sampling, random assignment, both, or neither involved in this study? Explain.

    (b) What are the appropriate hypotheses for testing whether there is convincing evidence that the proportion of all youths 8 to 12 years old who will tell an adult about a gun in the room is greater for those children shown a gun safety video than for those children shown a car safety video?

    (c) The test statistic is $z = 4.167$. What does this value mean in context?

    (d) Is there convincing evidence to reject the null hypothesis at the 0.01 significance level? Explain.

    (e) A 99% confidence interval for the difference in proportions (gun safety video viewers minus car safety video viewers) based on this data is (0.09342, 0.36676). Interpret this interval in context.

[23]"Minorities who 'Whiten' Job Resumes Get More Interviews," *Working Knowledge,* Harvard Business School, May 17, 2017.
[24]*https://jamanetwork.com/journals/jamapediatrics/article-abstract/2807325*

(f) Is this confidence interval consistent with the test decision in part (d)? Explain.

(g) The above two videos both featured a police chief in full uniform. A second experiment where both videos featured a cartoon figure called Eddie Eagle showed no statistical difference between the proportions of the two groups telling an adult about seeing a gun in the playroom. Give a possible explanation.

(h) Another study involved children in homes with a gun in the home and children in homes without a gun in the home. Those children living in homes with a gun in the home were more likely to tell an adult about seeing a gun in the playroom. Give a possible explanation.

6. Law Enforcement Assisted Diversion (LEAD) is a prebooking diversion program whereby low-level drug and prostitution offenders are redirected to community-based services instead of jail and prosecution.[25] There were 318 offenders who participated in a pilot program in which they were assigned to either the LEAD program ($n = 203$) or control (i.e., booking as usual; $n = 115$). All 318 offenders were deemed eligible for participation in the LEAD program.

(a) The average yearly criminal and legal system costs of those participating in the LEAD program were $4,763. For the control group, the average was $11,695. (Assume standard deviations of $875 and $2,125, respectively.) Is this statistically significant evidence that the LEAD program has lower costs than the usual booking program?

(b) After participation in the program, the LEAD group showed significantly fewer average number of days per year in jail than the control group ($P < 0.05$). Which error, Type I or Type II, might have been committed and what would be a possible consequence?

(c) Even though it was felt that both the LEAD and the control groups were representative of the targeted offender population, a nonrandomized design was used. This may be of concern. What is the purpose of *random assignment* in such studies?

(d) Using your answer from part (a), make an argument in favor of the LEAD program to a city council whose members are unfamiliar with statistical arguments.

7. Local activists and community leaders have long maintained that local law enforcement "overpolice" people of color in Berkeley, California. One claim is that people of color are being stopped at a higher rate than their population would indicate and that the stops are often for no legitimate reasons. A study[26] of the Police Department found that during the period from January 26 through August 12, 2015, 4,658 civilians were stopped. Of that total, 1,710 were White, 1,423 were African American, and 543 were Hispanic. Of those stopped, 652 of the Whites, 942 of the African Americans, and 306 of the Hispanics were released without arrest or citation.

According to 2013 census data, the population of Berkeley was 116,774, of which 64,412 were White, 10,076 were African American, and 11,600 were Hispanic.

(a) Sketch side-by-side bar charts to illustrate the population percentage distribution and the percentage distribution of who was stopped by race/ethnicity.

(b) Is there statistically significant evidence that African Americans are being stopped at a higher rate than their population would indicate? (Assume that we have a representative sample of common practices for years.)

(c) Is there statistically significant evidence of a relationship between race/ethnicity (White, African American, Hispanic) and disposition (arrest/citation or not)? (Assume that we have a representative sample of common practices for years.)

[25]"LEAD: Reducing the Role of Criminalization in Local Drug Control," *http://www.drugpolicy.org*
[26]*https://www.nbcbayarea.com/news/local/racial-profiling-rampant-in-berkeley-police-department-report/103331/*

8. Is there a relationship between inequality and health? The Equality Trust, a nonprofit advocacy group, has created an Index of Health and Social Problems (HSP Index)[27] to combine several troublesome indicators (including lower life expectancy, higher infant mortality, and weak social mobility) into a single variable that can be used to describe the overall "health" of a society. A measure of income inequality is the ratio of the income of the top 20% to the bottom 20% in a country. Computer output of a regression analysis of HSP Index versus Income ratio in 21 industrialized countries is as follows:

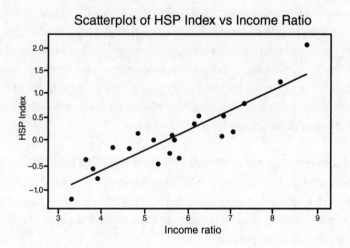

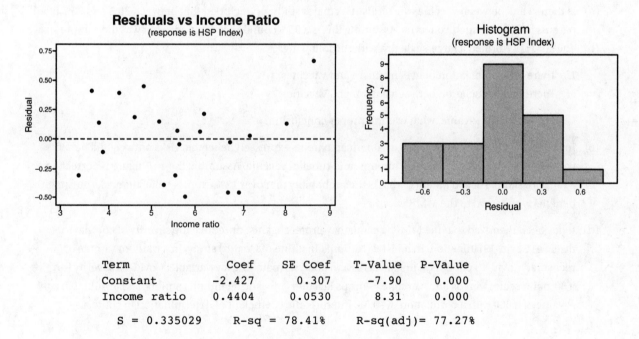

Term	Coef	SE Coef	T-Value	P-Value
Constant	-2.427	0.307	-7.90	0.000
Income ratio	0.4404	0.0530	8.31	0.000

S = 0.335029   R-sq = 78.41%   R-sq(adj)= 77.27%

(a) Find a 95% confidence interval for the slope of the regression line.
(b) Is there sufficient evidence that the higher the income ratio (that is, the higher the income inequality in a country), the greater the HSP Index (that is, the greater the health and social problems of the country)?
(c) Assuming that a lower HSP Index is desirable, what can be concluded about working to lessen the Income ratio (that is, to lessen the income inequality in a country)?

[27] http://inequality.org/inequality-health/#sthash.H77M9bp9.dpuf

9. According to the Bureau of Justice Statistics, 1 out of every 35 adults in the United States is under correctional supervision (probation, parole, jail, or prison).[28] According to a 2014 Human Rights Watch report, "tough-on-crime" laws have filled U.S. prisons with mostly nonviolent offenders.[29]

   (a) A previous study[30] on the population of local jails showed that 60.6% were people too poor to make bail or pay fines. More specifically, 34.4% were convicted males, 5.0% convicted females, 53.4% unconvicted males, and 7.2% unconvicted females. Suppose a current random sample of local jail populations counts 81 convicted males, 5 convicted females, 110 unconvicted males, and 4 unconvicted females. Is there sufficient evidence that the population distribution has changed from the findings of the earlier study?

   (b) Advocacy groups are pushing for counties across the country to reform their local criminal justice systems to make them fairer. The groups argue that these reforms are needed to stop increasing costs, reduce overcrowding, and improve outdated facilities. Local jails hold nearly 700,000 inmates on any given day, with small counties holding 44% of the total.[31] More data are very much needed. How would you proportionately choose 500 inmates to interview and study?

10. Immunizations are a cornerstone of the world's efforts to protect people from infectious diseases. After a 1998 research paper falsely claimed a link between MMR (measles, mumps, rubella) vaccines and ASD (autism spectrum disorder), vaccination rates dropped sharply.[32] The World Health Organization estimated that in 2018, over 140,000 people worldwide died of measles. In 2019, the U.S. experienced a large, multistate measles outbreak, involving mostly unvaccinated children.

   (a) Is there a link between vaccines and autism? A major study[33] concluded that between 30% and 36% of all parents have some such concerns. Assuming this is a 95% confidence interval, what was the sample size?

   (b) The study performed a large-scale test with the following:

   H_0: There is no association between autism and vaccines.
   H_a: There is an association between autism and vaccines.

   Given a very large P-value, what was the proper conclusion?

   (c) In a 2020 Pew Institute Study, 10% of American parents expressed belief that risks outweighed benefits with regard to the MMR (measles, mumps, and rubella) vaccine. Assuming the 10% figure is correct, in a random sample of 500 parents, what is the probability that over 12% express belief that risks outweigh benefits with regard to the MMR vaccine?

   (d) It has been theorized that the MMR vaccine may protect against or reduce the severity of coronavirus disease (Covid-19) infection. In an NIH (National Institute of Health) study, in a random sample of 50 recovered Covid-19 patients, a linear regression analysis matching a y-variable, Covid-19 severity (on a 30-point scale), versus an x-variable, mumps IgG titer (measuring mumps antibodies), resulted in a P-value of 0.000 with a correlation of –0.58. Interpret the coefficient of determination in context.

[28]"Correctional Populations in the United States," Bureau of Justice Statistics, 2013.
[29]"Nation Behind Bars: A Human Rights Solution," Human Rights Watch, May 2014.
[30]Bureau of Justice Statistics, 2012.
[31]Pearson, Jake, "Study: Smaller Counties Driving U.S. Jail Population Growth," December 2015.
[32]https://www.sciencedaily.com/releases/2012/06/120604142726.htm
[33]DeStefano, Price, and Weintraub, "Increasing Exposure to Antibody-Stimulating Proteins and Polysaccharides in Vaccines Is Not Associated with Risk of Autism," 2013.

11. There is a strong consensus among scientists that global surface temperatures have increased markedly in recent years and that the trend is caused by emissions of greenhouse gases. However, in the popular media, especially in the United States, there is much debate over whether there is any problem—and even if there is a problem, the debate continues over whether humans are responsible. (China and the U.S. are the two largest producers of greenhouse gases.)

(a) Given the following graph, describe the relationship between Temperature Anomaly (deviation) and Time.

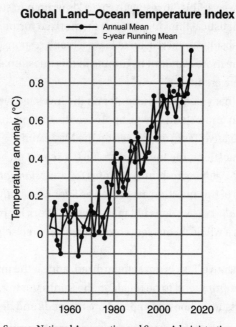

Source: National Aeronautics and Space Administration,
Goddard Institute for Space Studies

(b) In a Pew Global Attitudes Project survey in a random sample of 2,029 Indian people, 1,319 expressed a great deal of worry about global warming. In a random sample of 2,180 Chinese people, 436 expressed a great deal of worry about global warming. Is the difference statistically significant?

(c) Suppose we wish to conduct a survey among the student body of your high school and calculate a confidence interval for the proportion who worry a great deal about global warming. Comment about bias and direction of bias, from the wording of the following survey:

> King Charles III has consistently taken a strong stance criticizing both climate change deniers and corporate lobbyists by likening the Earth to a dying patient. Do you have a great deal of worry about global warming?

The answers for this quiz can be found beginning on page 525.

The Investigative Task: Free-Response Question 6

The last question on the exam is called the Investigative Task. This free-response question counts for one-eighth of the total grade on the AP Statistics exam. The purpose of this question is to test your ability to take what you've learned throughout the course and apply that knowledge in a novel way. The topic of this question often includes material not only unfamiliar to students but also to their teachers.

This question always has several parts. Very often the first parts are straightforward, coming directly out of standard concepts of the curriculum. These are followed by parts that will challenge you to think creatively and require you to use learned concepts in new ways. This "investigative part" is written in such a way as to be accessible to students who have learned sound statistical thinking, not just memorized formulas and procedures.

There is no reason to panic and absolutely no reason to leave a blank response. All students should be able to receive some credit by answering the more standard beginning to this question. Following this with something "reasonable" will earn you even more credit.

It is not possible to list all topics that you might encounter on the Investigative Task. However, working through the following examples will help you understand how to approach the task on the exam. This should give you the confidence to do well on the Investigative Task you will see on the May exam.

Just as the other free-response questions, the Investigative Task is scored on a 0 to 4 scale with 1 point for a *minimal* response, 2 points for a *developing* response, 3 points for a *substantial* response, and 4 points for a *complete* response. Individual parts of this question are scored as E for *essentially* correct, P for *partially* correct, and I for *incorrect*. Note that *essentially* correct does not mean *perfect*. Work is also graded holistically—that is, your complete response is considered as a whole whenever scores do not fall precisely on an integral value on the 0 to 4 scale.

The scoring rubric for the Investigative Task is not standardized, and so the minimal requirements change depending on the question. These rubrics tend to be tough on the things the Investigative Task is intended to address and lenient on peripheral issues. Choosing appropriate methods and demonstrating good communication are the keys to high scores.

Following are three Investigative Tasks with carefully explained solutions illustrating possible thought processes for approaching such problems. Following these is Quiz 29 with seven more Investigative Tasks for which you're on your own. (There are answers in the Appendix.)

> Example 1

The R&D division of an automobile company has developed a new fuel additive that they hope will increase gas mileage. For a pilot study, they send 30 cars using the new additive on a road trip from Detroit to Los Angeles. From previous studies, they know that without the additive, the cars will average 32.0 mpg with a standard deviation of $\sigma = 2.5$ mpg. (Assume σ remains the same with or without the additive.)

(a)   What is the parameter of interest, and what are the null and alternative hypotheses?

Think: The interest is in gas mileage, and there is mention of the "average" mpg. So the parameter is the mean mpg using the new fuel additive. The parameter is always something about a whole population, so be sure to use μ, not $\bar{x}$. The null hypothesis is some specific claim about this parameter, and the old mean mpg is given to be a specific 32.0 mpg. The alternative hypothesis will have an inequality. Since the R&D division hopes to increase gas mileage, the alternative must be > 32.0.

> **TIP**
>
> You've already practiced so many hypotheses problems that you should be able to name the parameter and the hypotheses in your sleep!

(b)   What are Type I and Type II errors in this context? Give a possible consequence of each.

Think: A Type I error is when you mistakenly reject a true null hypothesis. In this case, if the null hypothesis was true, that would mean the cars really get a mean of 32.0 mpg. If you reject this, you mistakenly think the mean mpg is > 32.0. A consequence is that you'll waste money on this new additive, thinking it will improve gas mileage when in reality it won't.

Think: A Type II error is when you mistakenly fail to reject a false null hypothesis. In this case, if the null hypothesis is false, that would mean the cars really get a mean mpg > 32.0. If you fail to reject the null hypothesis claim of 32.0, you don't think the additive helps when it really does.

(c)   The company decides to work with a level of significance of $\alpha = 0.05$. What is the probability that a Type I error will be committed?

Think: That's easy. The significance level, or α-risk, gives the cutoff score for rejecting the null hypothesis, but it's based on the assumption that the null hypothesis is true. Rejecting a true null hypothesis is what we call a Type I error.

(d)   What is the null hypothesis rejection region? In other words, what critical value of the test statistic $\bar{x}$ will lead the company to conclude that there is statistically significant evidence that the new fuel additive increases gas mileage?

Think: You are given the population standard deviation ($\sigma = 2.5$), so you can use a normal distribution for the sampling distribution. The mean is μ and standard deviation is $\frac{\sigma}{\sqrt{n}}$. The 0.05 in a tail of the normal curve corresponds to a z-score of 1.645, and the critical value is 1.645 standard deviations from the mean.

(e)   The R&D statistician plots the following *power curve* showing the "performance" that can be expected from this hypothesis test. That is, for each possible true value of the population parameter, the probability of rejecting the null hypothesis is shown.

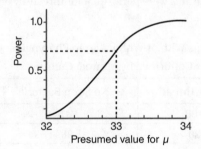

Explain why the graph intersects the pictured vertical axis above $\mu = 32$ at a y-value of 0.05.

Think: If the true value is 32.0, it's actually the null hypothesis. However, you just explained in part (c) that the probability of rejecting a true null hypothesis is 0.05!

(f)   What is the meaning of the point (33, 0.7) in terms of a Type II error?

Think: A Type II error is when you mistakenly fail to reject a false null hypothesis. What does that have to do with the given graph? According to the graph, (33, 0.7) signifies that if the true mean is 33, the probability of rejecting it is 0.7. Then the probability of not rejecting it must be $1 - 0.7 = 0.3$. Not rejecting it is exactly what the Type II error is all about.

✓ **Solutions**

(a)   The parameter of interest is μ, the mean gas mileage in mpg for cars using the new fuel additive.

$$H_0: \mu = 32.0 \text{ and } H_a: \mu > 32.0$$

(b)   Type I error: The additive doesn't improve gas mileage, but the company mistakenly thinks it does. Consequence: The company invests in something that has no value.

Type II error: The additive does improve gas mileage, but the company mistakenly thinks it doesn't. Consequence: The company doesn't invest in a potentially lucrative product.

(c)   0.05 (the probability of a Type I error is the same as the level of significance).

(d)   With the population standard deviation $\sigma = 2.5$ known and with $n \geq 30$, the sampling distribution of $\bar{x}$ is roughly normal with mean $\mu_{\bar{x}} = \mu = 32.0$ and standard deviation $\sigma_{\bar{x}} = \frac{2.5}{\sqrt{30}} = 0.4564$. Then invNorm(0.95) = 1.645, and the critical value is $32.0 + 1.645(0.4564) = 32.751$, so any $\bar{x} \geq 32.751$ will lead the company to conclude that there is statistically significant evidence that the new fuel additive increases gas mileage.

(e) $H_0: \mu = 32.0$, so if the true mean was 32, the probability of rejecting the null hypothesis is simply the level of significance, which is $\alpha = 0.05$.

(f)   If the true mean gas mileage using the additive was 33, the probability of correctly rejecting the null hypothesis is 0.7. So, the probability of committing a Type II error, that is, of mistakenly failing to reject the false null hypothesis, is $1 - 0.7 = 0.3$.

❭ **Example 2**

When you have a sore throat, your doctor has two major options to diagnose strep pharyngitis: a *rapid antigen test* to analyze the bacteria in your throat quickly (10–15 minutes) or a much more accurate *throat culture* that tests to see what grows on a Petri dish (takes 1–2 days).

In a random sample of 500 children brought into a pediatrician's office with sore throat symptoms, both rapid antigen tests and throat cultures were performed for this study with the following summary counts:

TIP

Nothing new is here. However, you probably should think for a moment about how to calculate conditional probabilities from a two-way table.

Patients With or Without Strep Pharyngitis
as Confirmed by a Throat Culture

		Strep Throat	No Strep Bacteria
Results of Rapid Antigen Test	Positive Test	180	15
	Negative Test	20	285

(a)   *Sensitivity* is defined as the probability of a positive test given that the subject has the disease. What was the sensitivity of this study?

Think: The phrase "given that" indicates this is a conditional probability. For such probabilities from a two-way table, you should first calculate row and column totals, writing them directly on the question page.

Patients With or Without Strep Pharyngitis
as Confirmed by a Throat Culture

		Strep Throat	No Strep Bacteria	*Totals*
Results of Rapid Antigen Test	Positive Test	180	15	*195*
	Negative Test	20	285	*305*
	Totals	*200*	*300*	*500*

You're "given that" the subject has the disease (strep throat in this case). So, the first column of numbers is indicated. Note that 200 subjects had strep throat, and 180 of these had a positive test.

(b)   *Specificity* is defined as the probability of a negative test given that the subject is healthy. What was the specificity of this study?

Think: Here you're given that the subject is healthy; there is "no strep throat." So, the second column of numbers is indicated. Note that 300 subjects did not have strep throat, and 285 of these had a negative test.

(c)   A valuable tool for assessing the value of such diagnostic tests is the *positive diagnostic likelihood ratio* (LR^+). It gives the ratio of the probability a positive test result will be observed in a diseased person compared with the probability that the same result will be observed in a healthy person.

$$LR^+ = \frac{\text{sensitivity}}{1 - \text{specificity}}$$

Calculate LR^+ in this study. Explain why the larger the value of LR^+ is, the more useful the test is.

TIP

"Explain" is a key word in this problem— underline it and don't forget to answer it!

Think: To calculate the LR^+, it looks as if you simply plug in the answers from parts (a) and (b). However, the expository part of the question involves some thought. LR^+ is a fraction. So, it is larger when either its numerator is larger or its denominator is smaller. The numerator, sensitivity, is larger when there is a greater probability of a positive test when a child does have strep, and this clearly makes the test more useful. The denominator, $1 -$ specificity, is smaller when specificity is larger and closer to 1. (As a probability, it can't be greater than 1.) Thus, there is a greater probability of a negative test when a child is healthy, and this too clearly makes the test more useful.

(d)   Another valuable tool is the *negative diagnostic likelihood ratio* (LR^-). It gives the ratio of the probability a negative test result will be observed in a diseased person compared with the probability that the same result will be observed in a healthy person. Give an expression for LR^- in terms of *sensitivity* and *specificity*.

This type of problem has appeared frequently. In other words, you are given a "new" statistic and use simulation and a dotplot to estimate the *P*-value.

(e)   Suppose in one such sample study, $LR^+ = 14.7$. To determine whether or not this is sufficient evidence that the population LR^+ is below the desired value of 15.0, 100 samples from a population with a known LR^+ of 15.0 are generated and the resulting simulated values of LR^+ are shown in the dotplot:

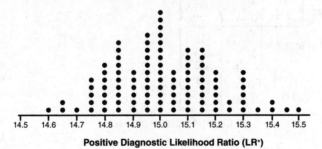

Positive Diagnostic Likelihood Ratio (LR⁺)

Based on this dotplot and the sample $LR^+ = 14.7$, is there convincing evidence that the population LR^+ is below the desired value of 15.0? Explain.

TIP

When the question says to use two givens, in this case the dotplot and the sample LR^+, that's a very strong hint on what you must base your answer!

Think: There are 4 dots out of 100 that correspond to values of 14.7 or less. However, this is really a *P*-value for the sample LR^+. With a *P*-value of 0.04, the LR^+ value of 14.7 is statistically significant at the 5% significance level. We have enough evidence to reject the null hypothesis of $LR^+ = 15.0$.

✓ Solutions

(a) $P(\text{positive test} \mid \text{strep}) = \dfrac{180}{200} = 0.90$

(b) $P(\text{negative test} \mid \text{healthy}) = \dfrac{285}{300} = 0.95$

(c) $\text{LR}^+ = \dfrac{0.90}{1 - 0.95} = 18.0$

A positive test given a child with strep and a negative test given a healthy child are both desired outcomes, so higher values for both sensitivity and specificity are good. Higher values of sensitivity, the numerator of LR^+, lead to greater values of LR^+. Higher values of specificity give lower values of $1 -$ specificity, the denominator of LR^+, again leading to greater values of LR^+.

TIP

It is not enough simply to give 0.90 for an answer. You must show where the answer comes from. Writing $\dfrac{180}{200} = 0.90$ is sufficient.

(d)   Since sensitivity is the probability of a positive test given that the subject has the disease, $1 -$ sensitivity gives the probability of a negative test given that the subject has the disease. The probability that the same result (a negative test) will be observed in a healthy person is the definition of "specificity." The answer is

$$\text{LR}^- = \frac{1 - \text{sensitivity}}{\text{specificity}}$$

(e)   The estimated P-value is the proportion of the simulated statistics that are less than or equal to the sample statistic of 14.7. Counting values in the dotplot, 4 out of 100, gives an estimated P-value of 0.04. With this small of a P-value, $0.04 < 0.05$, there is convincing evidence that the population LR^+ is below the desired value of 15.0.

❯ Example 3

A grocery chain executive in charge of picking locations for future developments is interested in developing a function to predict weekly sales (in \$1,000) based on store size (in 1,000 ft^2), median family income (in \$1,000) in the neighborhood, median family size in the neighborhood, and number of competitor stores in the neighborhood. Data gathered on 15 existing stores yield the following scatterplot matrix:

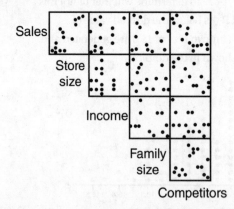

(a)   The scatterplot matrix can be used to determine which variables seem to be the best predictors of sales. Explain which variables seem to be the best predictors of sales and the relationship between sales and those variables.

Think: What is going on? There are lots of scatterplots. In fact, you are given 10 of them with 5 different variables. You're asked about predicting "sales," so sales must be the response variable. "Sales" is to the left of the top row. So, the top row of 4 scatterplots must be sales against the other 4 variables. Do any of these 4 scatterplots look as if they could be used

TIP

Wow, this question starts right off with a brand-new, complex picture! If you have no idea how to interpret this, simply move on. You may still be able to make sense out of later problem parts!

for prediction, that is, do any of them look like there is a linear relationship? Yes, it occurs in the first and fourth scatterplots along the top! The first seems to indicate some kind of positive association between sales and store size. The fourth seems to indicate some kind of negative association between sales and competitors.

(b)   Individual regressions of sales on each of the possible predictors give:

```
Predictor    Coef    SE Coef      T       P
Constant  -115.51     51.74    -2.23   0.044
St Size    5.8801     0.9808    6.00   0.000

Predictor    Coef    SE Coef      T       P
Constant    66.69     86.15     0.77   0.453
Income       3.859     2.691     1.43   0.175

Predictor    Coef    SE Coef      T       P
Constant   214.19     91.31     2.35   0.036
Fam Size   -11.58     36.71    -0.32   0.758

Predictor    Coef    SE Coef      T       P
Constant   374.64     33.41    11.21   0.000
Comp       -24.777     4.109   -6.03   0.000
```

> **TIP**
> Even though this full table looks unfamiliar, you should recognize each of the 4 parts as a standard linear regression summary.

We will not use any variable whose t-test P-value is over 0.15 and will first use the variable whose t-test P-value is least. Under these guidelines, which, if any, variables will be discarded and which variable will be used first? Explain.

Think: The question is about t-tests and P-values, so this must be about hypothesis tests for regression. You're given a generic computer printout that seems to give separate information about using each of the 4 possible predictors. From working with simpler computer outputs, remember that the line with the predictor first gives the slope, then the SE of the slope, followed by the t- and P-values from a linear regression test. You're told to throw out the predictor variables with P over 0.15. You see that Income has $P = 0.175$ and Fam Size has $P = 0.758$. (Looking back, this gives you confidence that your answer to part (a) is correct!) You're asked which predictor variable has the smallest P, but both St Size and Comp have $P = 0.000$ rounded to 3 decimals. How can you tell which is smaller? Well, you know the P-values come from the t-values, and you see that Comp has the larger t-value in absolute value ($6.03 > 6.00$), so it must have the smaller P if you were given more decimal places!

(c)   Multiple regression of sales on the two variables, store size and number of competitors, gives:

```
Predictor    Coef    SE Coef      T       P
Constant   122.79     94.44     1.30   0.218
St Size      3.342     1.200     2.78   0.017
Comp       -14.213     5.051    -2.81   0.016

S = 37.3598    R-Sq = 84.0%
```

> **TIP**
> It seems like you're simply adding another variable to the standard linear regression analysis—which is not unreasonable!

What sales does this predict for a store of 50,000 ft^2 with 8 competitors?

Think: With a linear regression generic computer printout, you know that the "Coef" column gives the "Constant" and under it the slope or coefficient of the independent variable in the formula for predicting the response variable. Here, though, you have an extra row under the Constant, which means two predictor variables, "St Size" and "Comp." The constant in the formula must be 122.79, and the coefficients of "St Size" and "Comp" must be 3.342 and -14.213, respectively.

(d)   What proportion of the variation in weekly sales is explained by its linear relationship with store size and number of competitors?

Think: This sounds like the definition of the "coefficient of determination" or R^2. Sure enough, "R-Sq" is in the computer output!

(e)   Assuming normality of residuals and using the computer output, estimate the probability that the weekly sales will be at least $50,000 over what is predicted by the regression line.

Think: You're being asked about the difference between an actual weekly sale and what is predicted, and this is what you know is a "residual." What is the distribution of the residuals? The question says the distribution is roughly normal. Remember that the sum and thus the mean of the residuals is always 0. By looking at the computer output, you see that the standard deviation of the residuals is $s = 37.3598$. So, you can simply use normalcdf on your calculator to find the probability!

✓ Solutions

(a)   "Store size" and "number of competitor stores" seem to be the best predictors of "weekly sales." The first scatterplot in the top row indicates a strong positive association between weekly sales and store size. The last scatterplot in the top row indicates a strong negative association between weekly sales and number of competitor stores.

(b)   The variables "median family income" and "median family size" with t-test P-values of 0.175 and 0.758, respectively, both over 0.15, should be removed. Although both "store size" and "number of competitor stores" have t-test P-values of 0.000 to three decimal places, "number of competitor stores" will be used first as it has the smaller P-value because it has the larger t-statistic in absolute value, 6.03 versus 6.00.

(c) $122.79 + 3.342(50) - 14.213(8) = 176.186$ thousand or $176,186

(d)   0.84. To be safe, I'll write it out: The proportion of the variation in weekly sales explained by its linear relationship with store size and number of competitors is 0.84.

(e)   The sum and thus the mean of the residuals is always 0. The standard deviation of the residuals is estimated with $s = 37.3598$. With a z-score of $\frac{50 - 0}{37.3598} = 1.338$ and a roughly normal distribution, we have $P(z > 1.338) \approx 0.090$.

Quiz 29

1. A basic problem in ecology is to estimate the number of animals in a wildlife population. Suppose a wildlife management team captures and tags a random sample of 108 springboks in a Southwest African wildlife preserve. The springboks are released and given time to mingle with the population. One month later, a second random sample capture is made of 80 springboks, among which 12 are noted to be tagged from the original capture.

   (a) Assuming that the proportion of marked individuals within the second sample is equal to the proportion of marked individuals in the whole population, estimate the total number of springboks in the wildlife preserve.

   (b) A formula for variance of this estimate is $\text{var}(N) = \dfrac{ac(a-b)(c-b)}{b^3}$, where N = estimate of total population size, a = number of animals originally captured and tagged, b = number of tagged animals that are recaptured, and c = number of animals in the second sample. What is the standard deviation for the population size estimate above?

   The sampling distribution for estimates of N calculated as above is approximately normal under the condition that both samples are random and both $b \geq 10$ and $c - b \geq 10$.

   (c) Determine a 90% confidence interval for the total population size of springboks in this wildlife preserve.

   The above analysis assumes that there is no change to the population during the investigation (the population is closed).

   (d) Suppose that some of the tagged springboks were killed by hunters or escaped from the preserve before they could rejoin the herd. Would this result in an underestimation or an overestimation of the herd? Explain.

   (e) Suppose that the detection device missed some of the wire tags on the springboks in the second random sample. Would this result in an underestimation or an overestimation of the herd? Explain.

2. When operating at full capacity, a stainless steel bottle filler can fill twelve 64-ounce containers per minute. Once a day, a random set of twelve containers is picked, the containers are weighed, and if $\hat{w} = \dfrac{\text{Max} + \text{Min}}{2}$ varies significantly from 64 ounces, the machine is stopped and adjustments are performed. (The distribution of $\hat{w}$ values is a sampling distribution for a parameter W.) To explore this quality control procedure further, the machine is perfectly calibrated, and 100 random samples of size 12 bottles are taken from the assembly line. For each of these 100 samples, the $\hat{w}$ measurement is calculated and shown plotted below.

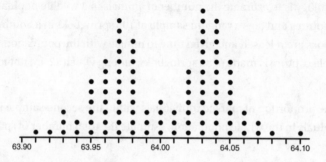

(a) Describe the above sampling distribution.

(b) Assuming a median of 64.005 and quartiles of 63.97 and 64.04, are there any outliers in the above distribution? Explain.

(c) When a sample of size 12 is picked, what are the null and alternative hypotheses being tested?

(d) One day, the random sample is {63.8, 63.84, 63.87, 63.88, 63.91, 63.95, 63.97, 63.97, 63.98, 64.0, 64.02, 64.04}. Based on the dotplot above, is this sufficient evidence at the 5% significance level to conclude that the machine needs adjustment? Justify your answer.

3. In a random sample of 10 years, the number of people who died by becoming tangled in their bedsheets (deaths per year) versus per capita consumption of cheese (pounds per year) in the United States is illustrated in the following scatterplot:

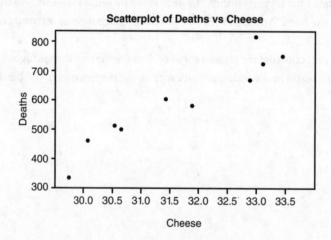

Scatterplot of Deaths vs Cheese

Below is the computer regression output from the 10 data points:

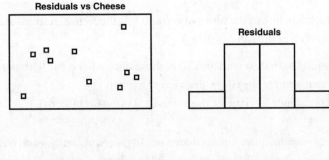

Predictor	Coef	SE Coef	T	P
Constant	-2977.3	427.6	-6.96	0.000
Cheese	113.13	13.56	8.35	0.000

S = 50.0007   R-Sq = 89.7%   R-Sq(adj) = 88.4%

(a) Discuss how conditions for inference are met.

(b) During the year when the per capita consumption of cheese was 30.1 pounds, there were 456 deaths from people becoming tangled in their bedsheets. Calculate and interpret the residual for this point.

(c) Determine a 95% confidence interval for the regression slope, and interpret it in context.

(d) What kind of cause-and-effect conclusion is appropriate? Explain.

(e) Using the computer output, give a rough estimate of the probability that the number of deaths in a year is at least 75 over what is predicted by the regression line.

4. Probiotics are microorganisms that are believed to give health benefits, especially with regard to healthy digestion, when consumed. One brand advertises a strength of 30 billion CFU (colony-forming units) per capsule. A company performs a quality check of 100 randomly chosen capsules. The 100 CFU values are arranged in order, the percentile rank of each is noted, and the standard normal value (z-score) corresponding to those percentile ranks is found. (For example, if a value had a percentile rank of 75%, then the corresponding z-score would be invNorm(0.75) = 0.6745.) The resulting plot of z-scores versus CFU values (in billions) is shown below, together with summary statistics:

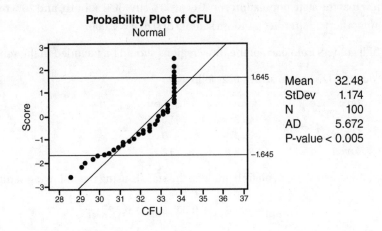

Mean	32.48
StDev	1.174
N	100
AD	5.672
P-value	< 0.005

(a) What is the *range* of CFU values?

(b) What is the 5th percentile of the data? Explain.

The diagonal (sloping) straight line in the plot is the graph of a theoretical normal distribution with the same mean and standard deviation of the sample data.

(c) In a normal distribution with the mean and SD of these data, what is the 95th percentile? Explain.

(d) What shape will a histogram of the CFU values have? Explain.

(e) Determine a 95% confidence interval for the mean CFU value (in billions).

5. On their website, a candy manufacturer claims there are 210 pieces of candy, each weighing 1 gram, in their 7-ounce bags. A machine is used to fill the bags. For quality control, a randomly chosen bag is weighed each day to test the hypotheses:

H_0: the machine is putting 210 pieces of candy into each 7-ounce bag.

H_a: the machine is putting fewer than 210 pieces of candy into each 7-ounce bag.

(a) In this context, what would be a Type I error, and what would be a possible consequence of making a Type I error?

(b) In this context, what would be a Type II error, and what would be a possible consequence of making a Type II error?

Assume that the candy weights are roughly normally distributed with a mean of 1 gram and a standard deviation of 0.05 grams. Let W represent the total weight of candy in a 7-ounce bag.

(c) Describe the distribution of W if a selected 7-ounce bag contains exactly 210 pieces of candy.

If the total weight of candy in the daily randomly chosen bag is 209.5 grams or less, the machine is shut down for inspection and possible recalibration.

(d) If a bag contains 210 pieces, what is the probability that the machine will be shut down?

(e) If a bag contains 209 pieces, what is the probability that the machine will be shut down?

(f) What is the statistical name for the probability calculated in (e)?

6. At the end of the school year, a state administers a mathematics exam to a sample of fifth-grade students. Census results show that the state population consists of 50% city, 30% suburb, and 20% rural. The state decides to use a proportionate stratified random sample of size $n = 2,000$.

(a) How many fifth graders from each of the three regions should be included in the sample?

Partial computer output of the exam scores is as follows:

Variable	Mean	StDev
City	74.3	10.2
Suburb	80.4	9.3
Rural	69.8	12.1

As an estimator for the mean score of all fifth graders, the state is using the following sample statistic:

$$\bar{x}_{overall} = 0.5\bar{x}_{city} + 0.3\bar{x}_{suburb} + 0.2\bar{x}_{rural}$$

(b) Calculate the sample statistic $\bar{x}_{overall}$ for these data.

(c) Using part (a) and the above computer output, calculate the three standard errors, one for each region, $SE(\bar{x}_{city})$, $SE(\bar{x}_{suburb})$, and $SE(\bar{x}_{rural})$.

(d) Calculate the standard error, $SE(\bar{x}_{overall})$, of the sampling distribution of $\bar{x}_{overall}$.

(e) Assuming all conditions for inference are satisfied and using z as an approximation for t (given the very large sample size), determine a 99% confidence interval for the mean score of all fifth graders if they were to take this exam.

7. The ADA (American Dental Association) recommends that you should brush your teeth for 2 minutes, but over 90 percent of people using a standard toothbrush don't brush that long. A dentist is interested in estimating the mean brushing time of her patients who use electric toothbrushes. Fifteen patients were selected at random, and a histogram of their self-reported brushing times is shown below.

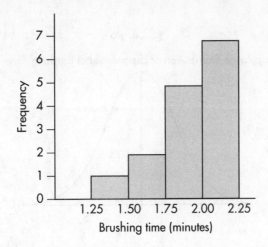

(a) What is an example of possible response bias in this context?

(b) Based on the histogram, why might it not be appropriate to use a one-sample t-interval to estimate the mean brushing time for all the dentist's patients who use electric toothbrushes?

If x represents a patient's brushing time, an exponential transformation of the patient's brushing time is given by 10^x. The table below gives some summary statistics for the original brushing times and the respective exponential-transformed brushing times.

	Brushing time, x	10^x
Mean	1.943	99.24
Median	2.000	100.0
Standard deviation	0.247	44.35

A histogram of the 15 exponential-transformed data values is shown below.

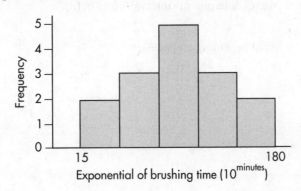

(c) The conditions for inference are met for the exponential-transformed data. Construct and interpret a 95 percent confidence interval for the mean of the exponential of the brushing times.

(d) The mean of the exponential-transformed data is $99.24(10^{\text{minutes}})$. This can be converted back to minutes by calculating $\log(99.24) = 2.00$ minutes. Convert the endpoints of your interval in (c) back to minutes to obtain a new interval.

Graph 1 below shows the population distribution of the exponential of brushing times in (10^{minutes}), which is approximately normal with the mean, μ. Graph 2 shows the result of converting the population distribution in Graph 1 back to the population of brushing times, in minutes. The median of the distribution is shown in each graph.

Graph 1

Population Distribution of Exponential of Brushing Time

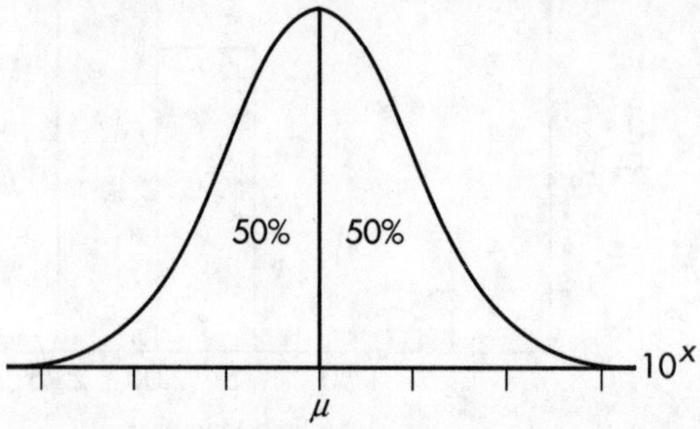

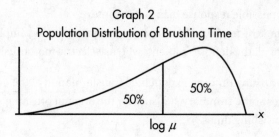

Graph 2

Population Distribution of Brushing Time

(e) Consider the parameter $\log \mu$ in Graph 2.

   (i) How does the parameter $\log \mu$ compare with the median of the population distribution of brushing times?

   (ii) How does the parameter $\log \mu$ compare with the mean of the population distribution of brushing times?

(f) Using your results from part (e), interpret the interval found in (d).

The answers for this quiz can be found beginning on page 530.

50 Misconceptions

Misconceptions About Graphs

Misconception #1: In a boxplot, the median is in the interquartile range (IRQ) box.

Fact: The IQR is a number that represents the length of the box. The box itself is not the IQR.

Misconception #2: A symmetric boxplot indicates an approximately normal distribution.

Fact: Many differently shaped distributions can have identical boxplots; in general, it is very difficult to say anything about the shape of a distribution from its boxplot.

Misconception #3: When comparing two distributions, it is sufficient to list the center, spread, shape, and unusual features of each distribution.

Fact: *Comparative* language must be used to state which distribution has the greater center and which has the greater spread.

Misconceptions About Linear Regression

Misconception #4: When $r = 0$, there is no relationship between the variables.

Fact: Correlation measures only linearity. When $r = 0$, there is no LINEAR relationship between the variables, but there might be a strong NONLINEAR relationship between these variables.

Misconception #5: When $r = 0.3$, then 30 percent of the variables are closely related.

Fact: When $r^2 = 0.3$, then 30 percent of the variation in the dependent variable can be explained by the linear relationship with the independent variable.

Misconception #6: When $r = 1$, there is a perfect cause-and-effect relationship between the variables.

Fact: Correlation shows association, not cause and effect.

Misconception #7: A correlation close to 1 means that a linear model will give the best fit to the data.

Fact: Curved data can also have a correlation near 1.

Misconception #8: The slope of a linear regression line gives the change in the dependent variable, y, for each unit change in the independent variable, x.

Fact: The slope gives the "predicted," "estimated," "expected," or "average" change in the y-variable for each one unit increase in the x-variable.

Misconception #9: If $\hat{y} = a + bx$, then we can algebraically solve to get $\hat{x} = -\frac{a}{b} + \frac{y}{b}$.

Fact: The variables x and $\hat{y}$ are not the same as $\hat{x}$ and y. The equation that predicts $\hat{y}$ from x does not correctly predict $\hat{x}$ from y.

Misconception #10: If $\hat{y} = a + bx$ and $\hat{x} = c + dy$ are derived from the same set of data, then b and d are reciprocals.

Fact: $b = r\frac{s_y}{s_x}$ and $d = r\frac{s_x}{s_y}$ are not reciprocals.

Misconceptions About Collecting Data

Misconception #11: An experiment cannot be double-blind if anyone knows who gets which treatment.

> **Fact:** Someone always knows, but as long as this person doesn't interact with the subjects or measure the response, the experiment can still be double-blind.

Misconception #12: There is no difference between a stratified random sample and a randomized block design experiment.

> **Fact:** Although the ideas are similar, as both involve forming groups of similar objects, stratified sampling occurs when taking a sample from a population, while blocking occurs before assigning units to treatments in an experiment.

Misconception #13: Blocks and treatment groups are basically the same.

> **Fact:** Blocks are not formed at random but are formed by grouping together similar units. Each block contains units with a certain characteristic. Treatment groups are formed at random with the goal of making the different groups as similar as possible, other than the treatment received.

Misconceptions About Probability

Misconception #14: The complement rule gives that $0.3^C = 0.7$ or $P(E)^C = 0.7$.

> **Fact:** Probabilities do not have complements; events do. The "C" should appear in the superscript of the event E; that is, $P(E^C) = 1 - 0.3 = 0.7$.

Misconception #15: If E and F are mutually exclusive, they are independent.

> **Fact:** If E and F are mutually exclusive, $P(E \cap F) = 0$ and $P(E|F) = 0$. If E and F are independent, $P(E \cap F) = P(E)P(F)$ and $P(E|F) = P(E)$. "Mutually exclusive" and "independent" are two very different concepts.

Misconception #16: The probability a student taking AP Statistics owns a graphing calculator is the same as the probability a student who owns a graphing calculator is taking AP Statistics.

> **Fact:** $P(E|F) \neq P(F|E)$. This can be seen as $P(E|F) = \dfrac{P(E \cap F)}{P(F)}$ while $P(F|E) = \dfrac{P(E \cap F)}{P(E)}$.

Misconception #17: If a fair coin comes up "heads" five times in a row, by the "law of averages," the probability that the next toss is "tails" is greater than 0.5.

> **Fact:** There is no "law of averages" or "law of small numbers." By the "law of large numbers," in the long run, the proportion of heads will tend toward 0.5. However, coins have no memories, and the probability the next toss comes up "tails" is still 0.5.

Misconception #18: Writing normalcdf(20, 30, 35, 15) = 0.211 will receive full credit on a Normal calculation question, and writing binompdf(10, 0.2, 3) = 0.201 will receive full credit on a Binomial calculation question.

> **Fact:** If using "calculator-speak," every input must be identified. For example, you may write normalcdf(lower bd = 20, upper bd = 30, mean = 35, SD = 15) = 0.211, and you may write binompdf($n = 10$, $p = 0.2$, $x = 3$) = 0.201.

Misconception #19: If a random process is repeated long enough, eventually the expected value will occur.

> **Fact:** The expected value of a random process may never occur. It is simply a long-run average.

Misconception #20: The expected value of a random variable will always equal one of the possible values of that variable.

> **Fact:** There is no reason why $\sum xP(x)$ should equal one of the x-values. In fact, $\sum xP(x)$ often equals some nonwhole number. Do NOT round the expected value to a whole number after calculation.

Misconceptions About Sampling Distributions

Misconception #21: The larger the sample size, the closer the sample distribution is to a normal distribution.

> **Fact:** The larger the sample size, the closer the sample distribution is to the *population* distribution; the larger the sample size, the closer the *sampling* distribution of the sample mean is to a normal distribution.

Misconception #22: The larger the sample size, the closer the sampling distribution is to the population distribution.

> **Fact:** The larger the sample size, the closer the sampling distribution of the sample mean is to a normal distribution.

Misconception #23: If the sample size is large, the sampling distribution has the same mean as the population and has a standard deviation that is close to the standard deviation of the population.

> **Fact:** No matter what the sample size, the mean of the sampling distribution of $\bar{x}$ equals the population mean, $\mu_{\bar{x}} = \mu$, but the standard deviation of the sampling distribution of $\bar{x}$ equals the population standard deviation divided by the square root of the sample size, $\sigma_{\bar{x}} = \frac{\sigma}{\sqrt{n}}$.

Misconception #24: The larger the sample size, the closer the distribution of any statistic will approximate a normal distribution.

> **Fact:** The *central limit theorem* states that for sufficiently large sample sizes, the sampling distribution of the sample mean can be approximated by a normal distribution. This is not true, for example, for the sampling distribution of statistics such as the sample *maximum* or the sample *range*.

Misconception #25: Accuracy and precision are the same thing.

> **Fact:** Accuracy means that a statistic has low or no bias, while precision means that a statistic has low variability.

Misconception #26: The *central limit theorem* has universal validity; that is, it can always be applied to give normal approximations to appropriate sampling distributions.

> **Fact:** There is a critical condition regarding a large enough sample size that must be satisfied (for the sampling distribution of $\bar{x}$, we use $n \geq 30$).

Misconceptions About Confidence Intervals and Levels

Misconception #27: If (35, 40) is a 95% confidence interval of a population mean, there is a 95% probability that the true mean is between 35 and 40.

> **Fact:** The probability that the population mean is between 35 and 40 is either 0 or 1 depending upon whether or not it is in the interval (35, 40).

Misconception #28: If (35, 40) is a 95% confidence interval of a population mean, 95% of the subjects are between 35 and 40.

> **Fact:** The confidence interval is about a population mean, not about individual values.

Misconception #29: If (35, 40) is a 95% confidence interval of a population mean, 95% of sample means from all such samples will be between 35 and 40.

> **Fact:** (35, 40) is one specific interval derived from one sample mean and has nothing to do with other samples and other sample means.

Misconception #30: Interpreting a confidence interval is the same as interpreting a confidence level.

> **Fact:** When constructing a confidence interval, the confidence interval must be interpreted in context. Only interpret the confidence level when specifically asked, for example, "What does 95% confidence mean?"

Misconception #31: In general, increasing the sample size makes us more confident.

Fact: In general, the confidence level, for example, 95%, doesn't change. Increasing the sample size while keeping the same confidence level gives a narrower interval, that is, a *more precise* estimate.

Misconception #32: For the same data, a 90% confidence interval is wider than a 95% confidence interval.

Fact: The 95% confidence interval is wider. If we want higher confidence, we must accept a wider, less precise interval.

Misconception #33: The *Normal/Large Sample condition* can be simplified to saying that the sample must have a normal distribution.

Fact: A finite sample cannot have a normal distribution. When samples are too small for the *central limit theorem* to apply, we look to see if the sample is unimodal with no outliers and no extreme skewness in order to conclude that it is not unreasonable to say "the sample came from a roughly normal population."

Misconceptions About Hypothesis Testing

Misconception #34: H_0 and H_a are alternative conclusions.

Fact: Hypotheses testing is a decision-making process in order to reject or fail to reject a null hypothesis, H_0. You are not deciding that one of H_0 or H_a is true. Your conclusion should be either that there *is* sufficient evidence to reject H_0 (that is, there *is* convincing evidence in support of H_a) or that there is *not* sufficient evidence to reject H_0 (that is, there is *not* convincing evidence in support of H_a).

Misconception #35: The hypotheses can refer to either the population or the sample.

Fact: The hypotheses **always** refer to the population, never to the sample.

Misconception #36: The null hypothesis and the failure-to-reject region are the same thing.

Fact: The null hypothesis is a claim about a population parameter. The failure-to-reject region is an interval of values for the sample statistic that would indicate insufficient evidence to reject the given null hypothesis claim.

Misconception #37: The choice of alternative hypothesis depends on the sample statistic, that is, whether the sample statistic is positive or negative.

Fact: Both the null and alternative hypotheses are decided before the data are collected. One should never decide if a test is one-sided (and in which direction) or two-sided based upon the data.

Misconception #38: The significance level α is the probability that the null hypothesis is true given that it has been rejected.

Fact: The significance level is a fixed value we use to decide how small a *P*-value is required to be in order to reject the null hypothesis.

Misconception #39: The significance level α is the probability that the alternative hypothesis is true.

Fact: The significance level is a fixed value we use to decide how small a *P*-value is required to be in order to reject the null hypothesis.

Misconception #40: A hypotheses test, if correctly performed, establishes the truth of one of the two hypotheses, either the null or the alternative.

Fact: Truth is never established. The conclusion is simply whether or not there is sufficient evidence to reject the null hypothesis.

Misconception #41: Statistical significance and practical significance are the same.

Fact: A statistically significant result might have no practical significance, and a practically significant result might turn out to *not* be statistically significant.

Misconception #42: The probability of a Type I error plus the probability of a Type II error should equal 1.

Fact: The probability of a Type I error, α, is a fixed value giving the threshold value of the hypothesis test, while the probability of a Type II error, β, is different for each possible alternative parameter value.

Misconception #43: A well-planned test of significance should result in a statement either that the null hypothesis is true or that it is false.

Fact: Tests of significance are designed only to measure the strength of evidence against the null hypothesis. They do not establish truth.

Misconceptions About the *P*-Value

Misconception #44: If the *P*-value is 0.16, the probability that the null hypothesis is correct is 0.16.

Fact: The *P*-value of a test is the conditional probability of obtaining a result as extreme or more extreme as the one obtained assuming the null hypothesis is true.

Misconception #45: When the null hypothesis is rejected, it is because it is not true.

Fact: The null hypothesis is rejected when the *P*-value is small. The small *P*-value indicates there is sufficient evidence to doubt the null hypothesis, but it is always possible that we mistakenly reject a true null hypothesis.

Misconception #46: A very large *P*-value provides convincing evidence that the null hypothesis is true.

Fact: A very large *P*-value simply says that there is not sufficient evidence to reject the null hypothesis; it does not say that the null hypothesis is true and does not say that we can accept the null hypothesis.

Misconception #47: The *P*-value is the probability that the observed event has happened by chance.

Fact: This misses that the *P*-value is a conditional probability. The *P*-value is the probability of obtaining a statistic as extreme or more extreme than the one actually observed, by chance alone, when assuming that the null hypothesis is true.

Misconceptions About χ^2-Tests

Misconception #48: The chi-square statistic, $\Sigma \dfrac{(\text{observed} - \text{expected})^2}{\text{expected}}$, can be calculated using either counts or proportions.

Fact: The chi-square statistic is **always** calculated using integers for the observed counts, and never using proportions for either the observed or expected cells.

Misconception #49: For chi-square goodness-of-fit tests, the null hypothesis is that some observed distribution is the same as some claimed distribution.

Fact: The hypotheses **never** refer to the sample. The null hypothesis is that the *population* distribution is the same as some claimed distribution.

Misconception #50: Expected counts should always be rounded to integer values.

Fact: While observed counts are always integers, the expected counts need not be, and students may lose points with this rounding.

50 Common Errors on the AP Exam

Common Errors: Exploratory Analysis

1. When asked to draw a graph, many students forget *labeling* and *keys*, which are critical for stemplots, and *labeling* and *scaling*, which are critical for histograms, boxplots, and scatterplots. For example, if asked to draw parallel boxplots, be sure to label which is which. Proper labeling is more important than getting every point in exactly the right place.

2. When asked to describe a distribution from a graph such as a histogram or dotplot, students address "center, spread, shape, and unusual features" and forget to mention context.

3. When describing a distribution given only a boxplot, students may try to conclude shape like approximate normality or even symmetry; however, shape cannot be concluded from a boxplot.

4. When describing center or spread given only a histogram, definitive language like "the median is 120" will be penalized. One needs to use language that conveys uncertainty, such as "the median is approximately 120" or "the median is in the 115–130 interval."

5. When claiming a value is an outlier or claiming that there are no outliers, some calculation is necessary to communicate how the decision was made, otherwise credit is lost.

6. When asked to *compare* distributions, it is not enough to address center, spread, shape, and unusual features, in context, separately for each distribution. In addition, the characteristics of center and spread must be *compared* using comparative expressions like "greater than" or "less than" or "equal to."

7. When mentioning skew, students sometimes get the direction wrong. A distribution is skewed in the direction the tail goes, not in the direction where the peak is.

8. When using words with a statistical meaning, incorrect use will be penalized. For example, in statistics, "range" refers to a single number, the maximum minus the minimum. Writing "the data range from 50 to 100" will be marked wrong. You can say that all the data are in the interval from 50 to 100.

9. When combining independent events, variances add, but it is a mistake to add standard deviations.

10. When writing the equation of a regression line, students forget to include the "hat" on the y-variable (or use the word "expected") and forget to define the variables.

11. When asked if a linear model is appropriate, students need to refer to the residual plot, not to whether or not the correlation is close to ± 1.

12. When a correlation is close to ± 1, students incorrectly think correlation implies causation.

13. When explaining the slope of a regression line, students need to use a word like "predicted," "estimated," "expected," or "average" for the change in the y-variable for each one unit increase in the x-variable. Lack of such a nondeterministic word will be penalized.

14. When asked to describe a scatterplot, students address "form, direction, and strength" but forget context.

15. When interpreting a residual, students address the distance between observed and predicted but forget the *direction* of the error, that is, whether the predicted value is an underestimate or overestimate. Remember that the residual is calculated as the actual minus the predicted, not the other way around.

Common Errors: Collecting Data

16. When describing how to select a sample using a random integer generator or a random integer table, students forget to explicitly state that repeated integers are to be ignored and forget to state that the selected integers then correspond to the subjects to be chosen for the sample.

17. When asked about a possible source of bias, students name the bias but are penalized for not clearly describing the problem caused by the bias and giving consequences in context.

18. When looking at data collection, students confuse *stratification* and *blocking*. While both involve dividing into homogeneous groups, stratification is a sampling method while blocking is an experimental design.

19. When looking at data collection, students confuse *stratified* and *cluster* sampling. Stratified sampling involves homogeneous groups and random sampling from each strata, while cluster sampling involves heterogeneous groups and a random sample of clusters.

20. When analyzing experimental design, students confuse *treatment groups* and *blocks*. A treatment group is a randomly formed collection of individuals to which a treatment is applied. A block is a group of similar experimental units that is then randomly allocated into each of the treatment groups. The goal is for treatment groups to be as similar as possible to each other (other than the treatment), while blocks are chosen to be very different from each other.

21. When looking at data collection, students fail to understand that *random selection* and *random assignment* have different purposes. Random selection in sampling is used to generalize to a population, while random assignment in experiments is used to minimize the effect of confounding variables.

22. When looking at data collection, students may correctly state what variable is a confounding variable but are penalized for not explaining *why* the variable is a confounding variable in the context of the problem.

Common Errors: Probability

23. When calculating probabilities, minor arithmetic errors might not be penalized; however, reporting that a probability is a negative number or a number greater than 1 will always be penalized.

24. When studying two events, students confuse mutually exclusive and independent events. E and F are mutually exclusive if $P(E \cap F) = 0$. E and F are independent if $P(E \cap F) = P(E)P(F)$.

25. When using the conditional probability formula, students should show the numerator and denominator of the probability, not just a reduced fraction and not just a decimal answer.

26. When interpreting probabilities, there is a "law of large numbers" (in the long run, proportions will tend toward a specific probability); however, students incorrectly think there is a "law of averages" or a "law of small numbers."

27. When calculating probabilities, students confuse "intersection" and "given that."

28. When calculating probabilities, students confuse $P(E|F)$ with $P(F|E)$.

29. When given a straightforward probability problem, students sometimes incorrectly try to turn it into a full-blown inference problem.

30. When using the normal distribution for probability calculations, students forget to state that they are using a normal distribution and forget to give the values of the two parameters, the mean μ and the standard deviation σ.

31. When using the binomial distribution for probability calculations, students forget to state that they are using a binomial distribution and forget to give the values of the two parameters, the number of trials n and the probability of success p.

32. When using the geometric distribution for probability calculations, students forget to state that they are using a geometric distribution and forget to give the value of the parameter p, the probability of success.

33. When calculating probabilities, students forget to clearly identify the *boundary* and *direction* for the outcome of interest.

34. When using "calculator-speak" (often not recommended) like normalcdf(55, 60, 65, 5) or binompdf(25, 0.3, 8), students forget to state what every number refers to.

35. When calculating expected values, rounding to integers will often be penalized. Expected values do not have to be integers!

36. When describing a distribution, students are unclear as to whether they are describing a sample distribution, a population distribution, or a sampling distribution.

Common Errors: Inference

37. When beginning an inference procedure, students sometimes mistakenly refer to the *sample* instead of the *population* in stating hypotheses. There is nothing to hypothesize about the sample since the sample results are known exactly.

38. When communicating which procedure is being used, it is generally easier to use *words* rather than writing out a *formula*. In free-response inference questions, more mistakes are made when trying to describe using formulas than when using words.

39. When defining a parameter, do so in context and refer to the population (not mistakenly to the sample!).

40. When a formula is used, it is often safer to start with numbers plugged in and not to include the symbolic formula, because including an incorrect symbol will be penalized.

41. When concluding an inference procedure, students confuse interpreting a confidence *interval* and interpreting a confidence *level*.

42. When concluding an inference procedure, students sometimes mistakenly refer to the sample instead of the population in giving the conclusion. For example, students mistakenly refer to a confidence interval of people who gave a particular answer to a survey question rather than to the people who would have answered in a certain way (only those in the sample actually "answered").

43. When analyzing an inference procedure, students confuse *parameters* and *statistics*. A parameter is a fixed, usually unknowable, truth about a population such as p, μ, σ, or β, while statistics are measurable values such as $\hat{p}$, $\bar{x}$, s, or b coming from samples. We use sample statistics to make inferences about population parameters.

44. When conducting an inference procedure, students incorrectly use the expression to "prove" or "accept" the null or alternative hypothesis. Conclusions should always be that there is or is not "convincing evidence" of some claim. (Often, the alternative hypothesis is the claim.)

45. When concluding a confidence interval or hypothesis test question about the population mean, students forget to include the word "mean" in the conclusion.

46. When performing an inference procedure, there is no reason to conduct both a significance test and a confidence interval unless specifically asked to. Students will be graded on whichever is the weaker response.

47. When concluding a hypothesis test, students are often incomplete in their conclusions, forgetting to involve linkage to a P-value (a direct comparison of the P-value to alpha), or forgetting context, or forgetting to refer to the parameter (such as "true proportion" or "true mean").

48. When asked about interpretations of errors on multiple-choice questions, students confuse the definitions of Type I and Type II errors.

49. When asked which type of error might have been committed based on the result of a hypothesis test, students often are unable to pick between Type I or Type II.

50. When asked to explain or analyze possible errors, students often cannot provide or explain possible consequences of Type I and Type II errors.

50 AP Exam Hints, Advice, and Reminders

1. There is no penalty for wrong answers on the multiple-choice section, so **answer every question**.

2. For free-response questions, let the graders understand *what* you are doing, *why* you are doing it, and *how* you are doing it. **Communication** is just as important as statistical knowledge! Graders want to give you credit—help them! Don't make the readers guess at what you are doing.

3. Read each of the multiple-choice and free-response questions *carefully* while underlining key words and phrases. Be sure you understand **exactly what you are being asked to do, find, or explain**.

4. *Check assumptions.* **Don't just state them!** Be sure the assumptions to be checked are stated correctly. Verifying assumptions and conditions means more than simply listing them with check marks—you must show work or give some reason to confirm verification.

5. Learn and practice how to read *generic computer output.* Linear regression questions are often very straightforward from such output. Also note that when answering questions using computer output, you usually will **not** need to use all the information given.

6. Show where answers come from. *Naked* or *bald answers* will receive little or **no** credit! If you are using a formula, start with showing the numbers substituted rather than writing the formula with symbols, which is easy to mess up. Also, don't give more than one solution to the same problem—you will receive credit only for the weaker one.

7. Sketch, with labeled axes and numbered scales when possible, any graph to which you refer. This includes histograms, boxplots, stemplots, scatterplots, residuals plots, normal probability plots, or any other kind of graph. It is not enough to simply say, "I did a normal probability plot of the residuals on my calculator and it looked linear."

8. **Use proper terminology!** For example, the language of experiments is different from the language of observational studies—you shouldn't mix up *blocking* and *stratification.* Know what *confounding* means and when it is proper to use this term. Words like *bias, correlation, normal, power, range, skew,* and even *statistic* have specific statistical meanings and should not be used colloquially. Don't use statistical vocabulary unless you use it correctly.

9. **Avoid** "calculator-speak"! For example, do not simply write "2-SampTTest..." or "binomcdf...." There are lots of calculators out there, each with its own abbreviations. Some abbreviated function notation can be referenced IF the parameters used are identified. For example, normalcdf(left bd $= 55$, right bd $= 60$, mean $= 45$, SD $= 15) = 0.0938$ is okay to write.

10. When part of a question says to use results from previous parts of the question, you must explicitly refer to the previous parts in your answer.

11. Write null and alternative hypotheses in terms of the population parameter. If the question refers to sample data, don't automatically parrot the stem of the problem in your hypotheses.

12. When describing a one-variable distribution, you need to mention context, but you need only to mention it once, either with regard to shape, center, or spread.

13. Approach each problem systematically. Some problems look scary on first reading but are not overly difficult and are surprisingly straightforward. Other questions might take you beyond the scope of the AP curriculum; however, remember that they will be phrased in ways that you should be able to answer them based on what you have learned in your AP Statistics class.

14. Read through all six free-response questions, underlining key points. Go back and answer the questions you think are easiest (this will usually, but not always, include Question 1). Then tackle the heavily weighted Problem 6 for a while. Finally, try the remaining problems. Hopefully you will have time at the end to work further on Problem 6!

15. Be sure that your methods, reasoning, and calculations are clear to the reader and that the explanations and conclusions are given *in the context of the problem.* Answers do not have to be in paragraph form. Symbols and algebra are fine.

16. Show calculations carefully. A wrong answer due to a computational error might still result in full credit.

17. For any part of a multistep problem that you can't answer but a solution is needed in a later step, make up a reasonable answer. Later steps might be easy, so don't give up.

18. State the values that you are substituting into any formula that you use. You do not have to write down the actual formula.

19. Don't waste time erasing. Cross out wrong answers and draw arrows to help the reader follow your work. You are given lots of space to answer, but be succinct—there's no reason to fill in all the space you are given. When you have answered the question, stop writing!

20. Read carefully and recognize that sometimes very different hypothesis tests or very different approaches are required in different parts of the same problem.

21. Realize that there may be several reasonable approaches to a given problem. In such a case, pick the one with which you are most comfortable or the one you feel will require the least amount of time to complete.

22. Realize that there may not be one clear, correct answer. Any of several different answers may receive full credit if they are backed up with well-reasoned arguments and explanations.

23. Punching a long list of data into the calculator does not show statistical knowledge. Be sure it is necessary!

24. Remember that uniform and symmetric are not the same, and that not all symmetric unimodal distributions are normal.

25. Boxplots show center (median), variability (IQR and range), and outliers but not shape. Differently shaped distributions can have identical boxplots.

26. Use comparative words like "more than," "less than," "equal to," "same as," or "different from" when comparing distributions. A "laundry list" of *each* distribution's shape, center, spread, and unusual features is not enough.

27. Do not refer to the "law of averages" or "law of small numbers" since there are no such things. There *is* a "law of large numbers," which refers to the relative frequency of an event becoming closer to the true probability of the event as the experiment is performed more and more times.

28. Comment on whether or not there are nonlinear patterns, increasing or decreasing spread, and whether the residuals are small or large compared with the different x-values when describing residual plots. "Randomly scattered" is not the same as "half below and half above."

29. Know how to *interpret* the slope, the y-intercept, and the coefficient of determination in context when using a calculator or computer printout to find a regression line.

30. Think about the smallest collection of things to which a single treatment is applied when identifying experimental units.

31. Understand that there are two types of control groups—experimental units that receive a placebo and experimental units that receive no treatment at all.

32. Explain how you will "randomly assign" subjects to treatment groups when asked to design an experiment. For example, you must state that you will use and how you will use a random number table or a random number generator on a calculator.

33. Follow these three steps when using a random number generator to pick a sample: (1) number the subjects; (2) define the interval of values from which integers are being selected, indicate that repeats are to be ignored, and note the stopping point; and (3) state that the sample is the group of subjects associated with the selected integers.

34. Understand that the expected value of a random variable does *not* mean the value that is expected to occur. In fact, it may have probability 0 of ever occurring. It is simply a long-run average that is somewhere between the lowest and highest values of the random variable.

35. Know the difference between a *sample distribution* and a *sampling distribution*. Do not state that the sample distribution will be close to normal because it is large! The larger the sample, the closer the *sample distribution* is to the population distribution. On the other hand, the larger the sample size, the closer the *sampling distribution* of the sample mean is to a normal distribution (due to the central limit theorem).

36. Look carefully at the original data scatterplot together with the residual plot as they often give critical insights in understanding a regression question.

37. Determine for inference problems whether a variable is *categorical* (leading to proportions or chi-square) or *quantitative* (leading to means or linear regression), whether you are given *raw data* or *summary statistics*, whether there is a *single population* of interest or *two populations* being compared, and, in the case of comparison, whether there are *independent samples* or a *paired comparison*.

38. Always define parameters! For example: Let p be the proportion of the population of all AP students who plan to major in statistics when they go to college.

39. Use clear subscripts if there are two populations. For example, $\mu_{\text{new drug}}$ and $\mu_{\text{old drug}}$.

40. Follow four steps when answering a confidence interval problem: (1) identify the procedure by name or formula (usually name is easier); (2) check conditions (do not just state them); (3) calculate the confidence interval (calculator software does this quickly); and (4) state a conclusion with reference to the confidence level, the parameter, and the population, and use context.

> **NOTE**
>
> In general with inference, do not show work with formulas as it is easy to mess this up.

41. Follow five steps when answering a hypothesis test problem: (1) state the hypotheses in terms of population parameters (don't forget to define the parameters); (2) state the test by name or formula (usually by name is easier); (3) check conditions (do not just state them); (4) calculate the test statistic and associated P-value (calculator software does this quickly); and (5) state a decision based upon the P-value and write a conclusion *in context* about the alternative hypothesis with reference to the parameter and the population.

42. Follow five steps when answering a chi-square problem: (1) state the hypotheses (both H_0 and H_a in context); (2) name the test (goodness-of-fit, independence, or homogeneity), which can often be inferred from the way the hypotheses are stated; (3) check the conditions, which requires showing the expected values (they can be copied from the calculator); (4) report the resulting chi-square value, the P-value, and *df*, the number of degrees of freedom; and (5) give a conclusion in context with linkage to the P-value.

43. Communicate that there is convincing evidence to support the *alternative hypothesis* if the decision is to reject the null hypothesis.

44. Know which type of error, Type I or Type II, might have been committed based on the conclusion to a hypothesis test, and be able to give a possible reasonable consequence of committing that error.

45. Use nondeterministic language like "on average," "estimated," "expected," and "predicted." Use words like "approximately" often, especially in front of "normal."

46. When asked to calculate a confidence interval, you need a conclusion in context about the *interval; however,* you do not need to interpret the confidence *level* unless specifically asked.

47. When performing a significance test, you need to *calculate* a P-value, but you do not need to *interpret* the P-value unless specifically asked.

48. Describe both an advantage of one thing and a disadvantage of the other if asked why one option is better than another.

49. In a yes or no question, be sure to answer yes or no, in addition to your explanation.

50. Follow directions on the Investigative Task (Free-Response Question 6). It often has a flow leading to an overall idea.

PART 4
Practice Tests

ANSWER SHEET
Practice Test 1

1. Ⓐ Ⓑ Ⓒ Ⓓ Ⓔ   11. Ⓐ Ⓑ Ⓒ Ⓓ Ⓔ   21. Ⓐ Ⓑ Ⓒ Ⓓ Ⓔ   31. Ⓐ Ⓑ Ⓒ Ⓓ Ⓔ

2. Ⓐ Ⓑ Ⓒ Ⓓ Ⓔ   12. Ⓐ Ⓑ Ⓒ Ⓓ Ⓔ   22. Ⓐ Ⓑ Ⓒ Ⓓ Ⓔ   32. Ⓐ Ⓑ Ⓒ Ⓓ Ⓔ

3. Ⓐ Ⓑ Ⓒ Ⓓ Ⓔ   13. Ⓐ Ⓑ Ⓒ Ⓓ Ⓔ   23. Ⓐ Ⓑ Ⓒ Ⓓ Ⓔ   33. Ⓐ Ⓑ Ⓒ Ⓓ Ⓔ

4. Ⓐ Ⓑ Ⓒ Ⓓ Ⓔ   14. Ⓐ Ⓑ Ⓒ Ⓓ Ⓔ   24. Ⓐ Ⓑ Ⓒ Ⓓ Ⓔ   34. Ⓐ Ⓑ Ⓒ Ⓓ Ⓔ

5. Ⓐ Ⓑ Ⓒ Ⓓ Ⓔ   15. Ⓐ Ⓑ Ⓒ Ⓓ Ⓔ   25. Ⓐ Ⓑ Ⓒ Ⓓ Ⓔ   35. Ⓐ Ⓑ Ⓒ Ⓓ Ⓔ

6. Ⓐ Ⓑ Ⓒ Ⓓ Ⓔ   16. Ⓐ Ⓑ Ⓒ Ⓓ Ⓔ   26. Ⓐ Ⓑ Ⓒ Ⓓ Ⓔ   36. Ⓐ Ⓑ Ⓒ Ⓓ Ⓔ

7. Ⓐ Ⓑ Ⓒ Ⓓ Ⓔ   17. Ⓐ Ⓑ Ⓒ Ⓓ Ⓔ   27. Ⓐ Ⓑ Ⓒ Ⓓ Ⓔ   37. Ⓐ Ⓑ Ⓒ Ⓓ Ⓔ

8. Ⓐ Ⓑ Ⓒ Ⓓ Ⓔ   18. Ⓐ Ⓑ Ⓒ Ⓓ Ⓔ   28. Ⓐ Ⓑ Ⓒ Ⓓ Ⓔ   38. Ⓐ Ⓑ Ⓒ Ⓓ Ⓔ

9. Ⓐ Ⓑ Ⓒ Ⓓ Ⓔ   19. Ⓐ Ⓑ Ⓒ Ⓓ Ⓔ   29. Ⓐ Ⓑ Ⓒ Ⓓ Ⓔ   39. Ⓐ Ⓑ Ⓒ Ⓓ Ⓔ

10. Ⓐ Ⓑ Ⓒ Ⓓ Ⓔ   20. Ⓐ Ⓑ Ⓒ Ⓓ Ⓔ   30. Ⓐ Ⓑ Ⓒ Ⓓ Ⓔ   40. Ⓐ Ⓑ Ⓒ Ⓓ Ⓔ

Practice Test 1

Section I: Questions 1–40

SPEND 90 MINUTES ON THIS PART OF THE EXAM.

> **DIRECTIONS:** The questions or incomplete statements that follow are each followed by five suggested answers or completions. Choose the response that best answers the question or completes the statement.

1. The following is a histogram of 87 home sale prices (in thousands of dollars) in one community:

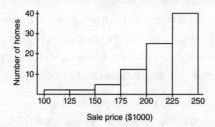

   Sale price ($1000)

   Which of the following could be the median of the sale prices?

   (A) $162,500
   (B) $175,000
   (C) $187,500
   (D) $212,500
   (E) $237,500

2. Which of the following is most useful in establishing cause-and-effect relationships?

   (A) A complete census
   (B) A least squares regression line showing high correlation
   (C) A simple random sample
   (D) A well-designed, well-conducted survey incorporating chance to ensure a representative sample
   (E) An experiment

3. The average height of ancient Egyptian men was 5′6″. Assuming a roughly normal distribution of heights, with a standard deviation of 2.5″, would Pharaoh Sa-Nakht's height of 6′1″ be considered an outlier using our standard definition of outliers? (invNorm(0.75) = 0.6745)

   (A) No, because $6'1'' < 5'6'' + 3(2.5'')$
   (B) Yes, because $6'1'' > 5'6'' + 0.6745(2.5'')$
   (C) Yes, because $6'1'' > 5'6'' + 2(0.6745)(2.5'')$
   (D) Yes, because $6'1'' > 5'6'' + 3(0.6745)(2.5'')$
   (E) Yes, because $6'1'' > 5'6'' + 4(0.6745)(2.5'')$

4. To determine the average cost of running for a congressional seat, a simple random sample of 50 politicians is chosen and the politicians' records examined. The cost figures show a mean of $1,825,000 with a standard deviation of $132,000. Which of the following is the best interpretation of a 90% confidence interval estimate for the average cost of running for office?

(A) Of all politicians running for a congressional seat, 90% spend between $1,794,000 and $1,856,000.

(B) Of all politicians running for a congressional seat, 90% spend a mean dollar amount that is between $1,794,000 and $1,856,000.

(C) We are 90% confident that politicians running for a congressional seat spend between $1,794,000 and $1,856,000.

(D) We are 90% confident that politicians running for a congressional seat spend a mean dollar amount between $1,794,000 and $1,856,000.

(E) We are 90% confident that in the chosen sample, the mean dollar amount spent running for a congressional seat is between $1,794,000 and $1,856,000.

5. In one study on the effect that eating meat products has on weight level, a simple random sample (SRS) of 500 subjects who admitted to eating meat at least once a day had their weights compared with those of an independent SRS of 500 people who claimed to be vegetarians. In a second study, an SRS of 500 subjects was served at least one meat meal per day for 6 months, while an independent SRS of 500 others were chosen to receive a strictly vegetarian diet for 6 months, with weights compared after 6 months. Which of the following is a true statement?

(A) The first study is a controlled experiment, while the second is an observational study.

(B) The first study is an observational study, while the second is a controlled experiment.

(C) Both studies are controlled experiments.

(D) Both studies are observational studies.

(E) Each study is part controlled experiment and part observational study.

6. A plumbing contractor obtains 60% of her boiler circulators from a company whose defect rate is 0.005 and the rest from a company whose defect rate is 0.010. If a circulator is defective, what is the probability that it came from the first company?

(A) $(0.6)(0.005)$

(B) $(0.6)(0.005) + (0.4)(0.010)$

(C) $\dfrac{0.6}{(0.6)(0.005) + (0.4)(0.010)}$

(D) $\dfrac{0.005}{(0.6)(0.005) + (0.4)(0.010)}$

(E) $\dfrac{(0.6)(0.005)}{(0.6)(0.005) + (0.4)(0.010)}$

7. A kidney dialysis center periodically checks a sample of its equipment and performs a major recalibration if readings are sufficiently off target. Similarly, a fabric factory periodically checks the sizes of towels coming off an assembly line and halts production if measurements are sufficiently off target. In both situations, we have the null hypothesis that the equipment is performing satisfactorily. For each situation, which is the *more* serious concern, a Type I or Type II error?

(A) Dialysis center: Type I error;
    towel manufacturer: Type I error

(B) Dialysis center: Type I error;
    towel manufacturer: Type II error

(C) Dialysis center: Type II error;
    towel manufacturer: Type I error

(D) Dialysis center: Type II error;
    towel manufacturer: Type II error

(E) This is impossible to answer without making an expected value judgment between human life and accurate towel sizes.

8. A coin is weighted so that the probability of heads is 0.75. The coin is tossed 10 times, and the number of heads is noted. This procedure is repeated a total of 50 times, and the number of heads is recorded each time. What kind of distribution has been simulated?

   (A) The sampling distribution of the sample proportion with $n = 10$ and $p = 0.75$
   (B) The sampling distribution of the sample proportion with $n = 50$ and $p = 0.75$
   (C) The sampling distribution of the sample mean with $\bar{x} = 10(0.75)$ and $\sigma = \sqrt{10(0.75)(0.25)}$
   (D) The binomial distribution with $n = 10$ and $p = 0.75$
   (E) The binomial distribution with $n = 50$ and $p = 0.75$

9. During the years 1886 through 2019, there were an average of 9.0 tropical cyclones per year, of which an average of 5.4 became hurricanes. Assuming that the probability of any cyclone becoming a hurricane is independent of what happens to any other cyclone, if there are five cyclones in one year, what is the probability that at least three become hurricanes?

   (A) $\binom{5}{3}(0.6)^3$

   (B) $\binom{5}{3}(0.6)^3(0.4)^2$

   (C) $(0.6)^3(0.4)^2 + (0.6)^4(0.4) + (0.6)^5$

   (D) $\binom{5}{3}(0.6)^3(0.4)^2 + \binom{5}{4}(0.6)^4(0.4) + (0.6)^5$

   (E) $1 - [(0.4)^5 + 5(0.6)(0.4)^4 + 10(0.6)^2(0.4)^3 + 10(0.6)^3(0.4)^2]$

10. Given the probabilities $P(A) = 0.3$ and $P(B) = 0.2$, what is the probability of the union $P(A \cup B)$ if A and B are mutually exclusive? If A and B are independent? If B is a subset of A?

    (A) 0.44, 0.5, 0.2
    (B) 0.44, 0.5, 0.3
    (C) 0.5, 0.44, 0.2
    (D) 0.5, 0.44, 0.3
    (E) 0, 0.5, 0.3

11. A social science researcher is testing the claim that 69 percent of Americans use social media every day. In a survey involving a simple random sample (SRS) of 1,100 Americans, 770 say that they use social media every day. With $H_0: p = 0.69$ and $H_a: p \neq 0.69$, what is the value of the test statistic?

    (A) $z = \dfrac{0.7 - 0.69}{\sqrt{1{,}100(0.69)(1 - 0.69)}}$

    (B) $z = 2\dfrac{0.7 - 0.69}{\sqrt{1{,}100(0.69)(1 - 0.69)}}$

    (C) $z = \dfrac{0.7 - 0.69}{\sqrt{\dfrac{(0.69)(1 - 0.69)}{1{,}100}}}$

    (D) $z = 2\dfrac{0.7 - 0.69}{\sqrt{\dfrac{(0.69)(1 - 0.69)}{1{,}100}}}$

    (E) $z = 2\dfrac{0.7 - 0.69}{\sqrt{1{,}100(0.7)(1 - 0.7)}}$

12. A medical research team tests for tumor reduction in a sample of patients using three different dosages of an experimental cancer drug. Which of the following is true?

    (A) There are three explanatory variables and one response variable.
    (B) There is one explanatory variable with three levels of response.
    (C) Tumor reduction is the only explanatory variable, but there are three response variables corresponding to the different dosages.
    (D) There are three levels of a single explanatory variable.
    (E) Each explanatory level has an associated level of response.

13. The graph below shows cumulative proportion plotted against age for a population.

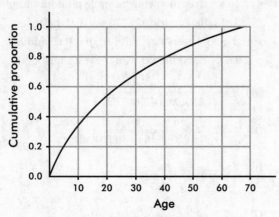

The median is approximately what age?

(A) 17

(B) 30

(C) 35

(D) 40

(E) 45

14. Mr. Bee's statistics class had a standard deviation of 11.2 on a standardized test, while Mr. Em's class had a standard deviation of 5.6 on the same test. Which of the following is the most reasonable conclusion concerning the two classes' performance on the test?

(A) Mr. Bee's class is less heterogeneous than Ms. Em's.

(B) Ms. Em's class is more homogeneous than Mr. Bee's.

(C) Mr. Bee's class performed twice as well as Ms. Em's.

(D) Ms. Em's class did not do as well as Mr. Bee's.

(E) Mr. Bee's class had the higher mean, but this may not be statistically significant.

15. A college admissions officer is interested in comparing the SAT Math scores of high school applicants who have and have not taken AP Statistics. She randomly pulls the files of five applicants who took AP Statistics and five applicants who did not, and proceeds to run a t-test to compare the mean SAT Math scores of the two groups. Which of the following is a necessary assumption?

(A) The population variances from each group are known.

(B) The population variances from each group are unknown.

(C) The population variances from the two groups are equal.

(D) The population of all SAT scores from each group is roughly normally distributed.

(E) The samples must be independent simple random samples, and for each sample, np and $n(1-p)$ must both be at least 10.

16. The appraised values of houses in a city have a mean of $195,000 with a standard deviation of $23,000. Because of a new teachers' contract, the school district needs an extra 10% in funds compared with the previous year. To raise this additional money, the city instructs the assessment office to raise all appraised house values by $5,000. What will be the new standard deviation of the appraised values of houses in the city?

(A) $23,000

(B) $25,300

(C) $28,000

(D) $30,300

(E) $30,800

17. A psychologist hypothesizes that scores on an aptitude test are normally distributed with a mean of 70 and a standard deviation of 10. In a random sample of size 100, scores are distributed as in the table below. What is the χ^2-statistic for a goodness-of-fit test?

Score:	Below 60	60−70	70−80	Above 80
Number of people:	10	40	35	15

(A) $\dfrac{(10-16)^2}{10} + \dfrac{(40-34)^2}{40} + \dfrac{(35-34)^2}{35} + \dfrac{(15-16)^2}{15}$

(B) $\dfrac{(10-16)^2}{16} + \dfrac{(40-34)^2}{34} + \dfrac{(35-34)^2}{34} + \dfrac{(15-16)^2}{16}$

(C) $\dfrac{(10-25)^2}{25} + \dfrac{(40-25)^2}{25} + \dfrac{(35-25)^2}{25} + \dfrac{(15-25)^2}{25}$

(D) $\dfrac{(10-25)^2}{10} + \dfrac{(40-25)^2}{40} + \dfrac{(35-25)^2}{35} + \dfrac{(15-25)^2}{15}$

(E) $\dfrac{(10-25)^2}{16} + \dfrac{(40-25)^2}{34} + \dfrac{(35-25)^2}{34} + \dfrac{(15-25)^2}{16}$

18. A talk show host recently reported that in response to his on-air question, 82% of the more than 2,500 email messages received through his publicized address supported the death penalty for anyone convicted of selling drugs to children. What does this show?

(A) The survey is meaningless because of voluntary response bias.

(B) No meaningful conclusion is possible without knowing something more about the characteristics of his listeners.

(C) The survey would have been more meaningful if he had picked a random sample of the 2,500 listeners who responded.

(D) The survey would have been more meaningful if he had used a control group.

(E) This was a legitimate sample, randomly drawn from his listeners, and of sufficient size to be able to conclude that most of his listeners support the death penalty for such a crime.

19. Define a new measurement as the difference between the 60th and 40th percentile scores in a population. This measurement will give information concerning

(A) central tendency.
(B) variability.
(C) symmetry.
(D) skewness.
(E) clustering.

20. Suppose X and Y are random variables with $E(X) = 37$, $\text{var}(X) = 5$, $E(Y) = 62$, and $\text{var}(Y) = 12$. What are the expected value and variance of the random variable $X + Y$?

(A) $E(X+Y) = 99$, $\text{var}(X+Y) = 8.5$
(B) $E(X+Y) = 99$, $\text{var}(X+Y) = 13$
(C) $E(X+Y) = 99$, $\text{var}(X+Y) = 17$
(D) $E(X+Y) = 49.5$, $\text{var}(X+Y) = 17$
(E) There is insufficient information to answer this question.

21. Which of the following can affect the value of the correlation r?

(A) A change in measurement units
(B) A change in which variable is called x and which is called y
(C) Adding the same constant to all values of the x-variable
(D) All of the above can affect the r-value.
(E) None of the above can affect the r-value.

22. The American Medical Association (AMA) wishes to determine the percentage of obstetricians who are considering leaving the profession because of the rapidly increasing number of lawsuits against obstetricians. The AMA would like an answer to within ±3% at the 95% confidence level. Which of the following should be used to find the sample size (n) needed?

(A) $1.645\sqrt{\dfrac{0.5}{n}} \leq 0.03$

(B) $1.645\sqrt{\dfrac{0.5}{n}} \leq 0.06$

(C) $1.645\sqrt{\dfrac{(0.5)(0.5)}{n}} \leq 0.06$

(D) $1.96\sqrt{\dfrac{0.5}{n}} \leq 0.03$

(E) $1.96\sqrt{\dfrac{(0.5)(0.5)}{n}} \leq 0.03$

23. Suppose you did 10 independent tests of the form $H_0: \mu = 25$ versus $H_a: \mu < 25$, each at the $\alpha = 0.05$ significance level. What is the probability of committing a Type I error and incorrectly rejecting a true H_0 with at least one of the 10 tests?

(A) 0.05
(B) 0.40
(C) 0.50
(D) 0.60
(E) 0.95

24. Which of the following sets has the smallest standard deviation? Which has the largest?

I. 1, 2, 3, 4, 5, 6, 7
II. 1, 1, 1, 4, 7, 7, 7
III. 1, 4, 4, 4, 4, 4, 7

(A) Smallest SD: I; largest SD: II
(B) Smallest SD: II; largest SD: III
(C) Smallest SD: III; largest SD: I
(D) Smallest SD: II; largest SD: I
(E) Smallest SD: III; largest SD: II

25. A researcher plans a study to examine long-term confidence in the U.S. economy among the adult population. She obtains a simple random sample of 30 adults as they leave a Wall Street office building one weekday afternoon. All but two of the adults agree to participate in the survey. Which of the following conclusions is correct?

(A) Proper use of chance as evidenced by the simple random sample makes this a well-designed survey.
(B) The high response rate makes this a well-designed survey.
(C) Selection bias makes this a poorly designed survey.
(D) A voluntary response study like this gives too much emphasis to persons with strong opinions.
(E) Lack of anonymity makes this a poorly designed survey.

26. Which among the following would result in the narrowest confidence interval?

(A) Small sample size and 95% confidence
(B) Small sample size and 99% confidence
(C) Large sample size and 95% confidence
(D) Large sample size and 99% confidence
(E) Large sample size with either 95% or 99% confidence

27. In a random sample of 500 working adults, heights x (in inches) and annual salaries y (in dollars) are tabulated. When the least-squares regression analysis is performed, the correlation is 0.38. Which of the following is the correct way to label the correlation?

(A) 0.38
(B) 0.38 dollars per inch
(C) 0.38 inches per dollar
(D) 0.38 dollar inches
(E) 0.38 inch dollars

28. There are two games involving flipping a fair coin. In the first game, you win a prize if you can throw between 45% and 55% heads. In the second game, you win if you can throw more than 80% heads. For each game, would you rather flip the coin 30 times or 300 times?

    (A) 30 times for each game
    (B) 300 times for each game
    (C) 30 times for the first game and 300 times for the second
    (D) 300 times for the first game and 30 times for the second
    (E) The outcomes of the games do not depend on the number of flips.

29. City planners are trying to decide among various parking plan options ranging from more on-street spaces to multilevel facilities to spread-out small lots. Before making a decision, they wish to test the downtown merchants' claim that shoppers park for an average of only 47 minutes in the downtown area. The planners have decided to tabulate parking durations for 225 shoppers and to reject the merchants' claim if the sample mean exceeds 50 minutes. If the merchants' claim is wrong and the true mean is 51 minutes, what is the probability that the random sample will lead to a mistaken failure to reject the merchants' claim? Assume that the standard deviation in parking durations is 27 minutes.

    (A) $P\left(z < \dfrac{50 - 47}{27/\sqrt{225}}\right)$

    (B) $P\left(z > \dfrac{50 - 47}{27/\sqrt{225}}\right)$

    (C) $P\left(z < \dfrac{50 - 51}{27/\sqrt{225}}\right)$

    (D) $P\left(z > \dfrac{50 - 51}{27/\sqrt{225}}\right)$

    (E) $P\left(z > \dfrac{51 - 47}{27/\sqrt{225}}\right)$

30. Consider the following parallel boxplots indicating the starting salaries (in thousands of dollars) for blue-collar and white-collar workers at a particular production plant:

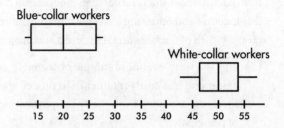

    Which of the following is a correct conclusion?

    (A) The ranges of the distributions are the same.
    (B) In each distribution, the mean is equal to the median.
    (C) Each distribution is symmetric.
    (D) Each distribution is roughly normal.
    (E) The distributions are outliers of each other.

31. The mean Law School Aptitude Test (LSAT) score for applicants to a particular law school is 650 with a standard deviation of 45. Suppose that only applicants with scores above 700 are considered. What percentage of the applicants considered have scores below 740? (Assume the scores are approximately normally distributed.)

    (A) 13.3%
    (B) 17.1%
    (C) 82.9%
    (D) 86.7%
    (E) 97.7%

32. If all the other variables remain constant, which of the following will increase the power of a hypothesis test?

    I. Increasing the sample size
    II. Increasing the significance level
    III. Increasing the probability of a Type II error

    (A) I only
    (B) II only
    (C) III only
    (D) I and II only
    (E) I, II, and III

33. A researcher planning a survey of school principals in a particular state has lists of the school principals employed in each of the 125 school districts. The procedure is to obtain a random sample of principals from each of the districts rather than grouping all the lists together and obtaining a sample from the entire group. Which of the following is a correct conclusion?

(A) This is a simple random sample obtained in an easier and less costly manner than procedures involving sampling from the entire population of principals.

(B) This is a cluster sample in which the population was divided into heterogeneous groups called clusters.

(C) This is an example of systematic sampling, which gives a reasonable sample as long as the original order of the list is not related to the variables under consideration.

(D) This is an example of proportional sampling based on sizes of the school districts.

(E) This is a stratified sample, which may give comparative information that a simple random sample wouldn't give.

34. A simple random sample is defined by

(A) the method of selection.

(B) examination of the outcome.

(C) both of the above.

(D) how representative the sample is of the population.

(E) the size of the sample versus the size of the population.

35. A teacher randomly assigns students to seat locations and then notes final test scores versus assigned row. Some computer output of the regression analysis follows.

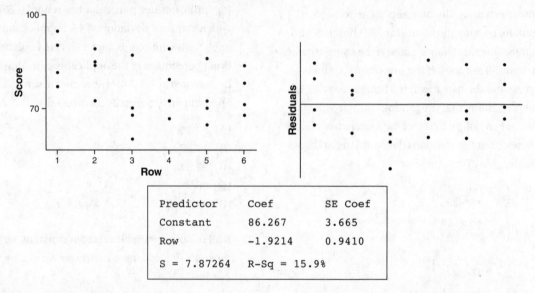

Predictor	Coef	SE Coef
Constant	86.267	3.665
Row	−1.9214	0.9410

S = 7.87264   R-Sq = 15.9%

Which of the following is an appropriate test statistic for testing the null hypothesis that the slope of the regression line is less than 0?

(A) $\dfrac{0 - (-1.9214)}{0.9410}$

(B) $\dfrac{-1.9214 - 0}{0.9410}$

(C) $\dfrac{-1 - (-1.9214)}{0.9410}$

(D) $\dfrac{0 - (-1.9214)}{7.87264}$

(E) $\dfrac{-1.9214 - 0}{7.87264}$

PRACTICE TEST 1

36. Which of the following is a true statement about hypothesis testing?

(A) If there is sufficient evidence to reject a null hypothesis at the 10% level, there is sufficient evidence to reject it at the 5% level.

(B) Whether to use a one- or a two-sided test is typically decided after the data are gathered.

(C) If a hypothesis test is conducted at the 1% level, there is a 1% chance of rejecting the null hypothesis.

(D) The probability of a Type I error plus the probability of a Type II error always equals 1.

(E) The power of a test concerns its ability to detect an alternative hypothesis.

37. A population is normally distributed with mean 25. Consider all samples of size 10. The variable $\dfrac{\bar{x} - 25}{\frac{s}{\sqrt{10}}}$

(A) has a normal distribution.

(B) has a t-distribution with $df = 10$.

(C) has a t-distribution with $df = 9$.

(D) has neither a normal distribution nor a t-distribution.

(E) has either a normal or a t-distribution depending on the characteristics of the population standard deviation.

38. A correlation of 0.6 indicates that the percentage of variation in y that is explained by the variation in x is how many times the percentage indicated by a correlation of 0.3?

(A) 2

(B) 3

(C) 4

(D) 6

(E) There is insufficient information to answer this question.

39. As reported in a national CNN poll, 43% of high school students expressed fear about going to school. Which of the following best describes what is meant by the poll having a margin of error of 5%?

(A) It is likely that the true proportion of high school students afraid to go to school is between 38% and 48%.

(B) Five percent of the students refused to participate in the poll.

(C) Between 38% and 48% of those surveyed expressed fear about going to school.

(D) There is a 0.05 probability that the 43% result is in error.

(E) If similar size polls were repeatedly taken, they would be wrong about 5% of the time.

40. In a high school of 1,650 students, 132 have personal investments in the stock market. To estimate the total stock investment by students in this school, two plans are proposed. Plan I would sample 30 students at random, find a confidence interval estimate of their average investment, and then multiply both ends of this interval by 1,650 to get an interval estimate of the total investment. Plan II would sample 30 students at random from among the 132 who have investments in the market, find a confidence interval estimate of their average investment, and then multiply both ends of this interval by 132 to get an interval estimate of the total investment. Which is the better plan for estimating the total stock market investment by students in this school?

(A) Plan I

(B) Plan II

(C) Both plans use random samples and so will produce equivalent results.

(D) Neither plan will give an accurate estimate.

(E) The resulting data must be seen to evaluate which is the better plan.

Section II: Part A

Questions 1–5

SPEND ABOUT 65 MINUTES ON THIS PART OF THE EXAM.
PERCENTAGE OF SECTION II GRADE—75

> **DIRECTIONS:** You must show all work and indicate the methods you use. You will be graded on the correctness of your methods and on the accuracy of your results and explanations.

1. A new anti-spam software program is field tested on 1,000 emails, and the results are summarized in the following table (positive test = program labels email as spam):

	Spam	Legitimate	
Positive test	205	90	295
Negative test	45	660	705
	250	750	1,000

Using the above empirical results, determine the following probabilities.

(a) (i) What is the *predictive value* of the test? That is, what is the probability that an email is spam and the test is positive?

(ii) What is the *false-positive* rate? That is, what is the probability of testing positive given that the email is legitimate?

(b) (i) What is the *sensitivity* of the test? That is, what is the probability of testing positive given that the email is spam?

(ii) What is the *specificity* of the test? That is, what is the probability of testing negative given that the email is legitimate?

(c) (i) Given a random sample of five legitimate emails, what is the probability that the program labels at least one as spam?

(ii) Given a random sample of five spam emails, what is the probability that the program correctly labels at least three as spam?

(d) (i) If 35% of the incoming emails from one source are spam, what is the probability that an email from that source will be labeled as spam?

(ii) If an email from this source is labeled as spam, what is the probability it really is spam?

2. When students do not understand an author's vocabulary, they often miss subtleties of meaning in the text; thus, learning vocabulary is recognized as one of the most important skills in all subject areas. A school system is planning a study to see which of two programs is more effective in ninth grade. Program A involves teaching students to apply morphemic analysis, the process of deriving a word's meaning by analyzing its meaningful parts, such as roots, prefixes, and suffixes. Program B involves teaching students to apply contextual analysis, the process of inferring the meaning of an unfamiliar word by examining the surrounding text.

   (a) Explain the purpose of incorporating a control group in this experiment.
   (b) To comply with informed consent laws, parents will be asked their consent for their children to be randomly assigned to be taught in one of the new programs or by the local standard method. Students of parents who do not return the consent forms will continue to be taught by the local standard method. Explain why these students should not be considered part of the control group.
   (c) Given 90 students randomly selected from ninth-grade students with parental consent, explain how to assign them to the three groups, Program A, Program B, and control, for a completely randomized design.

3. In the 1960s and 1970s, college students were in the front line of the Civil Rights and anti-war movements. Such activism is one of the most powerful mechanisms for social change. A study looked at the association between social justice advocacy as measured on the ACT activity scale and political interest level as measured on the 1–10 POL scale. The researchers hoped to predict advocacy from political interest. The statistical summary of the data from a random sample of 134 graduate students at a Midwestern university is shown below.

Variable	Mean	SD
ACT	18.92	4.85
POL	5.48	2.24
r = 0.49		

   (a) What is b, the slope of the regression line? Interpret it in context.
   (b) What is the equation of the least squares regression line?
   (c) Suppose a student has a POL of 4 with an ACT of 15. What is the residual? Interpret it in context.

4. Cumulative frequency graphs of the heights (in inches) of athletes playing college baseball (A), basketball (B), and football (C) are given below:

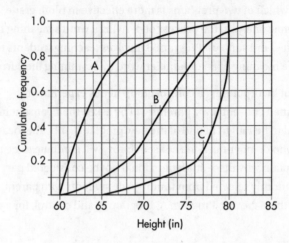

(a) Compare the distributions of heights of athletes playing the three college sports.
(b) What is the probability that a randomly picked college basketball player is under 71 inches in height?
(c) If a recruiter randomly picks college basketball players, what is the probability that the first player under 71 inches in height will be the third player picked?

5. Cumulative exposure to nitrogen dioxide (NO_2) is a major risk factor for lung disease in tunnel construction workers. In a 2004 study, researchers compared cumulative exposure to NO_2 (in parts per million per year, ppm/yr) for a random sample of drill and blast workers and for an independent random sample of concrete workers. Summary statistics are shown below.

Tunnel Worker Activity	Sample Size	Mean Cumulative NO_2 Exposure (ppm*/yr)	Standard Deviation in Cumulative NO_2 Exposure (ppm*/yr)
Drill and blast	115	4.1	1.8
Concrete	69	4.8	2.4

(a) Is random sampling, random assignment, both, or neither involved in this study?
(b) Find a 95% confidence interval estimate for the difference in mean cumulative NO_2 exposure for drill and blast workers and for concrete workers.
(c) Using this confidence interval, is there evidence of a difference in mean cumulative NO_2 exposure for drill and blast workers and for concrete workers? Explain.
(d) It turns out that both data sets were somewhat skewed. Does this invalidate your analysis? Explain.

Section II: Part B

Question 6

SPEND ABOUT 25 MINUTES ON THIS PART OF THE EXAM.
PERCENTAGE OF SECTION II GRADE—25

6. A physical education teacher compared balance between male and female students at her school. Two independent samples were randomly selected, one of nine men and one of nine women. Each student was timed as to how long (in seconds) he or she could balance on one foot with eyes closed.

Balance times (in seconds)									
Men	19	13	35	20	48	19	24	21	17
Women	42	23	38	19	27	30	34	46	33

(a) Create a back-to-back stemplot to display these data.

(b) Compare the two distributions.

(c) To test whether or not the mean balance times of men and women are different, a two-sample t-test was performed with a resulting t-statistic of -1.816 and P-value of 0.0889. At a 10 percent significance level, what should be concluded?

Also of interest is whether men (M) or women (W) have more variability in their balance times. To test the hypotheses $H_0: \sigma_M = \sigma_W$ versus $H_a: \sigma_M \neq \sigma_W$, the test statistic $\frac{s_M}{s_W}$ (the ratio of the sample standard deviations) will be used. To investigate the sampling distribution of this test statistic when the population standard deviations are equal, a simulation is conducted. Two independent samples of size 9 are selected from the same population, and the ratio of the sample standard deviations is calculated. A histogram of 500 simulated values of $\frac{s_M}{s_W}$ is shown below.

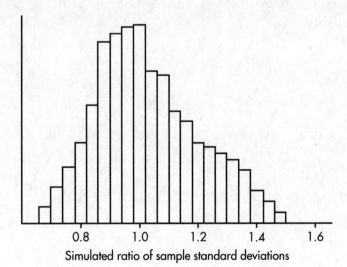

Simulated ratio of sample standard deviations

(d) Explain why the median of the above ratios appears to be about 1.0.

Summary statistics for the original data are below.

	Sample Size	Sample Mean	Sample Standard Deviation
Men	9	24.00	10.85
Women	9	32.44	8.76

(e) Using the summary statistics above and the simulation histogram above, is there convincing evidence of a difference in variability of balance times between men and women?

STOP   If there is still time remaining, you may check your work on this section.

Answer Explanations

Section I

1. **(D)** With 87 home sale prices, the median is between the 43rd and 44th values when placed in either ascending or descending order. There appear to be around 40 values between \$225,000 and \$250,000 and 25 more values between \$200,000 and \$225,000. So, both the 43rd and 44th values are in the 200 to 225 interval in the histogram.

2. **(E)** Regression lines show association, not causation. Surveys suggest relationships, which experiments can help to show to be cause and effect.

3. **(E)** $\text{invNorm}(0.75) = 0.6745$, so $Q_3 = 5'6'' + 0.6745(2.5'')$, $Q_1 = 5'6'' - 0.6745(2.5'')$, and $\text{IQR} = Q_3 - Q_1 = 2(0.6745)(2.5'')$. On the higher side, outliers are values greater than $Q_3 + 1.5(\text{IQR}) = [5'6'' + 0.6745(2.5'')] + 1.5[2(0.6745)(2.5'')] = 5'6'' + 4(0.6745)(2.5'') = 6'0.745''$ and $6'1'' > 6'0.745''$.

4. **(D)** Calculator software gives (1,794,000, 1,856,000). We are 90% confident that the population mean is within the interval calculated using the data from the sample.

5. **(B)** The first study was observational because the subjects were not chosen for treatment; their current eating habits were simply noted. In the second study, two treatments were applied (meat meals and vegetarian meals).

6. **(E)**

$$P(\text{def}) = P(\text{1st} \cap \text{def}) + P(\text{2nd} \cap \text{def})$$

$$= (0.6)(0.005) + (0.4)(0.010)$$

$$P(\text{1st} \mid \text{def}) = \frac{P(\text{1st} \cap \text{def})}{P(\text{def})} = \frac{(0.6)(0.005)}{(0.6)(0.005) + (0.4)(0.010)}$$

7. **(C)** At the dialysis center the more serious concern would be a Type II error, which is that the equipment is not performing correctly yet the check does not pick this up; however, at the towel manufacturing plant the more serious concern would be a Type I error, which is that the equipment is performing correctly yet the check causes a production halt.

8. **(D)** There are two possible outcomes (heads and tails), with the probability of heads always 0.75 (independent of what happened on the previous toss), and we are interested in the number of heads in 10 tosses. Thus, this is a binomial model with $n = 10$ and $p = 0.75$. Repeating this over and over (in this case 50 times) simulates the resulting binomial distribution.

9. **(D)** In a binomial distribution with $n = 5$ and probability of success $p = \frac{5.4}{9.0} = 0.6$, the probability of at least 3 successes is $P(3 \text{ successes}) + P(4 \text{ successes}) + P(5 \text{ successes})$:
$$= \binom{5}{3}(0.6)^3(0.4)^2 + \binom{5}{4}(0.6)^4(0.4) + (0.6)^5$$

10. **(D)** If A and B are mutually exclusive, then $P(A \cap B) = 0$, and so $P(A \cup B) = 0.3 + 0.2 - 0 = 0.5$. If A and B are independent, then $P(A \cap B) = P(A)P(B)$, and so $P(A \cup B) = 0.3 + 0.2 - (0.3)(0.2) = 0.44$. If B is a subset of A, then $A \cup B = A$, and so $P(A \cup B) = P(A) = 0.3$.

11. **(C)** The standard deviation of the test statistic is $\sigma_{\hat{p}} = \sqrt{\frac{p_0(1 - p_0)}{n}} = \sqrt{\frac{(0.69)(1 - 0.69)}{1,100}}$ and $\hat{p} = \frac{770}{1,100} = 0.7$. Then $z = \frac{\hat{p} - p_0}{\sigma_{\hat{p}}}$
$$= \frac{0.7 - 0.69}{\sqrt{\frac{(0.69)(1 - 0.69)}{1,100}}}.$$ Since this is a two-sided test, the P-value will be twice the tail probability of the test statistic; however, the test statistic itself is not doubled.

12. **(D)** Dosage is the only explanatory variable, and it is being tested at three levels. Tumor reduction is the single response variable.

13. **(A)** The median corresponds to a cumulative proportion of 0.5.

14. **(B)** Standard deviation is a measure of variability. The less variability, the more homogeneity there is.

15. **(D)** Since the sample sizes are small, the samples must come from roughly normally distributed populations. While the samples should be independent simple random samples, np and $n(1 - p)$ refer to conditions for tests involving sample proportions, not means.

16. **(A)** Adding the same constant to all values in a set will increase the mean by that constant but will leave the standard deviation unchanged.

17. **(B)** With about 68% of the values within 1 standard deviation of the mean, the expected numbers for a normal distribution are as follows:

16	34	34	16

$$\chi^2 = \sum \frac{(\text{observed} - \text{expected})^2}{\text{expected}}$$

$$= \frac{(10 - 16)^2}{16} + \frac{(40 - 34)^2}{34} + \frac{(35 - 34)^2}{34} + \frac{(15 - 16)^2}{16}$$

18. **(A)** This is a good example of voluntary response bias, which often overrepresents strong or negative opinions. The people who chose to respond were very possibly the parents of children facing drug problems or people who had had bad experiences with drugs being sold in their neighborhoods. There is very little chance that the 2,500 respondents were representative of the population. Knowing more about his listeners or taking a sample of the sample would not have helped.

19. **(B)** The range (difference between largest and smallest values), the interquartile range $(Q_3 - Q_1)$, and this difference between the 60th and 40th percentile scores all are measures of variability, or how spread out is the population or a subset of the population.

20. **(E)** It is true that $E(X + Y) = E(X) + E(Y) = 37 + 62 = 99$; however, without *independence,* we cannot determine $\text{var}(X + Y)$ from the information given.

21. **(E)** The correlation coefficient r is not affected by changes in units, by which variable is called x or y, by adding the same number to all values of a variable, or by multiplying all values of a variable by the same positive number.

22. **(E)** The critical z-score for a 95% confidence interval is `invNorm(0.975)` = 1.96. With unknown p, we use $p = 0.5$ in our calculation of n.

23. **(B)** With $\alpha = 0.05$, the probability of committing a Type I error is 0.05, and the probability of not committing a Type I error is 0.95. $P(\text{at least one Type I error in 10 tests}) = 1 - P(\text{no Type I errors in 10 tests}) = 1 - (0.95)^{10} = 0.40$.

24. **(E)** Note that all three sets have the same mean and the same range. The third set has most of its values concentrated right at the mean, while the second set has most of its values concentrated far from the mean.

25. **(C)** People coming out of a Wall Street office building are a very unrepresentative sample of the adult population, especially given the question under consideration. Using chance and obtaining a high response rate will not change the selection bias and make this into a well-designed survey. This is a convenience sample, not a voluntary response sample.

26. **(C)** Larger samples (so $\frac{\sigma}{\sqrt{n}}$ is smaller) and less confidence (so the critical z or t is smaller) both result in smaller intervals.

27. **(A)** The correlation r is unit-free.

28. **(D)** The probability of throwing heads is 0.5. By the law of large numbers, the more times you flip the coin, the more the relative frequency tends to become closer to this probability. With fewer tosses there is more chance for wide swings in the relative frequency.

29. **(C)** $\sigma_{\bar{x}} = \frac{\sigma}{\sqrt{n}} = \frac{27}{\sqrt{225}}$. If the true mean parking duration is 51 minutes, the normal curve should be centered at 51. The critical value of 50 has a z-score of $\frac{50 - 51}{27/\sqrt{225}}$. The city planners will reject the merchants' claim if the sample mean is greater than 50 and so will fail to reject if less than 50.

30. **(A)** The overall lengths (between tips of whiskers) are the same, and so the ranges are the same. Just because the min and max are equidistant from the median, and Q_1 and Q_3 are equidistant from the median, does not imply that a distribution is symmetric or that the mean and median are equal. Even if a distribution is symmetric, this does not imply that it is roughly normal. Particular values, not distributions, may be outliers.

31. **(C)** Critical z-scores are $\frac{700 - 650}{45} = 1.11$ and $\frac{740 - 650}{45} = 2$ with right tail probabilities of 0.1335 and 0.0228, respectively. The percentage below 740 given that the scores are above

    700 is $\frac{0.1335 - 0.0228}{0.1335} = 82.9\%$. Or

    $\frac{\text{normalcdf}(700, 740, 650, 45)}{\text{normalcdf}(700, 1000000, 650, 45)} = 0.82928.$

32. **(D)** There is a different probability of Type II error for each possible correct value of the population parameter, and 1 minus this probability is the power of the test against the associated correct value. Increasing the sample size will make the standard deviations smaller and thus can decrease the probabilities of Type I and Type II errors and so increase the power. Increasing the significance level α (making it easier to reject the null) will lower the probability of a Type II error and thus also increase the power.

33. **(E)** Stratified samples are often easier and less costly to obtain and also make comparative data available. In this case, responses can be compared among various districts. This is not a simple random sample because all possible sets of the required size do not have the same chance of being picked. For example, a set of principals all from just half the school districts has no chance of being picked to be the sample. This is not a cluster sample in that there is no reason to believe that each school district resembles the population as a whole, and furthermore, there was no random sample taken of the school districts. This is not systematic sampling as the districts were not put in some order with every nth district chosen.

34. **(A)** A simple random sample can be any size and may or may not be representative of the population. It is a method of selection in which every possible sample of the desired size has an equal chance of being selected.

35. **(B)** With H_0: $\beta = 0$ and H_a: $\beta < 0$, $t = \frac{b - 0}{SE(b)}$ $= \frac{-1.9214 - 0}{0.9410}$.

36. **(E)** If the P-value is less than 0.10, it does not follow that it is less than 0.05. Decisions such as whether a test should be one- or two-sided are made before the data are gathered. If $\alpha = 0.01$, there is a 1% chance of rejecting the null hypothesis *if* the null hypothesis is true. There is one probability of a Type I error, the significance level, while there is a different probability of a Type II error associated with each possible correct alternative, so the sum does not equal 1. When given a true alternative, power is the probability of rejecting the false null hypothesis.

37. **(C)** If the population is normally distributed, then $\frac{\bar{x} - \mu}{\frac{\sigma}{\sqrt{n}}}$ is normally distributed, but $\frac{\bar{x} - \mu}{\frac{s}{\sqrt{n}}}$ has a t-distribution with $df = n - 1$.

38. **(C)** A correlation of 0.6 explains $(0.6)^2$, or 36%, of the variation in y, while a correlation of 0.3 explains only $(0.3)^2$, or 9%, of the variation in y.

39. **(A)** When using a measurement from a sample, we are never able to say *exactly* what a population proportion is; rather, we always say we have a certain *confidence* that the population proportion lies in a particular *interval*. In this case, that interval is 43% ± 5% or between 38% and 48%.

40. **(B)** With Plan I, the expected number of students with stock investments is only $30\left(\frac{132}{1,650}\right) = 2.4$ out of 30. Plan II allows an estimate to be made using a full 30 investors.

Section II: Part A

1. (a) (i) $P(\text{spam} \cap \text{positive}) = \frac{205}{1,000} = 0.205$

   (ii) $P(\text{positive} \mid \text{legitimate}) = \frac{90}{750} = 0.12$

   (b) (i) $P(\text{positive} \mid \text{spam}) = \frac{205}{250} = 0.82$

   (ii) $P(\text{negative} \mid \text{legitimate}) = \frac{660}{750} = 0.88$

   (c) (i) $1 - (0.88)^5 = 0.4723$

   (ii) $10(0.82)^3(0.18)^2 + 5(0.82)^4(0.18) + (0.82)^5 = 0.9563$
       Or $1 - \text{binomcdf}(5, 0.82, 2) = 1 - 0.0437 = 0.9563$

   (d) (i) $P(\text{positive}) = P(\text{positive} \cap \text{spam}) + P(\text{positive} \cap \text{legitimate}) =$
       $P(\text{spam})P(\text{positive} \mid \text{spam}) + P(\text{legitimate})P(\text{positive} \mid \text{legitimate}) =$
       $(0.35)(0.82) + (0.65)(0.12) = 0.365$

   (ii) $P(\text{spam} \mid \text{positive}) = \dfrac{P(\text{spam} \cap \text{positive})}{P(\text{positive})} = \dfrac{(0.35)(0.82)}{0.365} = 0.7863$

SCORING

There are two probabilities to calculate in each Part (a)−(d). Each Part is essentially correct for both probabilities correctly calculated and is partially correct for one probability correctly calculated. In Parts that use results from previous parts, full credit is given for correctly using the results of the earlier Part, whether that earlier calculation was correct or not. For credit for Part (d), a correct methodology must also be shown.

Count partially correct answers as one-half an essentially correct answer.

4	**Complete Answer**	Four essentially correct answers.
3	**Substantial Answer**	Three essentially correct answers.
2	**Developing Answer**	Two essentially correct answers.
1	**Minimal Answer**	One essentially correct answer.

Use a holistic approach to decide a score totaling between two numbers.

---

2. (a) A control group would allow the school system to compare the effectiveness of each of the new programs to the local standard method currently being used.

   (b) Parents who fail to return the consent form are a special category who may well place less priority on education. The effect of using their children may distort results since their children could be placed only in the control group.

   (c) Assign each student a unique number 01−90. Using a random number table or a random number generator, pick numbers between 01 and 90, throwing out repeats. The students corresponding to the first 30 such numbers picked will be assigned to Program A, the next 30 picked to Program B, and the remaining to the control group.

SCORING

Part (a) is essentially correct if the purpose is given for using a control group in this study. Part (a) is partially correct if a correct explanation for the use of a control group is given but not in the context of this study.

Part (b) is essentially correct for a clear explanation in context. Part (b) is partially correct if the explanation is weak.

Part (c) is essentially correct if randomization is used correctly and the method is clear. Part (c) is partially correct if randomization is used but the method is not clearly explained.

4	**Complete Answer**	All three parts essentially correct.
3	**Substantial Answer**	Two parts essentially correct and one part partially correct.
2	**Developing Answer**	Two parts essentially correct OR one part essentially correct and one or two parts partially correct OR all three parts partially correct.
1	**Minimal Answer**	One part essentially correct OR two parts partially correct.

3. (a) $b = r\left(\frac{s_y}{s_x}\right) = 0.49\left(\frac{4.85}{2.24}\right) = 1.0609$. For each unit increase on the POL political interest scale, the expected increase in the ACT activism scale is 1.0609.

(b) The least squares regression line passes through $(\bar{x}, \bar{y})$. Thus, $a + 1.0609(5.48) = 18.92$, which gives $a = 13.11$, and the equation is $\widehat{ACT} = 13.11 + 1.0609(POL)$.

(c) $13.11 + 1.0609(4) = 17.35$. The residual = observed − predicted = $15 - 17.35 = -2.35$. Thus, the regression model overestimated the ACT activity score for this student by 2.35.

SCORING

Part (a) is essentially correct for a correct calculation of the slope, an interpretation of the slope in context, and the use of nondeterministic language (such as "expected" or "predicted"). Part (a) is partially correct for two of these three components.

Part (b) is essentially correct for a correct equation, using a "hat" for the dependent variable or the word "Predicted," and defining the variables. Part (b) is partially correct for two of these three components.

Part (c) is essentially correct for a correct calculation of the residual, referring to context in a correct interpretation, and referring to direction (such as an "overestimate"). Part (c) is partially correct for two of these three components.

4	**Complete Answer**	All three parts essentially correct.
3	**Substantial Answer**	Two parts essentially correct and one part partially correct.
2	**Developing Answer**	Two parts essentially correct OR one part essentially correct and one or two parts partially correct OR all three parts partially correct.
1	**Minimal Answer**	One part essentially correct OR two parts partially correct.

4. (a) A complete answer compares shape, center, and spread and also mentions context.

Shape: The baseball players (A), for which the cumulative frequency plot rises steeply at first, include more shorter players, and thus the distribution is skewed to the right (toward the greater heights). The football players (C), for which the cumulative frequency plot rises slowly at first and then steeply toward the end, include more taller players, and thus the distribution is skewed to the left (toward the lower heights). The basketball players (B), for which the cumulative frequency plot rises slowly at each end and steeply in the middle, have a more bell-shaped distribution of heights.

Center: The medians correspond to relative frequencies of 0.5. Reading across from 0.5 and then down to the *x*-axis shows the median heights to be about 63.5 inches for baseball players, about 72.5 inches for basketball players, and about 79 inches for football players. Thus, the center of the baseball height distribution is the least, and the center of the football height distribution is the greatest.

Spread: The range of the football players is the smallest, $80 - 65 = 15$ inches, then comes the range of the baseball players, $80 - 60 = 20$ inches, and finally the range of the basketball players is the greatest, $85 - 60 = 25$ inches.

(b) Seventy-one inches corresponds to a probability of 0.4.

(c) We have a geometric probability with $p = 0.4$. Then $P(X = 3) = (0.6)^2(0.4) = 0.144$.

SCORING

This question is scored in three sections. Section 1 consists of part (a1). Section 2 consists of part (a2). Section 3 consists of parts (b) and (c).

Section 1 is essentially correct for correctly identifying which distribution is skewed left, skewed right, and more bell-shaped and for giving a correct justification based on the cumulative frequency plots. Section 1 is partially correct for correctly identifying which distribution is skewed left, skewed right, and more bell-shaped without giving a good explanation.

Section 2 is essentially correct for (1) correctly noting that the baseball players have the lowest median height, (2) correctly noting that the football players have the greatest median height, (3) correctly noting that the football players have the smallest range for their heights, and (4) correctly noting that the basketball players have the greatest range for their heights. Section 2 is partially correct for two or three of these four components correct.

Section 3 is essentially correct for (1) the correct probability 0.4 in (b), (2) identification of a geometric probability in (c), and (3) giving a correct numerical derivation of 0.144 in (c), and is partially correct for two of these three components correct.

4	Complete Answer	All three sections essentially correct.
3	Substantial Answer	Two sections essentially correct and one section partially correct.
2	Developing Answer	Two sections essentially correct OR one section essentially correct and one or two sections partially correct OR all three sections partially correct.
1	Minimal Answer	One section essentially correct OR two sections partially correct.

Lower a 4 to a 3, or a 3 to a 2, if context is never mentioned in either Section 1 or Section 2.

5. (a) Random sampling is involved as two independent random samples of workers is chosen, a random sample of drill and blast workers and an independent random sample of concrete workers. There is no random assignment of workers to positions.

(b) *Parameters:* Let μ_{db} represent the mean cumulative NO_2 exposure for the population of drill and blast workers. Let μ_c represent the mean cumulative NO_2 exposure for the population of concrete workers.
Procedure: Two-sample t-interval for $\mu_{db} - \mu_c$.
Checks: We check *independence* and *normality*. For independence, we are given independent random samples, and 115 and 69 are assumed less than 10% of all "drill and blast" and "concrete" workers, respectively. For normality, the sample sizes, 115 and 69, are both ≥ 30 so the CLT applies.
Mechanics: Calculator software (such as 2-SampTInt on the TI-84 or 2-SampletInterval on the Casio Prizm) gives $(-1.362, -0.0381)$.
Conclusion in context: We are 95% confident that the true difference (drill and blast minus concrete) in the mean cumulative NO_2 exposure for all drill and blast workers and all concrete workers is between -1.362 and -0.0381 ppm/yr.

(c) Zero is not in the above 95% confidence interval, so at the $\alpha = 0.05$ significance level, there is evidence to reject $H_0: \mu_{db} - \mu_c = 0$ in favor of $H_a: \mu_{db} - \mu_c \neq 0$. That is, there is evidence of a difference in mean cumulative NO_2 exposure for drill and blast workers and for concrete workers.

(d) With sample sizes this large, the central limit theorem applies and our analysis is valid.

SCORING

This question is scored in three sections. Section 1 consists of parts (a) and (b1). Section 2 consists of part (b2). Section 3 consists of parts (c) and (d).

Section 1 is essentially correct for (1) stating that random sampling and not random assignment are involved, (2) defining parameters, (3) naming the procedure, (4) noting independent random samples, and (5) noting the large sample sizes. Section 1 is partially correct for 3 or 4 of these 5 components.

Section 2 is essentially correct for (1) the correct interval, and in the conclusion reference to: (2) 95% confidence, (3) the difference in means, (4) the population rather than the samples, and (5) context. Section 2 is partially correct for 3 or 4 of these 5 components.

Section 3 is essentially correct for (1) noting that zero is not in the interval so the observed difference is significant, (2) stating this in context of the problem, and (3) relating the central limit theorem to the samples being large. Section 3 is partially correct for 2 of these 3 components.

4	**Complete Answer**	All three sections essentially correct.
3	**Substantial Answer**	Two sections essentially correct and one section partially correct.
2	**Developing Answer**	Two sections essentially correct OR one section essentially correct and one or two sections partially correct OR all three sections partially correct.
1	**Minimal Answer**	One section essentially correct OR two sections partially correct.

Section II: Part B

6. (a)

```
            Men     Women
                  0 |
           9 9 7 3 | 1 | 9
             4 1 0 | 2 | 3 7
                 5 | 3 | 0 3 4 8
                 8 | 4 | 2 6
```

3|1|9 represents a man's
time of 13 seconds and a
woman's time of 19 seconds

(b) The distribution of men's balance times is skewed right, while the distribution of women's balance times is more unimodal and roughly symmetric; neither distribution appear to have outliers. The median women's time, 33 seconds, is greater than the median men's time, 20 seconds. The men's times exhibit more variability than the women's times (range of men's times is $48 - 13 = 35$, while range of women's times is $46 - 19 = 27$).

(c) With this small of a P-value, $0.0889 < 0.10$, there is convincing evidence that the mean balance times of men and women are different.

(d) The simulation was conducted under the null hypothesis H_0: $\sigma_M = \sigma_W$, which gives $\frac{\sigma_M}{\sigma_W} = 1$. So, it is reasonable that the simulated sample ratios $\frac{s_M}{s_W}$ will be centered around 1.

(e) In the original samples, $\frac{s_M}{s_W} = \frac{10.85}{8.76} = 1.24$. The value of 1.24 in the distribution of simulated ratios does not fall far into the tail; thus, it could have well occurred by random sampling. So, no, there is not convincing evidence of a difference in variability of balance times between men and women.

SCORING

This question is scored in three sections. Section 1 consists of parts (a) and (b). Section 2 consists of parts (c) and (d). Section 3 consists of part (e).

Section 1 is essentially correct in (a) and (b) for (1) a correctly drawn back-to-back stemplot, (2) a key for the stemplot, and a comparison addressing (3) shape, (4) noting that the center of the distribution of women's times is greater than that of the men's, and (5) noting that the spread of the men's distribution is greater than the women's. Section 1 is partially correct for three or four out of the five components above.

Section 2 is essentially correct for (1) in (c) for a correct conclusion in context linked to the P-value and is partially correct if just missing context and (2) in (d) for correctly stating that the population standard deviations are equal if the null hypothesis is true, and so the ratios of the simulated standard deviations should be centered around 1. Section 2 is partially correct for one of these two components correct OR for both correct conclusions but weak explanations.

Section 3 is essentially correct in (e) for a correct conclusion with a clear explanation and is partially correct for a correct conclusion with a weak explanation.

4	**Complete Answer**	All three sections essentially correct.
3	**Substantial Answer**	Two sections essentially correct and one section partially correct.
2	**Developing Answer**	Two sections essentially correct OR one section essentially correct and one or two sections partially correct OR all three sections partially correct.
1	**Minimal Answer**	One section essentially correct OR two sections partially correct.

ANSWER SHEET
Practice Test 2

1. Ⓐ Ⓑ Ⓒ Ⓓ Ⓔ 11. Ⓐ Ⓑ Ⓒ Ⓓ Ⓔ 21. Ⓐ Ⓑ Ⓒ Ⓓ Ⓔ 31. Ⓐ Ⓑ Ⓒ Ⓓ Ⓔ
2. Ⓐ Ⓑ Ⓒ Ⓓ Ⓔ 12. Ⓐ Ⓑ Ⓒ Ⓓ Ⓔ 22. Ⓐ Ⓑ Ⓒ Ⓓ Ⓔ 32. Ⓐ Ⓑ Ⓒ Ⓓ Ⓔ
3. Ⓐ Ⓑ Ⓒ Ⓓ Ⓔ 13. Ⓐ Ⓑ Ⓒ Ⓓ Ⓔ 23. Ⓐ Ⓑ Ⓒ Ⓓ Ⓔ 33. Ⓐ Ⓑ Ⓒ Ⓓ Ⓔ
4. Ⓐ Ⓑ Ⓒ Ⓓ Ⓔ 14. Ⓐ Ⓑ Ⓒ Ⓓ Ⓔ 24. Ⓐ Ⓑ Ⓒ Ⓓ Ⓔ 34. Ⓐ Ⓑ Ⓒ Ⓓ Ⓔ
5. Ⓐ Ⓑ Ⓒ Ⓓ Ⓔ 15. Ⓐ Ⓑ Ⓒ Ⓓ Ⓔ 25. Ⓐ Ⓑ Ⓒ Ⓓ Ⓔ 35. Ⓐ Ⓑ Ⓒ Ⓓ Ⓔ
6. Ⓐ Ⓑ Ⓒ Ⓓ Ⓔ 16. Ⓐ Ⓑ Ⓒ Ⓓ Ⓔ 26. Ⓐ Ⓑ Ⓒ Ⓓ Ⓔ 36. Ⓐ Ⓑ Ⓒ Ⓓ Ⓔ
7. Ⓐ Ⓑ Ⓒ Ⓓ Ⓔ 17. Ⓐ Ⓑ Ⓒ Ⓓ Ⓔ 27. Ⓐ Ⓑ Ⓒ Ⓓ Ⓔ 37. Ⓐ Ⓑ Ⓒ Ⓓ Ⓔ
8. Ⓐ Ⓑ Ⓒ Ⓓ Ⓔ 18. Ⓐ Ⓑ Ⓒ Ⓓ Ⓔ 28. Ⓐ Ⓑ Ⓒ Ⓓ Ⓔ 38. Ⓐ Ⓑ Ⓒ Ⓓ Ⓔ
9. Ⓐ Ⓑ Ⓒ Ⓓ Ⓔ 19. Ⓐ Ⓑ Ⓒ Ⓓ Ⓔ 29. Ⓐ Ⓑ Ⓒ Ⓓ Ⓔ 39. Ⓐ Ⓑ Ⓒ Ⓓ Ⓔ
10. Ⓐ Ⓑ Ⓒ Ⓓ Ⓔ 20. Ⓐ Ⓑ Ⓒ Ⓓ Ⓔ 30. Ⓐ Ⓑ Ⓒ Ⓓ Ⓔ 40. Ⓐ Ⓑ Ⓒ Ⓓ Ⓔ

Practice Test 2

Section I: Questions 1–40

SPEND 90 MINUTES ON THIS PART OF THE EXAM.

> **DIRECTIONS:** The questions or incomplete statements that follow are each followed by five suggested answers or completions. Choose the response that best answers the question or completes the statement.

1. Suppose that the regression line for a set of data, $\hat{y} = 3 + bx$, passes through the point $(2, 7)$. If $\bar{x}$ and $\bar{y}$ are the sample means of the x- and y-values, respectively, which of the following is equal to $\bar{y}$?

   (A) $\bar{x}$
   (B) $-2 + \bar{x}$
   (C) $3 + \bar{x}$
   (D) $3 + 2\bar{x}$
   (E) $3 + 3.5\bar{x}$

2. A study is made to determine whether more hours of academic studying leads to a greater number of points scored by basketball players. In surveying 50 basketball players, it is noted that the 25 who claim to study the most hours have a higher point average than the 25 who study less. Based on this study, the coach begins requiring the players to spend more time studying. Which of the following is a correct statement?

   (A) While this study may have its faults, it still does prove causation.
   (B) There could well be a confounding variable responsible for the seeming relationship.
   (C) While this is a controlled experiment, the conclusion of the coach is not justified.
   (D) To get the athletes to study more, it would be more meaningful to have them put in more practice time on the court to boost their point averages, as higher point averages seem to be associated with more study time.
   (E) No proper conclusion is possible without somehow introducing *blinding*.

3. In a social experiment, students were asked to take a selfie and post it to social media. Unknown to the students, what the study was really about was counting how many selfies a student would take before feeling satisfied to post one. The standard deviation for number of selfies taken before posting was 4.3, and 30% of numbers of selfies taken were over 15. Assuming an approximately normal distribution for numbers of selfies taken before posting, what was the mean number?

   (A) $15 - 0.30(4.3)$
   (B) $15 + 0.30(4.3)$
   (C) $15 - 0.4756(4.3)$
   (D) $15 + 0.5244(4.3)$
   (E) $15 - 0.5244(4.3)$

4. Which of the following is a correct statement about correlation?

   (A) If the slope of the regression line is exactly 1, the correlation is exactly 1.
   (B) If the correlation is 0, the slope of the regression line is undefined.
   (C) Switching which variable is called x and which is called y changes the sign of the correlation.
   (D) The correlation r is equal to the slope of the regression line when z-scores for the y-variable are plotted against z-scores for the x-variable.
   (E) Changes in the measurement units of the variables may change the correlation.

5. Which of the following are affected by outliers?

   I. Mean
  II. Median
 III. Standard deviation
 IV. Range
  V. Interquartile range

(A) I and II only
(B) III, IV, and V only
(C) I and V only
(D) III and IV only
(E) I, III, and IV only

6. An engineer wishes to determine the quantity of heat being generated by a particular electronic component. She knows that the standard deviation is 2.4 and wants to be 99% sure of knowing the mean quantity to within ±0.6. Which of the following should be used to find the sample size (n) needed?

(A) $1.645\sqrt{\dfrac{2.4}{n}} \le 0.6$

(B) $1.96\left(\dfrac{2.4}{\sqrt{n}}\right) \le 0.6$

(C) $2.576\left(\dfrac{2.4}{\sqrt{n}}\right) \le 0.6$

(D) $2.326\sqrt{\dfrac{2.4}{n}} \le 0.6$

(E) $2.576\sqrt{\dfrac{2.4}{n}} \le 0.6$

7. A company that produces facial tissues continually monitors tissue strength. If the mean strength from sample data drops below a specified level, the production process is halted and the machinery inspected. Which of the following would result from a Type I error?

(A) Halting the production process when sufficient customer complaints are received.
(B) Halting the production process when the tissue strength is below specifications.
(C) Halting the production process when the tissue strength is within specifications.
(D) Allowing the production process to continue when the tissue strength is below specifications.
(E) Allowing the production process to continue when the tissue strength is within specifications.

8. Two possible wordings for a questionnaire on a proposed school budget increase are as follows:

  I. This school district has one of the highest per student expenditure rates in the state. This has resulted in low failure rates, high standardized test scores, and most students going on to good colleges and universities. Do you support the proposed school budget increase?

 II. This school district has one of the highest per student expenditure rates in the state. This has resulted in high property taxes, with many people on fixed incomes having to give up their homes because they cannot pay the school tax. Do you support the proposed school budget increase?

One of these questions showed that 58% of the population favor the proposed school budget increase, while the other question showed that only 13% of the population support the proposed increase. Which produced which result and why?

(A) The first showed 58% and the second 13% because of the lack of randomization as evidenced by the wording of the questions.
(B) The first showed 13% and the second 58% because of a placebo effect due to the wording of the questions.
(C) The first showed 58% and the second 13% because of the lack of a control group.
(D) The first showed 13% and the second 58% because of response bias due to the wording of the questions.
(E) The first showed 58% and the second 13% because of response bias due to the wording of the questions.

9. Suppose that for a certain Caribbean island, in any 3-year period the probability of a major hurricane is 0.25, the probability of water damage is 0.44, and the probability of both a hurricane and water damage is 0.22. What is the probability of water damage given that there is a hurricane?

(A) $0.25 + 0.44 - 0.22$

(B) $\dfrac{0.22}{0.44}$

(C) $0.25 + 0.44$

(D) $\dfrac{0.22}{0.25}$

(E) $0.25 + 0.44 + 0.22$

10. A union spokesperson is trying to encourage a college faculty to join the union. She would like to argue that faculty salaries are not truly based on years of service as most faculty believe. She gathers data and notes the following scatterplot of salary versus years of service.

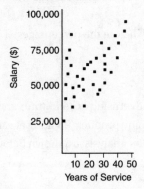

Which of the following *most* correctly interprets the overall scatterplot?

(A) The faculty member with the fewest years of service makes the lowest salary, and the faculty member with the most years of service makes the highest salary.

(B) A faculty member with more years of service than another has a higher salary than the other.

(C) There is a strong positive correlation with little deviation.

(D) There is no clear relationship between salary and years of service.

(E) While there is a strong positive correlation, there is a distinct deviation from the overall pattern for faculty with fewer than ten years of service.

11. Two random samples of students are chosen, the first set from those taking an AP Statistics class and the second set from those not taking an AP Statistics class. The following back-to-back stemplots compare the GPAs.

AP Statistics		No AP Statistics	
	1	89	Key: 1 | 8 means 1.8
97653	2	015688	
98775332110	3	133344777888	
1100	4		

Which of the following is true about the ranges and standard deviations?

(A) The first set has both a greater range and a greater standard deviation.

(B) The first set has a greater range, while the second has a greater standard deviation.

(C) The first set has a greater standard deviation, while the second has a greater range.

(D) The second set has both a greater range and a greater standard deviation.

(E) The two sets have equal ranges and equal standard deviations.

12. In a group of 10 scores, the largest score is increased by 40 points. What will happen to the mean?

(A) It will remain the same.

(B) It will increase by 4 points.

(C) It will increase by 10 points.

(D) It will increase by 40 points.

(E) There is not sufficient information to answer this question.

13. Suppose X and Y are random variables with $\mu_X = 32$, $\sigma_X = 5$, $\mu_Y = 44$, and $\sigma_Y = 12$. Given that X and Y are independent, what are the mean and standard deviation of the random variable $X + Y$?

(A) $\mu_{X+Y} = 76$, $\sigma_{X+Y} = 8.5$

(B) $\mu_{X+Y} = 76$, $\sigma_{X+Y} = 13$

(C) $\mu_{X+Y} = 76$, $\sigma_{X+Y} = 17$

(D) $\mu_{X+Y} = 38$, $\sigma_{X+Y} = 17$

(E) There is insufficient information to answer this question.

PRACTICE TEST 2

14. Suppose you toss a fair die three times and it comes up an even number each time. Which of the following is a true statement?

(A) By the law of large numbers, the next toss is more likely to be an odd number than another even number.

(B) Based on the properties of conditional probability, the next toss is more likely to be an even number given that three in a row have been even.

(C) Dice actually do have memories, and thus the number that comes up on the next toss will be influenced by the previous tosses.

(D) The law of large numbers tells how many tosses will be necessary before the percentages of evens and odds are again in balance.

(E) The probability that the next toss will again be even is 0.5.

15. A pharmaceutical company is interested in the association between advertising expenditures and sales for various over-the-counter products. A sales associate collects data on nine products, looking at sales (in $1,000) versus advertising expenditures (in $1,000). The results of the regression analysis are shown below.

```
Dependent variable: Sales

Variable        Coefficient      SE Coef          t-ratio          P
Constant         123.800          1.798            68.84            0.000
Advertising       12.633          0.378            33.44            0.000
R-Sq = 99.4%   R-Sq (adj) = 99.3%
s = 2.926 with 9-2 = 7 degrees of freedom
```

Which of the following gives a 90% confidence interval for the slope of the regression line?

(A) $12.633 \pm 1.415(0.378)$

(B) $12.633 \pm 1.895(0.378)$

(C) $123.800 \pm 1.414(1.798)$

(D) $123.800 \pm 1.895(1.798)$

(E) $123.800 \pm 1.645\left(\dfrac{1.798}{\sqrt{9}}\right)$

16. A school principal wanted to estimate the hours per month that parents spend volunteering at their children's school. The principal took repeated random samples of size 25 parents and estimated the mean and standard deviation of the sampling distribution to be 5.5 hours and 0.4 hours, respectively. Based on the estimates for the sampling distribution, which of the following provides the best estimates of the population parameters?

(A) Mean of 5.5 hours and standard deviation of 0.016 hours

(B) Mean of 5.5 hours and standard deviation of 0.08 hours

(C) Mean of 5.5 hours and standard deviation of 1 hour

(D) Mean of 5.5 hours and standard deviation of 2 hours

(E) Mean of 5.5 hours and standard deviation of 10 hours

17. Jonathan obtained a score of 80 on a statistics exam, placing him at the 90th percentile. Suppose 5 points are added to everyone's score. Jonathan's new score will be at the

(A) 80th percentile.

(B) 85th percentile.

(C) 90th percentile.

(D) 95th percentile.

(E) There is not sufficient information to answer this question.

18. To study the effect of music on piecework output at a clothing manufacturer, two experimental treatments are planned: day-long classical music for one group versus day-long light rock music for another. Which one of the following groups would serve best as a control for this study?

    (A) A third group for which no music is played
    (B) A third group that randomly hears either classical or light rock music each day
    (C) A third group that hears day-long R & B music
    (D) A third group that hears classical music every morning and light rock every afternoon
    (E) A third group in which each worker has earphones and chooses his or her own favorite music

19. Suppose H_0: $p = 0.6$ and the power of the test for the alternative $p = 0.7$ is 0.8. Which of the following is a valid conclusion?

    (A) The probability of committing a Type I error is 0.1.
    (B) If the alternative is true, the probability of failing to reject H_0 is 0.2.
    (C) The probability of committing a Type II error is 0.3.
    (D) All of the above are valid conclusions.
    (E) None of the above are valid conclusions.

20. The following is a histogram of the numbers of ties owned by bank executives.

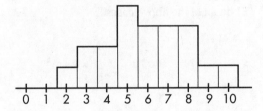

    Which of the following is a correct statement?

    (A) The median number of ties is five.
    (B) More than four executives own over eight ties each.
    (C) An executive is equally likely to own fewer than five ties or more than seven ties.
    (D) One tie is a reasonable estimate for the standard deviation.
    (E) Removing all the executives with three, nine, and ten ties may change the median.

21. A policewoman counts the number of motorists not wearing seat belts in a random sample of five drivers. Define the random variable of interest for the policewoman, and state how the random variable is distributed.

    (A) The random variable of interest is the proportion of motorists wearing seat belts in a random sample of five drivers. The distribution is binomial with parameters $n = 5$ and $p =$ probability a motorist is wearing a seat belt.
    (B) The random variable of interest is the number of motorists wearing seat belts in a random sample of five drivers. The distribution is binomial with parameters $n = 5$ and $p =$ probability a motorist is not wearing a seat belt.
    (C) The random variable of interest is the number of motorists not wearing seat belts in a random sample of five drivers. The distribution is binomial with parameters $n = 5$ and $p =$ probability a motorist is not wearing a seat belt.
    (D) The random variable of interest is the percent of motorists not wearing seat belts in a random sample of five drivers. The distribution is binomial with parameters $n = 5$ and $p =$ probability a motorist is wearing a seat belt.
    (E) The random variable of interest is the proportion of motorists not wearing seat belts in a random sample of five drivers. The distribution is binomial with parameters $n = 5$ and $p =$ probability a motorist is not wearing a seat belt.

22. Company I manufactures demolition fuses that burn an average of 50 minutes with a standard deviation of 10 minutes, while Company II advertises fuses that burn an average of 55 minutes with a standard deviation of 5 minutes. Which company's fuse is more likely to last at least 1 hour? Assume roughly normal distributions of fuse times.

    (A) Company I's, because of its greater standard deviation
    (B) Company II's, because of its greater mean
    (C) For both companies, the probability that a fuse will last at least 1 hour is 0.159.
    (D) For both companies, the probability that a fuse will last at least 1 hour is 0.841.
    (E) The problem cannot be solved from the information given.

23. Which of the following is *not* important in the design of experiments?

    (A) Control of confounding variables
    (B) Randomization in assigning subjects to different treatments
    (C) Use of a confounding variable to control the placebo effect
    (D) Replication of the experiment using sufficient numbers of subjects
    (E) All of the above are important in the design of experiments.

24. The travel miles claimed in weekly expense reports of the sales personnel at a corporation are summarized in the following boxplot.

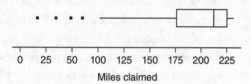

Miles claimed

    Which of the following is the most reasonable conclusion?

    (A) The mean and median numbers of travel miles are roughly equal.
    (B) The mean number of travel miles is probably greater than the median number.
    (C) Most of the claimed numbers of travel miles are in the [0, 200] interval.
    (D) Most of the claimed numbers of travel miles are in the [200, 240] interval.
    (E) The left and right whiskers contain the same number of values from the set of personnel travel mile claims.

25. For the least squares regression line, which of the following statements about residuals is true?

    (A) Influential scores have large residuals.
    (B) If the linear model is good, the number of positive residuals will be the same as the number of negative residuals.
    (C) The mean of the residuals is always zero.
    (D) If the correlation is 0, there will be a distinct pattern in the residual plot.
    (E) If the correlation is 1, there will not be a distinct pattern in the residual plot.

26. Four pairs of data are used in determining a regression line $\hat{y} = 3x + 4$. If the four values of the independent variable are 32, 24, 29, and 27, respectively, what is the mean of the four values of the dependent variable?

    (A) 68
    (B) 84
    (C) 88
    (D) 100
    (E) The mean cannot be determined from the given information.

27. According to one poll, 12% of the public favor legalizing all drugs. In a simple random sample of six people, what is the probability that at least one person favors legalization?

    (A) $6(0.12)(0.88)^5$
    (B) $(0.88)^6$
    (C) $1 - (0.88)^6$
    (D) $1 - 6(0.12)(0.88)^5$
    (E) $6(0.12)(0.88)^5 + (0.88)^6$

28. Sampling error (sampling variability) occurs

    (A) when interviewers make mistakes resulting in bias.
    (B) because a sample statistic is used to estimate a population parameter.
    (C) when interviewers use judgment instead of random choice in picking the sample.
    (D) when samples are too small.
    (E) in all of the above cases.

29. A telecommunications executive instructs an associate to contact 104 customers using their service to obtain their opinions in regard to an idea for a new pricing package. The associate notes the number of customers whose names begin with *A* and uses a random number table to pick four of these names. She then proceeds to use the same procedure for each letter of the alphabet and combines the $4 \times 26 = 104$ results into a group to be contacted. Which of the following is a correct conclusion?

(A) Her procedure makes use of chance.

(B) Her procedure results in a simple random sample.

(C) Each customer has an equal probability of being included in the survey.

(D) Her procedure introduces bias through *sampling error.*

(E) With this small a sample size, it is better to let the surveyor use the company's data banks to pick representative customers to be surveyed based on features such as gender, political affiliation, income level, race/ethnicity, age, and so on.

30. The graph below shows cumulative proportions plotted against GPAs for high school seniors.

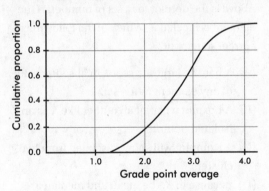

What is the approximate interquartile range?

(A) 0.85

(B) 2.25

(C) 2.7

(D) 2.75

(E) 3.1

31. PCB (polychlorinated biphenyl) contamination of a river by a manufacturer is being measured by amounts of the pollutant found in fish. A company scientist claims that the fish contain only 5 parts per million, but an investigator believes the figure is higher. The investigator catches six fish that show the following amounts of PCB (in parts per million): 6.8, 5.6, 5.2, 4.7, 6.3, and 5.4. In performing a hypothesis test with $H_0: \mu = 5$ and $H_a: \mu > 5$, what is the test statistic?

(A) $t = \dfrac{5.67 - 5}{0.763}$

(B) $t = \dfrac{5.67 - 5}{\sqrt{\dfrac{0.763}{5}}}$

(C) $t = \dfrac{5.67 - 5}{\sqrt{\dfrac{0.763}{6}}}$

(D) $t = \dfrac{5.67 - 5}{\left(\dfrac{0.763}{\sqrt{5}}\right)}$

(E) $t = \dfrac{5.67 - 5}{\left(\dfrac{0.763}{\sqrt{6}}\right)}$

32. Consider two sets, *A* and *B*, both roughly normally distributed with $n = 1,000$. Set *A* has mean 10 and standard deviation 2, while set *B* has mean 20 and standard deviation 1. Define two new sets, *X* and *Y*, where *X* is formed by combining all the values from *X* and *Y* into a larger set with $n = 2,000$, while *Y* is formed by adding each value in *A* to each value in *B* to form a set with $n = 1,000 \times 1,000$. Which of the following best describes the distributions of *X* and *Y*?

(A) Both are roughly normal with identical means.

(B) Both are roughly normal, but one has a mean twice the other.

(C) One is roughly normal, the other is bimodal, and they have identical means.

(D) One is roughly normal, the other is bimodal, and one has a mean twice the other.

(E) Both are bimodal with identical means.

33. In general, how does tripling the sample size change the confidence interval width?

    (A) It triples the interval width.
    (B) It divides the interval width by 3.
    (C) It multiples the interval width by 1.732.
    (D) It divides the interval width by 1.732.
    (E) This question cannot be answered without knowing the sample size.

34. Which of the following statements is *false*?

    (A) Like the normal distribution, the *t*-distributions are symmetric.
    (B) The *t*-distributions are lower at the mean and higher at the tails and so are more spread out than the normal distribution.
    (C) The greater the *df*, the closer the *t*-distributions are to the normal distribution.
    (D) The smaller the *df*, the better the 68-95-99.7 rule works for *t*-models.
    (E) The areas under all *t*-distribution curves are 1.

35. Consider the following three histograms:

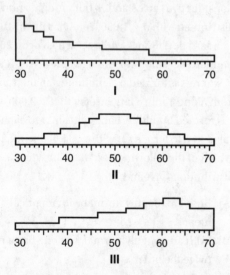

I

II

III

    Which of the following correctly orders the histograms from the one with the smallest proportion of data above its mean to the one with the largest proportion of data above its mean?

    (A) I, II, III
    (B) I, III, II
    (C) III, II, I
    (D) II, III, I
    (E) All three histograms have the same proportion of data above their respective means.

36. The sampling distribution of the sample mean is close to the normal distribution

    (A) only if both the original population has a normal distribution and *n* is large.
    (B) if the standard deviation of the original population is known.
    (C) if *n* is large, no matter what the distribution of the original population.
    (D) no matter what the value of *n* or what the distribution of the original population.
    (E) only if the original population is not badly skewed and does not have outliers.

37. What is the probability of a Type II error when a hypothesis test is being conducted at the 10% significance level ($\alpha = 0.10$)?

    (A) 0.05
    (B) 0.10
    (C) 0.90
    (D) 0.95
    (E) There is insufficient information to answer this question.

38.

Above is the dotplot for a set of numbers. One element is labeled *X*. Which of the following is a correct statement?

    (A) *X* has the largest *z*-score, in absolute value, of any element in the set.
    (B) A boxplot will plot an outlier like *X* as an isolated point.
    (C) A stemplot will show *X* isolated from two clusters.
    (D) Because of *X*, the mean and median are different.
    (E) The interquartile range (IRQ) is exactly half the range.

39. A 2019 survey of 500 adults concluded that 82% of adults support increasing the minimum wage by $5. Which of the following best describes what is meant by the poll having a margin of error of $\pm 3\%$?

(A) Three percent of those surveyed refused to participate in the poll.

(B) It would not be unexpected for 3% of adults to begin supporting an increase of the minimum wage by $5 or to stop supporting an increase of the minimum wage by $5.

(C) Between 395 and 425 of the adults surveyed responded that they support increasing the minimum wage by $5.

(D) If a similar survey of 500 adults were taken weekly, a 3% change in each week's results would not be unexpected.

(E) It is likely that between 79% and 85% of all adults support increasing the minimum wage by $5.

40.

| | College Plans | |
	Public	Private
Taking AP Statistics:	18	27
Not taking AP Statistics:	26	40

The above two-way table summarizes the results of a survey of a random sample of high school seniors conducted to determine if there is a relationship between whether or not a student is taking AP Statistics and whether the student plans to attend a public or a private college after graduation. Which of the following is the most reasonable conclusion about the relationship between taking AP Statistics and the type of college a student plans to attend?

(A) There appears to be no association since the proportion of AP Statistics students planning to attend public schools is almost identical to the proportion of students not taking AP Statistics who plan to attend public schools.

(B) There appears to be an association since the proportion of AP Statistics students planning to attend public schools is almost identical to the proportion of students not taking AP Statistics who plan to attend public schools.

(C) There appears to be an association since more students plan to attend private than public schools.

(D) There appears to be an association since fewer students are taking AP Statistics than are not taking AP Statistics.

(E) These data do not address the question of association.

STOP If there is still time remaining, you may check your work on this section.

PRACTICE TEST 2

Section II: Part A

Questions 1–5

SPEND ABOUT 65 MINUTES ON THIS PART OF THE EXAM.
PERCENTAGE OF SECTION II GRADE—75

> **DIRECTIONS:** You must show all work and indicate the methods you use. You will be graded on the correctness of your methods and on the accuracy of your results and explanations.

1. Ten volunteer male subjects are to be used for an experiment to study four substances (aloe, camphor, eucalyptus oil, and benzocaine) and a placebo with regard to itching relief. Itching is induced on a forearm with an itch stimulus (cowage), a drug is topically administered, and the duration of itching is recorded.

   (a) If the experiment is to be done in one sitting, how would you assign treatments for a completely randomized design?
   (b) If 5 days are set aside for the experiment, with one sitting a day, how would you assign treatments for a randomized block design where the subjects are the blocks?
   (c) What limits are there on generalization of any results?

2. A comprehensive study of more than 3,000 baseball games results in the graph below showing relative frequencies of runs scored by home teams.

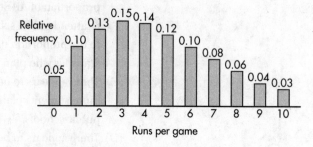

   (a) Calculate the mean and the median.
   (b) Between the mean and the median, is the one that is greater what was to be expected? Explain.
   (c) What is the probability that in 4 randomly selected games, the home team is shut out (scores no runs) in at least one of the games?
   (d) If X is the random variable for the runs scored per game by home teams, its standard deviation is 2.578. Suppose 200 games are selected at random, and $\bar{x}$, the mean number of runs scored by the home teams, is calculated. Describe the sampling distribution of $\bar{x}$.

3. A study is performed to explore the relationship, if any, between 24-hour urinary metabolite 3-methoxy-4-hydroxyphenylglycol (MHPG) levels and depression in bipolar patients. The MHPG level is measured in micrograms per 24 hours while the manic-depression (MD) scale used goes from 0 (manic delirium), through 5 (euthymic), up to 10 (depressive stupor). A partial computer printout of regression analysis with MHPG as the independent variable follows:

```
Average MHPG = 1243.1 with SD = 384.9
Average MD = 5.4 with SD = 2.875
95% confidence interval for slope b = (-0.0096, -0.0016)
```

(a) Calculate the slope of the regression line, and interpret it in the context of this problem.

(b) Find the equation of the regression line.

(c) Calculate and interpret the value of r^2 in the context of this problem.

(d) What does the correlation say about causation in the context of this problem?

4. An experiment is run to test whether daily stimulation of specific reflexes in young infants will lead to earlier walking. Twenty infants were recruited through a pediatrician's service and were randomly split into two groups of ten. One group received the daily stimulation, while the other was considered a control group. The ages (in months) at which the infants first walked alone were recorded.

<u>Ages in months at which first steps alone were taken</u>

With stimulation: 10, 12, 11, 10.5, 11, 11.5, 11.5, 11, 12, 11.5   Mean = 11.2, SD = 0.63246

Control: 10, 13, 12, 11, 11.5, 11.5, 12.5, 12, 11.5, 11   Mean = 11.6, SD = 0.84327

(a) Is random sampling, random assignment, both, or neither involved in this study?

(b) Is there statistical evidence that infants walk earlier with daily stimulation of specific reflexes?

(c) To what population, if any, can the conclusion be generalized?

5. Suppose that the volumes of liquid dispensed into a soft drink bottle follow a roughly normal distribution with a mean slightly above the advertised volume. Suppose further that the manufacturer wants to make a change that will result in fewer bottles weighing less than the advertised number of ounces.

(a) If they change only the advertised volume, should they make the advertised volume smaller or larger? Explain.

(b) If they change only the mean volume of liquid dispensed, should they make the mean smaller or larger? Explain.

(c) If they change only the standard deviation in volume of liquid dispensed, should they make the standard deviation smaller or larger? Explain.

Section II: Part B
Question 6

SPEND ABOUT 25 MINUTES ON THIS PART OF THE EXAM.
PERCENTAGE OF SECTION II GRADE—25

6. A candy manufacturer sells boxes advertised to contain 30 packs of candy per box. Occasionally a packing machine malfunctions and begins putting fewer than 30 packs in a box. A quality control inspector wants to determine whether or not a machine is functioning properly without opening up boxes. The inspector will weigh a box to test the following hypotheses.

H_0: The machine is putting 30 packs of candy in each box.

H_a: The machine is putting fewer than 30 packs of candy in each box.

(a) Describe what a Type I error would be, and describe a consequence of making a Type I error.
(b) Describe what a Type II error would be, and describe a consequence of making a Type II error.

The distribution of weights of packs of candy is approximately normal with a mean of 2 ounces and a standard deviation of 0.05 ounce. Assume the packs in a box are a random sample of packs produced.

Let the random variable C represent the total weight of a box of packs of candy, and assume the weight of the packaging is negligible.

(c) Describe the distribution of C if the selected box contains 30 packs.

The inspector decides to use the following rule. If the total weight of a box of packs of candy is 59 ounces or less, he will conclude the box contains fewer than 30 packs; otherwise he will conclude the box contains 30 packs.

(d) Suppose a box actually contains 30 packs. What is the probability the inspector mistakenly concludes that the box contains fewer than 30 packs?
(e) Suppose a box actually contains 29 packs. What is the probability the inspector mistakenly concludes that the box contains 30 packs?
(f) Based on your answers to (d) and (e), comment of the effectiveness of the inspector's rule choice.

PRACTICE TEST 2

Answer Explanations

Section I

1. **(D)** Since (2, 7) is on the line, $\hat{y} = 3 + bx$, we have $7 = 3 + 2b$ and $b = 2$. Thus, the regression line is $\hat{y} = 3 + 2x$. The point $(\overline{x}, \overline{y})$ is always on the regression line, so we have $\overline{y} = 3 + 2\overline{x}$.

2. **(B)** It could well be that conscientious students are the same ones who both study and do well on the basketball court. If students could be randomly assigned to study or not study, the results would be more meaningful. Of course, ethical considerations might make it impossible to isolate the confounding variable in this way.

3. **(E)** `invNorm(0.7)` $= 0.5244$ so $15 - \mu = 0.5244(4.3)$, which gives $\mu = 15 - 0.5244(4.3)$.

4. **(D)** The slope of the regression line and the correlation are related by $b = r\dfrac{S_y}{S_x}$. When using z-scores, the standard deviations s_x and s_y are 1. If $r = 0$, then $b = 0$. Switching which variable is x and which is y, or changing units, will not change the correlation.

5. **(E)** The median and interquartile range are specifically used when outliers are suspected of unduly influencing the mean, range, or standard deviation.

6. **(C)** `invNorm(0.995)` $= 2.576$ is the critical z-score for a 99% confidence interval, and $\sigma_{\overline{x}} = \dfrac{\sigma}{\sqrt{n}} = \dfrac{2.4}{\sqrt{n}}$.

7. **(C)** This is a hypothesis test with H_0: tissue strength is within specifications, and H_a: tissue strength is below specifications. A Type I error is committed when a true null hypothesis is mistakenly rejected.

8. **(E)** The wording of questions can lead to response bias. The neutral way of asking this question would simply have been, "Do you support the proposed school budget increase?"

9. **(D)**

$$P(\text{water} \mid \text{hurricane}) = \frac{P(\text{water} \cap \text{hurricane})}{P(\text{hurricane})} = \frac{0.22}{0.25}$$

10. **(E)** Although it is important to look for basic patterns, it is also important to look for deviations from these patterns. In this case, there is an overall positive correlation; however, those faculty with under ten years of service show little relationship between years of service and salary. Although (A) is a true statement, it does not give an overall interpretation of the scatterplot.

11. **(D)** The second set has a greater range, $3.8 - 1.8 = 2.0$ as compared with $4.1 - 2.3 = 1.8$, and with its skewness it also has a greater standard deviation.

12. **(B)** With $n = 10$, increasing $\sum x$ by 40 increases $\dfrac{\sum x}{n}$ by $\dfrac{40}{10} = 4$.

13. **(B)** Means can always be added, and in this example because of independence, variances can also be added. Thus, the new variance is $5^2 + 12^2 = 169$, and the new standard deviation is $\sqrt{169} = 13$.

14. **(E)** Dice have no memory, so the probability that the next toss will be an even number is 0.5 and the probability that it will be an odd number is 0.5. The law of large numbers says that as the number of tosses becomes larger, the proportion of even numbers tends to become closer to 0.5.

15. **(B)** The sample slope is 12.633, the standard error of the slope is 0.378, and the critical t-scores for 90% confidence with $df = 7$ are $\pm$`invT(0.95, 7)` $= \pm 1.895$. The confidence interval is $b \pm t^* SE(b) = 12.633 \pm (1.895)(0.378)$.

16. **(D)** The mean of the sampling distribution is equal to the population mean. The standard deviation of the sampling distribution is the population standard deviation divided by the square root of the sample size. In this case, $\sigma_{\overline{x}} = \dfrac{\sigma}{\sqrt{n}}$ gives $\dfrac{\sigma}{\sqrt{25}} = 0.4$, and thus the population standard deviation equals $\sigma = (0.4)\sqrt{25} = 2$.

17. **(C)** Percentile ranking is a measure of relative position. Adding 5 points to everyone's score will not change the relative positions.

18. **(A)** The control group should have experiences identical to those of the experimental groups except for the treatment under examination. They should not be given a new treatment.

19. **(B)** If the alternative $p = 0.7$ is true, the probability of failing to reject H_0 and thus committing a Type II error is 1 minus the power, that is, $1 - 0.8 = 0.2$.

20. **(C)** In histograms, relative area corresponds to relative frequency. The area from 1.5 to 4.5 (fewer than 5) appears to be the same as the area between 7.5 and 10.5 (more than 7). Five does not split the area in half, so 5 is not the median. Histograms such as these show relative frequencies, not actual frequencies. Given the spread, 1 is too small an estimate of the standard deviation. The area above 3 looks to be the same as the area above 9 and 10, so the median won't change.

21. **(C)** The policewoman is *counting*, not calculating a *proportion*. More specifically, she is counting the number of motorists *not* wearing seat belts rather than the number wearing seat belts. It seems reasonable to assume that for any given driver, the probability of not wearing a seat belt is the same and is independent from driver to driver. There is a fixed number, $n = 5$, of repetitions, and $p =$ probability a motorist is not wearing a seat belt.

22. **(C)** In both cases, 1 hour is one standard deviation from the mean with a right tail probability of 0.159.

23. **(C)** Control, randomization, and replication are all important aspects of well-designed experiments. We try to control confounding variables, not to use them to control something else.

24. **(D)** The median appears to be roughly 215, indicating that the interval [200, 240] probably has more than 50% of the values. While the shape of a distribution is difficult to discern from a boxplot, the data do appear to be skewed to the left, indicating that the mean is less than the median. While in a boxplot without outliers each whisker contains 25% of the values, this boxplot shows four outliers on the left, and so the left whisker has four fewer values than the right whisker.

25. **(C)** For the least squares regression line, the sum and thus the mean of the residuals are always zero. An influential score may have a small residual but still have a great effect on the regression line. If the correlation is 1, all the residuals would be 0, resulting in a very distinct pattern.

26. **(C)** $\bar{x} = \frac{32 + 24 + 29 + 27}{4} = 28$. Since $(\bar{x}, \bar{y})$ is a point on the regression line, $\bar{y} = 3(28) + 4 = 88$.

27. **(C)** $P(\text{at least } 1) = 1 - P(\text{none}) = 1 - (0.88)^6$.

28. **(B)** Different samples give different sample statistics, all of which are estimates for the same population parameter, and so error, called *sampling error* (also called *sampling variability*), is naturally present.

29. **(A)** While the associate does use chance, each customer would have the same chance of being selected only if the same number of customers had names starting with each letter of the alphabet. This selection does not result in a simple random sample because each possible set of 104 customers does not have the same chance of being picked as part of the sample. For example, a group of customers whose names all start with A will not be chosen. Sampling error (sampling variability), the natural variation inherent in a survey, is always present and is not a source of bias. Letting the surveyor have free choice in selecting the sample, rather than incorporating chance in the selection process, is a recipe for disaster!

30. **(A)** Corresponding to cumulative proportions of 0.25 and 0.75 are $Q_1 = 2.25$ and $Q_3 = 3.1$, respectively, and so the interquartile range is $3.1 - 2.25 = 0.85$.

31. **(E)** By putting the data in a list, a calculator gives $\bar{x} = 5.67$ and $s = 0.763$. Then $SE(\bar{x}) = \frac{s}{\sqrt{n}} = \frac{0.763}{\sqrt{6}}$ and the test statistic is $t = \frac{\bar{x} - x_0}{SE(\bar{x})} = \frac{5.67 - 5}{\left(\frac{0.763}{\sqrt{6}}\right)}$.

32. **(D)** The set $X = A \cup B$ will be bimodal with modes at the values 10 and 20. Since both A and B are of the same size, the mean of X will be $\frac{(10 + 20)}{2} = 15$. $Y = A + B$ will be roughly normal because both A and B are roughly normal. The mean of Y will be $10 + 20 = 30$, the sum of the means of A and B.

33. **(D)** Increasing the sample size by a multiple of d divides the interval width by $\sqrt{d}$.

34. **(D)** The t-distributions are symmetric; however, they are lower at the mean and higher at the tails and so are more spread out than the normal distribution. The greater the df, the closer the t-distributions are to the normal distribution. The 68-95-99.7 rule applies to the z-distribution and will work for t-models with very large df. All probability density curves have an area of 1 below them.

35. **(A)** Right-skewed distributions, as in I, generally have means greater than their medians, so less than 50% of the data are greater than the mean. Symmetric distributions, as in II, have means equal to their medians, so 50% of the data are greater than the mean. Left-skewed distributions, as in III, generally have means less than their medians, so more than 50% of the data are greater than the mean.

36. **(C)** This follows from the central limit theorem.

37. **(E)** There is a different Type II error for each possible correct value for the population parameter.

38. **(C)** X and the two clusters will be clearly visible in a stemplot of these data. X is close to the mean and so will have a z-score close to 0. While boxplots do show outliers as isolated points, outliers are from the mean, so X is not an outlier. In symmetric distributions, the mean and median are equal. The IQR here is close to the range.

39. **(E)** When using a measurement from a sample, we are never able to say *exactly* what a population proportion is; rather, we always say we have a certain *confidence* that the population proportion lies in a particular *interval*. In this case, that interval is 82% $\pm$ 3% or between 79% and 85%.

40. **(A)** Whether or not students are taking AP Statistics seems to have no relationship to which type of school they are planning to go to. Chi-square is close to 0.

Section II: Part A

1. (a) Number the volunteers 1 through 10. Use a random number generator to pick numbers between 1 and 10, throwing out repeats. The volunteers corresponding to the first two numbers chosen will receive aloe, the next two will receive camphor, the next two eucalyptus oil, the next two benzocaine, and the remaining two a placebo.

   (b) Each volunteer (the volunteers are "blocks") should receive all five treatments, one a day, with the time order randomized. For example, label aloe 1, camphor 2, eucalyptus oil 3, benzocaine 4, and the placebo 5. Then, for each volunteer, use a random number generator to pick numbers between 1 and 5, throwing away repeats. The order picked gives the day on which each volunteer receives each treatment.

   (c) Results cannot be generalized to women.

SCORING

Part (a) is essentially correct for giving a procedure that randomly assigns two volunteers to each of the five treatments. Part (a) is partially correct for giving a procedure that randomly assigns one of the treatments for each volunteer but may not result in *two* volunteers receiving each treatment.

Part (b) is essentially correct for giving a procedure assigns a random order for each of the volunteers to have all five treatments. Part (b) is partially correct for having each volunteer take all five treatments, one a day, but not clearly randomizing the time order.

Part (c) is essentially correct for stating that the results cannot be generalized to women and is incorrect otherwise.

4	Complete Answer	All three parts essentially correct.
3	Substantial Answer	Two parts essentially correct and one part partially correct.
2	Developing Answer	Two parts essentially correct OR one part essentially correct and one or two parts partially correct OR all three parts partially correct.
1	Minimal Answer	One part essentially correct OR two parts partially correct.

2. (a) $\bar{x} = \sum xp(x) = 0(0.05) + 1(0.10) + 2(0.13) + 3(0.15) + 4(0.14) + 5(0.12) + 6(0.10) + 7(0.08) + 8(0.06) + 9(0.04) + 10(0.03) = 4.27$.

With $(0.05 + 0.10 + 0.13 + 0.15) = 0.43$ below 4 runs and with $(0.12 + 0.10 + 0.08 + 0.06 + 0.04 + 0.03) = 0.43$ above 4 runs, the median must be 4.

(b) The mean is greater than the median, as was to be expected because the distribution is skewed to the right.

(c) This is a binomial with $n = 4$ and $p = 0.05$.

$P(\text{at least one shutout in 4 games}) = 1 - P(\text{no shutouts in the 4 games})$
$$= 1 - (1 - 0.05)^4 = 1 - (0.95)^4 = 0.1855$$

(d) The distribution of $\bar{x}$ is approximately normal with mean $\mu_{\bar{x}} = 4.27$ (from above) and standard deviation $\sigma_{\bar{x}} = \dfrac{2.578}{\sqrt{200}} = 0.1823$.

SCORING

This question is scored in three sections. Section 1 consists of parts (a) and (b). Section 2 consists of part (c). Section 3 consists of part (d).

Section 1 is essentially correct for correctly calculating the mean, correctly calculating the median, noting that the mean is greater than the median, and relating this to the skew. Section 1 is partially correct for two or three of these four elements.

Section 2 in Part (c) is essentially correct for recognizing this as a binomial probability calculation and making the correct calculation. Section 2 is partially correct for recognizing this as a binomial probability calculation but with an error such as $1 - (0.05)^4$ or $4(0.05)(0.95)^3$.

Section 3 in Part (d) is essentially correct for "approximately normal," $\mu_{\bar{x}} = 4.27$ and $\sigma_{\bar{x}} = 0.1823$. Section 3 is partially correct for two of these three answers.

4	**Complete Answer**	All three sections essentially correct.
3	**Substantial Answer**	Two sections essentially correct and one section partially correct.
2	**Developing Answer**	Two sections essentially correct OR one section essentially correct and one or two sections partially correct OR all three sections partially correct.
1	**Minimal Answer**	One section essentially correct OR two sections partially correct.

3. (a) The slope b is in the center of the confidence interval, so $b = \dfrac{-0.0096 - 0.0016}{2} = -0.0056$. In context, 0.0056 *estimates* the *average* decrease in the manic-depression scale score for each 1-microgram increase in the level of urinary MHPG. (Thus, high levels of MHPG are associated with increased mania, and conversely, low levels of MHPG are associated with increased depression.)

(b) Recalling that the regression line goes through the point $(\bar{x}, \bar{y})$ or using the AP exam formula page, $a = \bar{y} - b\bar{x} = 5.4 - (-0.0056)(1243.1) = 12.36$, and thus the equation of the regression line is $\widehat{MD} = 12.36 - 0.0056(MHPG)$ (or $\hat{y} = 12.36 - 0.0056x$, where x is the level of urinary MHPG in micrograms per 24 hours and y is the score on a 0–10 manic-depressive scale).

(c) Recalling that on the regression line, each one SD increase in the independent variable corresponds to an increase of r SD in the dependent variable (or from the AP exam formula page, $b = r\dfrac{S_y}{S_x}$),

$$r = b\frac{S_x}{S_y} = -0.0056\left(\frac{384.9}{2.875}\right) = -0.750$$

and $r^2 = 56.2\%$. Thus, 56.2% of the variation in the manic-depression scale level is accounted for by the linear regression model with urinary MHPG levels as the explanatory variable.

(d) Correlation never proves causation. It could be that depression causes biochemical changes, leading to low levels of urinary MHPG, or it could be that low levels of urinary MHPG cause depression, or it could be that some other variable (a confounding variable) simultaneously affects both urinary MHPG levels and depression.

SCORING

Part (a) is essentially correct if the slope is correctly calculated and correctly interpreted in context. Part (a) is partially correct if the slope is not correctly calculated but a correct interpretation is given using the incorrect value for the slope.

Part (b) is essentially correct if the regression equation is correctly calculated (using the slope found in Part (a)) and it is clear what the variables stand for. Part (b) is partially correct if the correct equation is found (using the slope found in Part (a)) but it is unclear what the variables stand for.

Part (c) is essentially correct if the coefficient of determination, r^2, is correctly calculated and correctly interpreted in context. Part (c) is partially correct if r^2 is not correctly calculated but a correct interpretation is given using the incorrect value for r^2.

Part (d) is essentially correct for noting that correlation never proves causation and referring to context. Part (d) is partially correct for a correct statement about correlation and causation but with no reference to context.

Count partially correct answers as one-half an essentially correct answer.

4	**Complete Answer**	Four essentially correct answers.
3	**Substantial Answer**	Three essentially correct answers.
2	**Developing Answer**	Two essentially correct answers.
1	**Minimal Answer**	One essentially correct answer.

Use a holistic approach to decide a score totaling between two numbers.

4. (a) There was random assignment into two groups (one group given daily stimulation and the other group acting as a control); however, there was no random sampling as volunteers were used.

   (b) *Hypotheses:* $H_0 : \mu_1 - \mu_2 = 0$ and $H_a : \mu_1 - \mu_2 < 0$

   where μ_1 = true mean number of months at which first steps alone are taken for infants receiving daily stimulation, and μ_2 = true mean number of months at which first steps are taken for infants not receiving daily stimulation. (Other notations are also possible, for example, $H_0 : \mu_1 = \mu_2$ and $H_a : \mu_1 < \mu_2$ or $H_0 : \mu_{\text{stimulation}} = \mu_{\text{control}}$ and $H_a : \mu_{\text{stimulation}} < \mu_{\text{control}}$.)

   Procedure: Two-sample *t*-test.

   Checks: We check *independence* and *normality*. For independence, we are given that the infants were randomly assigned to the two groups, and the two groups are independent. For normality, dotplots of the two groups show no outliers, no strong skew, and are roughly bell-shaped. So it is not unreasonable to assume the data come from a roughly normal distribution.

   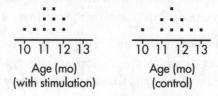

   Age (mo)   Age (mo)
   (with stimulation)   (control)

   Mechanics and *conclusion in context with linkage to the P-value:* Calculator software gives $t = -1.20$ and $P = 0.1234$. With this large of a *P*-value, $0.1234 > 0.05$, there is not sufficient evidence to reject H_0; that is, there is not convincing evidence that infants walk earlier with daily stimulation of specific reflexes. (There is not convincing evidence that the true mean number of months at which first steps alone are taken is less for infants receiving daily stimulation than for infants not receiving daily stimulation.)

   (c) No, the conclusion cannot be generalized to any group, not even to this pediatrician's service, because the original sample consists of volunteers who were not randomly selected from a population.

SCORING

This question is scored in three sections. Section 1 consists of parts (a) and (b1). Section 2 consists of part (b2). Section 3 consists of parts (b3) and (c).

Section 1 is essentially correct for (1) correctly answering that random assignment and not random sampling were used, (2) identifying the parameters, (3) stating the hypotheses, and (4) identifying the procedure, and is partially correct for 3 of these 4 components.

Section 2 is essentially correct for (1) mentioning random assignment, (2) noting that the two sample distributions are roughly bell-shaped, (3) finding the test statistic *t*, and (4) finding the *P*-value, and is partially correct for 3 of these 4 components.

Section 3 is essentially correct for (1) a correct conclusion, (2) linkage to the *P*-value, (3) answering in context, and (4) stating that this conclusion cannot be generalized because the sample consisted of volunteers, not a random sample, and is partially correct for 3 of these 4 components.

4	**Complete Answer**	All three sections essentially correct.
3	**Substantial Answer**	Two sections essentially correct and one section partially correct.
2	**Developing Answer**	Two sections essentially correct OR one section essentially correct and one or two sections partially correct OR all three sections partially correct.
1	**Minimal Answer**	One section essentially correct OR two sections partially correct.

5. (a) Smaller. To reduce the percentage of bottles weighing less than advertised, without changing the mean or standard deviation, the manufacturer would have to *decrease* the advertised volume, that is, make it further into the tail of the roughly normal distribution of volumes.

   (b) Larger. Decreasing the percentage of underweight bottles would require putting more volume in each bottle, so the mean of the distribution of volumes would need to *increase* so that the advertised volume is further into the tail of the roughly normal distribution of volumes.

   (c) Smaller. Decreasing the percentage of underweight bottles without changing the advertised volume or the mean would require a taller, narrower normal curve so that the advertised volume is further into the tail of the roughly normal distribution of volumes. Thus, the standard deviation of the volumes would need to *decrease*.

SCORING

Part (a) is essentially correct for stating that the advertised volume would have to be decreased and giving a correct explanation. Part (a) is partially correct for stating that the advertised volume would have to be decreased and giving a weak explanation. Part (a) is incorrect for stating that the advertised volume would have to be increased or for stating that the advertised volume would have to be decreased but giving no explanation or an incorrect explanation.

Part (b) is essentially correct for stating that the mean volume would have to be increased and giving a correct explanation. Part (b) is partially correct for stating that the mean volume would have to be increased and giving a weak explanation. Part (b) is incorrect for stating that the mean volume would have to be decreased or for stating that the mean volume would have to be increased but giving no explanation or an incorrect explanation.

Part (c) is essentially correct for stating that the standard deviation would have to be decreased and giving a correct explanation. Part (c) is partially correct for stating that the standard deviation would have to be decreased and giving a weak explanation. Part (c) is incorrect for stating that the standard deviation would have to be increased or for stating that the standard deviation would have to be decreased but giving no explanation or an incorrect explanation.

4	**Complete Answer**	All three parts essentially correct.
3	**Substantial Answer**	Two parts essentially correct and one part partially correct.
2	**Developing Answer**	Two parts essentially correct OR one part essentially correct and one or two parts partially correct OR all three parts partially correct.
1	**Minimal Answer**	One part essentially correct OR two parts partially correct.

Section II: Part B

6. (a) A Type I error will be committed if the machine is correctly putting 30 packs of candy into each box; however, based on the sample, the inspector concludes that the machine is putting fewer than 30 packs of candy into each box. A consequence could be that the production process is stopped to fix the machine when, in reality, nothing is wrong with the machine.

    (b) A Type II error will be committed if the machine is actually putting fewer than 30 packs of candy into each box; however, based on the sample, the inspector concludes that the machine is functioning properly and is putting 30 packs of candy into each box. A consequence could be that the machine is not fixed even though it is malfunctioning, so it will continue to put fewer than 30 packs into each box.

    (c) The sampling distribution of C is approximately normal with mean $30 \times 2 = 60$ and standard deviation $\sqrt{30} \times 0.05 = 0.27386$.

    (d) $P(C < 59) = P\left(z < \frac{59 - 60}{0.27386}\right) = 0.000130$. The probability that the inspector will mistakenly conclude that the box contains fewer than 30 packs when the box really does contain 30 packs is 0.000130.

    (e) Now we have a roughly normal distribution with $\mu = 29 \times 2 = 58$ and $\sigma = \sqrt{29} \times 0.05 = 0.26926$. $P(C > 59) = P\left(z > \frac{59 - 58}{0.26926}\right) = 0.000102$. The probability that the inspector will mistakenly conclude that the box contains 30 packs when the box really does contain 29 packs is 0.000102.

    (f) The inspector's rule choice is very effective. The probability he commits a Type I error and mistakenly concludes that the machine is putting fewer than 30 packs of candy into each box when it is actually putting the correct 30 packs into each box is only 0.000130. The probability he commits a Type II error and mistakenly concludes that the machine is putting 30 packs of candy into each box when it is actually putting 29 packs into each box is only 0.000102.

SCORING

This question is scored in four sections. Section 1 consists of parts (a) and (b). Section 2 consists of parts (c) and (d). Section 3 consists of part (e). Section 4 consists of part (f).

Section 1 is essentially correct for a correct description of a Type I error, a correct consequence of a Type I error, a correct description of a Type II error, and a correct consequence of a Type II error. Section 1 is partially correct for two or three correct out of the four steps above.

Section 2 is essentially correct for a correct description of the shape of the sampling distribution as approximately normal, a correct calculation of the mean of the sampling distribution, and a correct calculation of the standard deviation of the sampling distribution all in (c), and a correct calculation of the probability in (d). Section 2 is partially correct for three out of these four steps correct.

Section 3 is essentially correct for a correct description of the shape of the sampling distribution as approximately normal, a correct calculation of the mean of the sampling distribution, a correct calculation of the standard deviation of the sampling distribution, and a correct calculation of the probability all in (e). Section 3 is partially correct for two or three out of these four steps correct.

Section 4 is essentially correct for the correction conclusion about the effectiveness of the inspector's rule choice with clear justification linked to both (d) and (e). Section 4 is partially correct if the justification is weak.

Count essentially correct answers as one point and partially correct answers as one-half point.

4	**Complete Answer**	Four points
3	**Substantial Answer**	Three points
2	**Developing Answer**	Two points
1	**Minimal Answer**	One point

Use a holistic approach to decide a score totaling between two numbers, deciding whether to score up or down depending on the strength of the response and communication.

ANSWER SHEET
Practice Test 3

1. Ⓐ Ⓑ Ⓒ Ⓓ Ⓔ   11. Ⓐ Ⓑ Ⓒ Ⓓ Ⓔ   21. Ⓐ Ⓑ Ⓒ Ⓓ Ⓔ   31. Ⓐ Ⓑ Ⓒ Ⓓ Ⓔ

2. Ⓐ Ⓑ Ⓒ Ⓓ Ⓔ   12. Ⓐ Ⓑ Ⓒ Ⓓ Ⓔ   22. Ⓐ Ⓑ Ⓒ Ⓓ Ⓔ   32. Ⓐ Ⓑ Ⓒ Ⓓ Ⓔ

3. Ⓐ Ⓑ Ⓒ Ⓓ Ⓔ   13. Ⓐ Ⓑ Ⓒ Ⓓ Ⓔ   23. Ⓐ Ⓑ Ⓒ Ⓓ Ⓔ   33. Ⓐ Ⓑ Ⓒ Ⓓ Ⓔ

4. Ⓐ Ⓑ Ⓒ Ⓓ Ⓔ   14. Ⓐ Ⓑ Ⓒ Ⓓ Ⓔ   24. Ⓐ Ⓑ Ⓒ Ⓓ Ⓔ   34. Ⓐ Ⓑ Ⓒ Ⓓ Ⓔ

5. Ⓐ Ⓑ Ⓒ Ⓓ Ⓔ   15. Ⓐ Ⓑ Ⓒ Ⓓ Ⓔ   25. Ⓐ Ⓑ Ⓒ Ⓓ Ⓔ   35. Ⓐ Ⓑ Ⓒ Ⓓ Ⓔ

6. Ⓐ Ⓑ Ⓒ Ⓓ Ⓔ   16. Ⓐ Ⓑ Ⓒ Ⓓ Ⓔ   26. Ⓐ Ⓑ Ⓒ Ⓓ Ⓔ   36. Ⓐ Ⓑ Ⓒ Ⓓ Ⓔ

7. Ⓐ Ⓑ Ⓒ Ⓓ Ⓔ   17. Ⓐ Ⓑ Ⓒ Ⓓ Ⓔ   27. Ⓐ Ⓑ Ⓒ Ⓓ Ⓔ   37. Ⓐ Ⓑ Ⓒ Ⓓ Ⓔ

8. Ⓐ Ⓑ Ⓒ Ⓓ Ⓔ   18. Ⓐ Ⓑ Ⓒ Ⓓ Ⓔ   28. Ⓐ Ⓑ Ⓒ Ⓓ Ⓔ   38. Ⓐ Ⓑ Ⓒ Ⓓ Ⓔ

9. Ⓐ Ⓑ Ⓒ Ⓓ Ⓔ   19. Ⓐ Ⓑ Ⓒ Ⓓ Ⓔ   29. Ⓐ Ⓑ Ⓒ Ⓓ Ⓔ   39. Ⓐ Ⓑ Ⓒ Ⓓ Ⓔ

10. Ⓐ Ⓑ Ⓒ Ⓓ Ⓔ   20. Ⓐ Ⓑ Ⓒ Ⓓ Ⓔ   30. Ⓐ Ⓑ Ⓒ Ⓓ Ⓔ   40. Ⓐ Ⓑ Ⓒ Ⓓ Ⓔ

Practice Test 3

Section I: Questions 1–40

SPEND 90 MINUTES ON THIS PART OF THE EXAM.

DIRECTIONS: The questions or incomplete statements that follow are each followed by five suggested answers or completions. Choose the response that best answers the question or completes the statement.

---

1. Which of the following is a true statement?

   (A) While properly designed experiments can strongly suggest cause-and-effect relationships, a complete census is the only way of establishing such a relationship.

   (B) If properly designed, observational studies can establish cause-and-effect relationships just as strongly as can properly designed experiments.

   (C) Controlled experiments are often undertaken later to establish cause-and-effect relationships first suggested by observational studies.

   (D) A useful approach to overcome bias in observational studies is to increase the sample size.

   (E) In an experiment, the control group is a self-selected group who choose not to receive a designated treatment.

2. Two classes take the same exam. Suppose a certain score is at the 40th percentile for the first class and at the 80th percentile for the second class. Which of the following is the most reasonable conclusion?

   (A) Students in the first class generally scored higher than students in the second class.

   (B) Students in the second class generally scored higher than students in the first class.

   (C) A score at the 20th percentile for the first class is at the 40th percentile for the second class.

   (D) A score at the 50th percentile for the first class is at the 90th percentile for the second class.

   (E) One of the classes has twice the number of students as the other.

3. In an experiment, the control group should receive

    (A) treatment opposite that given to the experimental group.
    (B) the same treatment given to the experimental group without knowing they are receiving the treatment.
    (C) a procedure identical to that given to the experimental group except for receiving the treatment under examination.
    (D) a procedure identical to that given to the experimental group except for a random decision on receiving the treatment under examination.
    (E) none of the procedures given to the experimental group.

4. In a simple random sample (SRS) of 625 families who do not live near any chemical plant, 10 had children with leukemia. In an SRS of 412 families living near chemical plants, 15 had children with leukemia. A 90% confidence interval of the difference in proportions is reported to be -0.0204 ± 0.0173. What is a proper conclusion?

    (A) The interval is invalid because probabilities cannot be negative.
    (B) The interval is invalid because it does not contain zero.
    (C) Families living near chemical plants are approximately 2.04% more likely to have children with leukemia than families who do not live near any chemical plant.
    (D) Ninety percent of families living near chemical plants are approximately 2.04% more likely to have children with leukemia than families who do not live near any chemical plant.

    (E) We are 90% confident that the difference in proportions between families who do not live near any chemical plant having children with leukemia and families who live near chemical plants having children with leukemia is between -0.0377 and -0.0031.

5. In a large geographic region, the proportional demographics include 14% African American, 5% Asian, 75% White and 6% Hispanic. The number of times characters of different ethnicities appeared in 400 TV commercials aired in this region during one randomly chosen week are as follows:

	African American	Asian	White	Hispanic
Observed frequency	48	10	315	27

What is the chi-square statistic for a goodness-of-fit test?

(A) $\dfrac{(48-56)^2}{48} + \dfrac{(10-20)^2}{10} + \dfrac{(315-300)^2}{315}$
$+ \dfrac{(27-24)^2}{27}$

(B) $\dfrac{(48-56)^2}{56} + \dfrac{(10-20)^2}{20} + \dfrac{(315-300)^2}{300}$
$+ \dfrac{(27-24)^2}{24}$

(C) $\dfrac{(48-100)^2}{100} + \dfrac{(10-100)^2}{100} + \dfrac{(315-100)^2}{100}$
$+ \dfrac{(27-100)^2}{100}$

(D) $\dfrac{(0.12-0.14)^2}{0.12} + \dfrac{(0.025-0.05)^2}{0.025}$
$+ \dfrac{(0.7875-0.75)^2}{0.7875} + \dfrac{(0.0675-0.06)^2}{0.0675}$

(E) $\dfrac{(0.12-0.14)^2}{0.14} + \dfrac{(0.025-0.05)^2}{0.05}$
$+ \dfrac{(0.7875-0.75)^2}{0.75} + \dfrac{(0.0675-0.06)^2}{0.06}$

6. A computer manufacturer sets up three locations to provide technical support for its customers. Logs are kept, noting whether or not calls about problems are solved successfully. Data from a sample of 1000 calls are summarized in the following table:

	Location			
	1	**2**	**3**	**Total**
Problem solved	325	225	150	700
Problem not solved	125	100	75	300
Total	450	325	225	1,000

Assuming there is no association between location and whether or not a problem is resolved successfully, what is the expected number of successful calls (problem solved) from location 1?

(A) $\dfrac{(325)(450)}{700}$

(B) $\dfrac{(325)(700)}{450}$

(C) $\dfrac{(325)(450)}{1,000}$

(D) $\dfrac{(325)(700)}{1,000}$

(E) $\dfrac{(450)(700)}{1,000}$

7. Which of the following statements about the correlation coefficient is true?

(A) The correlation coefficient and the slope of the regression line may have opposite signs.

(B) A correlation of 1 indicates a perfect cause-and-effect relationship between the variables.

(C) Correlations of $+0.87$ and -0.87 indicate the same degree of clustering around the regression line.

(D) Correlation applies equally well to quantitative and categorical data.

(E) A correlation of 0 shows little or no association between two variables.

8. Suppose X and Y are random variables with $E(X) = 780$, $\text{var}(X) = 75$, $E(Y) = 430$, and $\text{var}(Y) = 25$. Given that X and Y are independent, what is the variance of the random variable $X - Y$?

(A) $75 - 25$

(B) $75 + 25$

(C) $\sqrt{75 - 25}$

(D) $\sqrt{75 + 25}$

(E) $\sqrt{75} - \sqrt{25}$

9. What is the proper order of size for the probabilities of the following three events?

I. A randomly selected person over 80 years old is registered to vote.

II. A randomly selected person is over 80 years old and registered to vote.

III. A randomly selected person who is registered to vote is over 80 years old.

(A) $P(\text{I}) < P(\text{II}) < P(\text{III})$

(B) $P(\text{I}) > P(\text{II}) > P(\text{III})$

(C) $P(\text{I}) < P(\text{III}) < P(\text{II})$

(D) $P(\text{I}) > P(\text{III}) > P(\text{II})$

(E) $P(\text{II}) < P(\text{I}) < P(\text{III})$

10. A histogram of the cholesterol levels of all employees at a large law firm is as follows:

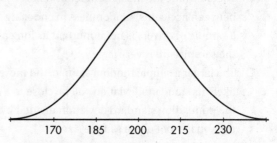

Which of the following is the best estimate of the standard deviation of this distribution?

(A) $\dfrac{230 - 170}{6} = 10$

(B) 15

(C) $\dfrac{200}{6} = 33.3$

(D) $230 - 170 = 60$

(E) $245 - 155 = 90$

11. A soft drink dispenser can be adjusted to deliver any fixed number of ounces. If the machine is operating with a standard deviation in delivery equal to 0.3 ounce, what should be the mean setting so that a 12-ounce cup will overflow less than 1% of the time? Assume an approximately normal distribution for ounces delivered.

   (A) $12 - 0.99(0.3)$ ounces
   (B) $12 - 2.326(0.3)$ ounces
   (C) $12 - 2.576(0.3)$ ounces
   (D) $12 + 2.326(0.3)$ ounces
   (E) $12 + 2.576(0.3)$ ounces

12. An insurance company wishes to study the number of years drivers in a large city go between automobile accidents. They plan to obtain and analyze the data from a sample of drivers. Which of the following is a true statement?

   (A) A reasonable time-and-cost-saving procedure would be to use systematic sampling on an available list of all AAA (Automobile Association of America) members in the city.
   (B) A reasonable time-and-cost-saving procedure would be to randomly choose families and include all drivers in each of these families in the sample.
   (C) To determine the mean number of years between accidents, randomness in choosing a sample of drivers is not important as long as the sample size is very large.
   (D) The larger a simple random sample, the more likely its standard deviation will be close to the population standard deviation divided by the square root of the sample size.
   (E) None of the above are true statements.

13. The probability that a person will show a certain gene-transmitted trait is 0.8 if the father shows the trait and 0.06 if the father doesn't show the trait. Suppose that the children in a certain community come from families in 25% of which the father shows the trait. Given that a child shows the trait, what is the probability that her father shows the trait?

   (A) $(0.25)(0.8)$
   (B) $(0.25)(0.8) + (0.75)(0.06)$
   (C) $\dfrac{0.25}{(0.25)(0.8) + (0.75)(0.06)}$
   (D) $\dfrac{0.8}{(0.25)(0.8) + (0.75)(0.06)}$
   (E) $\dfrac{(0.25)(0.8)}{(0.25)(0.8) + (0.75)(0.06)}$

14. Given an experiment with $H_0: \mu = 10$, $H_a: \mu > 10$, and a sample mean of 11, which of the following increases if n is discovered to have been greater than originally thought?

   I. The P-value
   II. The probability of a Type II error
   III. The power of the test

   (A) I only
   (B) II only
   (C) III only
   (D) II and III
   (E) None will increase.

15. If all the values of a data set are the same, all of the following must equal zero except for which one?

   (A) Mean
   (B) Standard deviation
   (C) Variance
   (D) Range
   (E) Interquartile range

16. Peer relationships are an important part of the socialization process of children and require the use of many complex social skills. Substantial evidence suggests that children with poor peer relationships are more likely to be truant, repeat grade levels, and drop out of school. In a random sample of 15 fifth-grade students, peer acceptance as measured on a sociometric scale showed a mean of 2.72 with a standard deviation of 0.15. Which of the following gives a 90% confidence interval for the mean peer acceptance of all fifth-grade students?

(A) $2.72 \pm 1.645\dfrac{0.15}{\sqrt{14}}$

(B) $2.72 \pm 1.753\dfrac{0.15}{\sqrt{14}}$

(C) $2.72 \pm 1.761\dfrac{0.15}{\sqrt{14}}$

(D) $2.72 \pm 1.753\dfrac{0.15}{\sqrt{15}}$

(E) $2.72 \pm 1.761\dfrac{0.15}{\sqrt{15}}$

17. A company has 1,000 employees evenly distributed throughout five assembly plants. A sample of 30 employees is to be chosen as follows. Each of the five managers will be asked to place the 200 time cards of their respective employees into a bag, shake them up, and randomly draw out six names. The six names from each plant will be put together to make up the sample. Will this method result in a simple random sample of the 1,000 employees?

(A) Yes, because every employee has the same chance of being selected.

(B) Yes, because every plant is equally represented.

(C) Yes, because this is an example of stratified sampling, which is a special case of simple random sampling.

(D) No, because the plants are not chosen randomly.

(E) No, because not every group of 30 employees has the same chance of being selected.

18. Given that $P(E) = 0.32$, $P(F) = 0.15$, and $P(E \cap F) = 0.048$, which of the following is a correct conclusion?

(A) The events E and F are both independent and mutually exclusive.

(B) The events E and F are neither independent nor mutually exclusive.

(C) The events E and F are mutually exclusive but not independent.

(D) The events E and F are independent but not mutually exclusive.

(E) The events E and F are independent, but there is insufficient information to determine whether or not they are mutually exclusive.

19. Researchers surveyed a random sample of 55 adults in neighborhood A and 64 adults in neighborhood B from populations of over 1,000 adults in each neighborhood. The sampled individuals were asked what their families anticipated spending on Halloween. Let $\bar{x}_A$ be the calculated mean spending of the 55 adults in neighborhood A, and let $\bar{x}_B$ be the calculated mean spending of the 64 adults in neighborhood B. Which of the following is the best explanation for why the sampling distribution of $\bar{x}_A - \bar{x}_B$ can be modeled with a normal distribution?

(A) There are at least 30 people in each population.

(B) The sample sizes are sufficiently large.

(C) The population distributions are assumed to be roughly normal.

(D) The sample distributions are assumed to be unimodal and roughly symmetric.

(E) The two sample standard deviations are assumed to be roughly equal.

20. Consider the following back-to-back stemplot:

	0	348
	1	01256
843	2	29
65210	3	2557
92	4	
7552	5	6
	6	1458
6	7	09
8541	8	
90	9	

Key: 2 |5| 6 represents a value of 52 on the left and 56 on the right

Which of the following is a correct statement?

(A) The distributions have the same mean.

(B) The distributions have the same median.

(C) The interquartile range of the distribution to the left is 20 greater than the interquartile range of the distribution to the right.

(D) The distributions have the same variance.

(E) None of the above is correct.

21. Which of the following is a correct statement?

(A) A study results in a 99% confidence interval estimate of (34.2, 67.3). This means that in about 99% of all samples selected by this method, the sample means will fall between 34.2 and 67.3.

(B) A high confidence level may be obtained no matter what the sample size.

(C) The central limit theorem is most useful when drawing samples from normally distributed populations.

(D) The sampling distribution for a mean has standard deviation $\frac{\sigma}{\sqrt{n}}$ only when n is sufficiently large (typically one uses $n \geq 30$).

(E) The center of any confidence interval is the population parameter.

22. A fair coin is flipped 6 times. Consider the following 3 outcomes where H refers to a toss landing heads, and T refers to a toss landing tails:

 I.   The result in order is HHHTTT.

 II. The result in order is HTTHHT.

 III. The result includes 3 Hs and 3 Ts.

What is a comparison of the three probabilities?

(A) $P(\text{I}) < P(\text{II}) < P(\text{III})$

(B) $P(\text{II}) < P(\text{I}) < P(\text{III})$

(C) $P(\text{III}) < P(\text{I}) < P(\text{II})$

(D) $P(\text{I}) = P(\text{II}) < P(\text{III})$

(E) $P(\text{III}) < P(\text{I}) = P(\text{II})$

23. Suppose two events, E and F, have nonzero probabilities p and q, respectively. Which of the following is impossible?

(A) $p + q > 1$

(B) $p - q < 0$

(C) $\frac{p}{q} > 1$

(D) E and F are neither independent nor mutually exclusive.

(E) E and F are both independent and mutually exclusive.

24. An inspection procedure at a manufacturing plant involves picking four items at random and accepting the whole lot if at least three of the four items are in perfect condition. If in reality 90% of the whole lot are perfect, what is the probability that the lot will be accepted?

(A) $(0.9)^4$

(B) $1 - (0.9)^4$

(C) $4(0.9)^3(0.1)$

(D) $0.1 - 4(0.9)^3(0.1)$

(E) $4(0.9)^3(0.1) + (0.9)^4$

25. A town has one high school, which buses students from urban, suburban, and rural communities. Which of the following samples is recommended in studying attitudes toward tracking of students in honors, regular, and below-grade classes?

(A) Convenience sample

(B) Simple random sample

(C) Stratified sample

(D) Systematic sample

(E) Voluntary response sample

26. Suppose there is a correlation of $r = 0.9$ between number of hours per day students study and GPAs. Which of the following is a reasonable conclusion?

   (A) 90% of students who study receive high grades.
   (B) 90% of students who receive high grades study a lot.
   (C) 90% of the variation in GPAs can be explained by variation in number of study hours per day.
   (D) 10% of the variation in GPAs cannot be explained by variation in number of study hours per day.
   (E) 81% of the variation in GPAs can be explained by variation in number of study hours per day.

27. To determine the average number of children living in single-family homes, a researcher picks a simple random sample of 50 such homes. However, even after one follow-up visit, the interviewer is unable to make contact with anyone in 8 of these homes. Concerned about nonresponse bias, the researcher picks another simple random sample and instructs the interviewer to keep trying until contact is made with someone in 8 more homes for a total of 50 homes. The average number of children is determined to be 1.73. Is this estimate probably too low or too high?

   (A) Too low, because of undercoverage bias
   (B) Too low, because convenience samples overestimate average results
   (C) Too high, because of undercoverage bias
   (D) Too high, because convenience samples overestimate average results
   (E) Too high, because voluntary response samples overestimate average results

28. The graph below shows cumulative proportions plotted against land values (in dollars per acre) for farms on sale in a rural community.

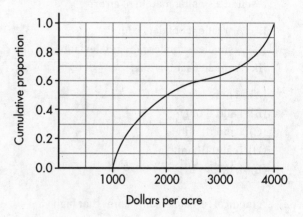

What is the median land value?

   (A) $2,000
   (B) $2,250
   (C) $2,500
   (D) $2,750
   (E) $3,000

29. An experiment is to be conducted to determine whether taking fish oil capsules or garlic capsules has more of an effect on cholesterol levels. In past studies, it was noted that daily exercise intensity (low, moderate, high) is associated with cholesterol level, but average sleep length ($< 5, 5-8, > 8$ hours) is not associated with cholesterol level. This experiment should be done

   (A) by blocking on exercise intensity.
   (B) by blocking on sleep length.
   (C) by blocking on cholesterol level.
   (D) by blocking on capsule type.
   (E) without blocking.

PRACTICE TEST 3

30. A confidence interval estimate is determined from the monthly grocery expenditures in a random sample of *n* families. Which of the following will result in a smaller margin of error?

   I.   A smaller confidence level
   II. A smaller sample standard deviation
   III. A smaller sample size

   (A) II only
   (B) I and II only
   (C) I and III only
   (D) II and III only
   (E) I, II, and III

31. A medical research team claims that high vitamin C intake increases endurance. In particular, 1,000 milligrams of vitamin C per day for a month should add an average of 4.3 minutes to the length of maximum physical effort that can be tolerated. Army training officers believe the claim is exaggerated and plan a test on a simple random sample of 400 soldiers in which they will reject the medical team's claim if the sample mean is less than 4.0 minutes. Suppose the standard deviation of added minutes is 3.2. If the true mean increase is only 4.2 minutes, what is the probability that the officers will fail to reject the false claim of 4.3 minutes?

   (A) $P\left(z < \dfrac{4.0 - 4.3}{\left(\dfrac{3.2}{\sqrt{400}}\right)}\right)$

   (B) $P\left(z > \dfrac{4.0 - 4.3}{\left(\dfrac{3.2}{\sqrt{400}}\right)}\right)$

   (C) $P\left(z < \dfrac{4.3 - 4.2}{\left(\dfrac{3.2}{\sqrt{400}}\right)}\right)$

   (D) $P\left(z > \dfrac{4.3 - 4.2}{\left(\dfrac{3.2}{\sqrt{400}}\right)}\right)$

   (E) $P\left(z > \dfrac{4.0 - 4.2}{\left(\dfrac{3.2}{\sqrt{400}}\right)}\right)$

32. Consider the two sets $X = \{10, 30, 45, 50, 55, 70, 90\}$ and $Y = \{10, 30, 35, 50, 65, 70, 90\}$. Which of the following is false?

   (A) The sets have identical medians.
   (B) The sets have identical means.
   (C) The sets have identical ranges.
   (D) The sets have identical boxplots.
   (E) None of the above are false.

33. The weight of an aspirin tablet is 300 milligrams according to the bottle label. An FDA investigator weighs a simple random sample of seven tablets, obtains weights of 299, 300, 305, 302, 299, 301, and 303, and runs a hypothesis test of the manufacturer's claim. Which of the following gives the *P*-value of this test?

   (A) $P(t > 1.54)$ with $df = 6$
   (B) $2P(t > 1.54)$ with $df = 6$
   (C) $P(t > 1.54)$ with $df = 7$
   (D) $2P(t > 1.54)$ with $df = 7$
   (E) $0.5P(t > 1.54)$ with $df = 7$

34. A group of 20 male college students took an IQ test, and each participant underwent an MRI scan to measure the size of his brain (in cubic centimeters). The scatterplot and some regression analysis are shown.

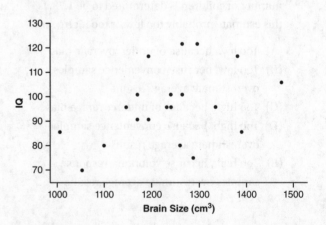

Predictor	Coef	SE Coef
Constant	38.47	31.80
Brain size	0.04986	0.02545

S = 13.4401   R-Sq = 17.6%

Which of the following is the *P*-value for an appropriate test statistic for testing the null hypothesis that the slope of the regression line is greater than 0?

(A) $P\left(z > \dfrac{0.04986}{0.02545}\right)$

(B) $P\left(t > \dfrac{0.04986}{0.02545}\right)$ with $t = 18$

(C) $P\left(t > \dfrac{0.04986}{0.02545}\right)$ with $t = 19$

(D) $P\left(t > \dfrac{0.04986}{13.4401}\right)$ with $t = 18$

(E) $P\left(t > \dfrac{0.04986}{13.4401}\right)$ with $t = 19$

35. Which of the following is *not* true with regard to contingency tables for chi-square tests for independence?

(A) Both variables are categorical.
(B) Observed frequencies should be whole numbers.
(C) Expected frequencies should be whole numbers.
(D) Expected frequencies in each cell should be at least 5, and to achieve this, one sometimes combines categories for one or the other or both of the variables.
(E) The expected frequency for any cell can be found by multiplying the row total by the column total and dividing by the table total.

36. Which of the following is a correct statement?

(A) The probability of a Type II error does not depend on the probability of a Type I error.
(B) In conducting a hypothesis test, it is possible to simultaneously make both a Type I and a Type II error.
(C) A Type II error will result if one incorrectly assumes the data are normally distributed.
(D) In medical disease testing with the null hypothesis that the patient is healthy, a Type I error is associated with a *false negative*; that is, the test incorrectly indicates that the patient is disease free.
(E) When you choose a significance level α, you're setting the probability of a Type I error to exactly α.

37. A random sample of 20 school districts looked at average EOC (end-of-course) test scores versus percent of students living in poverty. A scatterplot of the data is below.

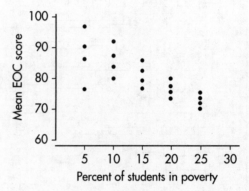

Which of the following would be revealed by a graph of the residuals versus percent of students in poverty?

(A) There is a negative linear relationship between the residuals and the percent in poverty.
(B) The sum of the residuals is less than 0.
(C) The sum of the residuals is greater than 0.
(D) The mean of the residuals is less than 0.
(E) The variation in the mean EOC scores is not the same across the percents in poverty.

38. Studies have shown that only 36% of young adults feel that the death penalty is applied fairly by courts. In a random sample of 50 young adults, what is the expected number of young adults that do not feel that the death penalty is applied fairly by courts?

(A) $50(0.36)$
(B) $50(0.64)$
(C) $50(0.36)(0.64)$
(D) $\sqrt{50(0.36)(0.64)}$
(E) $\sqrt{\dfrac{(0.36)(0.64)}{50}}$

39. The parallel boxplots below show monthly rainfall summaries for Liberia, West Africa.

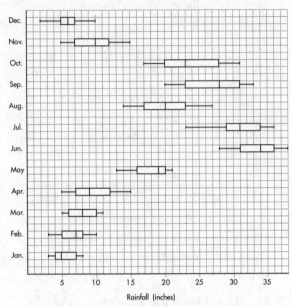

Rainfall (inches)

Which of the following months has the least variability as measured by *interquartile range*?

(A) January
(B) February
(C) March
(D) May
(E) December

40. In comparing the life expectancies of two models of refrigerators, the average years before complete breakdown of 10 model A refrigerators is compared with that of 15 model B refrigerators. The 90% confidence interval estimate of the difference is (6, 12). Which of the following is the most reasonable conclusion?

(A) The mean life expectancy of one model is twice that of the other.
(B) The mean life expectancy of one model is 6 years, while the mean life expectancy of the other is 12 years.
(C) The probability that the life expectancies are different is 0.90.
(D) The probability that the difference in life expectancies is greater than 6 years is 0.90.
(E) We should be 90% confident that the difference in life expectancies is between 6 and 12 years.

PRACTICE TEST 3

STOP   If there is still time remaining, you may check your work on this section.

Section II: Part A

Questions 1–5

SPEND ABOUT 65 MINUTES ON THIS PART OF THE EXAM.
PERCENTAGE OF SECTION II GRADE—75

> **DIRECTIONS:** You must show all work and indicate the methods you use. You will be graded on the correctness of your methods and on the accuracy of your results and explanations.

1. The Information Technology Services division at a university is considering installing a new spam filter software product on all campus computers to combat unwanted advertising and spyware. A sample of 60 campus computers was randomly divided into two groups of 30 computers each. One group of 30 was considered to be a control group, while each computer in the other group had the spam filter software installed. During a two-week period, each computer user was instructed to keep track of the number of unwanted spam emails received. The back-to-back stemplot below shows the distribution of such emails received for the control and treatment groups.

```
             Control        Treatment
       9 8 7 5    0    2 2 3 5 5 6 6 6 7 8 8 9 9 9
9 8 8 7 5 5 5 4 2 2 0    1    0 0 1 1 2 2 5 7 9
  8 7 5 4 4 3 2 1 1 0    2    0 2 4 6
         4 3 2 0 0    3    2 5
                      4    1
```

Key: 0|3|2 represents 30 and 32 spam emails for the control and treatment groups, respectively.

(a) Compare the distribution of spam emails from the control and treatment groups.
(b) The standard deviation of the numbers of emails in the control group is 8.1. How does this value summarize variability for the control group data?
(c) A researcher in Information Technology Services calculates a 95% confidence interval for the difference in mean number of spam emails received between the control group and the treatment group with the new software and obtains (1.5, 10.9). Assuming all conditions for a two-sample t-interval are met, comment on whether or not there is evidence of a difference in the means for the number of spam emails received during a two-week period by computers with and without the software.
(d) The computer users on campus fall into four groups: administrators, staff, faculty, and students. Explain why a researcher might decide to use blocking in setting up this experiment.

2. A game contestant flips three fair coins and receives a score equal to the absolute value of the difference between the number of heads and number of tails showing.

(a) Construct the probability distribution table for the possible scores in this game.
(b) Calculate the expected value of the score for a player.
(c) What is the probability that if a player plays this game three times, the total score will be exactly 3?
(d) Suppose a player wins a major prize if he or she can average a score of at least 2. Given the choice, should he or she try for this average by playing 10 times or by playing 15 times? Explain.

3. A sociologist is researching a possible link between poverty and lack of education in the United States. A least squares regression on "percent of population 25 years and over who did not graduate from high school" versus "percent of population below poverty level" among a random sample of 45 regions yields the following computer printout on the next page.

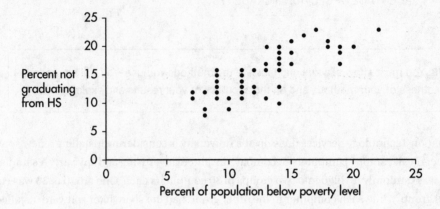

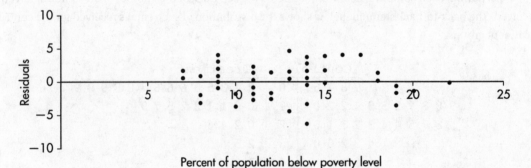

Percent of population below poverty level

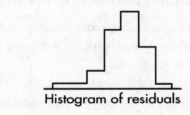

Histogram of residuals

	Coef	SE Coef	t	P
Intercept	4.1929	1.3394	3.1304	0.00148
Poverty	0.8691	0.1010	8.6050	0.00000

(a) Determine the equation of the least squares regression line.

(b) What is the y-intercept of the regression line? Interpret in context.

(c) Predict the percentage of students not graduating from high school who are from regions in which the percent of the population below the poverty line are 22% and 42%, and explain your confidence in both of your answers.

(d) Find the 90 percent confidence interval for the slope of the regression line, and interpret in context.

4. A study is proposed to compare two treatments for patients with significantly narrowed neck arteries. Some patients will be treated with surgery to remove built-up plaque, while others will be treated with stents to improve circulation. The response variable will be the proportion of patients who suffer a major complication such as a stroke or heart attack within one month of the treatment. The researchers decide to block on whether or not a patient has had a mini stroke in the previous year.

   (a) There are 1,000 patients available for this study, half of whom have had a mini stroke in the previous year. Explain a block design to assign patients to treatments, including how to make treatment assignments.

   (b) Give two methods other than blocking to increase the power of detecting a difference between using surgery versus stents for patients with this medical condition. Explain your choice of methods.

5. A point is said to have *high leverage* if it is an outlier in the *x*-direction, and a point is said to be *influential* if its removal sharply changes the regression line.

   (a) In the scatterplots below, compare points *A* and *B* with regard to having high leverage and with regard to being influential.

   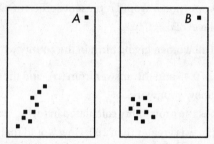

   (b) In the scatterplots below, compare points *C* and *D* with regard to residuals and influence on β, the slope of the regression line.

   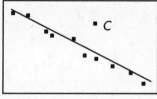 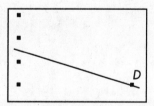

   (c) In the scatterplot below, compare the effect of removing point *E* or point *F* to that of removing both *E* and *F*.

   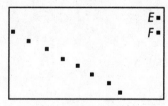

   (d) In the scatterplot below, compare the effect of removing point *G* or point *H* to that of removing both *G* and *H*.

   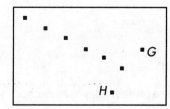

Section II: Part B

Question 6

SPEND ABOUT 25 MINUTES ON THIS PART OF THE EXAM.
PERCENTAGE OF SECTION II GRADE—25

6. It was recently reported that in the United States, 40.3 percent of all births are to unmarried mothers. A county health administrator is investigating whether births to unmarried women is higher in her county than in the national average. If so, she will propose additional funding to counsel unmarried mothers. A random sample of 100 births in the county will be looked at.

Let p represent the proportion of the population of women giving birth in the county who are unmarried. Consider the following hypotheses.

$H_0: p = 0.403$, $H_a: p > 0.403$

(a) Describe a Type II error in context and a possible consequence.

(b) What values of the sample proportion $\hat{p}$ would represent sufficient evidence to reject the null hypothesis at a significance level of $\alpha = 0.05$?

Suppose the actual proportion of all women giving birth in the county who are unmarried is 0.45.

(c) Using the actual proportion of 0.45 and the answer from (b), find the probability that the null hypothesis will be rejected. Show your work.

(d) What statistical term describes the probability calculated in (c)?

(e) Suppose the size of the sample was greater than 100. How would that affect the probability of rejecting the null hypothesis calculated in (c)? Explain.

Answer Explanations

Section I

1. **(C)** A complete census can give much information about a population, but it doesn't necessarily establish a cause-and-effect relationship among seemingly related population parameters. While the results of well-designed observational studies might suggest relationships, it is difficult to conclude that there is cause and effect without running a well-designed experiment. If bias is present, increasing the sample size simply magnifies the bias. The control group is selected by the researchers making use of chance procedures.

2. **(A)** In the first class, only 40% of the students scored below the given score, while in the second class, 80% scored below the same score.

3. **(C)** The control group should have experiences identical to those of the experimental groups except for the treatment under examination. The control group should not be given a new treatment.

4. **(E)** The negative sign comes about because we are dealing with the difference of proportions. The confidence interval estimate means that we have a certain *confidence* that the difference in population proportions lies in a particular *interval*. Note that $-0.0204 - 0.0173 = -0.0377$ and $-0.0204 + 0.0173 = -0.0031$.

5. **(B)** The expected values are $0.14(400) = 56$; $0.05(400) = 20$; $0.75(400) = 300$; and $0.06(400) = 24$. Then
$$\chi^2 = \sum \frac{(\text{observed} - \text{expected})^2}{\text{expected}} = \frac{(48 - 56)^2}{56}$$
$$+ \frac{(10 - 20)^2}{20} + \frac{(315 - 300)^2}{300} + \frac{(27 - 24)^2}{24}.$$

6. **(E)** The proportion of successful calls (problem solved) is $\frac{700}{1,000}$, so $\frac{700}{1,000}(450)$ is the expected number of calls from location 1 that are successful. Alternatively, the proportion of calls from location 1 is $\frac{450}{1,000}$, so $\frac{450}{1,000}(700)$ gives the expected number of successful calls from location 1.
$$\left[\text{Or, Expected} = \frac{(\text{column total})(\text{row total})}{(\text{grand total})} \right.$$
$$\left. = \frac{(450)(700)}{1,000}. \right]$$

7. **(C)** The slope and the correlation always have the same sign. Correlation shows association, not causation. Correlation does not apply to categorical data. Correlation measures linear association, so even with a correlation of 0, there may be very strong nonlinear association.

8. **(B)** If two random variables are independent, the mean of the difference of the two random variables is equal to the difference of the two individual means; however, the variance of the difference of the two random variables is equal to the *sum* of the two individual variances.

9. **(D)** $P(\text{I}) = P(\text{vote}|\text{over } 80) = \frac{P(\text{vote} \cap \text{over } 80)}{P(\text{over } 80)}$, $P(\text{II}) = P(\text{vote} \cap \text{over } 80)$, and $P(\text{III}) = P(\text{over } 80|\text{vote}) = \frac{P(\text{vote} \cap \text{over } 80)}{P(\text{vote})}$. $P(\text{I}) > P(\text{III})$ because the denominator in (I) is smaller than the denominator in (III). $P(\text{II})$ is the smallest because (I) and (III) have denominators that are less than one.

10. **(B)** The markings, spaced 15 apart, clearly look like the standard deviation spacings associated with a roughly normal curve. The curve seems to be the steepest above the points 185 and 215, and this too indicates that the standard deviation is 15.

11. **(B)** With a right tail having probability 0.01, the critical z-score is $\texttt{invNorm}(0.99) = 2.326$. Thus, $\mu + 2.326(0.3) = 12$, giving $\mu = 12 - 2.326(0.3)$.

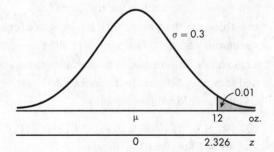

12. **(E)** There is no reason to think that AAA members are representative of the city's drivers. Family members may have similar driving habits, and the independence condition would be violated. Random selection is important regardless of the sample size. The larger a random sample, the closer its standard deviation will be to the population standard deviation.

13. **(E)**

$$P\left(\frac{\text{child}}{\text{shows}}\right) = P\left(\frac{\text{father}}{\text{shows}} \cap \frac{\text{child}}{\text{shows}}\right)$$

$$+ P\left(\frac{\text{father}}{\text{doesn't}} \cap \frac{\text{child}}{\text{shows}}\right)$$

$$= (0.25)(0.8) + (0.75)(0.06)$$

$$P\left(\frac{\text{father}}{\text{shows}} \middle| \frac{\text{child}}{\text{shows}}\right) = \frac{P\left(\frac{\text{father}}{\text{shows}} \cap \frac{\text{child}}{\text{shows}}\right)}{P\left(\frac{\text{child}}{\text{shows}}\right)}$$

$$= \frac{(0.25)(0.8)}{(0.25)(0.8) + (0.75)(0.06)}$$

14. **(C)** With a greater n, the standard error is smaller, resulting in the P-value and the probability of a Type II error both being less. Power equals one minus the probability of a Type II error, so it will be greater.

15. **(A)** The mean equals the common value of all the data elements. The other terms all measure variability, which is zero when all the data elements are equal.

16. **(E)** $SE(\bar{x}) = \frac{s}{\sqrt{n}} = \frac{0.15}{\sqrt{15}}$ and with $df = n - 1 = 15 - 1 = 14$, the critical t-scores are $\pm\texttt{invT(0.95, 14)} = \pm 1.761$. The 90% confidence interval is

$$\bar{x} \pm t^*(SE(\bar{x})) = 2.72 \pm 1.761\left(\frac{0.15}{\sqrt{15}}\right)$$

17. **(E)** In a simple random sample, every possible group of the given size has to be equally likely to be selected, and this is not true here. For example, with this procedure it is impossible for the employees in the final sample to all be from a single plant. This method is an example of stratified sampling, but stratified sampling does not result in simple random samples.

18. **(D)** $(0.32)(0.15) = 0.048$, so $P(E \cap F) = P(E)P(F)$ and, thus, E and F are independent. $P(E \cap F) \neq 0$, so E and F are not mutually exclusive.

19. **(B)** Both sample sizes are at least 30, so by the central limit theorem, the sampling distribution of $\bar{x}_A - \bar{x}_B$ can be modeled with a normal distribution.

20. **(D)** One set is a shift of 20 units from the other, so they have different means and medians, but they have identical shapes and thus the same variability, including IQR, standard deviation, and variance.

21. **(B)** The wider the confidence interval, the higher the confidence level, so one may obtain as high a confidence level as one wishes if a very wide confidence interval is accepted. A 99% confidence interval estimate means that in about 99% of all samples selected by this method, the population mean will be included in the confidence interval. The central limit theorem applies to any population, no matter if it is normally distributed or not. The sampling distribution for a mean always has standard deviation $\frac{\sigma}{\sqrt{n}}$; large enough sample size n refers to the closer the distribution will be to a normal distribution. The center of a confidence interval is the sample statistic, not the population parameter.

22. **(D)** $P(\text{I}) = P(\text{II}) = (0.5)^6$, while $P(\text{III}) = {}_6C_3(0.5)^3(0.5)^3 = 20(0.5)^6$

23. **(E)** Independent implies $P(E \cap F) = P(E)P(F)$, while mutually exclusive implies $P(E \cap F) = 0$. So, to be both independent and mutually exclusive in this example would mean $pq = 0$, which is impossible because p and q are given to be nonzero.

24. **(E)** In a binomial with $n = 4$ and $p = 0.9$, $P(\text{at least 3 successes}) = P(\text{exactly 3 successes}) + P(\text{exactly 4 successes}) = 4(0.9)^3(0.1) + (0.9)^4$.

25. **(C)** In stratified sampling, the population is divided into representative groups, and random samples of persons from each group are chosen. In this case, it might well be important to have sufficient numbers from each group and to be able to consider separately the responses from each of the three groups—urban, suburban, and rural.

26. **(E)** The coefficient of determination, r^2, indicates the percentage of variation in the response variable y that is explained by variation in the explanatory variable x.

27. **(C)** It is most likely that the homes at which the interviewer had difficulty finding someone home were homes with fewer children living in them. Replacing these homes with other randomly picked homes will most likely replace homes with fewer children with homes with more children.

28. **(A)** The median corresponds to the 0.5 cumulative proportion.

29. **(A)** Blocking divides the subjects into groups of similar individuals, in this case individuals with similar exercise habits, and runs the experiment on each separate group. This controls the known effect of variation in exercise level on cholesterol level.

30. **(B)** The margin of error varies directly with the critical z-value and directly with the standard deviation of the sample but inversely with the square root of the sample size.

31. **(E)** $\sigma_{\bar{x}} = \frac{3.2}{\sqrt{400}}$. With a true mean increase of 4.2, the z-score for 4.0 is $\frac{4.0 - 4.2}{\left(\frac{3.2}{\sqrt{400}}\right)}$ and the officers fail to reject the claim if the sample mean has a z-score greater than this.

32. **(E)** Both have 50 for their means and medians, both have a range of $90 - 10 = 80$, and both have identical boxplots, with first quartile 30 and third quartile 70.

33. **(B)** Since we are not told that the investigator suspects that the average weight is over 300 mg or is under 300 mg and since a tablet containing too little or too much of a drug clearly should be brought to the manufacturer's attention, this is a two-sided test. Thus, the P-value is twice the tail probability obtained (using the t-distribution with $df = n - 1 = 6$). Note that while t can be calculated by putting the data in a list and using T-Test, this is unnecessary here because every answer choice involves "$t > 1.54$."

34. **(B)** With $H_0: \beta = 0$ and $H_a: \beta > 0$, $t = \frac{b - 0}{SE(b)} = \frac{0.04986 - 0}{0.02545}$. $df = n - 2 = 20 - 2 = 18$, so the P-value equals $P\left(t > \frac{0.04986}{0.02545}\right)$ with $t = 18$.

35. **(C)** The expected frequencies, as calculated by $\frac{(\text{row total}) \times (\text{column total})}{\text{table total}}$, may not be whole numbers.

36. **(E)** We reject H_0 when the P-value falls below α, and when H_0 is true this rejection will happen precisely with probability α. The probabilities of Type I and Type II errors are related; for example, lowering the Type I error increases the probability of a Type II error. A Type I error can be made only if the null hypothesis is true, while a Type II error can be made only if the null hypothesis is false. In medical testing, with the usual null hypothesis that the patient is healthy, a Type I error is that a healthy patient is diagnosed with a disease, that is, a *false positive*.

37. **(E)** The variation in the mean EOC scores is larger for lower percent poverty and smaller for higher percent poverty, and this will be shown by residuals with larger absolute values and then smaller absolute values. The sum and the mean of the residuals is always exactly 0, and the residual plot will never show either a positive or a negative linear relationship.

38. **(B)** The probability of an adult feeling that the death penalty is not applied fairly by the courts is $1 - 0.36 = 0.64$. The expected value of a binomial with $n = 50$ and $p = 0.64$ is $np = 50(0.64)$.

39. **(E)** A boxplot gives a five-number summary: smallest value, 25th percentile (Q_1), median, 75th percentile (Q_3), and largest value. The interquartile range is given by $Q_3 - Q_1$, or the total length of the two "boxes," ignoring the "whiskers."

40. **(E)** When using a measurement from a sample, we are never able to say *exactly* what a population mean is; rather, we always say we have a certain *confidence* that the population mean lies in a particular *interval*.

Section II: Part A

1. (a)   A complete answer compares shape, center, and spread and mentions content.

   Shape: The control group distribution is somewhat bell-shaped and symmetric, while the treatment group distribution is somewhat skewed right (skewed toward the higher values).
   Center: The center of the control group distribution is around 20, which is greater than the center of the treatment group distribution, which is somewhere around 10 to 12.
   Spread: The spread of the control group distribution, 5 to 34, is less than the spread of the treatment group distribution, which is 2 to 41.

   (b)   For computers in the control group (no spam software), the number of spam emails received varies an "average" amount of 8.1 from the mean number of spam emails received in the control group.

   (c)   Since the 95% confidence interval for the difference does not contain zero, the researcher can conclude the observed difference in mean numbers of spam emails received between the control group and the treatment group that received spam software is statistically significant.

   (d)   It may well be that the four groups—administrators, staff, faculty, and students—are each exposed to different kinds of spam email risks, and possibly the software will be more or less of a help to each group. In that case, the researcher should in effect run four separate experiments on the homogeneous groups, called blocks. Even though results from the different blocks will be combined at the end, some conclusions may be more specific.

SCORING
This question is scored in three sections. Section 1 consists of part (a). Section 2 consists of parts (b) and (c). Section 3 consists of part (d).
Section 1 is essentially correct for correctly comparing shape, center, and spread. Section 1 is partially correct for correctly comparing two of the three features. Lower an E to a P or a P to an I if context is never mentioned.
Section 2 is essentially correct if (1) standard deviation is explained correctly in the context of this problem and (2) for noting that zero is not in the interval so the observed difference, in context, is significant. Section 2 is partially correct for one of these two parts correct.
Section 3 is essentially correct if the purpose of blocking is correctly explained in the context of this problem. Section 3 is partially correct if the general purpose of blocking is correctly explained but not in the context of this problem.

4	**Complete Answer**	All three sections essentially correct.
3	**Substantial Answer**	Two sections essentially correct and one section partially correct.
2	**Developing Answer**	Two sections essentially correct OR one section essentially correct and one or two sections partially correct OR all three sections partially correct.
1	**Minimal Answer**	One section essentially correct OR two sections partially correct.

2. (a) Listing the eight possibilities: {HHH, HHT, HTH, HTT, THH, THT, TTH, TTT} clearly shows that the only possibilities for the absolute value of the difference are 1 with probability $\frac{6}{8} = 0.75$ and 3 with probability $\frac{2}{8} = 0.25$. The table is

Absolute difference	Probability
1	0.75
3	0.25

(b) $E = \sum xP(x) = 1(0.75) + 3(0.25) = 1.5$.

(c) Since the only possible scores for each game are 1 and 3, the only way to have a total score of 3 in three games is to score 1 in each game. The probability of this is $(0.75)^3 = 0.421875$ [or $\left(\frac{3}{4}\right)^3 = \frac{27}{64}$].

(d) The more times the game is played, the closer the average score will be to the expected value of 1.5. The player does not want to average close to 1.5, so he or she should prefer playing 10 times rather than 15 times.

SCORING

This question is scored in three sections. Section 1 consists of part (a). Section 2 consists of parts (b) and (c). Section 3 consists of part (d).

Section 1 is essentially correct for the correct probability distribution table. Section 1 is partially correct for one minor error.

Section 2 is essentially correct for (1) the correct calculation of expected value based on the answer given in Part (a) and (2) for the correct probability calculation with some indication of where the answer is coming from. Section 2 is partially correct for one of these two parts correct.

Section 3 is essentially correct for choosing 10 and giving a clear explanation. Section 3 is partially correct for choosing 10 and giving a weak explanation. Section 3 is incorrect for choosing 10 with no explanation or with an incorrect explanation.

4	Complete Answer	All three sections essentially correct.
3	Substantial Answer	Two sections essentially correct and one section partially correct.
2	Developing Answer	Two sections essentially correct OR one section essentially correct and one or two sections partially correct OR all three sections partially correct.
1	Minimal Answer	One section essentially correct OR two sections partially correct.

3. (a) Predicted % not graduating HS = 4.1929 + 0.8691(% below poverty line)

(b) The y-intercept is 4.1929. In regions with 0% of the population below the poverty line, the average (or predicted) percent not graduating from high school is 4.2%.

(c) $4.1929 + 0.8691(22) = 23.3131$ and $4.1929 + 0.8691(42) = 40.6951$.

For regions with 22% of the population below the poverty line, the predicted percent not graduating from high school is 23.3%. For regions with 42% of the population below the poverty line, the predicted percent not graduating from high school is 40.7%. The 22% input is within the domain of the given sample data, while the 42% input is not. So, while we have confidence in the 23.3% calculation, we have very little confidence in the 40.7% calculation because it is an extreme extrapolation.

(d) *Procedure:* Confidence interval for the slope of the regression line.

Check conditions: We are given a random sample, we assume that the sample of 45 regions is less than 10% of all regions, the scatterplot looks roughly linear, there is no major pattern in the residual plot, and the histogram of the residuals is unimodal and roughly symmetric.

Mechanics: With $df = n - 2 = 43$, the critical t-scores are $\pm$invT(0.95, 43) $= \pm1.681$. Then $0.8691 \pm 1.681(0.1010) = 0.8691 \pm 0.1698$.

Conclusion in context: We are 90% confident that the true slope of the regression line linking percent not graduating from high school to percent below the poverty line is between 0.6993 and 1.0389. That is, we are 90% confident that for each additional 1 percent of the population below the poverty line, the predicted percent of the population not graduating from high school goes up between 0.6993 and 1.0389.

SCORING

This question is scored in three sections. Section 1 consists of parts (a) and (b). Section 2 consists of part (c). Section 3 consists of part (d).

Section 1, Parts (a) and (b), is essentially correct for (1) a correct equation, (2) defining the variables, (3) correct interpretation for the intercept, and (4) using nondeterministic language in both parts ("predicted" or "hat," "average"). Section 1 is partially correct for two or three of these four components correct.

Section 2, Part (c), is essentially correct for (1) two correct calculations, (2) answers in context, and (3) a clear explanation of why there is more confidence in the first calculation and is partially correct for two of these three components correct.

Section 3, Part (d), is essentially correct for (1) naming the procedure, (2) referencing all the listed conditions, (3) a correct interval, and (4) the conclusion referencing the confidence level, context, the population (such as "true"), and the parameter "slope" and is partially correct for two or three of the four components correct.

4	**Complete Answer**	All three sections essentially correct.
3	**Substantial Answer**	Two sections essentially correct and one section partially correct.
2	**Developing Answer**	Two sections essentially correct OR one section essentially correct and one or two sections partially correct OR all three sections partially correct.
1	**Minimal Answer**	One section essentially correct OR two sections partially correct.

4. (a) One block consists of the 500 patients who have had a mini stroke in the previous year, and a second block consists of the 500 patients who have not. In one block, number the patients 1 through 500, and then using a random number generator, pick integers between 1 and 500, throwing out repeats, until 250 unique integers are picked. Assign surgery to those patients with numbers corresponding to the generated integers, and assign stents to the remaining 250 patients. Repeat for the other block.

   (b) One method is to increase the sample size, resulting in a reduction in the standard error of the sampling distribution, which in turn increases the probability of rejecting the null hypothesis if it is false. A second method is to increase α, the significance level, which in turn also increases the probability of rejecting the null hypothesis if it is false. In either case, there is increased probability of detecting any difference in the proportions of patients suffering major complications between the surgery and stent recipients.

SCORING

Part (a) is essentially correct for (1) labeling the two groups of 500, (2) numbering the patients in each group, (3) using a random number generator or a random digit table to pick 250 unique integers, and (4) choosing the patients corresponding to the chosen integers for one of the treatments. Part (a) is partially correct for 3 of these 4 components.

Part (b) is essentially correct for (1) mentioning sample size, (2) explaining how increasing sample size increases power, (3) mentions increasing the significance level, and (4) explaining how increasing the significance level increases power. Part (b) is partially correct for 2 or 3 of these 4 components.

4	Complete Answer	Both parts essentially correct.
3	Substantial Answer	One part essentially correct and the other part partially correct.
2	Developing Answer	One part essentially correct and the other part incorrect OR both parts partially correct.
1	Minimal Answer	One part partially correct and the other part incorrect.

---

5. (a) Points A and B both have high leverage; that is, both their x-coordinates are outliers in the x-direction. However, point B is influential (its removal sharply changes the regression line), while point A is not influential (it appears to lie close to or directly on the regression line, so its removal will not change the line).

   (b) Point C lies off the regression line. So, its residual is much greater than that of point D, whose residual is 0 (point D lies on the regression line). However, the removal of point C will very minimally affect the slope of the regression line, if at all, while the removal of point D dramatically affects the slope of the regression line.

   (c) Removal of either point E or point F minimally affects the regression line, while removal of both has a dramatic effect.

   (d) Removing either point G or point H will definitely affect the regression line (pulling the line toward the remaining of the two points). However, removing both will have little, if any, effect on the line.

SCORING

Each of Parts (a), (b), (c), and (d) has two components and is scored essentially correct for both components correct and partially correct for one component correct.

Count essentially correct answers as one point and partially correct answers as one-half point.

4	Complete Answer	4 points
3	Substantial Answer	3 points
2	Developing Answer	2 points
1	Minimal Answer	1 point

Use a holistic approach to decide a score totaling between two numbers, deciding whether to score up or down depending on the strength of the response and communication.

Section II: Part B

6. (a) A Type II error is mistakenly failing to reject a false null hypothesis. In this situation, it would happen if the proportion of all women giving birth in the county who are unmarried is greater than 0.403, but the sample proportion does not provide sufficient evidence that it is. A possible consequence would be that needed additional funding to counsel unmarried mothers is not provided.

   (b) We first note that the standard deviation is $\sqrt{\dfrac{p(1-p)}{n}} = \sqrt{\dfrac{(0.403)(0.597)}{100}} = 0.04905$.
   This is a one-sided test, and so the critical z-score is $\mathtt{invNorm(0.95)} = 1.645$.
   Then $\dfrac{\hat{p} - 0.403}{0.04905} > 1.645$ gives $\hat{p} > 0.4837$.

   (c) If the true population proportion is 0.45, the sampling distribution of $\hat{p}$ is approximately normal with mean 0.45 and standard deviation $\sqrt{\dfrac{(0.45)(0.55)}{100}} = 0.04975$. Then $P(\hat{p} > 0.4837) = P\left(z > \dfrac{0.4837 - 0.45}{0.04975}\right) = 0.249$.

   (d) This is called the power of the test.

   (e) If the sample size is increased, then, while the rejection region is still $z > 1.645$, the sampling distribution of $\hat{p}$ will have a smaller standard deviation. Thus, the minimum value of $\hat{p}$ for which we would reject H_0 would be lower, and so the probability of rejecting H_0 would be greater.

SCORING

This question is scored in four sections. Section 1 consists of part (a). Section 2 consists of part (b). Section 3 consists of parts (c) and (d). Section 4 consists of part (e).

Section 1 is essentially correct in (a) for a correct explanation of a Type II error in context and including a possible consequence. Section 1 is partially correct for one out of the two steps above.

Section 2 is essentially correct in (b) for calculating $\sigma_{\hat{p}}$, stating the equation $\dfrac{\hat{p} - 0.403}{0.04905} > 1.645$ to be solved, and correctly solving for $\hat{p}$ and is partially correct for two out of these three steps correct.

Section 3 is essentially correct in (c) and (d) for the correct calculation of the new standard deviation $\sigma_{\hat{p}}$ correctly calculating the probability $P(\hat{p} > 0.4837)$, and correctly recognizing this to be the power of the test and is partially correct for two out of these three steps correct.

Section 4 is essentially correct in (e) for concluding that the probability of rejecting H_0 would be greater and giving a good explanation and is partially correct for concluding that the probability of rejecting H_0 would be greater and giving a weak explanation.

Count essentially correct answers as one point and partially correct answers as one-half point.

4	**Complete Answer**	Four points
3	**Substantial Answer**	Three points
2	**Developing Answer**	Two points
1	**Minimal Answer**	One point

Use a holistic approach to decide a score totaling between two numbers, deciding whether to score up or down depending on the strength of the response and communication.

ANSWER SHEET
Practice Test 4

1. Ⓐ Ⓑ Ⓒ Ⓓ Ⓔ   11. Ⓐ Ⓑ Ⓒ Ⓓ Ⓔ   21. Ⓐ Ⓑ Ⓒ Ⓓ Ⓔ   31. Ⓐ Ⓑ Ⓒ Ⓓ Ⓔ

2. Ⓐ Ⓑ Ⓒ Ⓓ Ⓔ   12. Ⓐ Ⓑ Ⓒ Ⓓ Ⓔ   22. Ⓐ Ⓑ Ⓒ Ⓓ Ⓔ   32. Ⓐ Ⓑ Ⓒ Ⓓ Ⓔ

3. Ⓐ Ⓑ Ⓒ Ⓓ Ⓔ   13. Ⓐ Ⓑ Ⓒ Ⓓ Ⓔ   23. Ⓐ Ⓑ Ⓒ Ⓓ Ⓔ   33. Ⓐ Ⓑ Ⓒ Ⓓ Ⓔ

4. Ⓐ Ⓑ Ⓒ Ⓓ Ⓔ   14. Ⓐ Ⓑ Ⓒ Ⓓ Ⓔ   24. Ⓐ Ⓑ Ⓒ Ⓓ Ⓔ   34. Ⓐ Ⓑ Ⓒ Ⓓ Ⓔ

5. Ⓐ Ⓑ Ⓒ Ⓓ Ⓔ   15. Ⓐ Ⓑ Ⓒ Ⓓ Ⓔ   25. Ⓐ Ⓑ Ⓒ Ⓓ Ⓔ   35. Ⓐ Ⓑ Ⓒ Ⓓ Ⓔ

6. Ⓐ Ⓑ Ⓒ Ⓓ Ⓔ   16. Ⓐ Ⓑ Ⓒ Ⓓ Ⓔ   26. Ⓐ Ⓑ Ⓒ Ⓓ Ⓔ   36. Ⓐ Ⓑ Ⓒ Ⓓ Ⓔ

7. Ⓐ Ⓑ Ⓒ Ⓓ Ⓔ   17. Ⓐ Ⓑ Ⓒ Ⓓ Ⓔ   27. Ⓐ Ⓑ Ⓒ Ⓓ Ⓔ   37. Ⓐ Ⓑ Ⓒ Ⓓ Ⓔ

8. Ⓐ Ⓑ Ⓒ Ⓓ Ⓔ   18. Ⓐ Ⓑ Ⓒ Ⓓ Ⓔ   28. Ⓐ Ⓑ Ⓒ Ⓓ Ⓔ   38. Ⓐ Ⓑ Ⓒ Ⓓ Ⓔ

9. Ⓐ Ⓑ Ⓒ Ⓓ Ⓔ   19. Ⓐ Ⓑ Ⓒ Ⓓ Ⓔ   29. Ⓐ Ⓑ Ⓒ Ⓓ Ⓔ   39. Ⓐ Ⓑ Ⓒ Ⓓ Ⓔ

10. Ⓐ Ⓑ Ⓒ Ⓓ Ⓔ   20. Ⓐ Ⓑ Ⓒ Ⓓ Ⓔ   30. Ⓐ Ⓑ Ⓒ Ⓓ Ⓔ   40. Ⓐ Ⓑ Ⓒ Ⓓ Ⓔ

Practice Test 4

Section I: Questions 1–40

SPEND 90 MINUTES ON THIS PART OF THE EXAM.

DIRECTIONS: The questions or incomplete statements that follow are each followed by five suggested answers or completions. Choose the response that best answers the question or completes the statement.

---

1. A company wishes to determine the relationship between the number of days spent training employees and their performances on a job aptitude test. Collected data result in a least squares regression line, $\hat{y} = 12.1 + 6.2x$, where x is the number of training days and $\hat{y}$ is the predicted score on the aptitude test. Which of the following statements best interprets the slope and y-intercept of the regression line?

   (A) The predicted base score on the test is 12.1, and for every day of training one would expect, on average, an increase of 6.2 on the aptitude test.
   (B) The predicted base score on the test is 6.2, and for every day of training one would expect, on average, an increase of 12.1 on the aptitude test.
   (C) The mean number of training days is 12.1, and for every additional 6.2 days of training one would expect, on average, an increase of one unit on the aptitude test.
   (D) The mean number of training days is 6.2, and for every additional 12.1 days of training one would expect, on average, an increase of one unit on the aptitude test.
   (E) The mean number of training days is 12.1, and for every day of training one would expect, on average, an increase of 6.2 on the aptitude test.

2. To survey the opinions of the students at your high school, a researcher plans to select every twenty-fifth student entering the school in the morning. Assuming there are no absences, will this result in a simple random sample of students attending your school?

   (A) Yes, because every student has the same chance of being selected.
   (B) Yes, but only if there is a single entrance to the school.
   (C) Yes, because the 24 out of every 25 students who are not selected will form a control group.
   (D) Yes, because this is an example of systematic sampling, which is a special case of simple random sampling.
   (E) No, because not every sample of the intended size has an equal chance of being selected.

3. Consider a hypothesis test with $H_0: \mu = 70$ and $H_a: \mu < 70$. Which of the following choices of significance level and sample size results in the greatest power of the test when $\mu = 65$?

   (A) $\alpha = 0.05, n = 15$
   (B) $\alpha = 0.01, n = 15$
   (C) $\alpha = 0.05, n = 30$
   (D) $\alpha = 0.01, n = 30$
   (E) There is no way of answering without knowing the strength of the given power.

4. In 2018, it was estimated that there were roughly 554,000 homeless people in the United States. The average shelter stay for homeless families with kids is 435 days. Assume a skewed left distribution with a standard deviation of 85 days. Consider random samples of size 100 taken from the distribution with the mean length of stay, $\bar{x}$, recorded for each sample. Which of the following is the best description of the sampling distribution of $\bar{x}$?

(A) Skewed left with mean 435 days and standard deviation 0.85 days

(B) Skewed left with mean 435 days and standard deviation 8.5 days

(C) Skewed left with mean 435 days and standard deviation 85 days

(D) Approximately normal with mean 435 days and standard deviation 0.85 days

(E) Approximately normal with mean 435 days and standard deviation 8.5 days

5. The label on a package of cords claims that the breaking strength of a cord is 3.5 pounds, but a hardware store owner believes the real value is less. She plans to test 36 such cords; if their mean breaking strength is less than 3.25 pounds, she will reject the claim on the label. If the standard deviation for the breaking strengths of all such cords is 0.9 pounds, what is the probability of mistakenly rejecting a true claim?

(A) $P\left(z < \dfrac{3.25 - 3.5}{0.9}\right)$

(B) $P\left(z > \dfrac{3.25 - 3.5}{0.9}\right)$

(C) $P\left(z < \dfrac{3.25 - 3.5}{\left(\dfrac{0.9}{\sqrt{36}}\right)}\right)$

(D) $2P\left(z < \dfrac{3.25 - 3.5}{\left(\dfrac{0.9}{\sqrt{36}}\right)}\right)$

(E) $P\left(z > \dfrac{3.25 - 3.5}{\left(\dfrac{0.9}{\sqrt{36}}\right)}\right)$

6. The graph below shows cumulative proportions plotted against numbers of employees working in midsized retail establishments.

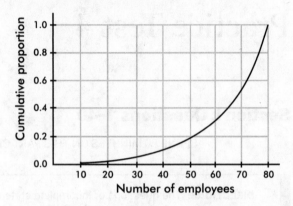

What is the approximate interquartile range?

(A) 18

(B) 35

(C) 57

(D) 68

(E) 75

7. For which of the following probability assignments are events E and F independent?

(A) $P(E \cap F^c) = 0.2$, $P(E \cap F) = 0.0$, $P(E^c \cap F) = 0.2$

(B) $P(E \cap F^c) = 0.3$, $P(E \cap F) = 0.3$, $P(E^c \cap F) = 0.3$

(C) $P(E \cap F^c) = 0.4$, $P(E \cap F) = 0.1$, $P(E^c \cap F) = 0.1$

(D) $P(E \cap F^c) = 0.4$, $P(E \cap F) = 0.2$, $P(E^c \cap F) = 0.1$

(E) $P(E \cap F^c) = 0.5$, $P(E \cap F) = 0.05$, $P(E^c \cap F) = 0.1$

8. To study the effect of alcohol on reaction time, subjects were randomly selected and given three beers to consume. Their reaction time to a simple stimulus was measured before and after drinking the alcohol. Which of the following is a correct statement?

(A) This study was an observational study.
(B) Lack of blinding makes this a poorly designed study.
(C) The placebo effect is irrelevant in this type of study.
(D) This study was an experiment with no controls.
(E) This study was an experiment in which the subjects were used as their own controls.

9. Suppose that the regression line for a set of data, $\hat{y} = a + 7x$ passes through the point $(-2, 4)$. If $\bar{x}$ and $\bar{y}$ are the sample means of the x- and y-values, respectively, then which of the following is equal to $\bar{y}$?

(A) $\bar{x}$
(B) $2 + \bar{x}$
(C) $-4 + \bar{x}$
(D) $7\bar{x}$
(E) $18 + 7\bar{x}$

10. Which of the following is a *false* statement about simple random samples?

(A) A sample must be reasonably large to be properly considered a simple random sample.
(B) Inspection of a sample will give no indication of whether or not it is a simple random sample.
(C) Attributes of a simple random sample may be very different from attributes of the population.
(D) Every element of the population has an equal chance of being picked.
(E) Every sample of the desired size has an equal chance of being picked.

11. In a random sample of 26 men, each with an approximate weight of 160 pounds, Blood Alcohol Concentration (BAC) was measured against number of 12-ounce beers consumed in a one-hour period. Some computer output of the regression analysis follows.

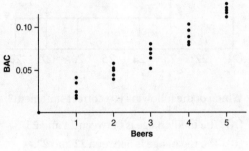

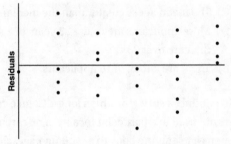

Predictor	Coef	SE Coef	T	P
Constant	0.002131	0.003711	0.57	0.571
Beers	0.023048	0.001141	20.20	0.000

S = 0.008739 R-Sq = 94.4%. R-Sq(adj) = 94.2%

Which of the following gives a 99% confidence interval for the slope of the regression line?

(A) $0.023048 \pm 2.787\left(\dfrac{0.001141}{\sqrt{26}}\right)$

(B) $0.023048 \pm 2.787(0.001141)$

(C) $0.023048 \pm 2.797(0.001141)$

(D) $0.023048 \pm 2.787\left(\dfrac{0.008739}{\sqrt{26}}\right)$

(E) $0.023048 \pm 2.797\left(\dfrac{0.008739}{\sqrt{26}}\right)$

12. The following is a histogram of ages of people applying for a particular high school teaching position.

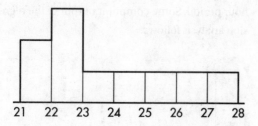

Which of the following is a correct statement?

(A) The median age is between 24 and 25.

(B) The mean age is between 22 and 23.

(C) The mean age is greater than the median age.

(D) More applicants are under 23 years of age than are over 23.

(E) There are a total of 10 applicants.

13. To conduct a survey of which long-distance carriers are used in a particular locality, a researcher opens a telephone book to a random page, closes his eyes, puts his finger down on the page, and then calls the next 75 names. Which of the following is a correct statement?

(A) The procedure results in a simple random sample.

(B) While the survey design does incorporate chance, the procedure could easily result in selection bias.

(C) This is an example of cluster sampling with 75 clusters.

(D) This is an example of stratified sampling with 26 strata.

(E) Given that the researcher truly keeps his eyes closed, this is a good example of blinding.

14. Which of the following is *not* true about *t*-distributions?

(A) There are different *t*-distributions for different values of *df* (degrees of freedom).

(B) *t*-distributions are bell-shaped and symmetric.

(C) *t*-distributions always have mean 0 and standard deviation 1.

(D) *t*-distributions are more spread out than the normal distribution.

(E) The larger the *df* value, the closer the distribution is to the normal distribution.

15. Suppose the probability that a person picked at random has lung cancer is 0.035 and the probability that the person both has lung cancer and is a heavy smoker is 0.014. Given that someone picked at random has lung cancer, what is the probability that the person is a heavy smoker?

(A) $0.035 - 0.014$

(B) $0.035 + 0.014$

(C) $0.035 + 0.014 - (0.035)(0.014)$

(D) $\dfrac{0.035}{1 - 0.014}$

(E) $\dfrac{0.014}{0.035}$

16. Suppose computer science graduates earn an average starting salary of $75,000 with a standard deviation of $12,000. What is the probability that a randomly selected computer science graduate has a starting salary less than $100,000 if it is known that his or her starting salary is over $80,000? Assume a roughly normal distribution of starting salaries of computer science graduates.

(A) $\dfrac{P(z < 100,000)}{P(z > 80,000)}$

(B) $\dfrac{P(80,000 < z < 100,000)}{P(z > 80,000)}$

(C) $\dfrac{P\left(z < \dfrac{100,000}{12,000}\right)}{P\left(z > \dfrac{80,000}{12,000}\right)}$

(D) $\dfrac{P\left(\dfrac{80,000}{12,000} < z < \dfrac{100,000}{12,000}\right)}{P\left(z > \dfrac{80,000}{12,000}\right)}$

(E) $\dfrac{P\left(\dfrac{80,000 - 75,000}{12,000} < z < \dfrac{100,000 - 75,000}{12,000}\right)}{P\left(z > \dfrac{80,000 - 75,000}{12,000}\right)}$

17. There are 4,900,000 people in a large city. A nutritionist wishes to determine the proportion of people in the city who like tofu. She wants to be 95% confident of the answer to within ± 0.014. Which of the following is closest to the sample size she should survey?

(A) 490

(B) 4,900

(C) 49,000

(D) 490,000

(E) 4,900,000

18. Suppose the correlation between two variables is $r = 0.19$. What is the new correlation if 0.23 is added to all values of the x-variable, every value of the y-variable is doubled, and the two variables are interchanged?

(A) 0.84

(B) 0.42

(C) 0.19

(D) −0.19

(E) −0.84

19. Two commercial flights per day are made from a small county airport. The airport manager tabulates the number of on-time departures for a sample of 200 days.

Number of on-time departures	0	1	2
Observed number of days	10	80	110

What is the χ^2 statistic for a goodness-of-fit test that the distribution is binomial with probability equal to 0.8 that a flight leaves on time?

(A) $\dfrac{(10-8)^2}{8} + \dfrac{(80-64)^2}{64} + \dfrac{(110-128)^2}{128}$

(B) $\dfrac{(10-8)^2}{10} + \dfrac{(80-64)^2}{80} + \dfrac{(110-128)^2}{110}$

(C) $\dfrac{(10-10)^2}{10} + \dfrac{(80-30)^2}{30} + \dfrac{(110-160)^2}{160}$

(D) $\dfrac{(10-10)^2}{10} + \dfrac{(80-30)^2}{80} + \dfrac{(110-160)^2}{110}$

(E) $\dfrac{(10-66)^2}{10} + \dfrac{(80-67)^2}{80} + \dfrac{(110-67)^2}{110}$

20. A company has a choice of three investment schemes. Option I gives a sure $25,000 return on investment. Option II gives a 50% chance of returning $50,000 and a 50% chance of returning $10,000. Option III gives a 5% chance of returning $100,000 and a 95% chance of returning nothing. Which option should the company choose?

(A) Option I if the company needs at least $20,000 to pay off an overdue loan

(B) Option II if the company wants to maximize expected return

(C) Option III if the company needs at least $80,000 to pay off an overdue loan

(D) All of the above answers are correct.

(E) Because of chance, it really doesn't matter which option the company chooses.

21. Suppose $P(X) = 0.35$ and $P(Y) = 0.40$. If $P(X|Y) = 0.28$, what is $P(Y|X)$?

(A) $\dfrac{(0.28)(0.35)}{0.40}$

(B) $\dfrac{(0.28)(0.40)}{0.35}$

(C) $\dfrac{(0.35)(0.40)}{0.28}$

(D) $\dfrac{0.28}{0.40}$

(E) $\dfrac{0.28}{0.35}$

22. To test whether extensive exercise lowers the resting heart rate, a study is performed by randomly selecting half of a group of volunteers to exercise 1 hour each morning, while the rest are instructed to perform no exercise. Is this study an experiment or an observational study?

(A) An experiment with a control group and blinding

(B) An experiment with blocking

(C) An observational study with comparison and randomization

(D) An observational study with little if any bias

(E) None of the above statements are correct.

23. The waiting times for a new roller coaster ride are approximately normally distributed with a mean of 35 minutes and a standard deviation of 10 minutes. If there are 150,000 riders the first summer, which of the following is the shortest time interval associated with 100,000 riders?

    (A) 0 to 31.7 minutes
    (B) 31.7 to 39.3 minutes
    (C) 25.3 to 44.7 minutes
    (D) 25.3 to 35 minutes
    (E) 39.3 to 95 minutes

24. A medical researcher, studying the effective durations of three over-the-counter pain relievers, obtains the following boxplots from three equal-sized groups of patients, one group using each of the pain relievers.

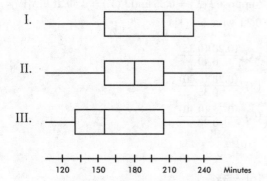

    Which of the following is a correct statement with regard to comparing the effective durations in minutes of the three pain relievers?

    (A) All three have the same interquartile range.
    (B) More patients had over 210 minutes of pain relief in the (I) group than in either of the other two groups.
    (C) More patients had over 240 minutes of pain relief in the (I) group than in either of the other two groups.
    (D) More patients had less than 120 minutes of pain relief in the (III) group than in either of the other two groups.
    (E) The durations of pain relief in the (II) group form a roughly normal distribution.

25. People at a high school were surveyed as to whether they were students or teachers and what their favorite ice cream flavor was, excluding vanilla and chocolate. The results are displayed in the following mosaic plot.

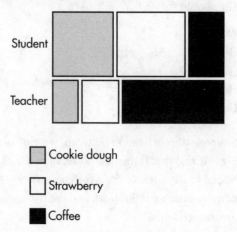

    Based on this plot, which of the following is a true statement?

    (A) The number of students answering "coffee" is greater than the number of teachers answering "coffee."
    (B) More teachers answered either "cookie dough" or "strawberry" than students who answered "cookie dough."
    (C) Of those who answered "strawberry," a greater proportion were teachers than students.
    (D) There were more students than teachers in the survey.
    (E) More people chose cookie dough than strawberry.

26. Suppose that 54% of the graduates from your high school go on to 4-year colleges, 20% go on to 2-year colleges, 19% find employment, and the remaining 7% search for a job. If a randomly selected student is not going on to a 2-year college, what is the probability he or she will be going on to a 4-year college?

    (A) 0.460
    (B) 0.540
    (C) 0.630
    (D) 0.675
    (E) 0.730

27. A congresswoman is interested in the proportion p of registered voters in her district who favor legalizing medical marijuana. Forty-four percent of a simple random sample of 750 registered voters in her district favor legalizing medical marijuana. What is the midpoint for a 95% confidence interval estimate of p?

    (A) 0.025
    (B) 0.05
    (C) 0.418
    (D) 0.44
    (E) p

28. Random samples of size n are drawn from a population. The mean of each sample is calculated, and the standard deviation of this set of sample means is found. Then the procedure is repeated, this time with samples of size $4n$. How does the standard deviation of the second group compare with the standard deviation of the first group?

    (A) It will be the same.
    (B) It will be twice as large.
    (C) It will be four times as large.
    (D) It will be half as large.
    (E) It will be one-quarter as large.

29. Leech therapy is used in traditional medicine for treating localized pain. In a double-blind experiment on 50 patients with osteoarthritis of the knee, half are randomly selected to receive injections of leech saliva, while the rest receive a placebo. Pain levels 7 days later among those receiving the saliva show a mean of 19.5, while pain levels among those receiving the placebo show a mean of 25.6 (higher numbers indicate more pain). Partial calculator output is shown below.

```
  2-SampTTest
μ1<μ2
t=-3.939503313
df=43.43159286
x̄1=19.5
x̄2=25.6
Sx1=4.5
Sx2=6.3
n1=25
n2=25
```

Which of the following is a correct conclusion?

(A) After 7 days, the mean pain level with the leech treatment is significantly lower than the mean pain level with the placebo at the 0.01 significance level.

(B) After 7 days, the mean pain level with the leech treatment is significantly lower than the mean pain level with the placebo at the 0.05 significance level but not at the 0.01 level.

(C) After 7 days, the mean pain level with the leech treatment is significantly lower than the mean pain level with the placebo at the 0.10 significance level but not at the 0.05 level.

(D) After 7 days, the mean pain level with the leech treatment is not significantly lower than the mean pain level with the placebo at the 0.10 significance level.

(E) The proper test should be a one-sample t-test on a set of differences.

30. A survey was conducted to determine the percentage of parents who would support raising the legal driving age to 18. The results were stated as 67% with a margin of error of ±3%. What is meant by ±3%?

(A) Three percent of the population were not surveyed.

(B) In the sample, the percentage of parents who would support raising the driving age is between 64% and 70%.

(C) The percentage of the entire population of parents who would support raising the driving age is between 64% and 70%.

(D) It is unlikely that the given sample proportion result could be obtained unless the true percentage was between 64% and 70%.

(E) Between 64% and 70% of the population were surveyed.

31. It is estimated that 30% of all cars parked in a metered lot outside city hall receive tickets for meter violations. In a random sample of 5 cars parked in this lot, what is the probability that at least one receives a parking ticket?

(A) $1 - (0.3)^5$
(B) $1 - (0.7)^5$
(C) $5(0.3)(0.7)^4$
(D) $5(0.3)^4(0.7)$
(E) $5(0.3)^4(0.7) + 10(0.3)^3(0.7)^2 + 10(0.3)^2(0.7)^3 + 5(0.3)(0.7)^4 + (0.7)^5$

32. Data are collected on income levels x versus number of bank accounts y. Summary calculations give $\bar{x} = 32,000$, $s_x = 11,500$, $\bar{y} = 1.7$, $s_y = 0.4$, and $r = 0.42$. What is the slope of the least squares regression line of number of bank accounts on income level?

(A) $\frac{(0.42)(0.4)}{11,500}$
(B) $\frac{(0.42)(11,500)}{0.4}$
(C) $\frac{(1.7)(11,500)}{0.42}$
(D) $\frac{(1.7)(0.42)}{11,500}$
(E) $\frac{(0.42)(11,500)}{1.7}$

33. Given that the sample has a standard deviation of zero, which of the following is a true statement?

(A) The standard deviation of the population is also zero.
(B) The sample mean and sample median are equal.
(C) The sample may have outliers.
(D) The population has a symmetric distribution.
(E) All samples from the same population will also have a standard deviation of zero.

34. In one study, half of a class were instructed to watch exactly 1 hour of television per day, the other half were told to watch 5 hours per day, and then their class grades were compared. In a second study, students in a class responded to a questionnaire asking about their television usage and their class grades.

(A) The first study was an experiment without a control group, and the second was an observational study.
(B) The first study was an observational study, and the second was a controlled experiment.
(C) Both studies were controlled experiments.
(D) Both studies were observational studies.
(E) Each study was part controlled experiment and part observational study.

35. The table below shows data that were collected from 125 middle school children. It indicates their gender and whether or not they believe that their father could overpower a sumo wrestler.

Believe father can overpower a sumo wrestler			
	Yes	No	Total
Boy	41	30	71
Girl	31	23	54
Total	72	53	125

One of these children is picked at random. Which of the following correctly interprets mutually exclusive events represented by the table?

(A) Believing one's father can overpower a sumo wrestler and not believing one's father can overpower a sumo wrestler
(B) Believing one's father can overpower a sumo wrestler and being a boy

(C) Believing one's father can overpower a sumo wrestler and being a girl

(D) Not believing one's father can overpower a sumo wrestler and being a boy

(E) Not believing one's father can overpower a sumo wrestler and being a girl

36. In leaving for school on an overcast April morning, you make a judgment on the null hypothesis: the weather will remain dry. What would the results be of Type I and Type II errors?

(A) Type I error: get drenched
Type II error: needlessly carry around an umbrella

(B) Type I error: needlessly carry around an umbrella
Type II error: get drenched

(C) Type I error: carry an umbrella, and it rains
Type II error: carry no umbrella, and weather remains dry

(D) Type I error: get drenched
Type II error: carry no umbrella, and weather remains dry

(E) Type I error: get drenched
Type II error: carry an umbrella, and it rains

37. The mean thrust of a certain model jet engine is 9,500 pounds. Concerned that a production process change might have lowered the thrust, an inspector tests a sample of units, calculating a mean of 9,350 pounds with a z-score of -2.46 and a P-value of 0.0069. Which of the following is the most reasonable conclusion?

(A) 99.31% of the engines produced under the new process will have a thrust under 9,350 pounds.

(B) 99.31% of the engines produced under the new process will have a thrust under 9,500 pounds.

(C) 0.69% of the time, an engine produced under the new process will have a thrust over 9,500 pounds.

(D) There is convincing evidence to conclude that the new process is producing engines with a mean thrust under 9,350 pounds.

(E) There is convincing evidence to conclude that the new process is producing engines with a mean thrust under 9,500 pounds.

38. In a group of 10 third graders, the mean height is 50 inches with a median of 47 inches, while in a group of 12 fourth graders, the mean height is 54 inches with a median of 49 inches. What is the median height of the combined group?

(A) 48 inches

(B) 52 inches

(C) $\dfrac{10(47) + 12(49)}{22}$ inches

(D) $\dfrac{10(50) + 12(54)}{22}$ inches

(E) With the information provided, a specific value for the median height of the combined group cannot be determined.

39. A reading specialist in a large public school system believes that the more time students spend reading, the better they will do in school. She plans a middle school experiment in which a simple random sample (SRS) of 30 eighth graders will be assigned four extra hours of reading per week, an SRS of 30 seventh graders will be assigned two extra hours of reading per week, and an SRS of 30 sixth graders with no extra assigned reading will be a control group. After one school year, the mean GPAs from each group will be compared. Is this a good experimental design?

(A) Yes, because simple random samples were used.

(B) No, because while this design may point out an association between reading and GPA, it cannot establish a cause-and-effect relationship.

(C) No, because without blinding, there is a strong chance of a placebo effect.

(D) No, because any conclusion would be flawed because of blocking bias.

(E) No, because grade level is a confounding variable.

40. A study at 35 large city high schools gives the following back-to-back stemplot of the percentages of students who say they have tried alcohol.

School Year 2016–17		School Year 2019–20
	0	
2	1	
9	2	
6 6 3	3	1 8
4 3 1 1	4	0 2 9
9 9 8 6 5 3 2 2 0	5	3 3 4 6
9 8 7 6 6 5 1 1	6	1 2 2 2 7
5 4 3 2 2	7	0 1 3 3 5 5 6 7 8 9
5 4 0	8	2 3 4 5 8 8 9
0	9	0 1 1 2

Key: 1|4|0 represents 41% for the 2016–17 school year and 40% for the 2019–20 school year

Which of the following does *not* follow from the above data?

(A) In general, the percentage of students trying alcohol seems to have increased from 2016–17 to 2019–20.

(B) The median alcohol percentage among the 35 schools increased from 2016–17 to 2019–20.

(C) The spread between the lowest and highest alcohol percentages decreased from 2016–17 to 2019–20.

(D) For both school years in most of the 35 schools, most of the students said they had tried alcohol.

(E) The percentage of students trying alcohol increased in each of the schools between 2016–17 and 2019–20.

STOP   If there is still time remaining, you may check your work on this section.

Section II: Part A

Questions 1–5

SPEND ABOUT 65 MINUTES ON THIS PART OF THE EXAM.
PERCENTAGE OF SECTION II GRADE—75

DIRECTIONS: You must show all work and indicate the methods you use. You will be graded on the correctness of your methods and on the accuracy of your results and explanations.

1. A guidance counselor at a large college is interested in interviewing students about their college experience. After talking with a local AP Statistics teacher, the guidance counselor plans the following three-stage sampling procedure.

   (1) The counselor will obtain separate lists of the four groups of students on the college campus: freshmen, sophomores, juniors, and seniors.
   (2) Each of the four groups takes a number of large classes together. The counselor will use a random number generator to pick three of these classes for each of the four groups.
   (3) In each of the chosen classes, the counselor will pick every fifth student entering the classroom.

   (a) The first stage above represents what kind of sampling procedure? Give an advantage in using it in this context.
   (b) The second stage above represents what kind of sampling procedure? Give an advantage in using it in this context.
   (c) The third stage above represents what kind of sampling procedure? Give a *disadvantage* in using it in this context.

2. A laboratory is testing the concentration level in milligrams per milliliter for the active ingredient found in a pharmaceutical product. In a random sample of five vials of the product, the concentrations were measured at 2.46, 2.57, 2.70, 2.64, and 2.54 mg/mL.

   (a) Determine a 95% confidence interval for the mean concentration level in milligrams per milliliter for the active ingredient found in this pharmaceutical product.
   (b) Explain in words what effect an increase in confidence level would have on the width of the confidence interval.
   (c) Suppose a concentration above 2.70 milligrams per milliliter is considered dangerous. What conclusion is justified by your answers to (a) and (b)?

3. An instructor takes an anonymous survey and notes exam score, hours studied, and gender for the first exam in a large college statistics class ($n = 250$). A resulting regression model is

$$\widehat{Score} = 50.90 + 9.45(Hours) + 4.40(Gender)$$

where *Gender* takes the value 0 for men and 1 for women.

(a) Provide an interpretation in context for each of the three numbers appearing in the above model formula.

(b) Sketch the separate prediction lines for men and women resulting from using 0 or 1 in the above model.

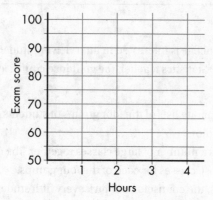

Looking at the data separately by gender results in the following two regression models:

For men: $\widehat{Score} = 51.8 + 8.5\ (Hours)$

For women: $\widehat{Score} = 54.4 + 10.4\ (Hours)$

(c) What comparative information do the slope coefficients from these two models give that does not show in the original model?

4. A high school guidance counselor compared the grade point averages of a random sample of students who have pulled at least one all-nighter (a night of total sleep deprivation) and a random sample of students who never have pulled an all-nighter. She calculated the following statistics:

	Sample Size	Sample Mean	Sample SD
At least one all-nighter	58	3.01	0.31
No all-nighters	43	3.14	0.26

(a) What are the explanatory and response variables?

(b) Is this an observational study or an experiment? Explain.

(c) A two-sample *t*-test with $H_a : \mu_1 < \mu_2$ yields a *P*-value of 0.012, where μ_1 is the mean GPA among the population of all students who have pulled at least one all-nighter, and μ_2 is the mean GPA among the population of all students who never have pulled an all-nighter. Is it appropriate to draw a cause-and-effect conclusion between pulling an all-nighter and having a lower GPA? Explain why or why not.

(d) Identify a (potential) confounding variable in this study. Describe how it could provide an alternative explanation for why students who pulled an all-nighter have a smaller mean GPA than students who have not.

5. In the famous doll experiment, researchers studied racial awareness and child development by investigating whether preschool African Americans would tend to select the white doll when offered a choice of playing with a white or black one. In one study, 18 out of 24 preschool African Americans chose the white doll.

(a) Describe the null hypothesis in words.

(b) Suppose you were to perform a simulation analysis to investigate whether the observed result provides strong evidence that preschool African Americans prefer to play with white, rather than black, dolls. Which of the following graphs could most reasonably have come from the simulation? Explain.

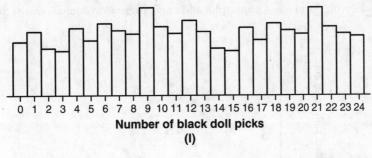

Number of black doll picks
(I)

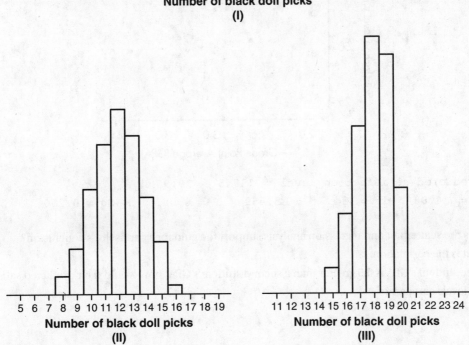

Number of black doll picks
(II)

Number of black doll picks
(III)

(c) Based on the correct graph, which is closest to the P-value of this test: 1.000, 0.500, 0.100, 0.050, or 0.001?

(d) What is the conclusion based on this test and the simulation?

Section II: Part B
Question 6

SPEND ABOUT 25 MINUTES ON THIS PART OF THE EXAM.
PERCENTAGE OF SECTION II GRADE—25

6. A college guidance counselor obtains a survey from a random sample of 40 students who live in the college dormitories. He concludes that students with higher GPAs (on a 4-point scale) appear to be less satisfied (on a 100-point scale) with dorm life. A scatterplot and linear regression analysis shows the following computer output:

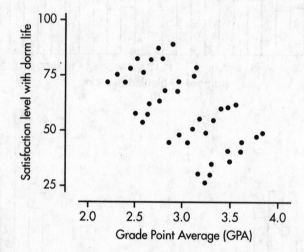

```
Predicted satisfaction level = 117.5 - 19.93(GPA)
S = 14.80    df = 38    t = -3.559    P = 0.001    R-Sq = 0.250
```

(a) Does the scatterplot and regression analysis support the guidance counselor's conclusion?
(b) What is the correlation, *r*?
(c) If one student's GPA is 0.5 greater than a second student's GPA, how would their predicted satisfaction levels compare?

A student who had taken AP Statistics in high school reads that the survey was obtained from a stratified random sample. She analyzes the data and produces the following revised scatterplot:

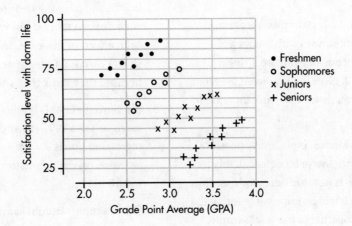

(d) If one student's GPA is 0.5 greater than a second student's GPA, how would their predicted satisfaction levels compare if the students are both sophomores?

(e) Based on the student's analysis, how should the counselor's conclusion be modified to give a better description of the relationship between GPA and satisfaction level for the students in the sample?

STOP If there is still time remaining, you may check your work on this section.

Answer Explanations

Section I

1. **(A)** The *y*-intercept, 12.1, gives the average aptitude test score for employees with 0 days training, and this can be considered to be a "base score." The slope, 6.2, gives the predicted average increase in the *y*-variable for each unit increase in the *x*-variable.

2. **(E)** For a simple random sample, every possible group of the given size has to be equally likely to be selected, and this is not true here. For example, with this procedure it will be impossible for all the early arrivals to be together in the final sample. This procedure is an example of systematic sampling (although the starting point should be randomly selected), but systematic sampling does not result in simple random samples.

3. **(C)** Power $= 1 - \beta$, and β is smallest when both the significance level α and the sample size n are greater.

4. **(E)** By the central limit theorem, the sampling distribution of $\bar{x}$ is approximately normal with mean equal to the population mean and with standard deviation equal to the population standard deviation divided by the square root of the sample size. In this example, $\mu_{\bar{x}} = \mu = 435$ and $\sigma_{\bar{x}} = \frac{\sigma}{\sqrt{n}} = \frac{85}{\sqrt{100}} = 8.5$.

5. **(C)** We have $H_0: \mu = 3.5$ and $H_a: \mu < 3.5$. Then $\sigma_{\bar{x}} = \frac{0.9}{\sqrt{36}}$ and the *z*-score of 3.25 is $\frac{3.25 - 3.5}{\left(\frac{0.9}{\sqrt{36}}\right)}$.

6. **(A)** The cumulative proportions of 0.25 and 0.75 correspond to $Q_1 = 57$ and $Q_3 = 75$, respectively, and so the interquartile range is $75 - 57 = 18$.

7. **(C)** Events *E* and *F* are independent if $P(E \cap F) = P(E)P(F)$. Here, $P(E) = 0.4 + 0.1 = 0.5$ and $P(F) = 0.1 + 0.1 = 0.2$. Then $P(E)P(F) = (0.5)(0.2) = 0.1 = P(E \cap F)$.

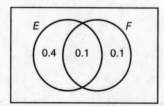

8. **(E)** In experiments on people, subjects can be used as their own controls, with responses noted before and after the treatment. However, with such designs there is always the danger of a placebo effect. In this case, subjects might well have slower reaction times after drinking the alcohol because they think they should.

9. **(E)** Since $(-2, 4)$ is on the line $\hat{y} = a + 7x$, we have $4 = a + 7(-2)$ and $a = 18$. Thus, the regression line is $\hat{y} = 18 + 7x$. The point $(\bar{x}, \bar{y})$ is always on the regression line, and so we have $\hat{y} = 18 + 7\bar{x}$.

10. **(A)** A simple random sample can be any size.

11. **(C)** The critical *t*-values with $df = 26 - 2 = 24$ and 0.005 in each tail are $\pm \text{invT}(0.995, 24) = \pm 2.797$. Thus, we have $b \pm t^* \times SE(b) = 0.023048 \pm 2.797(0.001141)$.

12. **(C)** When the distribution is skewed to the right, the mean is usually greater than the median. Half the area is on either side of 23, so 23 is the median. With half the area to each side of 23, half the applicants' ages are to each side of 23. Histograms such as this show relative frequencies, not actual frequencies.

13. **(B)** While the procedure does use some element of chance, all possible groups of size 75 do not have the same chance of being picked, so the result is not a simple random sample. There is a real chance of selection bias. For example, a number of relatives with the same last name and all using the same long-distance carrier might be selected.

14. **(C)** While *t*-distributions do have mean 0, their standard deviations are greater than 1.

15. **(E)**

$$P(\text{smoker}|\text{cancer}) = \frac{P(\text{smoker} \cap \text{cancer})}{P(\text{cancer})}$$

$$= \frac{0.014}{0.035}$$

16. **(E)** The critical z-scores are $\dfrac{80{,}000 - 75{,}000}{12{,}000}$ and $\dfrac{100{,}000 - 75{,}000}{12{,}000}$. The probability of being less than \$100,000 given that the starting salary is over \$80,000 is

$$\dfrac{P(80{,}000 < x < 100{,}000)}{P(x > 80{,}000)} =$$

$$\dfrac{P\left(\dfrac{80{,}000 - 75{,}000}{12{,}000} < z < \dfrac{100{,}000 - 75{,}000}{12{,}000}\right)}{P\left(z > \dfrac{80{,}000 - 75{,}000}{12{,}000}\right)}$$

17. **(B)**

$$1.96\sqrt{\dfrac{(0.5)(0.5)}{n}} \le 0.014 \text{ gives}$$

$$\sqrt{n} \ge \dfrac{1.96(0.5)}{0.014} = 70 \text{ and } n = 4{,}900.$$

18. **(C)** The correlation coefficient is not changed by adding the same number to every value of one of the variables, by multiplying every value of one of the variables by the same positive number, or by interchanging which are the x- and y-variables.

19. **(A)** The binomial distribution with $n = 2$ and $p = 0.8$ is $P(0) = (0.2)^2 = 0.04$, $P(1) = 2(0.2)(0.8) = 0.32$, and $P(2) = (0.8)^2 = 0.64$, resulting in expected numbers of $0.04(200) = 8$, $0.32(200) = 64$, and $0.64(200) = 128$. Thus,

$$X^2 = \sum \dfrac{(\text{observed} - \text{expected})^2}{\text{expected}}$$

$$= \dfrac{(10 - 8)^2}{8} + \dfrac{(80 - 64)^2}{64} + \dfrac{(110 - 128)^2}{128}$$

20. **(D)** Option I guarantees that the \$20,000 loan will be paid off. Option II gives the highest expected return: $(50{,}000)(0.5) + (10{,}000)(0.5) = 30{,}000$, which is greater than 25,000 and is also greater than $(100{,}000)(0.05) = 5000$. Option III provides the only chance of paying off the \$80,000 loan. The moral is that the highest expected value is not automatically the "best" answer.

21. **(B)** $P(X|\,Y) = \dfrac{P(X \cap Y)}{P(Y)}$ gives that $P(X \cap Y)$ $= P(X|Y)\,P(Y) = (0.28)(0.40)$.

Then $P(Y|\,X) = \dfrac{P(X \cap Y)}{P(X)} = \dfrac{(0.28)(0.40)}{0.35}$.

22. **(E)** This study is an experiment because a treatment (extensive exercise) is imposed. There is no blinding because subjects clearly know whether or not they are exercising. There is no blocking because subjects are not divided into blocks before random assignment to treatments. For example, blocking would have been used if subjects had been separated by gender or age before random assignment to exercise or not.

23. **(C)** From the shape of the normal curve, the answer is in the middle. We have $\dfrac{100{,}000}{150{,}000} = \dfrac{2}{3}$. The middle two-thirds (with $\dfrac{1}{6}$ in each tail) is between z-scores of $\pm\texttt{invNorm(5/6)} = \pm 0.97$, and $35 \pm 0.97(10)$ gives $(25.3, 44.7)$.

24. **(B)** More than 25% of the patients in the (I) group had over 210 minutes of pain relief, which is not the case for the other two groups. The interquartile range is the length of the box, so they are not all equal. There is no way to positively conclude a normal distribution from a boxplot.

25. **(D)** In mosaic plots, the area of a box is proportional to the count corresponding to that box. The vertical axis in this plot indicates the proportion of students versus teachers. The horizontal axis in this plot indicates the proportion of people choosing each of the three ice cream flavors for students and for teachers. Choice (D) is true because the three boxes corresponding to students have a total area greater than the three boxes corresponding to teachers, or noting that along the vertical axis, "Student" has a greater length than "Teacher."

26. **(D)** $\dfrac{0.54}{0.54 + 0.19 + 0.07} = 0.675$

27. **(D)** The midpoint of the confidence interval is the sample proportion $\hat{p} = 0.44$.

28. **(D)** $\sigma_{\bar{x}} = \dfrac{\sigma}{\sqrt{n}}$, so the standard deviation of sample means is inversely related to the square root of the sample size. Thus, increasing the sample size by a multiple of d^2 divides the standard deviation of the set of sample means by d.

29. **(A)** With $df = 43.43$ and $t = -3.94$, tcdf gives that the P-value is $0.000146 < 0.01$. With a P-value this small, there is sufficient evidence to reject H_0; that is, there is convincing evidence that the mean pain level with the leech treatment is lower than the mean pain level with the placebo.

30. **(D)** While the sample proportion is between 64% and 70% (more specifically, it is 67%), this is not the meaning of $\pm 3\%$. While the percentage of the entire population is likely to be between 64% and 70%, this is not known for certain.

31. **(B)** This is a binomial distribution with $n = 5$ and $p = 0.3$. The probability that a car does not receive a ticket is $1 - 0.3 = 0.7$, the probability that none of the five cars receives a ticket is $(0.7)^5$, and thus the probability that at least one receives a ticket is $1 - (0.7)^5$.

32. **(A)** $b = r\dfrac{s_y}{s_x} = 0.42\dfrac{0.4}{11,500} = \dfrac{(0.42)(0.4)}{11,500}$

33. **(B)** If the standard deviation of a set is zero, all the values in the set are equal. The mean and median would both equal this common value and so would equal each other. If all the values are equal, there are no outliers. Just because the sample happens to have one common value, there is no reason for this to be true for the whole population. Statistics from one sample can be different from statistics from any other sample.

34. **(A)** The first study is an experiment with two treatment groups (1 hour and 5 hours of television per night) and no control group. The second study is observational; the researcher simply noted the students' self-reported responses.

35. **(A)** Based on the table, children can either believe their father can overpower a sumo wrestler or not believe their father can overpower a sumo wrestler. Because the events cannot occur at the same time, the events are mutually exclusive.

36. **(B)** A Type I error means that the null hypothesis is correct (the weather will remain dry) but you reject it (thus you needlessly carry around an umbrella). A Type II error means that the null hypothesis is wrong (it will rain) but you fail to reject it (thus you get drenched).

37. **(E)** If the sample statistic is far enough away from the claimed population parameter, we say that there is sufficient evidence to reject the null hypothesis. In this case the null hypothesis is that $\mu = 9,500$. The P-value is the probability of obtaining a sample statistic as extreme as, or more extreme than, the one obtained if the null hypothesis is assumed to be true. The smaller the P-value, the more significant the difference between the null hypothesis and the sample results. With $P = 0.0069$, there is strong evidence to reject H_0.

38. **(E)** There are $10 + 12 = 22$ students in the combined group. In ascending order, where are the two middle scores? At least 5 third graders and 6 fourth graders have heights less than or equal to 49 inches, so at most 11 students have heights greater than or equal to 49. Thus, the median is less than or equal to 49. At least 5 third graders and 6 fourth graders have heights greater than or equal to 47 inches, so at most 11 students have heights less than or equal to 47. Thus, the median is greater than or equal to 47. All that can be said about the median of the combined group is that it is between 47 and 49 inches.

39. **(E)** Good experimental design aims to give each group the same experiences except for the treatment under consideration. Thus, all three SRSs should be picked from the same grade level.

40. **(E)** The stemplot does not indicate what happened for any individual school.

Section II: Part A

1. (a) This is an example of *stratified sampling*, where the groups (freshmen, sophomores, juniors, and seniors) are strata. The advantage is that the counselor will ensure that the final sample will represent all four groups of students, each of which might have different views of their college experience.

   (b) This an example of *cluster sampling*, where each class (cluster) resembles the overall group from which it is selected. The advantage is that using these clusters is much more practical than trying to sample from among the large groups (freshman, sophomores, juniors, and seniors).

   (c) This is an example of *systematic sampling*, which is quicker and easier than many other procedures. A possible disadvantage occurs if ordering is related to the variable under consideration. For example, in this study, if the order students enter the classroom is related to their views of their college experience, the counselor could end up with a nonrepresentative sample.

SCORING
Part (a) is essentially correct for identifying the procedure and giving a correct advantage in context. Part (a) is partially correct for one of these two elements.
Part (b) is essentially correct for identifying the procedure and giving a correct advantage in context. Part (b) is partially correct for one of these two elements.
Part (c) is essentially correct for identifying the procedure and giving a correct disadvantage in context. Part (c) is partially correct for one of these two elements.

4	**Complete Answer**	All three parts essentially correct.
3	**Substantial Answer**	Two parts essentially correct and one part partially correct.
2	**Developing Answer**	Two parts essentially correct OR one part essentially correct and one or two parts partially correct OR all three parts partially correct.
1	**Minimal Answer**	One part essentially correct OR two parts partially correct.

2. (a) *Parameter:* Let μ represent the mean concentration level for the active ingredient found in the population of vials of this pharmaceutical product.

 Procedure: A one-sample *t*-interval for the mean.

 Checks: We are given that this is a random sample, $n = 5$ is less than 10% of all vials of this product, and a dotplot

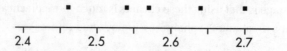

 makes the nearly normal condition not unreasonable.

 Mechanics: Putting the data in a List, calculator software gives (2.4674, 2.6966).

 Conclusion in context: We are 95% confident that the true mean concentration level for the active ingredient found in this pharmaceutical product is between 2.467 and 2.697 milligrams per milliliter.

> **TIP**
>
> Use TInterval on the TI-84 or 1-Sample tInterval on the Casio Prizm.

(b) For higher confidence, we must accept a wider interval of plausible values.

(c) Since the whole confidence interval (2.467, 2.697) is below the critical 2.7, at the 95% confidence level the mean concentration is at a safe level. However, with 2.697 so close to 2.7, based on the statement in (b), if the confidence level is raised we are no longer confident that the mean concentration is at a safe level.

SCORING

This question is scored in three sections. Section 1 consists of part (a1). Section 2 consists of part (a2). Section 3 consists of parts (b) and (c).

Section 1 is essentially correct for identifying the parameter, naming the procedure, and checking all three conditions, and is partially correct for three or four of these five components.

Section 2 is essentially correct for correct mechanics and a correct conclusion in context, and is partially correct for one of these two components.

Section 3 is essentially correct for (1) a correct statement in Part (b), (2) a correct conclusion in context for the 95% interval, and (3) a correct conclusion in context for the case where the confidence is raised. Section 3 is partially correct for two of these three components correct.

4	**Complete Answer**	All three sections essentially correct.
3	**Substantial Answer**	Two sections essentially correct and one section partially correct.
2	**Developing Answer**	Two sections essentially correct OR one section essentially correct and one or two sections partially correct OR all three sections partially correct.
1	**Minimal Answer**	One section essentially correct OR two sections partially correct.

3. (a) The value 50.90 estimates the average score of men who spend 0 hours studying. The value 9.45 estimates the average increase in score for each additional hour of study time. For any fixed number of hours of study time, the value 4.40 estimates the average number of points that women score higher than men.

(b)

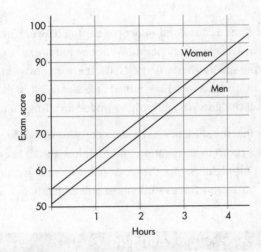

(c) For each additional hour of study time, the scores of men increase an average of 8.5 while those of women increase an average of 10.4. Additional hours of study time appear to benefit women more than men, something that does not show in the original model.

SCORING

Part (a) is essentially correct for correctly interpreting all three numbers in context and partially correct for correctly interpreting two of the three numbers.

Part (b) is essentially correct for correct graphs, clearly parallel and labeled as to which is which. Part (b) is partially correct for correct parallel graphs but missing labels or for labeled, parallel graphs with incorrect y-intercepts.

Part (c) is essentially correct for correctly interpreting both slopes and for some comparative statement about additional hours of study time appearing to benefit women more than men. Part (c) is partially correct for correctly interpreting both slopes but failing to make a comparative statement or for making a correct comparative statement without interpreting the slopes.

4	**Complete Answer**	All three parts essentially correct.
3	**Substantial Answer**	Two parts essentially correct and one part partially correct.
2	**Developing Answer**	Two parts essentially correct OR one part essentially correct and one or two parts partially correct OR all three parts partially correct.
1	**Minimal Answer**	One part essentially correct OR two parts partially correct.

4. (a) The categorical explanatory variable is whether or not the student pulled at least one all-nighter. The quantitative response variable is the student's grade point average (GPA).

   (b) This is an observational study, because the students decided for themselves whether or not to pull an all-nighter. They were not assigned, randomly or otherwise, to pull an all-nighter or not to pull an all-nighter.

   (c) No, it is not appropriate to draw a cause-and-effect conclusion between pulling an all-nighter and having a lower GPA because this was an observational study, not a randomized experiment. An appropriate conclusion is that because of the small P-value, $0.012 < 0.05$, there is convincing evidence that the mean GPA among the population of all students who have pulled at least one all-nighter is lower than the mean GPA among the population of all students who never have pulled an all-nighter.

   (d) One reasonable answer is that the student's *study skills* constitute a confounding variable. Perhaps students with poor study skills resort to all-nighters, and their low grades are a consequence of their poor study skills rather than the all-nighters. Another reasonable answer is *coursework difficulty*, the argument being that more difficult coursework forces students to pull all-nighters and also leads to lower grades.

SCORING

This question is scored in three sections. Section 1 consists of parts (a) and (b). Section 2 consists of part (c). Section 3 consists of part (d).

Section 1 is essentially correct for stating that (1) the explanatory variable is whether or not the student pulled at least one all-nighter, (2) the response variable is the student's grade point average, (3) this is an observational study, and (4) the students decided for themselves whether or not to pull an all-nighter. Section 1 is partially correct for 3 of the 4 components.

Section 2 is essentially correct for answering that it is not appropriate to draw a cause-and-effect conclusion and explaining why it is not. Section 2 is partially correct for answering that it is not appropriate to draw a cause-and-effect conclusion and giving a weak explanation. Section 2 is incorrect for answering that it is appropriate to draw a cause-and-effect conclusion or for answering that it is not appropriate to draw a cause-and-effect conclusion but with no explanation or an incorrect explanation.

Section 3 is essentially correct for giving a reasonable confounding variable with a correct explanation. Section 3 is partially correct for giving a reasonable confounding variable with a weak explanation. Section 3 is incorrect for giving a poor or reasonable confounding variable with no explanation or an incorrect explanation.

4	**Complete Answer**	All three sections essentially correct.
3	**Substantial Answer**	Two sections essentially correct and one section partially correct.
2	**Developing Answer**	Two sections essentially correct OR one section essentially correct and one or two sections partially correct OR all three sections partially correct.
1	**Minimal Answer**	One section essentially correct OR two sections partially correct.

5. (a) The null hypothesis is that preschool African Americans have no preference as to playing with white dolls or black dolls.

   (b) If there was no preference, we would expect the distribution to be centered around 12, tapering off in each direction from there. Choice II is the only reasonable graph.

   (c) For the middle graph, very few of the simulations produced 18 or more black doll picks, so the P-value closest to the given choices is the smallest, 0.001.

   (d) With this small of a P-value, $0.001 < 0.05$, there is convincing evidence that the population of preschool African Americans tends to select the white doll when offered a choice of playing with a white or black one.

SCORING

This question is scored in three sections. Section 1 consists of parts (a) and (b). Section 2 consists of part (c). Section 3 consists of part (d).

Section 1 is essentially correct for (1) stating the correct null hypothesis, (2) picking choice II, and (3) explaining why the graph in choice II is the correct choice. Section 1 is partially correct for 2 of these 3 components.

Section 2 is essentially correct for (1) picking 0.001 and (2) giving a correct explanation for this choice. It is partially correct for picking 0.001 and giving a weak explanation. Section 2 is incorrect for not picking 0.001 or for picking 0.001 and not giving at least a weak explanation.

Section 3 is essentially correct for a correct conclusion referencing (1) the P-value, (2) context, and (3) the population. Section 3 is partially correct for a correct conclusion referencing 2 of these 3 components.

4	**Complete Answer**	All three sections essentially correct.
3	**Substantial Answer**	Two sections essentially correct and one section partially correct.
2	**Developing Answer**	Two sections essentially correct OR one section essentially correct and one or two sections partially correct OR all three sections partially correct.
1	**Minimal Answer**	One section essentially correct OR two sections partially correct.

Section II: Part B

6. (a) The scatterplot is generally negative, and in fact the slope is given to be a negative number. So, yes, this supports the guidance counselor's conclusion.

   (b) $r = -\sqrt{0.250} = -0.5$.

   (c) $(-19.93)(0.5) = -9.965$. So, the student with the greater GPA is predicted to have a satisfaction level 9.965 lower than the other student.

   (d) Two points on the sophomore regression line are approximately (2.5, 55) and (3.0, 75), so the sophomore student with the 0.5 higher GPA is predicted to have a $75 - 55 = 20$ higher satisfaction level.

   (e) Overall, higher grade point averages are associated with lower satisfaction with dorm life. However, within each class level (freshmen, sophomores, juniors, and seniors), students with higher grade point averages tend to have higher satisfaction with dorm life.

SCORING

This question is scored in three sections. Section 1 consists of parts (a) and (b). Section 2 consists of parts (c) and (d). Section 3 consists of part (e).

Section 1 is essentially correct for answering yes in (a), justifying by noting the negative slope, and correctly calculating the correlation in (b). Section 1 is partially correct for two out of the three steps above.

Section 2 is essentially correct for calculating the "9.965 lower" in (c) and the "20 higher" in (d) and is partially correct for one out of these two steps correct.

Section 3 is essentially correct if the response states there is a positive association for each class AND the response notes the overall negative association. Section 3 is partially correct if the response states there is a positive association for each class but does not note the overall negative association.

4	Complete Answer	All three sections essentially correct.
3	Substantial Answer	Two sections essentially correct and one section partially correct.
2	Developing Answer	Two sections essentially correct OR one section essentially correct and one or two sections partially correct OR all three sections partially correct.
1	Minimal Answer	One section essentially correct OR two sections partially correct.

PART 5
Appendices

Appendices

Graphical Displays

	Right-skewed	**Symmetric**	**Left-skewed**

Dotplots

   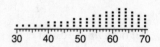

Stemplots

Right-skewed	Symmetric	Left-skewed
3 | 00000000022222224444446666668888	3 | 0246688	3 | 02468
4 | 000022244466688	4 | 000222444466666888888	4 | 00224466888
5 | 002244668	5 | 00000022222244444466666888	5 | 0002224446666688888
6 | 02468	6 | 000224468	6 | 0000002222224444446666668888
7 | 0	7 | 0	7 | 0000

Histograms

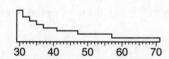   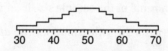   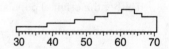

Cumulative Frequency Plots

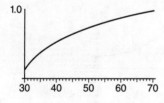   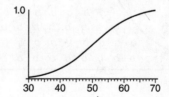   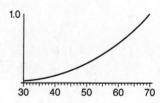

Boxplots

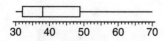   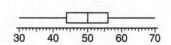   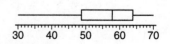

Sampling distribution of the sample proportion

$$\sigma_{\hat{p}} = \sqrt{\frac{p(1-p)}{n}}$$

If both $np \geq 10$ and $n(1-p) \geq 10$, then $\dfrac{\hat{p} - p}{\sqrt{\dfrac{p(1-p)}{n}}}$ has a roughly normal distribution.

Sampling distribution of the sample mean

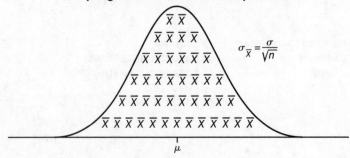

$$\sigma_{\overline{x}} = \frac{\sigma}{\sqrt{n}}$$

If either the original population is normal or the sample size is large enough ($n \geq 30$), then $\dfrac{\tilde{x} - \mu}{\sigma/\sqrt{n}}$ has a roughly normal distribution, and $\dfrac{\tilde{x} - \mu}{s/\sqrt{n}}$ has a roughly t-distribution with $df = n - 1$.

Sampling distribution of the sample slope

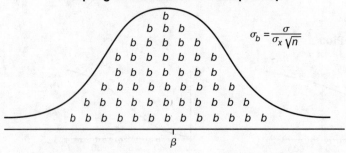

$$\sigma_b = \frac{\sigma}{\sigma_x \sqrt{n}}$$

If the distribution of residuals is normal, then $\dfrac{b - \beta}{\sigma/\sigma_x\sqrt{n}}$ has a roughly normal distribution, and $\dfrac{b - \beta}{s/s_x\sqrt{n-1}} = \dfrac{b - \beta}{SE(b)}$ has a roughly t-distribution with $df = n - 2$.

Guide to Inference

Confidence interval for proportions:

Procedure	Parameter	Statistic	Conditions	Formula	Calculators*
One-sample z-interval for a proportion	p	$\hat{p}$	■ Random sample ■ $n \leq 10\%N$ ■ $n\hat{p} \geq 10, n(1-\hat{p}) \geq 10$	$\hat{p} \pm z^* \sqrt{\dfrac{\hat{p}(1-\hat{p})}{n}}$	1-PropZInt 1-Props Z Interval Z-Int: 1π
Two-sample z-interval for a difference in proportions	$p_1 - p_2$	$\hat{p}_1 - \hat{p}_2$	■ Independent random samples ■ $n_1 \leq 10\%N_1, n_2 \leq 10\%N_2$ ■ $n_1\hat{p}_1 \geq 10, n_1(1-\hat{p}_1) \geq 10$ $n_2\hat{p}_2 \geq 10, n_2(1-\hat{p}_2) \geq 10$	$\hat{p}_1 - \hat{p}_2 \pm z^*$ $\sqrt{\dfrac{\hat{p}_1(1-\hat{p}_1)}{n_1} + \dfrac{\hat{p}_2(1-\hat{p}_2)}{n_2}}$	2-PropZInt 2-Prop Z Interval Z-Int: $\pi_1 - \pi_2$

Confidence interval for means:

Procedure	Parameter	Statistic	Conditions	Formula	Calculators*
One-sample t-interval for a mean	μ	$\bar{x}$	■ Random sample ■ $n \leq 10\%N$ ■ Population $\approx$ normal OR $n \geq 30$ OR sample shows no strong skewness or outliers	$\bar{x} \pm t^* \dfrac{s}{\sqrt{n}}$ with $df = n - 1$	TInterval 1-Sample t Interval T-Int: 1μ
Two-sample t-interval for a difference in means	$\mu_1 - \mu_2$	$\bar{x}_1 - \bar{x}_2$	■ Independent random samples ■ $n_1 \leq 10\%N_1, n_2 \leq 10\%N_2$ ■ For each sample, population $\approx$ normal OR $n \geq 30$ OR sample shows no strong skewness or outliers	$(\bar{x}_1 - \bar{x}_2) \pm t^* \sqrt{\dfrac{(s_1)^2}{n_1} + \dfrac{(s_2)^2}{n_2}}$ with df from calculator	2-SampTInt 2-Sample t Interval T-Int: $\mu_1 - \mu_2$
Paired data t-interval for a mean difference	μ_{diff}	$\bar{x}_{\text{diff}}$	■ Random sample of paired data ■ $n_{\text{diff}} \leq 10\%N_{\text{diff}}$ ■ Population of differences $\approx$ normal OR $n_{\text{diff}} \geq 30$ OR sample differences shows no strong skewness or outliers	$\bar{x}_{\text{diff}} \pm t^* \dfrac{s_{\text{diff}}}{\sqrt{n_{\text{diff}}}}$ with $df = n_{\text{diff}} - 1$	TInterval 1-Sample t Interval T-Int: 1μ

Confidence interval for slopes:

Procedure	Parameter	Statistic	Conditions	Formula	Calculators*
t-interval for a slope	β	b	■ Random sample ■ True relationship is linear ■ $n \leq 10\%N$ ■ Same y SD for all x-values ■ Plot of residuals $\approx$ normal OR $n \geq 30$	$b \pm t^*SE(b)$ with $df = n - 2$	LinRegTInt LinearReg t Interval Interval: Slope

Significance test for proportions:

Procedure	Null Hypothesis	Conditions	Formula	Calculators*
One-sample z-test for a proportion	$H_0: p = p_0$	■ Random sample ■ $n \leq 10\%N$ ■ $np_0 \geq 10, n(1-p_0) \geq 10$	$z = \dfrac{\hat{p} - p_0}{\sqrt{\dfrac{p_0(1-p_0)}{n}}}$	1-PropZTest 1-Prop Z Test Z-Test: 1π
Two-sample z-test for a difference in proportions	$H_0: p_1 - p_2 = 0$ OR $H_0: p_1 = p_2$	■ Independent random samples ■ $n_1 \leq 10\%N_1, n_2 \leq 10\%N_2$ ■ $n_1\hat{p}_c \geq 10, n_1(1-\hat{p}_c) \geq 10$ $n_2\hat{p}_c \geq 10, n_2(1-\hat{p}_c) \geq 10$ where $\hat{p}_c = \dfrac{x_1 + x_2}{n_1 + n_2}$	$z = \dfrac{(\hat{p}_1 - \hat{p}_2) - 0}{\sqrt{\dfrac{\hat{p}_c(1-\hat{p}_c)}{n_1} + \dfrac{\hat{p}_c(1-\hat{p}_c)}{n_2}}}$	2-PropZTest 2-Prop Z Test Z-Test: $\pi_1 - \pi_2$

Significance test for means:

Procedure	Null Hypothesis	Conditions	Formula	Calculators*
One-sample t-test for a mean	$H_0: \mu = \mu_0$	■ Random sample ■ $n \le 10\%N$ ■ Population $\approx$ normal OR $n \ge 30$ OR sample shows no strong skewness or outliers	$t = \dfrac{\bar{x} - \mu_0}{\frac{s}{\sqrt{n}}}$ with $df = n - 1$	T-Test 1-Sample tTest T-Test: $1\ \mu$
Two-sample t-test for a difference in means	$H_0: \mu_1 - \mu_2 = 0$ OR $H_0: \mu_1 = \mu_2$	■ Independent random samples ■ $n_1 \le 10\%N_1$, $n_2 \le 10\%N_2$ ■ For each sample, population $\approx$ normal OR $n \ge 30$ OR sample shows no strong skewness or outliers	$t = \dfrac{(\bar{x}_1 - \bar{x}_2) - 0}{\sqrt{\frac{(s_1)^2}{n_1} + \frac{(s_2)^2}{n_2}}}$ with df from calculator	2-SampTest 2-Sample tTest T-Test: $\mu_1 - \mu_2$
Paired data t-test for a mean difference	$H_0: \mu_{\text{diff}} = 0$	■ Random sample of paired data ■ $n_{\text{diff}} \le 10\%N_{\text{diff}}$ ■ Population of differences $\approx$ normal OR $n_{\text{diff}} \ge 30$ OR sample differences shows no strong skewness or outliers	$t = \dfrac{\bar{x}_{\text{diff}} - 0}{\frac{s_{\text{diff}}}{\sqrt{n_{\text{diff}}}}}$ with $df = n_{\text{diff}} - 1$	T-Test 1-Sample tTest T-Test: $1\ \mu$

Significance test for slopes:

Procedure	Null Hypothesis	Conditions	Formula	Calculators*
One-sample t-test for a slope	$H_0: \beta = 0$	■ Random sample ■ True relationship is linear ■ $n \le 10\%N$ ■ Same y SD for all x-values ■ Plot of residuals $\approx$ normal OR $n \ge 30$	$t = \dfrac{b - 0}{SE(b)}$ with $df = n - 2$	LinRegTTest LinearReg tTest Linear t test

Significance test for chi-square:

Procedure	Hypotheses	Conditions	Formula	Calculators*
χ^2-test for goodness-of-fit	H_0: The claimed distribution is correct. H_a: The claimed distribution is incorrect.	■ Random sample ■ $n \le 10\%N$ ■ All expected counts > 5	$\chi^2 = \sum \dfrac{(\text{observed} - \text{expected})^2}{\text{expected}}$ $df = \#$ of categories $- 1$	χ^2 GOF-Test χ^2 GOF Test χ^2 Test Goodness of Fit
χ^2-test for independence	H_0: Two variables are independent. H_a: Two variables are not independent.	■ Random sample ■ $n \le 10\%N$ ■ All expected counts > 5	$\chi^2 = \sum \dfrac{(\text{observed} - \text{expected})^2}{\text{expected}}$ $df = (\#$ of rows $- 1)(\#$ of columns $- 1)$	χ^2-Test χ^2 two-way Test χ^2 Test two-way test
χ^2-test for homogeneity	H_0: A distribution is the same across 2 or more populations. H_a: A distribution differs across 2 or more populations.	■ Random samples from each population ■ $n_i \le 10\%N_i$ each sample ■ All expected counts > 5	$\chi^2 = \sum \dfrac{(\text{observed} - \text{expected})^2}{\text{expected}}$ $df = (\#$ of rows $- 1)(\#$ of columns $- 1)$	χ^2-Test χ^2 two-way Test χ^2 Test two-way test

*Calculators refers to TI-84, Casio Prizm, and HP Prime

TEMPLATES

Following are 25 templates with examples to help you word your free-response answers. Remember, it is imperative that you communicate in your responses what you want the reader to understand. Correct calculations and answers will not give you full credit if you do not explain what the numbers mean within the context of the problem.

Describing a one-variable distribution

The distribution of (*context*) has a shape that is roughly (*symmetric/unimodal/bimodal/bell-shaped/skewed left/ skewed right/uniform*), (*center/mean/median*) of ____, (*range of ____ /IQR of ____ /spread from ____ to ____*), and unusual features of (*outliers, gaps, clusters*).

> *Example: In a competition among Gen X gamers, the distribution of Tetris scores was unimodal and roughly symmetric with a median of 160,000, a range of 175,000, a possible outlier at 260,000, and a gap between 240,000 and 260,000.*

Comparing one-variable distributions

The (*first distribution in context*) has a ____ shape that is (*similar/not similar*) to the (*second distribution in context*) that has a ____ shape, the (*first distribution in context*) has a (*larger/smaller/equal*) center to the (*second distribution in context*), the (*first distribution in context*) has a (*larger/smaller/equal*) variability to the (*second distribution in context*), and the (*distribution in context*) has (*outliers/gaps/clusters*) although the (*distribution in context*) (*also does/doesn't*).

> *Example: At the end of the 2018–2019 season, the distribution of wins among the 15 teams in the Eastern NBA Conference (EC) was roughly bell-shaped, which is different from the distribution of wins among the 15 teams in the Western NBA Conference (WC) that was roughly uniform with a low outlier. The WC distribution had a greater center, while the EC distribution had a greater spread. The WC distribution had an apparent outlier at 19 and a gap between 19 and 33, which was different from the EC distribution that had no apparent outliers or gaps.*

Interpreting the standard deviation

The mean distance between individual (*values in context*) and the mean (*context*) is approximately ____.

> *Example: Given that the number of Facebook friends of high school students has a mean of 13.5 and a standard deviation of 3.4, the mean distance between the number of Facebook friends of individual high school students and 13.5 is approximately 3.4.*

Interpreting a *z*-score

The (*value in context*) is ____ standard deviations (*above/below*) the mean (*value in context*).

> *Example: A student has a GPA, which is 0.85 standard deviations below the mean GPA of all students.*

Describing a scatterplot

There is a (*strong/moderate/weak*), (*positive/negative*), (*linear/nonlinear*) association between (*the explanatory variable* x *in context*) and (*the response variable* y *in context*).

> *Example: There is a moderate, negative, roughly linear association between the average number of minutes people sit during a day and their life expectancy in years.*

Interpreting the slope

The predicted increase in (*response variable* y *in context*) is ___ for each unit increase in (*explanatory variable* x *in context*).

> *Example: The predicted increase in number of calories burned is 6.55 for each additional minute walked at 4 mph.*

Interpreting the *y*-intercept

____ is the predicted value of the (*response variable* y *in context*) when the (*explanatory variable* x *in context*) is equal to 0.

> *Example: 0.142 is the predicted batting average for a high school player who spends 0 time in the batting cage.*

Interpreting *R*-squared, the coefficient of determination

____% of the variation in (*the y-variable in context*) is accounted for by the linear model relating (*the* y-*variable in context*) to (*the* x-*variable in context*).

> *Example: 38% of the variation in incidence of skin cancer is accounted for by the linear model relating incidence of skin cancer to the SPF (sun protection factor) value of sunscreen used.*

Picking a reasonable value for a correlation

Given a (*weak/moderate/strong*), (*positive/negative*), linear association between ____ and ____, a reasonable value for the correlation is (*a number between −1 and +1*).

> *Example: Given a strong, negative, linear association between resting heart rate and weekly hours of exercise, a reasonable value for the correlation is −0.83.*

Making a conclusion when the scatterplot shows a linear association and the residual plot has no definite pattern

Because the scatterplot shows a linear association and the residual plot shows a random scatter, it appears that a linear model relating (*the* y-*variable in context*) to (*the* x-*variable in context*) (*is*) reasonable.

> *Example: Because the scatterplot shows a linear association and the residual plot shows a random scatter, it appears that a linear model relating grades for research papers to length of the papers is reasonable.*

Making a conclusion when the residual plot has a definite pattern

Because the residual plot shows a definite pattern, a linear model relating (*the* y-*variable in context*) to (*the* x-*variable in context*) (*is not*) not appropriate.

> *Example: Because the residual plot shows a definite pattern, ∩-shaped, a linear model relating mpg to car speed is not appropriate.*

Describing how to use a random number generator to pick a simple random sample

Number the subjects 1, 2, 3, ..., *n*. Use a random number generator to generate integers between 1 and *n*, throwing out repeats, until *x* unique integers have been selected. The sample consists of the subjects with numbers corresponding to the *x* generated integers.

> *Example: To pick an SRS of 10 students from a classroom of 25 students, number the students from 1 through 25. Use a random number generator to generate integers between 1 and 25, throwing out repeats, until 10 unique integers have been selected. The sample consists of the 10 students with numbers corresponding to the 10 generated integers.*

Performing a normal distribution probability calculation

We have a normal distribution with $\mu = $ ____ and $\sigma = $ ____. $P(X > / < $ ____ $) = $ ____.
[You do not have to show whether you used a formula or calculator software.]

Example: We have a roughly normal distribution with mean $\mu = 25$ and standard deviation $\sigma = 10$. Then the probability a student's backpack weighs over 30 pounds is $P(X > 30) = 0.309$.

Performing a binomial distribution probability calculation

We have a binomial distribution with $n = $ ____ and $p = $ ____. $P(X = / > / < / \geq / \leq $ ____ $) = $ ____.
[You do not have to show whether you used a formula or calculator software.]

Example: We have a binomial distribution with $n = 25$ and $p = 0.63$. Then $P(X \geq 15) = 0.702$.

Performing a geometric distribution probability calculation

We have a geometric distribution with $p = $ ____. $P(X = / > / < / \geq / \leq $ ____ $) = $ ____.
[You do not have to show whether you used a formula or calculator software.]

Example: We have a geometric distribution with $p = 0.42$. Then $P(X \leq 3) = 0.805$.

Describing a sampling distribution

The sampling distribution for $(\hat{p} / \bar{x} / b)$ is approximately normal with mean $(p = $ ____ OR $\mu = $ ____ OR $\beta = $ ____ $)$ and standard deviation $(\sigma_{\hat{p}} = \sqrt{\dfrac{p(1-p)}{n}} = $ ____ OR $\sigma_{\bar{x}} = \dfrac{\sigma}{\sqrt{n}} = $ ____ OR $\sigma_b = $ ____ $)$.

Example: The sampling distribution of $\bar{x}$ is approximately normal with mean $\mu = 25$ and standard deviation $\sigma_{\bar{x}} = \dfrac{\sigma}{\sqrt{n}} = \dfrac{8}{\sqrt{45}} = 1.19$.

Checking conditions/assumptions for inference

We check *independence* (random sample and that the sample is less than 10% of the population) and normality (for proportions: $np \geq 10$ and $n(1-p) \geq 10$; for means: either the population is roughly normal OR the sample size $n \geq 30$ OR the sample distribution has no strong skewness or outliers).

Example (proportions): We check independence (we are given an SRS and $n = 125 < 10\%$ of all students) and normality ($n\hat{p} = (125)(0.61) = 76.25 \geq 10$ and $n(1-\hat{p}) = (125)(0.39) = 48.75 \geq 10$.

Example (means): We check independence (it is stated that the sample was randomly chosen and $n = 35 < 10\%$ of all teachers) and normality ($n = 35 \geq 30$).

Interpreting a confidence interval

We are ____ % confident that the true (*proportion/mean/slope in context*) is between ____ and ____.

Example: We are 95% confident that the true proportion of people who leave a tip at fast-food restaurants like McDonald's and Starbucks is between 0.13 and 0.19.

Interpreting a confidence level

If we take all possible samples of (*size in context*) and if a confidence interval for the (*proportion/mean/slope in context*) is constructed from each sample, then ____ % of these intervals will capture the true (*proportion/mean/slope in context*).

Example: If all possible samples of 40 smartphones are analyzed as to battery length and if a confidence interval for the mean battery length is constructed from each sample, then 90% of these intervals would capture the true mean battery length.

Interpreting a *P*-value

Assuming that (*the null hypothesis in context*) is true, there is a probability of ____ of getting a sample (*proportion/mean/slope*) as extreme as or more extreme than the one observed by chance alone in a random sample of (*size in context*).

> *Example: Assuming there is no linear relationship between smoking levels and risk of lung cancer, there is a probability of 0.0023 of getting a sample slope as extreme as or more extreme than the one observed by chance alone in a random sample of 100 smokers.*

Concluding a hypothesis test when rejecting the null hypothesis

Because the *P*-value is so small, $P =$ ____ $<$ (*significance level*), there is sufficient evidence to reject the null hypothesis. There is convincing evidence that (*alternative hypothesis in context*).

> *Example: With this small a P-value, P = 0.031 < 0.05, there is sufficient evidence to reject the null hypothesis. There is convincing evidence that less than 30% of all teenagers can name two current Supreme Court justices.*

Concluding a hypothesis test when failing to reject the null hypothesis

Because the *P*-value is so large, $P =$ ____ $>$ (*significance level*), there is not sufficient evidence to reject the null hypothesis. There is not convincing evidence that (*alternative hypothesis in context*).

> *Example: With this large a P-value, P = 0.18 > 0.05, there is not sufficient evidence to reject the null hypothesis. There is not convincing evidence that the mean age of all high school teachers is over 45.*

Interpretating a Type I error

Because the *P*-value was small, there was sufficient evidence to reject the (*null hypothesis in context*). However, the null hypothesis might have been true, in which case we committed a Type I error. A possible consequence would be ____.

> *Example: Because of a small P-value, there was sufficient evidence to reject the null hypothesis that a medication is not effective. However, if the medication really is not effective, a Type I error would have been committed. A possible consequence is that patients will not recover because they will continue to be treated with an ineffective medication.*

Interpretating a Type II error

Because the *P*-value was large, there was not sufficient evidence to reject the (*null hypothesis in context*). However, the null hypothesis might have been false, in which case we committed a Type II error. A possible consequence would be ____.

> *Example: Because of a large P-value, there is not sufficient evidence to reject the null hypothesis that a new teaching method fails to improve student performance. However, if the new teaching method really does improve student performance, a Type II error would have been committed. A possible consequence is that a school district will not adopt a new teaching method that would have helped their students.*

Interpreting power

There is a (*power*) probability of correctly rejecting the null hypothesis.

> *Example: With H_0: a medical device is operating correctly, there is a 0.92 probability of correctly recognizing that the medical device is not operating correctly.*

Answer Explanations for Quizzes 1–29

Quiz 1 (Pages 48–52)

Multiple-Choice

1. **(C)** Primary color is categorical (it takes values that are names), and width is quantitative (it takes values that can be measured).

2. **(D)** The minimum time is somewhere between 5 and 10 minutes but might not be exactly 5 minutes. Similarly, the maximum time is somewhere between 30 and 35 minutes. With 155 times, the middle time will be the 78th time if the times are arranged in order. There are 70 times less than 15 minutes and 40 times between 15 and 20 minutes, so the 78th time must be between 15 and 20 minutes.

3. **(C)** Stemplots are not used for categorical data sets, are too unwieldy to be used for very large data sets, and show every individual value. Stems should never be skipped over—gaps are important to see.

4. **(B)** Histograms give information about relative frequencies (relative areas correspond to relative frequencies) and may or may not have an axis with actual frequencies. Symmetric histograms can have any number of peaks. Choice of width and number of classes changes the appearance of a histogram. Stemplots clearly show outliers; however, in histograms outliers may be hidden in large class widths.

5. **(B)** The percentage of Generation X who preferred Company B is $\frac{31}{50} = 0.62 = 62\%$, while the percentage of Millennials who preferred Company B is $\frac{32}{75} = 0.427 = 42.7\%$.

6. **(B)** The number of students not choosing Comedy or Romance was $20 - (4 + 6)$ or $5 + 1 + 4$. Of these, $4 + 5$ chose Sci-Fi or Action for a proportion of $\frac{4 + 5}{20 - (4 + 6)}$.

7. **(A)** A cumulative relative frequency plot that rises slowly at first, then quickly in the middle, and finally slowly again at the end corresponds to a histogram with little area under the curve on the ends and much greater area in the middle.

8. **(B)** A histogram with little area under the curve early and much greater area later results in a cumulative relative frequency plot that rises slowly at first and then at a much faster rate later.

Free-Response Questions

1. (a)

(b) $\frac{3 + 11}{25} = \frac{14}{25} = 56\%$

(c) The answer would still be 56%.

2. (a) A complete answer mentions shape, center, spread, and unusual features (such as outliers) while including context within the description.

    Shape: Unimodal, skewed right, probable outlier at 10.
    Center: Around 2 or 3.
    Spread: From 0 to 10 goals were scored by each team.

    (b) If the player scored six goals on his own, then his/her team must have scored either 7 or 10 goals. Since the team lost, the only possible final score is 10 to 7.

    (c) No, there were six teams that scored exactly two goals, but there were only five teams that scored less than two goals. So, not all the two-goal teams could have won.

3. (a) The lowest winning percentage over the past 22 years was 46.0%.

    (b) A complete answer mentions shape, center, spread, and unusual features (such as outliers) while including context within the description.

    Shape: Two clusters, each somewhat bell-shaped.
    Center: Around 50%.
    Spread: The winning percentages were between 46.0% and 55.6%.

    (c) The team had more losing seasons (13) than winning seasons (9).

    (d) The cluster of winning percentages is further away from 50% than the cluster of losing percentages.

4. (a) 40% of the players averaged fewer than 20 points per game.

    (b) All the players averaged at least 3 points per game.

    (c) No players averaged between 5 and 7 points per game because the cumulative relative frequency was 10% for both 5 and 7 points.

    (d) Go over to the plot from 0.9 on the vertical axis and then down to the horizontal axis to result in 28 points per game.

    (e) Reading up to the plot and then over from 10 and from 20 shows that 0.25 of the players averaged under 10 points per game and 0.4 of the players averaged under 20 points per game.
    Thus, $0.4 - 0.25 = 0.15$ gives the proportion of players who averaged between 10 and 20 points per game.

Quiz 2 (Pages 68–73)

Multiple-Choice

1. **(D)** The distribution is clearly skewed right, so the mean is greater than the median and the ratio is greater than one.

2. **(D)** Multiplying every value in a set by the same constant (in this case, by $\frac{1}{60}$) multiplies both the mean and the standard deviation by the same constant. Standardized scores (the number of standard deviations from the mean) are unchanged and without units.

3. **(E)** Outliers are any values below $Q_1 - 1.5(IQR) = 25 - 1.5(38 - 25) = 5.5$ or above $Q_3 + 1.5(IQR) = 38 + 1.5(38 - 25) = 57.5$. With a minimum of $10 > 5.5$, there are no outliers on the low end; however, with a maximum of $60 > 57.5$, the maximum is an outlier and so are any other values falling between 57.5 and 60.

4. **(A)** Subtracting 10 from one value and adding 5 to two values leaves the sum of the values unchanged, so the mean will be unchanged. Exactly what values the outliers take will not change what value is in the middle, so the median will be unchanged.

5. **(C)** The median corresponds to the 0.5 cumulative proportion. The 0.25 and 0.75 cumulative proportions correspond to $Q_1 = 1.8$ and $Q_3 = 2.8$, respectively, and so the interquartile range is $2.8 - 1.8 = 1.0$.

6. **(A)** The standard deviation, $\sigma = \sqrt{\frac{\sum(x - \mu)^2}{n}}$, and the interquartile range, $IQR = Q_3 - Q_1$, can never be negative.

7. **(A)** The sum of the scores in one class is $20 \times 92 = 1{,}840$, while the sum in the other is $25 \times 83 = 2{,}075$. The total sum is $1{,}840 + 2{,}075 = 3{,}915$. There are $20 + 25 = 45$ students, and so the average score is $\frac{3{,}915}{45} = 87$.

8. **(B)** Removing a value that is greater than the average will always lower the average, and adding in a value that is less than the average will also always lower the average.

9. **(B)** There are 5 classes, and their average size is $\frac{150 + 25 + 25 + 25 + 25}{5} = \frac{250}{5} = 50$.

10. **(C)** Among the 250 students, there are 100 students in classes of size 25 and 150 students in a class of size 150. The average size of their history class is $\frac{100(25) + 150(150)}{250} = \frac{25{,}000}{250} = 100$.

Free-Response Questions

1. (a) Adding 10 to each value increases the mean by 10 but leaves measures of variability unchanged. So, the new mean is 340 hours while the range stays at 5,835 hours, the standard deviation remains at 245 hours, and the variance remains at $245^2 = 60{,}025 \text{ hr}^2$.

   (b) Increasing each value by 10% (multiplying by 1.10) will increase the mean to $1.1(330) = 363$ hours, the range to $1.1(5{,}835) = 6{,}418.5$ hours, the standard deviation to $1.1(245) = 269.5$ hours, and the variance to $(269.5)^2 = 72{,}630.25 \text{ hr}^2$. (Note that the variance increases by a multiple of $(1.1)^2$ and not by a multiple of 1.1.)

2. A complete answer compares shape, center, and spread and mentions context in at least one of the responses.
   Shape: The distribution of times to complete all tasks by females is skewed right (toward the higher values), whereas the distributions of times to complete all tasks by males is roughly bell-shaped.
   Center: The center of the distribution of female times (at around $2\frac{1}{4}$ minutes) is less than the center of the distribution of male times (at around $3\frac{3}{4}$ minutes).
   Spread: The spreads of the two distributions are roughly the same; the range of the female times $(5 - 1\frac{1}{2} = 3\frac{1}{2}$ minutes) equals the range of the male times $(5\frac{1}{2} - 2 = 3\frac{1}{2}$ minutes).

3. *z*-scores give the number of standard deviations from the mean, so
   $Q_1 = 300 - 0.7(25) = 282.5$ and $Q_3 = 300 + 0.7(25) = 317.5$.
   The interquartile range is $IQR = 317.5 - 282.5 = 35$, and $1.5(IQR) = 1.5(35) = 52.5$.
   The standard definition of outliers encompasses all values less than $Q_1 - 52.5 = 230$ and all values greater than $Q_3 + 52.5 = 370$.

4.   (a)   The Liberian median age lies in the 15–19 age interval because roughly 50% of the total bar lengths is above and below the 15–19 interval.

   (b)   Numerical values are needed to calculate means. Histograms such as these give numbers or proportions of values in specific intervals but do not give the actual values.

   (c)   Canada has more children younger than 10 years of age. There are about 1.2 million children younger than age 10 in Liberia (boys and girls) and roughly 3.5 million in Canada. (Note the difference in scales!)

   (d)   The population pyramid indicates that Canadian women live longer than men because all the higher age intervals show greater numbers of women than men.

   (e)   In the Liberian graph, the smaller 15–19 age group shows a definite break with the overall pattern. A plausible explanation is that a great number of children died in the civil war.

5.   A complete answer compares shape, center, and spread and also mentions context.

   Shape: The Sudan heights (for which the cumulative frequency plot rises slowly at each end and steeply in the middle) have a roughly bell-shaped distribution. The Rwanda heights (for which the cumulative frequency plot rises slowly at first and then steeply toward the end) have a distribution that is skewed to the left (toward the lower heights).

   Center: The medians correspond to relative frequencies of 0.5. Reading across from 0.5 and then down to the horizontal axis shows that the Sudan median height is less than the Rwanda median height.

   Spread: The range of the Sudan heights $(200 - 150 = 50$ cm$)$ is greater than the range of the Rwanda heights $(190 - 160 = 30$ cm$)$.

Quiz 3 (Pages 76–77)

Multiple-Choice

1.   **(E)**   Curve a appears to have a mean of 6 and a standard deviation of 2, while curve b appears to have a mean of 21 and also a standard deviation of 2. (It is important to look at the scale of each graph, not just the visual width.)

2.   **(D)**   Point E appears to be one standard deviation above the mean. $74.3 + 9.7 = 84.0$. Note that point E (and point C) are points where the slope is steepest.

3.   **(D)**   54.9 and 93.7 are two standard deviations below and above the mean, respectively. By the empirical rule, 95% of the data are in this interval.

4.   **(A)**   64.6 is one standard deviation below the mean. By the empirical rule, 68% of the data is between one standard deviation below and above the mean. This leaves 34% outside this interval and 17% in each tail.

5.   **(D)**   With a normal distribution, 16% in the tail to the right corresponds to one standard deviation above the mean. Then $\mu + 1(2.5\%) = 25\%$ gives $\mu = 22.5\%$.

6.   **(B)**   With bell-shaped data, the empirical rule applies. Given that the spread from 92 to 98 is roughly 6 standard deviations, one standard deviation is about 1.

Free-Response Questions

1. (a) $z = \dfrac{720 - 600}{120} = 1$, and for the normal model, the proportion of z-scores below 1 is $0.5 + 0.34 = 0.84$, or the 84th percentile.

   (b) Both \$360 and \$840 are \$240 from the mean of \$600, which makes \$360 two standard deviations below the mean and \$840 two standard deviations above. For the normal model, the proportion of z-scores between -2 and $+2$ is approximately 0.95 or 95%.

   (c) Under a normal curve, 99.7% of the values are within ± 3 standard deviations of the mean: $600 - 3(120) = 240$ and $600 + 3(120) = 960$. So, 99.7% of American teens spend between \$240 and \$960 on food per year.

2. No. In a normal distribution the mean and median should be equal or nearly equal, but here the mean, 6.4, is significantly greater than the median, 5. Furthermore, in a normal distribution the distance from the minimum to the median should be about the same as the distance from the median to the maximum. In this data set, however, the distance median − min = 5 − 0 = 5 is different from the distance max − median = 15 − 5 = 10. Both these facts suggest the distribution is skewed right.

Quiz 4 (Pages 85–87)

Multiple-Choice

1. **(E)** Of the 500 people surveyed, 250 were students, and $\dfrac{250}{500} = 0.5$, or 50%.

2. **(A)** Of the 500 people surveyed, 125 picked challenging as most important and were also teachers: $\dfrac{125}{500} = 0.25$, or 25%.

3. **(E)** There were 50 administrators, and 25 of them picked strict as most important: $\dfrac{25}{50} = 0.5$ or 50%.

4. **(E)** There were 210 people picking enthusiastic as most important, and 150 of them were students: $\dfrac{150}{210} = 0.714$, or 71.4%.

5. **(C)** The percentages of students, teachers, and administrators picking strict as most important were 20%, 12.5%, and 50%, respectively.

6. **(B)** Based on lengths of indicated segments, the percentage from the West who prefer country is the greatest.

7. **(E)** The given bar chart shows percentages, not actual numbers.

8. **(D)** In mosaic plots, the area of a box is proportional to the count corresponding to that box. Choice (D) is true because the three boxes corresponding to students with GPAs under 3.0 have a total area greater than the three boxes corresponding to students with GPAs 3.0 or higher, or note that along the vertical axis, "GPA under 3.0" has a greater length than "GPA 3.0 or higher."

Free-Response Questions

1. (a) Of the speeding pullovers who were given a ticket, 27 out of 93, or $\dfrac{27}{93}$, were Latino.

   (b) Of the speeding pullovers who were White, 84 out of 120, or $\dfrac{84}{120} = 0.70 = 70\%$, were given a warning.

   (c) Based upon the mosaic plot, there is an association between race/ethnicity and whether or not a ticket or warning is given to drivers pulled over for speeding at night. Whites are more likely to receive a warning, Blacks are equally as likely to receive a warning or a ticket, while Latinos are more likely to receive a ticket. Latinos were twice as likely to receive a ticket as Whites.

2. (a)

Program	Percentage of Men Accepted (%)	Percentage of Women Accepted (%)
A	62	82
E	28	24

Women seem to be favored in program A, while men seem to be slightly favored in program E.

(b) Overall, 564 out of 1,016 male applicants were accepted, for a 55.5% acceptance rate, while 184 out of 501 female applicants were accepted, for a 36.7% acceptance rate. This appears to contradict the results from part (a).

(c) You should tell the reporter that although it is true that the overall acceptance rate in these two programs for women is 36.7% compared with the 55.5% acceptance rate for men, in one program women have a much higher acceptance rate than men, while in the other program women have only a slightly lower acceptance rate than men. The reason behind this apparent paradox is that between these two programs, most men applied to program A, which had a high acceptance rate. However, most women applied to program E, which had a low acceptance rate.

Quiz 5 (Pages 104–109)

Multiple-Choice

1. **(D)** Residual = Measured − Predicted, so if the residual is negative, the predicted must be greater than the measured (observed).

2. **(A)** The correlation r is unit-free.

3. **(B)** The "Predictor" column indicates the independent variable, x, with its coefficient, b, to the right.

4. **(E)** $r = \sqrt{0.986} = 0.993$. (The sign is the same as the sign of the slope, which in this case is positive.)

5. **(C)** $\widehat{back} = 0.056 + 0.920(0.55) = 0.562$, and so the residual = $0.59 - 0.562 = 0.028$.

6. **(C)** The coefficient of determination r^2 gives the proportion of the y-variance that is accountable from a knowledge of the variability of x. In this case $r^2 = (0.632)^2 = 0.399$ or 39.9%.

7. **(C)** The correlation is not changed by adding the same number to every value of one of the variables, by multiplying every value of one of the variables by the same positive number, or by interchanging the x- and y-variables.

8. **(B)** The point (35, 14) appears to follow the linear trend of the rest of the data. The slope wouldn't change, but with another point added to the trend, r^2 will increase, and s, the measure of typical deviation from the line, will decrease.

9. **(D)** The word "negative" in the phrase "strong negative linear association" means that generally as one variable increases, the other variable decreases. Thus, regions with a higher percentage of seat belt usage tend to have lower numbers of highway deaths due to failure to wear seat belts. Choice (A) is an interpretation of the y-intercept. Correlation is a measure of the strength of a linear relationship but does not by itself explain the meaning of "positive" or "negative." While the overall pattern is of a negative association, anything can be true about two points on the scatterplot. Choice (E) is an interpretation of the slope.

10. **(A)** Slope $= 0.15\left(\dfrac{42,000}{1.3}\right) \approx 4,850$ and intercept $= 208,000 - 4,850(6.2) \approx 178,000$.

11. **(E)** A scatterplot readily shows that while the first three points lie on a straight line, the fourth point does not lie on this line. Thus, no matter what the fifth point is, all the points cannot lie on a straight line, and so r cannot be 1.

12. **(C)** The point X has low or no leverage because its x-value appears to be close to the mean x-value. The point X has a large residual because it is far above the line of best fit.

13. **(C)** Note that the smallest and largest x-values in X have y-values below the regression line, that is, with negative residuals, and so X corresponds to 2. Note that the three largest x-values in Z have y-values with positive residuals, and so Z corresponds to 3.

Free-Response Questions

1. (a) The correlation $r = \sqrt{0.8428} = 0.918$. It is positive because the slope of the regression line is positive.

    (b) The slope is 0.283, signifying that each unit increase in strength is associated with a predicted increase of 0.283 in speed.

    (c) The linear regression line passes through $(\bar{x}, \bar{y})$ so the mean speed $= 4.626 + 0.283(30.47) = 13.25$.

    (d) The formula $b = r\dfrac{s_y}{s_x}$ gives $0.283 = 0.918 \left(\dfrac{\text{SD of speed}}{13.72} \right)$. Then SD of speed $= \dfrac{0.283(13.72)}{0.918} = 4.23$.

    (e) The linear model predicts Thanos's speed to be $4.626 + 0.283(45) = 17.36$, so the model underestimates Thanos's speed by $18 - 17.36 = 0.64$.

    (f) $4.626 + 0.283(42) = 16.512$, so the point $(42, 16.5)$ is almost exactly on the regression line. Removing a point that follows the linear pattern will lower the correlation.

    (g) Removing the points corresponding to both Superman (with its high x-value) and the Penguin (with its low x-value) swings the line away from these points, in both cases leading to a decrease in the positive slope value.

2. (a) In scatterplot I, the points fall exactly on a downward sloping straight line, so $r = -1$.
       In scatterplot II, the isolated point is an influential point, and r is close to $+1$.
       In scatterplot III, the isolated point is also influential, and r is close to 0.

    (b) Inserting a point that lies directly on a straight line drawn through the three points will not change the correlation of $r = -1$.

3. (a) The association between stress level and debt for this sample of 25 college graduates is linear, moderate, and positive.

    (b) $19.376 + 0.42640(60) = 44.96$ and Observed − Predicted $= 25 - 44.96 = -19.96$. The residual value means that the stress level is 19.96 points less than what would be predicted for a student with a $60,000 debt.

    (c) Health is a better choice than income for including with debt in a regression model for predicting stress. The relatively strong association between health and residuals from regression of stress on debt shows that health will help explain the unexplained variation (the residuals) between stress and debt. The random scatter in the plot with income versus residuals from regression of stress on debt shows that income will not help explain the unexplained variation (the residuals) between stress and debt.

Quiz 6 (Pages 110–115)

Multiple-Choice

1. **(E)** The negative value of the slope (-2.84276) gives that, on average, the predicted average total SAT score of a school is 2.84 points lower for each one unit higher in the percentage of students taking the exam.

Choices (A) through (D) are incorrect for the following reasons. The variable column indicates the independent (explanatory) variable. The sign of the correlation is the same as the sign of the slope (negative here). In this example, the y-intercept is meaningless (predicted SAT result if no students take the exam). There can be a strong linear relation with high r^2-value but still a distinct pattern in the residual plot indicating that a nonlinear fit may be even stronger.

2. **(E)** A negative correlation shows a tendency for higher values of one variable to be associated with lower values of the other; however, given any two points, anything is possible.

3. **(E)** Since $(2, 5)$ is on the line $y = a + 3x$, we have $5 = a + 3(2)$ and $a = -1$. Thus, the regression line is $y = -1 + 3x$. The point $(\bar{x}, \bar{y})$ is always on the regression line, and so we have $\bar{y} = -1 + 3\bar{x}$.

4. **(C)** If the points lie on a straight line, $r = \pm 1$. Correlation has the formula $r = \dfrac{1}{n-1} \sum \left(\dfrac{x_i - \bar{x}}{s_x} \right) \left(\dfrac{y_i - \bar{y}}{s_y} \right)$, so x and y are interchangeable, and r does not depend on which variable is called x or y. However, since means and standard deviations can be strongly influenced by outliers, r too can be strongly affected by extreme values. Although $r = 0.75$ indicates a better fit with a linear model than $r = 0.25$ does, we cannot say that the linearity is threefold.

5. **(B)** Residuals can be any sign. In a good straight line fit, the residuals show a random pattern. The linear regression line minimizes, not maximizes, the sum of the squares of the residuals.

6. **(C)** Predicted winning percentage $= 44 + 0.0003(34,000) = 54.2$, and Residual $=$ Observed $-$ Predicted $= 55 - 54.2 = 0.8$.

7. **(B)** The slope and the correlation coefficient have the same sign. Multiplying every y-value by -1 changes this sign.

8. **(D)** A linear association means that as the explanatory variable (home size here) changes by a constant amount, the response variable (selling price here) also changes by a constant amount, on average. Unless there was perfect linear correlation, the points will not line up on a straight line. No distinct pattern in the residual plot simply means that there is not an obviously better model to be found, but it does not necessarily suggest that the data are linear. The coefficient of determination indicates something about the strength of the relationship but does not define linearity. Choice (E) is an interpretation of the slope, but again, not a definition of linearity.

9. **(E)** "Strong" means that the points in the related scatterplot fall close to the least squares regression line. In context, this means that the actual incidence of skin cancer at a given latitude will be very close to what is predicted by the least squares model. Association does not imply causation. Although (C) refers to the existence of an association, and (B) and (D) further refer to the direction of the association, none of these statements refer to the strength of the association.

10. **(E)** The least squares line passes through $(\bar{x}, \bar{y}) = (2, 4)$, and the slope b satisfies $b = r\dfrac{s_y}{s_x} = \dfrac{5r}{3}$. Since $-1 \leq r \leq 1$, we have $-\dfrac{5}{3} \leq b \leq \dfrac{5}{3}$.

11. **(B)** The mean equals the sum divided by the number of values; thus, if the sum is zero, so is the mean. However, if the sum of values equals zero, it does not follow that the middle value is zero.

12. **(B)** The point X has high leverage because its x-value is much greater than the mean x-value. Point X has a small residual because the regression line would pass close to it.

13. **(E)** Note that the four largest x-values in X have y-values above the regression line, that is, with positive residuals, and so X corresponds to 3. Note that the two points with largest x-values in Y have y-values above and below the regression line, one with a positive and one with a negative residual, so Y corresponds to 1.

Free-Response Questions

1. (a) The correlation coefficient is $r = \sqrt{0.746} = 0.864$. It is positive because the slope of the regression line is positive.

    (b) The slope is 1.106, signifying that each additional page raises a grade by 1.106, on average.

    (c) Including Mary's paper will lower the correlation coefficient because her result seems far off the regression line through the other points.

    (d) Including Mary's paper will swing the regression line down and lower the value of the slope.

2. (a)

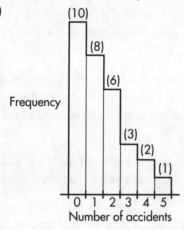

    (b) There is a roughly linear trend with daily accidents increasing during the month.

    (c) The daily number of accidents is strongly skewed to the right.

3. (a) The association between exam score and hours of sleep is *nonlinear* (or *curved*), weak (or moderate), and positive.

    (b) There appears to be very little, if any, association between exam score and hours of sleep for the 8 students who received at least 6 hours of sleep the night before the exam. (Because of the one very low score that also corresponds to hours of sleep just over 6, one could say that there is a very weak *positive* association.)

    (c) The student who scored 75 and slept for just over 2 hours the night before the exam has such a strong influence that that particular scatterplot point taken together with all the other points suggests a nonlinear form.

Quiz 7 (Pages 127–130)

Multiple-Choice

1. **(A)** This was an observational study because no treatment was imposed. It was retrospective because existing data from the past were examined.

2. **(E)** The wording "creating a level playing field" and "a right to express their individuality" are nonneutral and clearly leading phrasings.

3. **(D)** If there is bias, taking a larger sample just magnifies the bias on a larger scale. If there is enough bias, the sample can be worthless. Even when the subjects are chosen randomly, there can be bias due, for example, to nonresponse or to the wording of the questions. Convenience samples, like shopping mall surveys, are based on choosing individuals who are easy to reach, and they typically miss a large segment of the population. Voluntary response samples, like radio call-in surveys, are based on individuals who offer to participate, and they typically overrepresent persons with strong opinions.

4. **(E)** Each of the 50 states, with its own longtime past state standards and different regional culture, is considered a homogeneous stratum.

5. **(B)** Different samples give different sample statistics, all of which are estimates of a population parameter. Sampling variability (also called sampling error) relates to natural variation between samples, can never be eliminated, can be described using probability, and is generally smaller if the sample size is larger. Furthermore, it is not an error or mistake on anyone's part!

6. **(E)** In a simple random sample, every possible group of the given size has to be equally likely to be selected, and this is not true here. For example, with this procedure it would be impossible for all the players of one team to be together in the final sample. This procedure is an example of stratified sampling, but stratified sampling does not result in simple random samples.

7. **(C)** Surveying people coming out of any church results in a very unrepresentative sample of the adult population, especially given the question under consideration. Using chance and obtaining a high response rate will not change the selection bias and make this into a well-designed survey.

8. **(B)** Cluster samples work when each cluster roughly looks like the population as a whole. However, there is no reason to believe that each conference has similar graduation rates as any other conference. Some conferences may have different academic standards and different support levels for their student athletes. Picking two conferences may not at all be representative of the population of Division 1 college football programs. In general, cluster samples are known for ease, convenience, and less expense than other sampling methods. Homogeneity among schools in given conferences would relate to picking a stratified sample with the conferences as strata.

9. **(B)** Voluntary response bias results from surveys based on individuals who choose to participate. Examples include when people choose to call in to a radio station, to respond to a newspaper ad, to approach and answer a surveyor's questions, or to respond to an online website. The example in which people are randomly chosen from a voter registration list and all answer a survey would not involve voluntary response bias.

10. **(B)** Nonresponse bias is a potential problem whenever a survey does not obtain responses from everyone surveyed. Choice (A) has potential voluntary response bias, and choice (D) has potential undercoverage bias. However, choices (A), (C), (D), and (E) all achieved responses from everyone who was surveyed, so they do not have nonresponse bias.

Free-Response Questions

1. (a) This is an example of *stratified sampling,* where the chapters are strata. The advantage is that the student is ensuring that the final sample will represent the 12 different authors, who may well use different average word lengths.

    (b) This is an example of *cluster sampling,* where for each chapter the three chosen pages are clusters. It is reasonable to assume that each page (cluster) resembles the author's overall pattern. The advantage is that using these clusters is much more practical than trying to sample from among all an author's words.

    (c) This is an example of *systematic sampling,* which is quicker and easier than many other procedures. A possible disadvantage is that if ordering is related to the variable under consideration, this procedure will likely result in an unrepresentative sample. For example, in this study if an author's word length is related to word order in sentences, the student could end up with words of particular lengths.

2. (a) Method I results in cluster sampling (the rows are clusters) and can be implemented by numbering the rows 1 through 10, using a random number generator to randomly pick a number between 1 and 10, and using all 15 plots in that row. Method II results in stratified sampling (the columns are strata) and can be implemented by numbering the rows 1 through 10, using a random number generator to pick a number between 1 and 10 to pick a plot in the first column, repeating this process to pick a random plot in each of the 15 columns, and using the 15 plots resulting from picking one plot from each column.

    (b) Method II, a stratified sample (with the columns as strata) would be preferable here as each column is somewhat different, so picking a plot from each is meaningful. Method I, an attempt at cluster sampling with the rows as clusters, would work only if each of the rows was heterogeneous (representative of the whole area); this is clearly not true as, for example, the top and bottom rows have far fewer nests than the middle rows.

3. (a) People who attend the school's basketball game are probably more interested in supporting the athletic program than those who do not attend. This study will probably show that a higher proportion of people favor the increased spending than what is the true proportion.

    (b) Subscribers to a health magazine are more likely to believe in the benefits of eating organic than the general population, so the resulting proportion of believers in this study will be greater than it is in the general population.

    (c) The distributer could put any moldy strawberries on the bottom, out of sight! The proportion of accepted boxes will be higher than it should be.

    (d) Many patients will be reluctant to admit to their doctor that they are not following their diet instructions. So, the proportion who say they are following the instructions will be higher than the true proportion following the instructions.

    (e) Parents who are satisfied with the school system are probably less likely to take the time to respond to this voluntary response survey, while those who are unhappy with the education their kids are receiving are more likely to respond. So, the proportion of "unhappy" responses will probably be higher than the true proportion of parents who are unhappy.

Quiz 8 (Pages 139–141)

Multiple-Choice

1. **(A)** The main office at your school should be able to give you the class sizes of every math and English class. If need be, you can check with every math and English teacher.

2. **(C)** In the first study, the families were already in the housing units, while in the second study, one of two treatments was applied to each family.

3. **(D)** Octane is the only explanatory variable, and it is being tested at four levels. Miles per gallon is the single response variable.

4. **(D)** Using only a sample from the observations gives less information. It may well be that very bright students are the same ones who both choose to take AP Statistics and have high college GPAs. If students could be randomly assigned to take or not take AP Statistics, the results would be more meaningful. Of course, ethical considerations might make it impossible to isolate the confounding variable in this way.

5. **(E)** Association does not imply causation. It is likely that regular exercise and a person's age are both associated with greater heart strength, and either one could be the cause of the greater heart strength. Regular exercise and age are confounded. An experiment in which people are randomly assigned to exercise or not can prove a causal relationship between exercise and heart strength.

6. **(B)** Blocking divides the subjects into groups, such as men and women, or political affiliations, and thus reduces variation. That is, when we group similar individuals together into blocks and then randomize within the blocks, much of the variability due to the differences between the blocks is accounted for and so comparison of the treatment groups is clearer.

Free-Response Questions

1. (a) The explanatory variable is the frequency of substance abuse. The response variable is the age at death.

   (b) This is a prospective observational study. No treatments were assigned, and the researchers asked the rock and pop stars to report their frequency of substance abuse and later noted their ages at time of death.

   (c) No, this is an observational study, so no cause-and-effect conclusion is possible. It is possible that other variables are influencing the response. For example, rock and pop stars might engage in other forms of risk taking, such as poor eating and poor sleep patterns, which also might negatively affect life expectancy.

2. (a) The explanatory variable is flashing or not flashing "BUY POPCORN" on the screen for a fraction of a second. The response variable is the quantity of popcorn sold.

   (b) Every day for some specified period of time the manager could, look at the next digit on a random number table. If it is odd, flash the subliminal message all day on the screen; while if it is even, don't flash the message that day (randomization). Don't let the customers know what is happening (blinding) and don't let the clerks selling the popcorn know what is happening (double-blinding). Compare the quantity of popcorn bought by the treatment group, that is, by the people who receive the subliminal message, to the quantity bought by the control group, the people who don't receive the message (comparison).

3. (a) This cannot involve random assignment as it makes no sense to randomly assign teenagers to go to public or private schools. However, it would be good to use random sampling to select teenagers from each type of school to be asked about *Back to the Future* movies so as to be able to generalize to the population of public and private schools.

   (b) It is important for the company CEO to use random assignment to receive hourly pay raises or end-of-the-year bonuses because a cause-and-effect conclusion is of interest.

   (c) The sportscaster would probably want to include all the interleague games played by the Cubs and White Sox during the shortened 2020 season, and thus use neither random sampling nor random assignment.

4. (a) This was an experiment because treatments, which involved fitness trackers and incentives, were applied.

   (b) Treatments: (1) tracker and personal cash incentive, (2) tracker and charity cash incentive, (3) tracker alone, and (4) no tracker or cash incentive. Experimental units: the 800 volunteers. Response variable: change in blood pressure over a six-month period.

   (c) The "no fitness tracker or incentive" group was a *control group* whose purpose is to provide a baseline to determine if the activity tracker or cash incentive treatments had any effect on blood pressure change.

   (d) Assign each participant an integer from 1 to 800. Use a random integer generator to choose 600 unique integers between 1 and 800, throwing away repeats. The participants corresponding to the first 200 of these selected integers will be put in the first group, the participants corresponding to the second 200 of these selected integers will be put in the second group, the participants corresponding to the last 200 of these selected integers will be put in the third group, and the remaining 200 participants will be put into the fourth group.

   (e) These 800 participants were volunteers, who may not be representative of a larger population with regard to the effects of wearing fitness trackers or receiving cash incentives on changes in blood pressure. So it is not reasonable to generalize the results of this study to a larger population.

Quiz 9 (Pages 151–154)

Multiple-Choice

1. **(A)** 35 out of the entire population of 500 both have strep throat and tested positive.

2. **(A)** Of the 460 healthy people, 25 tested positive.

3. **(E)** Of the 40 people with strep throat, 35 tested positive.

4. **(E)** Of the 460 healthy people, 435 tested negative.

5. **(B)** The probability of the next child being a girl is independent of the sex of the previous children. Before she had any children, if the question had been about the probability of having eight girls in a row, then the answer would have been $(0.5)^8$, or about 1 in 256.

6. **(A)** The probability of a boy is given to be about 0.5. By the law of large numbers, the more births (Hospital B has more births per day), the more the relative frequency tends to become closer to this probability. With fewer births (Hospital A has fewer births per day), there is a greater chance of wide swings in the relative frequency.

7. **(D)** $P(exactly\ one) = P(\text{math: } yes)\,P(\text{physics: } no) + P(\text{math: } no)\,P(\text{physics: }yes) = (0.55)(0.7) + (0.3)(0.45)$. Note that the events "math and not physics" and "physics and not math" are mutually exclusive.

8. **(E)**

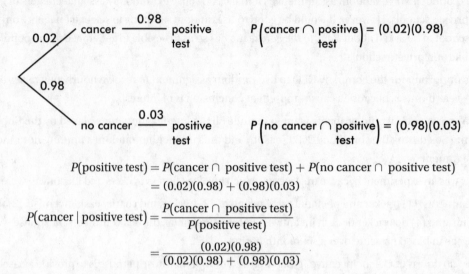

$$P(\text{positive test}) = P(\text{cancer} \cap \text{ positive test}) + P(\text{no cancer} \cap \text{ positive test})$$
$$= (0.02)(0.98) + (0.98)(0.03)$$

$$P(\text{cancer} \mid \text{positive test}) = \frac{P(\text{cancer} \cap \text{ positive test})}{P(\text{positive test})}$$

$$= \frac{(0.02)(0.98)}{(0.02)(0.98) + (0.98)(0.03)}$$

9. **(A)** We have $(21 - x) + x + (62 - x) + 15 = 90$, which gives $x = 8$. So there are 8 students who have at least two analog clocks and have at least three television sets in their homes. Because the events can occur at the same time, they are not mutually exclusive.

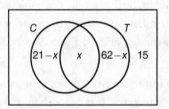

10. **(A)** Given that a person has a tattoo narrows our interest to

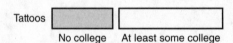

and people with at least some college are clearly over half of this group. Given that a person has taken a college course narrows our interest to

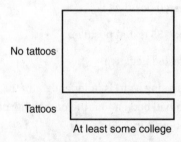

and people with tattoos are clearly less than half of this group.

11. **(D)** Events E and F are independent if $P(E \cap F) = P(E)P(F)$. Here, $P(E) = 0.4 + 0.1 = 0.5$ and $P(F) = 0.1 + 0.1 = 0.2$. Then $P(E)P(F) = (0.5)(0.2) = 0.1 = P(E \cap F)$.

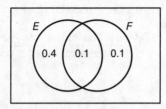

12. **(D)** $P(\text{I}) = P(\text{breakfast} \mid \text{senior}) = \dfrac{P(\text{breakfast} \cap \text{senior})}{P(\text{senior})}$

$P(\text{II}) = P(\text{breakfast} \cap \text{senior})$, and

$P(\text{III}) = P(\text{senior} \mid \text{breakfast}) = \dfrac{P(\text{breakfast} \cap \text{senior})}{P(\text{breakfast})}$

$P(\text{I}) > P(\text{III})$ because the denominator in (I) is smaller than the denominator in (III); that is, it is reasonable to assume that the set of all teenagers who are high school seniors is smaller than the set of all teenagers who eat breakfast.

$P(\text{II})$ is the smallest because (I) and (III) have denominators that are less than one.

Free-Response Questions

1. It's easiest to first sum the rows and columns:

	Years of experience			
	0–5	6–10	>10	
Less than 50 years old	80	125	20	225
More than 50 years old	10	75	50	135
	90	200	70	360

(a) $P(\text{age} < 50) = \dfrac{225}{360} = 0.625$

$P(\text{experience} > 10) = \dfrac{70}{360} = 0.194$

$P(\text{age} > 50 \cap \text{experience } 0\text{--}5) = \dfrac{10}{360} = 0.028$

(b) $P(\text{age} < 50 \mid \text{experience } 6\text{--}10) = \dfrac{125}{200} = 0.625$

(c) $P(\text{age} < 50) = \dfrac{225}{360} = 0.625$; however, $P(\text{age} < 50 \mid \text{experience} > 10) = \dfrac{20}{70} = 0.286$ and so they are not independent. $P(\text{age} > 50) = \dfrac{135}{360} = 0.375$ and also $P(\text{age} > 50 \mid \text{experience } 6\text{--}10) = \dfrac{75}{200} = 0.375$, and so these two are independent.

2. (a) She is assuming that acceptances to the two colleges are independent. This does not seem reasonable as most colleges use similar acceptance criteria.

(b) She is assuming that acceptances to the two colleges are mutually exclusive. This does not seem reasonable as students can simultaneously be accepted at both of the colleges.

3.   (a) $P((\text{smartphone} \geq 3 \text{ yrs}) \cap \text{Gen } X) = (0.26)(0.09) = 0.0234$

(b) $P(\text{smartphone} \geq 3 \text{ yrs}) = P(\text{Gen } X \cap (\text{smartphone} \geq 3 \text{ yrs})) + P(\text{Gen } Y \cap (\text{smartphone} \geq 3 \text{ yrs})) +$
$P(\text{Gen } Z \cap (\text{smartphone} \geq 3 \text{ yrs})) = 0.0234 + (0.38)(0.12) + (0.36)(0.15) = 0.123$

(c) $P(\text{Gen } X \,|\, (\text{smartphone} \geq 3 \text{ yrs})) = \dfrac{P(\text{Gen } X \cap (\text{smartphone} \geq 3 \text{ yrs}))}{P(\text{smartphone} \geq 3 \text{ yrs})} = \dfrac{0.0234}{0.123} = 0.190$

Quiz 10 (Pages 161–162)

Multiple-Choice

1. **(D)** $1{,}000(0.2) + 5{,}000(0.05) = 450$, and $800 - 450 = 350$

2. **(B)** Your expected winnings are only

$$\frac{1}{7{,}500{,}000}(1{,}000{,}000) + \frac{5}{7{,}500{,}000}(10{,}000) + \frac{20}{7{,}500{,}000}(100) = 0.14$$

3. **(D)** The mean is affected both by the multiplication and the addition of a constant, but the standard deviation is affected only by the multiplication of a constant. Thus, $\mu_Y = a + b\mu_X = 3 + 2(15) = 33$, and $\sigma_Y = |b|\sigma_X = 2(4) = 8$.

4. **(C)** Expected values and variances add. Thus,

$$E(\text{Thefts}) = E(X_1) + E(X_2) + \cdots + E(X_{45}) = 361 + 361 + \cdots + 361 = 45 \times 361 \text{ and}$$
$$\text{SD (Thefts)} = \sqrt{74^2 + 74^2 + \cdots + 74^2} = \sqrt{45 \times 74^2} = \sqrt{45} \times 74$$

5. **(C)** Means and variances add. Thus,

$$E(D + C) = E(D) + E(C), \; 212 = 98 + E(C), \; E(C) = 114$$
$$\text{var}(D + C) = \text{var}(D) + \text{var}(C), \; 39^2 = 25^2 + \text{var}(C),$$
$$\text{SD}(C) = \sqrt{39^2 - 25^2} = 29.93$$

6. **(B)** Means and variances add. Thus,

$$E(\text{Total}) = E(\text{Box}) + 50E(\text{Hole}), \; 16.0 = 1.0 + 50E(\text{Hole}), \; E(\text{Hole}) = 0.3$$
$$\text{var}(\text{Total}) = \text{var}(\text{Box}) + 50\text{var}(\text{Hole})$$
$$0.245^2 = 0.2^2 + 50\big(\text{SD}(\text{Hole})\big)^2, \; \text{SD}(\text{Hole}) = 0.02$$

Free-Response Questions

1.   (a) $E(S) = 0(0.2) + 1(0.5) + 2(0.3) = 1.1$

(b) $\text{var}(S) = (0 - 1.1)^2 (0.2) + (1 - 1.1)^2(0.5) + (2 - 1.1)^2(0.3) = 0.49$ and $\text{SD}(S) = \sqrt{0.49} = 0.7$

(c) $E(S + L) = E(S) + E(L) = 1.1 + 1.4 = 2.5$

(d) $\text{var}(S + L) = \text{var}(S) + \text{var}(L) = (0.7)^2 + (0.6)^2 = 0.85$ and $\text{SD}(S + L) = \sqrt{0.85} = 0.922$

2. (a) $F = 5C$ (the weight of coconut is being multiplied by a constant)

$E(F) = E(5C) = 5E(C) = 5(3.2) = 16$ ounces, and
$SD(F) = SD(5C) = 5SD(C) = 5(0.7) = 3.5$ ounces

(b) $T = C + C + C + C + C$ (total weight is the sum of the weights of 5 individual coconuts) $E(T) = E(C) + E(C) + E(C) + E(C) + E(C) = 5(3.2) = 16$ pounds and $\text{var}(T) = \text{var}(C) + \text{var}(C) + \text{var}(C) + \text{var}(C) + \text{var}(C) = 5(0.7)^2 = 2.45$ with $SD(T) = \sqrt{2.45} = 1.56552$

Quiz 11 (Pages 171–174)

Multiple-Choice

1. **(E)** This is a binomial with $n = 3$ and $p = 0.90$. Then $P(X \geq 2) = (0.9)^3 + 3(0.9)^2(0.1)$.

2. **(C)** This is a binomial with $n = 10$ and $p = 0.37$, and so the mean is $np = 10(0.37) = 3.7$.

3. **(B)** The sample has five adults, so $n = 5$, and the probability any of these adults experienced cyberbullying is given to be $\frac{1}{6}$.

4. **(C)** This is a binomial distribution with $n = 5$ and $p = \frac{2}{5} = 0.4$. Then, $P(X = 2) = \binom{5}{2}(0.4)^2(0.6)^3$.

5. **(D)** This is a geometric distribution with $p = 0.4$. Then, $P(X = 3) = (0.6)^2(0.4)$.

6. **(D)** This is a geometric distribution with $p = \frac{1}{20} = 0.05$. The mean $= \frac{1}{p} = \frac{1}{0.05} = 20$, and the standard deviation $= \sqrt{\frac{1-p}{p^2}} = \sqrt{\frac{1-0.05}{(0.05)^2}} = \sqrt{380}$.

7. **(D)** $P(\text{I}) = P(\text{II}) = (0.5)^5$, while $P(\text{III}) = {}_5C_2(0.5)^3(0.5)^2 = 10(0.5)^5$.

8. **(B)** The mean is 5, so $np = 5$. The standard deviation is 2 so $\sqrt{np(1-p)} = \sqrt{5(1-p)} = 2$. Solving for p results in $5(1-p) = 4, 1-p = 0.8$, and $p = 0.2$.

9. **(C)** Random variable F follows a binomial distribution with $n = 10$ and $p = 0.7$. Distribution F has mean $\mu_F = np = (10)(0.7) = 7$ and standard deviation $\sigma_F = \sqrt{np(1-p)} = \sqrt{10(0.7)(0.3)} = 1.45$. Answer choice (C) is the most reasonable simulation result.

10. **(D)** This is a binomial distribution, so the mean is $\mu = np = (50)(0.34) = 17$. The mean of a probability distribution is the long-run average of the values selected from many repetitions. Choices (A) and (B) refer to particular samples, not a long-run average. Choice (C) refers to a geometric probability; however, this is a binomial. Choice (E) is the interpretation of the standard deviation, not the mean.

11. **(C)** This is a geometric distribution, so the mean is $\mu = \frac{1}{p} = \frac{1}{0.7} = 1.43$. The mean of a probability distribution is the long-run average of the values selected from many repetitions. Choices (A) and (D) do not refer to long-run averages. Choice (B) does not refer to the calculated mean. Choice (E) is the interpretation of the standard deviation, not the mean.

Free-Response Questions

1. (a) We have a binomial with $n = 20$ and $p = 0.15$, so $P(X \geq 10) = 1 - \text{binomcdf}(20, 0.15, 9) = 0.00025$.

    [Or we can calculate $P(X \geq 10) = \binom{20}{10}(0.15)^{10}(0.85)^{10} + \binom{20}{11}(0.15)^{11}(0.85)^9 + \cdots + (0.15) = 0.00025$.]

    (b) If the probability of *Legionella* bacteria growing in an automatic faucet is 0.15, the probability of a result as extreme or more extreme than what was obtained in the Johns Hopkins study is only 0.00025. With such a low probability, there is strong evidence to conclude that the probability of *Legionella* bacteria growing in automatic faucets is greater than 0.15. That is, there is strong evidence that automatic faucets actually house more bacteria than the manual kind.

2. (a) $\dfrac{0.02}{0.11 + 0.02 + 0.014 + 0.006} = \dfrac{0.02}{0.15} = \dfrac{2}{15} = 0.133$

    (b) $P(\text{at least 2 tattoos}) = 0.02 + 0.014 + 0.006 = 0.04$. The student with tattoos can be the first, second, or third student, so we have $3(0.04)(0.85)^2 = 0.0867$.

    (c) $E(X) = 0(0.85) + 1(0.11) + 2(0.02) + 3(0.0014) + 4(0.006) = 0.216$. $\sigma = \sqrt{\sum(x - 0.216)^2\, P(x)} = \sqrt{0.365344} = 0.604$. The mean number of tattoos per college student is 0.216, and for any randomly selected college student, the number of tattoos will vary by about 0.604 from the mean of 0.216 tattoos.

3. (a) This is a binomial distribution with $n = 4$ and $p = 0.35$. Then $P(X = 2) = \text{binomialcdf}(4, 0.35, 1)$

    $= 0.3105$. [Or $P(X = 2) = \binom{4}{2}(0.35)^2(0.65)^2 = 0.3105$.]

    (b) Binomial, $n = 4$, $p = 0.35$, and $P(X \geq 2) = 1 - P(X \leq 1) = 1 - \text{binomialcdf}(4, 0.35, 1)] = 0.4370$.

    [Or $1 - [(0.65)^4 + 4(0.35)(0.65)^3] = 0.4370$.]

    (c) This is a geometric distribution with $p = 0.35$. $P(X = 2) = \text{geometpdf}(0.35, 2) = 0.2275$.

    [Or $(0.65)(0.35) = 0.2275$.]

    (d) Geometric, $p = 0.35$. Mean $= \dfrac{1}{p} = \dfrac{1}{0.35} = 2.8571$ and standard deviation $= \sqrt{\dfrac{1-p}{p^2}} = \sqrt{\dfrac{1-0.35}{(0.35)^2}} = 2.3035$. In random samples of teenagers, the mean number who must be questioned before one says that *The Hunger Games* is the best book series they have ever read is 2.8571. For any random sample of teenagers, the number who must be questioned before one says that *The Hunger Games* is the best book series they have ever read will vary about 2.3035 from the mean of 2.8571.

    (e) No, it would not be unusual, because the z-score of 5 is only $\dfrac{5 - 2.8571}{2.3035} = 0.93$, so 5 is less than one standard deviation from the mean.

Quiz 12 (Pages 185–188)

Multiple-Choice

1. **(A)** The area under any probability distribution is equal to 1. Many bell-shaped curves are *not* normal curves. The smaller the standard deviation of a normal curve, the higher and narrower the graph is. The mean determines the value around which the curve is centered; different means give different centers. Because of symmetry, the mean and median are identical for normal distributions.

2. **(A)** Probability for a normal model is calculated by the area under the normal curve within a given interval of X values. Since the point $X = 22$ does not represent an interval, $P(X = 22) = 0$.

3. **(E)** All normal distributions have the same percentage (about 95%) of their observations within two standard deviations of the mean.

4. **(B)** The z-score of 10 is $\frac{10 - 12.4}{1.2} = -2$. Then $P(X < 10) = P(z < -2)$.

5. **(E)** The z-scores of 3 and 4 are $\frac{3 - 3.4}{0.5}$ and $\frac{4 - 3.4}{0.5}$, respectively. Then $P(3 < X < 4) =$
$P\left(\frac{3 - 3.4}{0.5} < z < \frac{4 - 3.4}{0.5}\right)$.

6. **(E)** The critical z-score associated with 99% to the left is invNorm$(0.99) = 2.326$.

7. **(B)** The critical z-scores associated with the middle 95% are $\pm$invNorm$(0.975) = \pm 1.96$. Converting to raw scores gives $9{,}500 \pm 1.96(1{,}750)$.

8. **(B)** $\mu_{X+Y} = 150 + 120 = 270$ and $\sigma_{X+Y} = \sqrt{40^2 + 30^2} = 50$

9. **(C)** The critical z-score is invNorm$(0.85) = 1.036$. Then $\mu + 1.036(2) = 16$ gives $\mu = 16 - 1.036(2)$.

10. **(D)** Based on the table, 68% of the values are within one standard deviation of the mean, 95% are within two standard deviations of the mean, and 100% are within three standard deviations of the mean. These percentages are very close to the 68%-95%-99.7% of the empirical rule.

Free-Response Questions

1. (a) $P\left(z > \frac{x - 150}{25}\right) = 0.90$, so $\frac{x - 150}{25} = -1.282$ and $x = 117.95$ for a cutoff score of 118.

   (b) $P(x > 118 \mid x > 100) = \frac{P(x > 118)}{P(x > 100)} = \frac{0.900}{0.9772} = 0.921$

   (c) $1 - (0.921)^3 = 0.219$

2. (a) $P(X \geq 8.5) = 0.11 + 0.07 + 0.02 + 0.01 + 0.01 = 0.22$

   (b) $z = \frac{8.5 - 8.03}{0.49} = 0.9592$ and $P(X \geq 8.5) = $ normalcdf$(0.9592, \infty) = 0.1687$

   (c) The answers to (a) and (b) are different because the discrete model calculation sums the areas of the bars with a left boundary of 8.375 (8.5 is found between 8.375 and 8.625), whereas the continuous model calculation has a left boundary at exactly 8.5. Therefore, the lesser probability is found by using the continuous model from part (b).

   (d) Use the discrete model. The discrete model shows a bar centered directly over $X = 7.25$ and thus gives a probability of 0.06 for $X = 7.25$. However, the continuous model gives a probability of zero since there is no area above the single point $X = 7.25$.

   (e) Use the continuous model. The continuous model gives a probability of $X \geq 7.1$, which corresponds to the area under the curve to the right of 7.1. However, the relative frequency of 7.1 is not clear when using the discrete model because 7.1 does not correspond to any of the bars.

3. (a) If the probability distribution of each random variable is approximately normal, then the distribution of the difference of two independent random variables is also approximately normal. Let the random variable D be the difference in walking times (Steve $-$ Jan). Then $\mu_D = 30 - 25 = 5$, and $\sigma_D = \sqrt{5^2 + 4^2} = 6.403$. For Steve to arrive before Jan, the difference in walking times must be < 0.

$$P(D < 0) = P\left(z < \frac{0 - 5}{6.403}\right) = 0.217$$

   (b) If m is the number of minutes he should leave early, the new mean of the differences is $5 - m$, and we want $P\left(z < \frac{0 - (5 - m)}{6.403}\right) = 0.9$, which gives $\frac{-5 + m}{6.403} = 1.282$ and $m = 13.2$ minutes.

Quiz 13 (Pages 203–208)

Multiple-Choice

1. **(E)** Proportions come from binomial distributions, and a binomial distribution will be approximately normal if n is large enough. We check that np and $n(1 - p)$ are both ≥ 10. A sample size of at least 30 is a condition to be checked with regard to the sampling distribution of a sample mean. np and $\sqrt{np(1 - p)}$ refer to the mean and standard deviation of a binomial distribution.

2. **(E)** A biased estimator is one for which its mean is not equal to the value of the parameter being estimated.

3. **(C)** Because the population is approximately normal, normality can be assumed for both sampling distributions. Both will have mean 0.81. The standard deviation for the $n = 10$ sampling distribution is $\frac{0.24}{\sqrt{10}}$, which is greater than the standard deviation for the $n = 100$ sampling distribution which is $\frac{0.24}{\sqrt{100}}$.

4. **(D)** The variance for the sampling distribution of $\hat{p}$ equals $\frac{p(1 - p)}{n}$. A larger n in the denominator results in a smaller quotient, and $(0.1)(0.9) < (0.5)(0.5)$.

5. **(D)** The sample is given to be random, both $np = (400)(0.34) = 136 \geq 10$ and $n(1 - p) = (400)(0.66) = 264 \geq 10$, and our sample is clearly less than 10% of all people. So, the sampling distribution of $\hat{p}$ is approximately normal with mean 0.34 and standard deviation $\sigma_{\hat{p}} = \sqrt{\frac{(0.34)(0.66)}{400}}$. With z-scores of $\frac{0.30 - 0.34}{\sqrt{\frac{(0.34)(0.66)}{400}}}$ and $\frac{0.35 - 0.34}{\sqrt{\frac{(0.34)(0.66)}{400}}}$, the probability that the sample proportion is between 0.30 and 0.35 is

$$P(0.30) < X < 0.35) = P\left(\frac{0.30 - 0.34}{\sqrt{\frac{(0.34)(0.66)}{400}}} < z < \frac{0.35 - 0.34}{\sqrt{\frac{(0.34)(0.66)}{400}}}\right)$$

6. **(E)** We have a random sample that is less than 10% of the high school football population. With a sample size of 48, the central limit theorem applies, and the sampling distribution of $\bar{x}$ is approximately normal with mean $\mu_{\bar{x}} = 355$ and standard deviation $\sigma_{\bar{x}} = \frac{80}{\sqrt{48}}$. The z-scores of 340 and 360 are $\frac{340 - 355}{\frac{80}{\sqrt{48}}}$ and $\frac{360 - 355}{\frac{80}{\sqrt{48}}}$, respectively, and the probability of a sample mean between 340 and 360 is $P(340 < X < 360)$

$$= P\left(\frac{340 - 355}{\frac{80}{\sqrt{48}}} < z < \frac{360 - 355}{\frac{80}{\sqrt{48}}}\right).$$

7. **(E)** The sampling distribution of the sample mean will be roughly normal if the population is roughly normal or if the sample size is large. Because this sampling distribution is skewed, it is likely that neither of those conditions exist.

8. **(D)** The mean of the sampling distribution is equal to the population mean. The standard deviation of the sampling distribution is the population standard deviation divided by the square root of the sample size. In this case, $\sigma_{\bar{x}} = \frac{\sigma}{\sqrt{n}}$ gives $\frac{\sigma}{\sqrt{25}} = 4$ and thus the population standard deviation equals $\sigma = 4\sqrt{25} = 20$.

9. **(C)** The center of a sampling distribution for a population proportion equals the center of the distribution being sampled, so both distributions should be roughly centered at 0.14. The larger the sample size, the lesser the variability in the sampling distribution, so distribution B with a sample size of 200 will have less variability than distribution A with a sample size of 100.

10. **(B)** Both sample sizes are at least 30, so by the central limit theorem, the sampling distribution of $\bar{x}_M - \bar{x}_W$ can be modeled with a normal distribution.

11. **(A)** The sampling distribution of $\bar{x}$ has mean $\mu_{\bar{x}} = \mu = 50$ and standard deviation $\sigma_{\bar{x}} = \frac{\sigma}{\sqrt{n}} = \frac{10}{\sqrt{100}} = 1$.

With this large a sample size, $n = 100$, the CLT gives that the shape of the sampling distribution will be roughly normal regardless of what the original distribution is.

Free-Response Questions

1. (a) $\sigma_{\hat{p}} = \sqrt{\frac{p(1-p)}{n}} = \sqrt{\frac{(0.16)(0.84)}{80}} = 0.041$

   (b) Yes, the conditions for inference are met. First, it is given that a random sample is planned. Second, the sample size, $n = 80$, is clearly less than 10% of all game players. Third, $np = (80)(0.16) = 12.8 \geq 10$ and $nq = (80)(0.84) = 67.2 \geq 10$.

   (c) The sample size would be larger than 80, because $0.037 < 0.041$. In order for $\sigma_{\hat{p}}$ to be smaller, the denominator in $\sigma_{\hat{p}} = \sqrt{\frac{p(1-p)}{n}}$ must be larger.

> **TIP**
> On the exam, you can use expressions like `binomcdf` only if you note the parameters: `binomcdf (n=5, p=0.58,2)`.

2. (a) 0.58

   (b) This is a binomial with $n = 3$ and $p = 0.58$.
   $$P(X \geq 3) = \binom{5}{3}(0.58)^3(0.42)^2 + \binom{5}{4}(0.58)^4(0.42) + (0.58)^5 = 0.647$$
   or $1 - \text{binomcdf}(5,.58,2) = 0.647$.

   (c) The sample is given to be random, both $np = (350)(0.58) = 203 \geq 10$ and $n(1-p) = (350)(0.42) = 147 \geq 10$, and our sample is clearly less than 10% of all Americans. So, the sampling distribution of $\hat{p}$ is approximately normal with mean 0.58 and standard deviation $\sigma_{\hat{p}} = \sqrt{\frac{(0.58)(0.42)}{350}} = 0.0264$. With a z-score of $\frac{0.5 - 0.58}{0.0264} = -3.030$, the probability that the sample proportion is greater than 0.5 is $\text{normalcdf}(-3.030,1000) = 0.9988$.
   [Or $\text{normalcdf}(.5,1,.58,.0264) = 0.9988$.]

> **NOTE**
> You can also do a direct binomial calculation: $1 - \text{binomcdf}(350,0.58,174) = 0.9989$.

3. (a) $P(X > 4{,}250) = P\left(z > \frac{4{,}250 - 4{,}000}{125}\right) = P(z > 2) = 0.0228$.
   [Or $\text{normalcdf}(4250,10000,4000,125) = 0.0228$.]

   (b) The original population is approximately normal, so the sampling distribution of $\bar{x}$ is approximately normal with mean $\mu_{\bar{x}} = \mu = 4{,}000$ and standard deviation $\sigma_{\bar{x}} = \frac{\sigma}{\sqrt{n}} = \frac{125}{\sqrt{40}} = 19.764$.

   (c) $P(X < 3{,}950) = P\left(z < \frac{3{,}950 - 4{,}000}{19.764}\right) = P(z < -2.530) = 0.0057$.
   [Or $\text{normalcdf}(0,3950,4000,19.764) = 0.0057$.]

   (d) The answer to part (a) would be affected because it assumes an approximately normal population. The other answers would not be affected because for large enough n, the central limit theorem gives that the sampling distribution of $\bar{x}$ is roughly normal regardless of the distribution of the original population.

4. (a) Although shape cannot accurately be gauged from boxplots, a complete answer here must address center, spread, outliers, and mention context. The distribution of study hours per week for athletes at this college has a greater center (median) and a smaller spread (range) than the distribution of study hours per week for nonathletes at this college. The nonathlete distribution has three low end outliers while the athlete distribution has no outliers.

   (b) The value of $+3$ or above occurred in $7 + 10 + 4 + 1 = 22$ of the trials.

(c) The observed difference in median study hours was $21 - 18 = +3$. From (b), this occurred in 22 of the 110 trials giving a probability of $\frac{22}{110} = 0.2$.

(d) The observed difference in study hours was three hours, and this value or more will occur an estimated 20% of the time just by chance. Thus, there is not convincing statistical evidence that athletes participating in varsity sports at this college spend more hours per week with academic studies outside the classroom than do nonathletes.

Quiz 14 (Pages 219–221)

Multiple-Choice

1. **(D)** The relevant condition is that both $n\hat{p}$ and $n\hat{q} \geq 10$, and only choice (D) with $n\hat{p} = (75)(0.15) = 11.25$ and $n\hat{q} = (75)(0.85) = 63.75$ satisfies this.

2. **(B)** The critical z-scores with 0.01 in the tails are $\pm\texttt{invNorm(0.99)} = \pm 2.326$ and $SE(\hat{p}) = \sqrt{\dfrac{\hat{p}(1-\hat{p})}{n}} = \sqrt{\dfrac{(2/3)(1/3)}{60}}$. The confidence interval is $\hat{p} \pm z^* SE(\hat{p}) = \dfrac{2}{3} \pm 2.326\sqrt{\dfrac{(2/3)(1/3)}{60}}$.

3. **(D)** The sample size of 30 is greater than ten percent of the population, that is, $30 > (0.10)(250) = 25$. This violates the independence condition. Note that $n\hat{p} = 14 \geq 10$, and $n(1-\hat{p}) = 16 \geq 10$, which ensures that the sampling distribution of $\hat{p}$ is approximately normal.

4. **(E)** A confidence *level* of 95 percent gives the percent of all possible confidence intervals of the given size that will capture the population parameter. Choice (B) is a correct interpretation of the confidence *interval*.

5. **(E)** The margin of error is $\pm z^* SE(\hat{p})$, where $SE(\hat{p}) = \sqrt{\dfrac{(0.29)(0.71)}{1470}} = 0.011835$. Then $z^*(0.011835) = 0.03$, $z^* = 2.535$, and $\texttt{normalcdf(-2.535, 2.535)} = 0.9888$, or about 99%.

6. **(B)** $1.96\sqrt{\dfrac{(0.5)(0.5)}{n}} \leq 0.025$ gives $\sqrt{n} \geq \dfrac{1.96\,(0.5)}{0.025} = 39.2$ and $n \geq 1{,}537$. Note that the population size is not relevant here.

7. **(D)** The interval is a statement about how confident we are that the interval has captured the population parameter. The confidence interval is about the population, not the sample. The probability the true percentage is between 40% and 46% is either 1 or 0 depending upon whether it is or is not in the interval. Choice (E) is a correct interpretation of the confidence *level*, not the confidence *interval*.

Free-Response Questions

1. (a) The figure 93% comes from a sample, so it is a statistic.

    (b) The margin of error is proportional to $\dfrac{1}{\sqrt{n}}$. With such a very large sample size n, the margin of error will be very small.

    (c) This is a voluntary response survey, and so we should have very little confidence in the conclusion.

2. (a) *Parameter:* Let p represent the proportion of the population of American women between the ages of 45 and 64 who would say that breast cancer is the medical condition they fear most.
        Procedure: One-sample z-interval for a population proportion.
        Checks: We check *independence* and *normality*. For independence, we are given that this is a random sample, and clearly 500 is less than 10% of all American women between the ages of 45 and 64. For normality, $n\hat{p} = 305 \geq 10$ and $n(1-\hat{p}) = 195 \geq 10$.
        Mechanics: Calculator software gives (0.554, 0.666).
        [For instructional purposes, we note that $\hat{p} = \dfrac{305}{500} = 0.61$ and $0.61 \pm 2.576\sqrt{\dfrac{(0.61)(0.39)}{500}} = 0.61 \pm 0.056$ or (0.554, 0.666).]

Conclusion in context: We are 99% confident that the true proportion of all American women between the ages of 45 and 64 who would say that breast cancer is the medical condition they most fear is between 0.554 and 0.666.

(b) If all possible random samples of 500 women between the ages of 45 and 64 were surveyed, we would expect about 99% of the resulting confidence intervals to contain the true proportion of all American women between the ages of 45 and 64 who would say that breast cancer is the medical condition they fear most.

(c) No, the 3% value is not consistent with the confidence interval because 0.03 is not in the interval (0.554, 0.666).

Quiz 15 (Pages 222–224)

Multiple-Choice

1. **(E)** Both $n\hat{p}$ and $n(1-\hat{p})$ have to be ≥ 10, but $n\hat{p}$, the number of "successes," is only 8 in this example, so the normality condition is violated. The population size is immaterial because it is more than ten times the sample size.

2. **(D)** A confidence *level* of 90% gives the percent of all possible confidence intervals of the given size that will capture the population parameter. Choice (A) is a correct interpretation of the confidence *interval*.

3. **(C)** Because of sampling variability, samples and the resulting sample proportions differ. Thus, using a single $\hat{p}$ to estimate the true population proportion cannot give a one-point answer, but rather encompasses an interval of plausible values. The margin of error helps define this interval of plausible values around $\hat{p}$.

4. **(E)** There is no guarantee that 13.4 is anywhere near the interval, so none of the statements are true.

5. **(A)** The margin of error has to do with measuring chance variation but has nothing to do with faulty survey design. Sampling error (also called sampling variability) refers to the natural variation among samples and is quantified by the margin of error. As long as n is large, s is a reasonable estimate of σ; however, again this is not measured by the margin of error. (As seen in Unit 7, with t-scores there is a correction for using s as an estimate of σ.)

6. **(D)** The interval is a statement about how confident we are that the interval has captured the population parameter. The confidence interval is about the population, not the sample. The probability the true percentage is between 85% and 89% is either 1 or 0 depending upon whether it is or is not in the interval. Choice (E) is a correct interpretation of the confidence *level*, not the confidence *interval*.

7. **(C)** `invNorm(0.95)` $= 1.645$, $1.645\left(\frac{0.5}{\sqrt{n}}\right) \leq 0.04$, $\sqrt{n} \geq 20.563$, $n \geq 422.8$, so choose $n = 423$.

Free-Response Questions

1. (a) *Parameter:* Let p represent the proportion of the population of family members who contract H1N1 after an initial family member does in this state.
Procedure: One-sample z-interval for a population proportion.
Checks: We check *independence* and *normality*. For independence, we are given this is a random sample, and we assume that 876 is less than 10% of the number of all family members of H1N1 patients in the state. For normality, $n\hat{p} = 129 \geq 10$ and $n(1-\hat{p}) = 747 \geq 10$.
Mechanics: Calculator software gives (0.12757, 0.16695).
[For instructional purposes, we note that $\hat{p} = \frac{129}{876} = 0.147$ and $0.147 \pm 1.645\sqrt{\frac{(0.147)(0.853)}{876}} = 0.147 \pm 0.020$ or (0.127, 0.167).]

  Conclusion in context: We are 90% confident that the true proportion of all family members who contract H1N1 after an initial family member does in this state is between 0.127 and 0.167.

(b) Because $\frac{1}{8} = 0.125$ is not in the interval of plausible values for the population proportion, there *is* sufficient evidence that the true proportion of all family members who contract H1N1 after an initial family member does in this state is different from the 1 in 8 chance concluded in the published study.

(c) $0.147 \pm 2.576\sqrt{\frac{(0.147)(0.853)}{876}} = 0.147 \pm 0.031$ or $(0.116, 0.178)$, which does include 0.125. So, using a 99% confidence interval, there is *not* sufficient evidence that the true proportion of all family members who contract H1N1 after an initial family member does in this state is different from the 1 in 8 chance concluded in the published study.

2. (a) $z^* \, SE(\hat{p}) = 1.96\sqrt{\frac{(0.74)(0.26)}{1050}} = 0.0265$. So, the margin of error is $\pm 2.65\%$.

  (b) $0.74 \pm 0.0265 = (0.7135, 0.7265)$. We are 95% confident that between 71.35% and 76.65% of all smokers would like to give up smoking.

  (c) If all possible random samples of 1,050 smokers were surveyed, we would expect about 95% of the resulting confidence intervals to contain the true proportion of all smokers who would like to give up smoking.

  (d) Smoking is becoming more undesirable in society as a whole, so some smokers may untruthfully say they would like to stop.

  (e) To be more confident, we must accept a *wider* interval (the critical z-score would go from 1.96 to 2.576).

  (f) All other things being equal, the greater the sample size, the *smaller* the margin of error is (the standard error is inversely proportional to $\sqrt{n}$).

Quiz 16 (Pages 234–237)

Multiple-Choice

1. **(A)** Both the null hypothesis H_0 and the alternative hypothesis H_a are always about a population parameter, never about a sample statistic.

2. **(C)** If the null hypothesis is true, the P-value represents the probability of obtaining a sample proportion as extreme as or more extreme than the one obtained for the significance test, in the direction of the alternative hypothesis. In this case, the null hypothesis is that 70 percent of teens in the city pass their driving test on the first attempt; given the direction of the alternative hypothesis, "extreme as or more extreme" means "as large as or larger."

3. **(C)** A Type II error is a mistaken failure to reject a false null hypothesis or, in this case, a failure to realize that a person really does have ESP.

4. **(A)** This is a one-sided z-test, $H_0: p = 0.5$, $H_a: p > 0.5$, with $\sigma_{\hat{p}} = \sqrt{(0.5)(0.5)/565}$ and $\hat{p} = \frac{305}{565} = 0.540$. The P-value equals $P\left(z > \frac{\hat{p} - p_0}{\sigma_{\hat{p}}}\right) = P\left(z > \frac{0.54 - 0.50}{\sqrt{(0.5)(0.5)/565}}\right)$.

5. **(D)** $\hat{p}_1 = \frac{151}{500} = 0.302$, $\hat{p}_2 = \frac{345}{1{,}000} = 0.345$, $\hat{p}_c = \frac{151 + 345}{500 + 1{,}000} = 0.331$. This is a two-sided

test $(H_0\colon p_1 - p_2 = 0, H_a\colon p_1 - p_2 \neq 0)$ with $SE(\hat{p}_1 - \hat{p}_2) = \sqrt{\hat{p}_c(1 - \hat{p}_c)\left(\frac{1}{n_1} + \frac{1}{n_2}\right)} = $

$\sqrt{(0.331)(0.669)\left(\frac{1}{500} + \frac{1}{1{,}000}\right)}$. The *P*-value equals twice the tail probability (because this is a two-sided test),

so the *P*-value equals $2P\left(z < \dfrac{\hat{p}_1 - \hat{p}_2}{SE(\hat{p}_1 - \hat{p}_2)}\right) = 2P\left(z < \dfrac{0.302 - 0.345}{\sqrt{(0.331)(0.669)\left(\frac{1}{500} + \frac{1}{1{,}000}\right)}}\right)$.

6. **(B)** If the null hypothesis is far off from the true parameter value, there is a greater chance of rejecting the false null hypothesis and thus a smaller risk of a Type II error. Power is the probability that a Type II error is not committed, so a lower Type II error results in higher power.

7. **(B)** With a *P*-value greater than 0.05, the difference is not statistically significant at the 0.05 significance level, so a 95% confidence interval for the difference in population proportions will include both positive and negative values.

8. **(D)** The one-sample *z*-test for a population proportion results in a *P*-value (the probability of observing a value as extreme as or more extreme than $\hat{p}$ given that p_0 is true), which is a measure of the inconsistency between the claimed p_0 and the observed $\hat{p}$. It can be used to make a decision as to whether or not there is convincing evidence to reject the claimed p_0. Choice (C) concerns confidence intervals, not significance tests.

9. **(D)** The testers want to decrease the probability of a Type I error, that is, decrease the probability that the null hypothesis is true (the new software is not better than the old), but the test gives convincing evidence to reject the null hypothesis in favor of the alternative hypothesis (the new software is superior to the old). To make it more difficult to reject the null hypothesis, the significance level can be decreased (for example, going from a significance level of 0.05 to 0.01 makes it more difficult to reject the null hypothesis).

Free-Response Questions

1. (a)   *Parameters:* Let p_{US} represent the proportion of the population of new births in the United States that are to unmarried women. Let p_{UK} represent the proportion of the population of new births in the United Kingdom that are to unmarried women.

   Procedure: Two-sample *z*-interval for a difference between population proportions, $p_{US} - p_{UK}$, the difference in population proportions of new births to unmarried women in the United States and the United Kingdom.

   Checks: We check *independence* and *normality*. For independence, we are given these are random samples, they are clearly independent, and the sample sizes, 500 and 400, are less than 10% of new births in the United States and the United Kingdom, respectively. For normality, $n_{US}\hat{p}_{US} = (500)(0.412)$ $= 206 \geq 10$, $n_{US}(1 - \hat{p}_{US}) = (500)(0.588) = 294 \geq 10$, $n_{UK}\hat{p}_{UK} = (400)(0.465) = 186 \geq 10$, and $n_{UK}(1 - \hat{p}_{UK})$ $= (400)(0.535) = 214 \geq 10$.

   Mechanics: Calculator software (such as `2-PropZInt on the TI-84`) gives $(-0.1182, 0.01219)$. [For instructional purposes, we note that:

   $$(0.412 - 0.465) \pm 1.96\sqrt{\frac{(0.412)(0.588)}{500} + \frac{(0.465)(0.535)}{400}} = -0.053 \pm 0.065 \text{ or } (-0.118, 0.012).]$$

   Conclusion in context: We are 95% confident that the true difference in proportions, $p_{US} - p_{UK}$, of new births to unmarried women in the United States and the United Kingdom is between -0.118 and 0.012.

(b)   Because 0 is in the interval of plausible values for the difference of population proportions, this confidence interval does not support the belief by the UN health care statistician that the true proportions of new births to unmarried women are different in the United States and the United Kingdom.

2. (a)   *Parameter:* Let p represent the proportion of the population of AAUP members who own a Roth IRA.
Hypotheses: $H_0: p = 0.174$ and $H_a: p > 0.174$, where p is the proportion of all AAUP members who own Roth IRAs.
Procedure: A one-sample z-test for a population proportion.
Checks: Random sample (given), $np_0 = 750(0.174) = 130.5 \geq 10$ and $n(1 - p_0) = 750(0.826) = 619.5 \geq 10$ (where p_0 is the claimed 0.174) and the sample size, $n = 750$, is assumed to be less than 10% of all AAUP members.
Mechanics: Calculator software gives $z = 1.878$ and $P = 0.030$.

[For instructional purposes, we note that:

$$\hat{p} = \frac{150}{750} = 0.2,\ z = \frac{0.2 - 0.174}{\sqrt{\dfrac{(0.174)(0.826)}{750}}} = 1.878,\ \text{and the } P\text{-value} = P(z > 1.878) = 0.030.]$$

Conclusion in context with linkage to the P-value: With a P-value this small, $0.030 < 0.05$, there is sufficient evidence to reject H_0; that is, there is sufficient evidence that more than 17.4% of all AAUP members own Roth IRAs.

(b)   There was convincing evidence to reject the null hypothesis, so a Type I error might have been committed, that is, rejecting a true null hypothesis. If so, the AAUP would incorrectly think that more than 17.4% of their members own Roth IRAs and would not feel the need to recommend the Roth IRAs to their members when perhaps they should.

3. (a)   The sample was said to be selected randomly. There was no random assignment into ages or into who answered with what response.

(b) $H_0: p_{35} = p_{65}$ and $H_a: p_{35} \neq p_{65}$, where p_{35} is the proportion of all voters under age 35 who cannot name a single Supreme Court justice, and p_{65} is the proportion of all voters over age 65 who cannot name a single Supreme Court justice.

(c)   The two sample proportions (voters under 35 and over 65 who cannot name a single Supreme Court justice) are 4.43 standard deviations (or standard errors) apart.

(d)   With this large of a z-score, 4.43, the P-value will be very small, so there is sufficient evidence to reject the null hypothesis. That is, there is convincing evidence that the proportion of all voters under 35 years old who cannot name a single Supreme Court justice is different from the proportion of all voters over 65 years old who cannot name a single Supreme Court justice.

(e)   We are 99% confident that the proportion of all voters under age 35 who cannot name a single Supreme Court justice is greater than the proportion of all voters over age 65 who cannot name a single Supreme Court justice by between 0.04612 and 0.17388.

(f)   Since the confidence interval contains only positive values, that is, does not include zero as a plausible value for the difference in population proportions, the two procedures give consistent results.

4. (a)   The purpose of random assignment is to create two groups of subjects that are roughly equivalent except for the treatment, hydroxychloroquine or placebo, received. This allows for a cause-and-effect conclusion about hydroxychloroquine and a lowered susceptibility to Covid-19 if the difference between the two groups is statistically significant.

(b) $H_0: p_{hydro} = p_{placebo}$ and $H_a: p_{hydro} < p_{placebo}$, where p_{hydro} is the true probability of contracting Covid-19 after taking hydroxychloroquine as a postexposure prophylaxis for asymptomatic participants reporting a high-risk exposure to a confirmed Covid-19 contact, and $p_{placebo}$ is the true probability of contracting Covid-19 after taking a placebo for asymptomatic participants reporting a high-risk exposure to a confirmed Covid-19 contact.

(c)   Two-sample z-test for a difference of two population proportions.

- The treatments were randomly assigned.
- With $\hat{p}_c = \dfrac{49 + 58}{414 + 407} = 0.130$, we have $n_{hydro}\,\hat{p}_c = (414)(0.130) = 53.8$, $n_{hydro}(1 - \hat{p}_c) = (414)(0.870) = 360.2$, $n_{placebo}\,\hat{p}_c = (407)(0.130) = 52.9$, and $n_{hydro}(1 - \hat{p}_c) = (414)(0.870) = 360.2$ are all at least 10.

(d)   With this large of a P-value, $0.152 > 0.05$, there is not sufficient evidence to reject the null hypothesis; that is, there is not convincing evidence that hydroxychloroquine is any better than a placebo for preventing Covid-19 when used as a postexposure prophylaxis for asymptomatic participants reporting a high-risk exposure to a confirmed Covid-19 contact.

(e)   There was not convincing evidence to reject the null hypothesis, so a Type II error might have been committed, that is, failing to reject a false null hypothesis. If so, the researchers would not have recommended using hydroxychloroquine as a postexposure prophylaxis for asymptomatic participants reporting a high-risk exposure to a confirmed Covid-19 contact when, in reality, hydroxychloroquine is effective and should have been recommended.

Quiz 17 (Pages 238–241)

Multiple-Choice

1. **(B)** The critical z-scores are $\pm\texttt{invNorm(0.995)} = \pm 2.576$.

$$n_1 = 300 \quad n_2 = 400$$
$$\hat{p}_1 = 0.65 \quad \hat{p}_2 = 0.48$$
$$SE(\hat{p}_1 - \hat{p}_2) = \sqrt{\frac{\hat{p}_1(1 - \hat{p}_1)}{n_1} + \frac{\hat{p}_2(1 - \hat{p}_2)}{n_2}} = \sqrt{\frac{(0.65)(0.35)}{300} + \frac{(0.48)(0.52)}{400}}$$
$$(\hat{p}_1 - \hat{p}_2) \pm z^{*}\,SE(\hat{p}_1 - \hat{p}_2) = (0.65 - 0.48) \pm 2.576 \sqrt{\frac{(0.65)(0.35)}{300} + \frac{(0.48)(0.52)}{400}}$$

2. **(B)** If the null hypothesis is true, the P-value represents the probability of obtaining a sample proportion as extreme as or more extreme than the one obtained for the significance test, in the direction of the alternative hypothesis. In this case, the null hypothesis is that 15 percent of students at her college are willing to report cheating by other students. Given the direction of the alternative hypothesis, "extreme as or more extreme" means "as small as or smaller."

3. **(E)** The P-value is a conditional probability; in this case, there is a 0.032 probability of an observed difference in sample proportions as extreme as (or more extreme than) the one obtained if the null hypothesis is assumed to be true.

4. **(B)** The level of significance is defined to be the probability of committing a Type I error, that is, of mistakenly rejecting a true null hypothesis. Here it is 0.05, no matter what the sample size.

5. **(A)** With H_0: $p = \frac{2}{3}$ and H_a: $p > \frac{2}{3}$, 1PropZTest gives $P = 0.0062$. [For instructional purposes, we note that

$$\hat{p} = \frac{150}{200} = 0.75 \text{ and } z = \frac{0.75 - \frac{2}{3}}{\sqrt{\frac{\left(\frac{2}{3}\right)\left(\frac{1}{3}\right)}{200}}} = 2.5 \text{ and normalcdf}(2.5, 100) = 0.0062.] \text{ With this small of a } P\text{-value,}$$

$0.0062 < 0.01$, there is sufficient evidence to reject the null hypothesis; that is, there is convincing evidence in support of the sociologist's claim that over two-thirds of adult Whites have entirely White social networks without any minority presence.

6. **(C)** With a smaller α, that is, with a tougher standard to reject H_0, there is a greater chance of failing to reject a false null hypothesis; that is, there is a greater chance of committing a Type II error. Power is the probability that a Type II error is not committed, so a higher Type II error results in lower power.

7. **(D)** Look for the biggest difference between the proportions in the two groups (choices (A), (C), and (D) all compare 10% versus 30%), and when the difference is the same, the larger sample sizes will produce stronger evidence (no choice has larger sample sizes than (D)).

8. **(C)** If the two groups have the same proportion, then the risk ratio would be one. A risk ratio greater than one indicates that one group has a higher proportion than the other. Thus, a confidence interval consisting entirely of values greater than one provides convincing evidence that the proportions differ.

9. **(A)** The testers want to decrease the probability of a Type II error, that is, decrease the probability that the null hypothesis is false (the vaccine is effective), but the stringent test does not give convincing evidence to reject the null hypothesis in favor of the alternative hypothesis. To decrease the probability of a Type II error, the testers can increase the significance level (making it easier to reject a null hypothesis), thus increasing the probability of a Type I error. For example, going from a significance level of 0.05 to 0.10 makes it easier to reject the null hypothesis, increasing the probability of a Type I error, but decreasing the probability of a Type II error.

Free-Response Questions

1. (a) The P-value of 0.138 gives the probability of observing a sample proportion of GBS complications as great as (or greater than) the proportion found in the study if, in fact, the proportion of GBS complications is 0.000001.

   (b) Since $0.138 > 0.10$, there is not sufficient evidence to reject H_0; that is, there is not convincing evidence that under the new vaccine, the true proportion of GBS complications is greater than 0.000001 (one in a million).

   (c) The null hypothesis is not rejected, so there is the possibility of a Type II error, that is, of mistakenly failing to reject a false null hypothesis. A possible consequence is continued use of the vaccine with a higher rate of GBS complications than is acceptable.

2. *Parameters:* Let p_B represent the proportion of the population of high school boys who meet the recommended level of physical activity. Let p_G represent the proportion of the population of high school girls who meet the recommended level of physical activity.
   Hypotheses: H_0: $p_B - p_G = 0$ (or $p_B = p_G$) and H_a: $p_B - p_G > 0$ (or $p_B > p_G$).
   Procedure: Two-sample z-test for a difference of two proportions, $p_B - p_G$.
   Checks: We check *independence* and *normality*. For independence, we are given independent random samples, and the sample sizes, 850 and 580, are clearly less than 10% of the respective populations. For normality, with $\hat{p}_c = \frac{370 + 218}{850 + 580} = \frac{588}{1,430} = 0.411$, $n_B\hat{p}_c = (850)(0.411) = 349 \geq 10$, $n_B(1 - \hat{p}_c) = (850)(0.589) = 501 \geq 10$, $n_G\hat{p}_c = (580)(0.411) = 238 \geq 10$, and $n_G(1 - \hat{p}_c) = (580)(0.589) = 342 \geq 10$.

Mechanics: Calculator software (such as `2-PropZTest` on the TI-84) gives $z = 2.2427$ and $P = 0.012$.

[For instructional purposes, we note that

$$\hat{p}_B = \frac{370}{850} = 0.4353, \quad \hat{p}_G = \frac{218}{580} = 0.3759, \quad \hat{p}_c = \frac{370 + 218}{850 + 580} = \frac{588}{1,430} = 0.4112,$$

$$z = \frac{0.4353 - 0.3759}{\sqrt{(0.4112)(0.5888)\left(\frac{1}{850} + \frac{1}{580}\right)}} = 2.2427, \text{ and}$$

$$P = P(z > 2.2427) = 0.012.]$$

Conclusion in context with linkage to the P-value: With a *P*-value this small $(0.012 < 0.05)$, there is sufficient evidence to reject H_0; that is, there is convincing evidence that the proportion of all high school boys who meet the recommended level of physical activity is greater than the proportion of all high school girls who meet the recommended level of physical activity.

3. (a) The sample was said to be selected randomly. There was no random assignment into ages or into the responded answers.

    (b) $H_0: p_{18\text{-}25} = p_{65}$ and $H_a: p_{18\text{-}25} \neq p_{65}$ where $p_{18\text{-}25}$ is the proportion of all voters 18 to 25 years old who believe the false political claim, and p_{65} is the proportion of all voters over age 65 who believe the false political claim.

    (c) The two sample proportions (voters 18 to 25 years old and voters over 65 who believe the false political claim) are 1.834 standard deviations (or standard errors) apart.

    (d) This is a two-sided test with a *P*-value of $2 \times$ normalcdf(1.834, 1000) $= 2(0.0333) = 0.0666$. With this large of a *P*-value, $0.0666 > 0.05$, there is not sufficient evidence to reject the null hypothesis. That is, there is not convincing evidence that the proportion of all voters age 18 to 25 who believe the false political claim is different from the proportion of all voters over 65 years old who believe the false political claim.

    (e) We are 95% confident that the proportion of all voters age 18 to 25 who believe the false political claim is between 0.0012 less than and 0.06123 greater than the proportion of all voters over age 65 who believe the false political claim. We are 95% confident that the true difference in the proportions of 18- to 25-year-old voters and voters over 65 who believe the false political claim is between -0.0012 and 0.06123.

    (f) Since the confidence interval contains both negative and positive values, that is, it includes zero as a plausible value for the difference in population proportions, the two procedures give consistent results.

4. (a) Due to random sampling, the result can be generalized to the population from which the sample was drawn: All adults in Boulder, Colorado.

    (b) $P(\text{exercise addiction}) = \frac{151 + 48}{1,000} = 0.199$

    (c) $P(\text{exercise addiction} \mid \text{negative body image}) = \frac{68 + 26}{395} = \frac{94}{395} = 0.238$

    (d) The confidence interval calculation relies on the fact that the sampling distribution of $p_{neg} - p_{pos}$ is approximately normal. This will be the case if $n_{neg}\hat{p}_{neg}$, $n_{neg}(1 - \hat{p}_{neg})$, $n_{pos}\hat{p}_{pos}$, and $n_{pos}(1 - \hat{p}_{pos})$ are all at least 10.

    (e) We are 95% confident that the true difference between the proportion of people with severe exercise addiction among those in the population who have negative body image and the proportion of people with severe exercise addiction among those in the population who do not have negative body image is in the interval from 0.000815 to 0.0581. That is, we are 95% confident that the proportion of people with severe exercise addiction is between 0.0815 percentage points higher and 5.81 percentage points higher for those who have negative body image than those who do not have negative body image.

(f)   An example of a confounding variable could be a certain chemical imbalance in adults. This imbalance would cause both negativity in general (including negative thoughts about one's body) and a craving for the euphoria of endorphins (which are released during exercise).

Quiz 18 (Pages 250–252)

Multiple-Choice

1. **(A)** The *t*-distribution is bell-shaped (unimodal and symmetric) but lower at the mean, higher at the tails, and more spread out than the normal distribution. The larger the *df*, the closer the *t*-distribution is to the normal distribution, so as *df* increases, the area in the tails decreases.

2. **(D)** The women working in the mayor's office are a convenience sample, and the independence condition cannot be verified. The sample size, $n = 38$, is large enough for the central limit theorem to apply and given that the sampling distribution of $\bar{x}$ is approximately normal. The sample size of 38 can reasonably be assumed to be less than 10% of the women in the small city.

3. **(A)** Random assignment doesn't make sense here. The sample size, $n = 95$, is large enough for the central limit theorem to apply, so the population distribution does not have to be approximately normal. It is clear that the sample size of 95 students is less than 10% of all high school students. Given a random sample and less than 10% of the population, the independence condition is satisfied.

4. **(E)** The 95% refers to the method: 95% of all intervals obtained by this method will capture the true population parameter. Nothing is certain about any particular set of 20 intervals. For any particular interval, the probability that it captures the true parameter is 1 or 0 depending upon whether the parameter is or isn't in it.

5. **(A)** If you are willing to accept less confidence, that is, a smaller confidence level, you will have a smaller margin of error and a smaller confidence interval. The margin of error varies directly with the critical value (z^* or t^*) and directly with the standard deviation of the sample but inversely with the square root of the sample size. The value of the sample mean and the population size do not affect the margin of error.

6. **(C)** $df = n - 1 = 30 - 1 = 29$, $\texttt{invT(0.975, 29)} = 2.045$, and
$SE(\bar{x}) = \frac{s}{\sqrt{n}} = \frac{1.2}{\sqrt{30}}$, so the confidence interval is $\bar{x} \pm t^*_{n-1} SE(\bar{x}) = 28.5 \pm 2.045\left(\frac{1.2}{\sqrt{30}}\right)$.

7. **(B)** Although both methods will give estimates of the total quantity of sugary beverages consumed daily by students at this high school, Method II is better as it draws data from 30 students consuming sugary beverages. Method I could result in choosing a very small number, possibly even none at all, of students consuming sugary beverages to use for the overall estimation.

8. **(E)** The percent refers to how confident we are that the interval has captured the population mean. The confidence interval is about the population, not the sample. The probability the true mean difference is between 4.6 and 7.3 is either 1 or 0 depending on whether it is or is not in the interval. The confidence interval is about a population mean not about individual values.

9. **(E)** To divide the interval estimate by d without affecting the confidence level, multiply the sample size by a multiple of d^2. In this case, $4(50) = 200$.

10. **(D)** The formula for margin of error is $t^*\left(\frac{s}{\sqrt{n}}\right)$. Since s is given to be the same for both samples, the width of the confidence interval from X will be less than from Y if the sample sizes are the same but t^* is less (which will occur if the confidence level is less).

Free-Response Questions

1. (a) *Parameter:* the mean weight of all "11 oz" bags of Doritos tortilla chips.

   Procedure: A one-sample t-interval for a population mean.

   Checks: We check *independence* and *normality*. For independence, we are given a simple random sample, and $n = 10$ is less than 10% of all "11 oz" bags of Doritos tortilla chips. For normality, a dotplot of the sample data is roughly unimodal and symmetric and is free from strong skewness and outliers.

   Mechanics: Calculator software gives (11.08, 11.38).

   [For instructional purposes, we note that with $df = 10 - 1 = 9$, the critical t-score is $\text{invT}(0.975,9) = 2.262$. The confidence interval is $11.23 \pm 2.262\left(\frac{0.2058}{\sqrt{10}}\right) = 11.23 \pm 0.15$.]

   Conclusion in context: We are 95% confident that the mean weight of the population of all "11 oz" bags of Doritos tortilla chips is between 11.08 and 11.38 ounces.

   (b) Five of the bags in the sample have weights {11.1, 11.2, 11.2, 11.2, 11.3} in this confidence interval. Five out of ten gives an answer of 50%.

   (c) 50% is not close to 95%, but nothing went wrong. The confidence interval is estimating the mean weight, not weights of individual bags. There is no reason to expect that 95% of the individual weights to be within the confidence interval.

2. (a) The point estimate is the sample mean, $2,900,000.

   (b) $SE(\bar{x}) = \frac{s}{\sqrt{n}} = \frac{850,000}{\sqrt{30}} = 155,000$.

   (c) Yes, the conditions for inference are met: *independence* and *normality*. For independence, the sample is given to be random, and it is reasonable to assume that 30 is less than 10% of all white-collar crimes. For normality, the sample size, $n = 30$, is large enough for the central limit theorem to apply, thus giving the sampling distribution of the sample mean is approximately normal.

   (d) We are 95% confident that the mean amount involved in all white-collar crimes is between $2,580,000 and $3,222,000.

   (e) The confidence interval says nothing about the plausibility of a white-collar crime with a monetary amount of $2,000,000. A confidence interval provides an estimate of the population mean value but does not provide information about an individual value.

   (f) It is not plausible that the population mean is $2,000,000 because 2,000,000 is not in the interval from 2,580,000 to 3,222,000.

Quiz 19 (Pages 261–264)

Multiple-Choice

1. **(D)** Since couples are chosen, the samples of grandmothers and grandfathers are not independent, so a two-sample t-test cannot be performed. If the procedure would be to perform a one-sample t-test on the set of differences, then the conditions for inference would be met. Random assignment does not make sense here. The sample size, $n = 100$, is large enough for the central limit theorem to apply, so the population distribution need not be approximately normal. One hundred grandparents is clearly less than 10% of all grandparents.

2. **(D)** Since the sample size $n = 22 < 30$, the central limit theorem does not apply. In order to conclude approximate normality of the sampling distribution of $\bar{x}$, we either have to be told that the population is approximately normal, or failing that, that the sample distribution is roughly unimodal and symmetric so that it would not be unreasonable to assume the population is approximately normal. However, we are not given that the population is approximately normal, and we are told that the sample distribution is skewed right. Thus, the normality condition is violated.

3. **(C)** The population standard deviation is unknown, so this is a t-test. Medications having an effect shorter or longer than claimed should be of concern, so this is a two-sided t-test: H_a: $\mu \neq 58.4$, and $df = n - 1 = 40 - 1 = 39$.

4. **(A)** $SE(\bar{x}_1 - \bar{x}_2) = \sqrt{\dfrac{(s_1)^2}{n_1} + \dfrac{(s_2)^2}{n_2}} = \sqrt{\dfrac{(19.1)^2}{347} + \dfrac{(19.9)^2}{561}}$ and with such large sample sizes, the critical t-scores should be just slightly greater than the critical z-scores of ± 1.96 associated with 95% confidence. The confidence interval is $(\bar{x}_1 - \bar{x}_2) \pm t^* SE(\bar{x}_1 - \bar{x}_2) = (93.5 - 84.5) \pm 1.97 \sqrt{\dfrac{(19.1)^2}{347} + \dfrac{(19.9)^2}{561}}$.

5. **(E)** The test statistic in a significance test for a population mean is $t = \dfrac{(\bar{x} - \mu_0)}{SE(\bar{x})}$; that is, it is the number of standard errors the sample mean is away from the hypothesized mean. Be careful not to confuse the test statistic with the P-value.

6. **(D)** The P-value is the probability that, if many samples of size 50 were selected from the population, we would observe a sample mean of 29.4 gallons or more, given that the population mean is 28.2 gallons.

7. **(A)** With unknown population standard deviations, the t-distribution must be used, and $SE(\bar{x}_1 - \bar{x}_2) = \sqrt{\dfrac{s_1^2}{n_1} + \dfrac{s_2^2}{n_2}} = \sqrt{\dfrac{(0.3)^2}{10} + \dfrac{(0.2)^2}{10}}$. The critical t-score is $t = \dfrac{\bar{x}_1 - \bar{x}_2}{SE(\bar{x}_1 - \bar{x}_2)} = \dfrac{(4 - 4.8)}{\sqrt{\dfrac{(0.3)^2}{10} + \dfrac{(0.2)^2}{10}}}$.

8. **(C)** This is a hypothesis test with H_0: breaking strength is within specifications, and H_a: breaking strength is below specifications. A Type I error is committed when a true null hypothesis is mistakenly rejected. In this context, a Type I error would be committed if the breaking strength is within specifications, but because the sample wrongly indicates a drop below the specified level, the production process is unnecessarily halted.

9. **(B)** $P(\text{at least one Type I error}) = 1 - P(\text{no Type I errors}) = 1 - (0.99)^5 = 0.049$

10. **(C)** A larger sample size n reduces the standard deviation of the sampling distribution, resulting in a narrower sampling distribution. So, for the given sample statistic, the P-value is smaller, and the probabilities of mistakenly rejecting a true null hypothesis or mistakenly failing to reject a false null hypothesis are both decreased. Furthermore, a lower Type II error results in higher power.

11. **(B)**

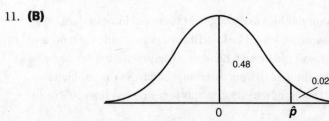

With 0.02 in the tail, a confidence interval over $0.48 + 0.48 = 0.96 = 96\%$ will contain 0.

Free-Response Questions

1. (a) The *P*-value of 0.0197 gives the probability of observing a sample mean of 1002.4 or greater if, in fact, this system results in a mean *E. coli* concentration of 1,000 MPN per 100 mL.

    (b) Since $0.0197 < 0.05$, there is sufficient evidence to reject H_0; that is, there is convincing evidence that the mean *E. coli* concentration is greater than 1,000 MPN per 100 mL.

    (c) With rejection of the null hypothesis, there is the possibility of a Type I error, that is, of mistakenly rejecting a true null hypothesis. Possible consequences are that sales of the system drop even though the system is doing what it claims or that the company performs an overhaul to fix the system even though the system is properly reducing the mean concentration of *E. coli* to 1,000 MPN per 100 mL.

2. (a) *Procedure:* A two-sample *t*-interval for $\mu_{NRW} - \mu_{RW}$, the difference in population means of BPA body concentrations in nonretail workers and retail workers.
    Checks: We check *independence* and *normality*. For independence, we are given random samples. It is reasonable to assume the samples are independent, and we assume the sample sizes are less than 10% of the populations. For normality, both sample sizes, $528 \geq 30$ and $197 \geq 30$, are large enough so that the CLT applies.
    Mechanics: Calculator software gives $(-0.9521, -0.7479)$ with $df = 332.3$.
    Conclusion in context: We are 99% confident that the difference in true means of BPA body concentrations in all nonretail and retail workers (nonretail mean minus retail mean) is between -0.75 and $-0.95 \, \mu g/L$.

    (b) Because 0 is not in the interval of plausible values for the difference of population means and the entire interval is negative, the interval does support the belief that retail workers carry higher amounts of BPA in their bodies than nonretail workers.

3. (a) Since couples are chosen, the samples of husbands' ages and wives' ages are not independent. So a two-sample *t*-interval cannot be found. If the procedure would be to calculate a one-sample *t*-interval on the set of differences, then the conditions would be met.

    (b) Because there is a clear relationship between husbands and wives in couples, the proper procedure is to form the set of 50 differences and to perform a one-sample *t*-test as follows.
    Parameter: Let μ_{diff} = the true mean difference in ages at wedding between husbands and wives (husbands minus wives).
    Hypotheses: $H_0: \mu_{diff} = 0$ and $H_a: \mu_{diff} > 0$.
    Procedure: A paired *t*-test, that is, a one-sample *t*-test on the set of differences.
    Checks: We check *independence* and *normality*. For independence, we are given a random sample, and $n = 50$ is less than 10% of all married couples. For normality, $n = 50$ is large enough for the CLT to apply.
    Mechanics: A *t*-test gives $t = 3.780$ with a *P*-value of 0.000213.
    [Or we can calculate $SE(\bar{x}_{diff}) = \dfrac{3.742}{\sqrt{50}} = 0.5292$ and $t = \dfrac{\bar{x}_{diff} - 0}{0.5292} = \dfrac{2}{0.5292} = 3.780$.
    With $df = n - 1 = 49$, the *P*-value is $P(t > 3.780) = tcdf(3.780, 1{,}000, 49) = 0.000213$.]
    Conclusion in context with linkage to the P-value: With this small of a *P*-value, $0.000213 < 0.05$, there is sufficient evidence to reject H_0. In other words, there is convincing evidence that the true mean age at wedding for husbands is greater than the true mean age at wedding for wives.

Quiz 20 (Pages 267–270)

Multiple-Choice

1. **(B)** For this day's sample, $G = \dfrac{\text{Max} + \text{Min}}{2} = \dfrac{35.55 + 33.51}{2} = 34.53$. A value of 34.53 or less occurred only twice in the 100 samples. Thus, the estimated P-value is $\dfrac{2}{100} = 0.02$. If the machinery was operating properly, a G measurement of 34.53 would be very unusual. The conclusion should be to recalibrate the machine.

2. **(E)** The P-value is a conditional probability, the probability of as extreme as (or more extreme) a result as the one obtained, given that the null hypothesis is true. This is a two-tailed test. In the situation, there were $12 + 7 + 1 = 20$ values on the lower end and $5 + 2 = 7$ values on the upper end that are at least as extreme as the $W = -0.7$ value. Thus, the estimated P-value is $\dfrac{20 + 7}{1,000} = 0.027$.

Free-Response Questions

1. (a) $H_0: \mu = 350$ and $H_a: \mu < 350$

   (b) The Normal/Large Sample condition is not met. There is no indication that the population of all battery cranking powers have a roughly normal distribution; the sample size, $n = 5$, is not large enough for the central limit theorem to apply; and since the sample data is not given, there is no way to determine if it is reasonable to assume that the sample data come from a roughly normal distribution.

   (c) The consumer group's random sample of 5 batteries had an average cranking power of only 320 CCA. The simulation indicates that the chance of picking 5 batteries with an average of 320 CCA or lower from a group of new batteries manufactured to have 350 CCA power is 4 out of 160 or 0.025. So, the estimated P-value is 0.025.

   (d) With a P-value this small, $0.025 < 0.05$, there is sufficient evidence to reject H_0; that is, there is convincing evidence that the true mean cranking power of these batteries is less than the company's claim of 350 CCA.

2. (a) $H_0: p_1 = p_2$ and $H_a: p_1 > p_2$, where p_1 is the proportion of all addicts taking desipramine who would not have relapsed after one year and p_2 is the proportion of all addicts taking lithium who would not have relapsed after one year.

   (b) $\dfrac{14}{25} - \dfrac{6}{25} = 0.32$

   (c) There are 3 values out of 500 that are 0.32 or greater, so the estimated P-value is $\dfrac{3}{500} = 0.006$.

   (d) With this small of a P-value, $0.006 < 0.05$, there is sufficient evidence to reject the null hypothesis; that is, there is convincing evidence that the antidepressant desipramine is more effective than the mood stabilizer lithium in treating cocaine addiction.

Quiz 21 (Pages 273–275)

Multiple-Choice

1. **(A)** The power is the probability of not committing a Type II error, that is, the power is the probability of correctly rejecting a false null hypothesis. In this context, a false null hypothesis means the air quality is poor, and thus, power is the probability of correctly giving a warning when the air quality is poor.

2. **(C)** The α-risk, also called the significance level of the test, is the probability of committing a Type I error and mistakenly rejecting a true null hypothesis.

3. **(D)** Power is the probability of not committing a Type II error. Here, $1 - 0.4 = 0.6$.

4. **(E)** Changing from a significance level of 0.05 to one of 0.01 makes it more difficult to reject the null hypothesis. Thus, there is a smaller probability of being able to reject a false null hypothesis. This gives a decrease in power and an increase in the probability of a Type II error.

5. **(A)** The null hypothesis is typically that no effect exists. Power, the probability of correctly rejecting a false null hypothesis, is thus the probability of correctly detecting an effect that exists. Choice (B) is a Type I error, and choice (C) is a Type II error.

6. **(A)** Power is the probability of correctly rejecting a false null hypothesis. All other things being equal, power will be greatest if the true mean is furthest from the mean claimed in the null hypothesis in the direction of the alternative.

7. **(E)** A larger sample size gives a smaller standard error and thus a greater probability of picking up that the null hypothesis is incorrect. Also, the further the true parameter is from the claimed null hypothesis about the parameter, the greater the probability of picking up that the null hypothesis is incorrect.

8. **(E)** With a greater significance level, 0.05 rather than 0.01, there is a greater chance of rejecting the null hypothesis. Also, a larger sample size gives a smaller standard error and thus a greater probability of picking up that the null hypothesis is incorrect.

Free-Response Question

1. (a) Given $N(81, 0.3)$, $P(80.5 < X < 81.5) = 0.904$.

   (b) The sampling distribution of $\bar{x}$ is approximately normal with mean $\mu_{\bar{x}} = \mu = 81$ and standard deviation $\sigma_{\bar{x}} = \frac{\sigma}{\sqrt{n}} = \frac{0.3}{\sqrt{5}} = 0.1342$.

   (c) A Type I error will be committed if the null hypothesis is true but is mistakenly rejected. Given $N(81, 0.1342)$, $P(X < 80.5) + P(X > 81.5) = 0.0002$.

   (d) A Type II error will be committed if the null hypothesis is false, but is mistakenly not rejected. Given $N(80.4, 0.1342)$, $P(80.5 < X < 81.5) = 0.228$.

Quiz 22 (Pages 290–292)

Multiple-Choice

1. **(C)** Picking separate samples from each of the schools and classifying according to one variable (perception of quality education) is a survey design that is most appropriately analyzed using a chi-square test of homogeneity of proportions.

2. **(E)** Tests for independence involve a single sample cross-classified on two variables, while tests for homogeneity involve samples from each of two or more populations to compare the distribution of some variable.

3. **(D)** With $df = (3 - 1)(5 - 1) = 8$, $P(\chi^2 > 13.95) = \chi^2\text{cdf}(13.95, 1000, 8) = 0.083$. Since $0.05 < 0.083 < 0.10$, there is sufficient evidence at the 10% significance level, but not at the 5% significance level, of a relationship between education level and sports interest.

4. **(D)** With $1 + 3 + 3 + 9 = 16$ and $n = 2,000$, the expected number of fruit flies for the first species is $\frac{3}{16}(2,000) = 375$.

5. **(E)** With $df = 3$, the P-value is $\chi^2\text{cdf}(5.998,1000,3) = 0.1117$. With a P-value this large, $0.1117 > 0.10$, there is not sufficient evidence of a difference in cafeteria food satisfaction among the class levels.

6. **(A)** The degrees of freedom are $df = (\text{rows} - 1)(\text{columns} - 1) = (3 - 1)(4 - 1) = 6$, and the P-value is $\chi^2\text{cdf}(12.7,1000,6) = 0.048$. With this small of a P-value, $0.048 < 0.05$, there is sufficient evidence to reject the null hypothesis. When the null hypothesis is rejected, there is the possibility of a Type I error, that is, the possibility of mistakenly rejecting a true null hypothesis.

Free-Response Questions

1. (a) *Hypotheses:* H_0: The colors of the sugar shells are distributed according to 35% cherry red, 10% vibrant orange, 10% daffodil yellow, 25% emerald green, and 20% royal purple, and H_a: The colors of the sugar shells are not distributed as claimed by the manufacturer. [Or H_0: $P_{CR} = 0.35$, $P_{VO} = 0.10$, $P_{DY} = 0.10$, $P_{EG} = 0.25$, $P_{RP} = 0.20$, and H_a: at least one proportion is different from this distribution.]

    > **NOTE**
    >
    > $\chi^2\text{GOF-Test}$ on the TI-84; CHI-GOF on the Casio Prizm; or $\chi^2\text{Test}$: Goodness-of-fit on the HP Prime.

    Procedure: χ^2 goodness-of-fit test.

    Checks: We are given a random sample, the data are measured as "counts," and all expected cells are greater than 5: 35% of 300 = 105, 10% of 300 = 30, 25% of 300 = 75, and 20% of 300 = 60. It can be assumed that the sample size of 300 is less than 10% of all the sweets produced by the manufacturer.

    Mechanics: A calculator gives $\chi^2 = \sum \dfrac{(\text{observed} - \text{expected})^2}{\text{expected}} = 6.689$, and with $df = 5 - 1 = 4$, $P = 0.153$.

    Conclusion in context with linkage to the P-value: With a P-value this large, $0.153 > 0.10$, there is not sufficient evidence to reject H_0; that is, at the 10% significance level there is not convincing evidence that the true distribution of colors is different from what is claimed by the manufacturer.

   (b) With $z = 1.876$, the tail probability is normalcdf$(1.876, \infty) = 0.0303$. This is a two-sided test so the P-value is $2(0.0303) = 0.0606$. With this small of a P-value, $0.0606 < 0.10$, there is convincing evidence to reject the null hypothesis; that is, at the 10% significance level there is convincing evidence that the proportion of the population of fruit-flavored sweets that are royal purple is different from 20%.

   (c) The answers to (a) and (b) are not consistent with each other. This shows that it is possible to have convincing evidence for one category to be different from a claim and not to have convincing evidence that the distribution as a whole is different from the full claim.

2. (a) The explanatory variable is the amount of lighting in the child's room before age two. The response variable is the child's eyesight diagnosis.

   (b) This is an observational study. Researchers did not assign the children to the amount of light in their rooms; they merely recorded this information.

   (c) $P(\text{farsighted} \mid \text{darkness}) = \dfrac{P(\text{farsighted} \cap \text{darkness})}{P(\text{darkness})} = \dfrac{40}{172}$

   $P(\text{darkness} \mid \text{farsighted}) = \dfrac{P(\text{darkness} \cap \text{farsighted})}{P(\text{farsighted})} = \dfrac{40}{91}$

(d) *Hypotheses:* H_0: A child's eyesight is independent of bedroom lighting condition, and H_a: A child's eyesight is not independent of bedroom lighting condition.

Procedure: A chi-square test of independence.

Checks: It is given that there was a random sample, the sample size of 479 is less than 10% of all children, the data are measured as "counts," and the expected counts are all at least 5:

21.45	49.19	66.36
14.25	32.68	44.08
39.30	90.13	121.57

Mechanics: Calculator software gives $\chi^2 = 56.51$ with $P = 0.000$ and $df = 4$.

Conclusion in context with linkage to the P-value: With this small of a P-value, $0.000 < 0.05$, there is strong evidence to reject H_0; that is, there is convincing evidence of a relationship between a child's eyesight and the bedroom lighting condition.

(e) Eyesight is hereditary, so children tend to have eyesight like that of their parents. Furthermore, near-sighted parents may tend to use more light in the child's room than other parents, perhaps so they can more easily check on the child during the night.

Quiz 23 (Pages 304–308)

Multiple-Choice

1. **(E)** The sampling distribution of b is a t-distribution with $df = n - 2$. An explanation of why $df = n - 2$ is beyond this course.

2. **(C)** The spread around the regression line; that is, the differences between the predicted scores (the estimates) and the actual scores, are measured with the residual standard deviation. Usually this value is labeled "S" in computer output.

3. **(D)** If the scatterplot is not roughly linear, a linear regression analysis should not be performed. If r and b are zero, then the linear regression line will simply be $\hat{y} = \bar{y}$. The point $(\bar{x}, \bar{y})$ is always on the regression line.

4. **(C)** The confidence interval is a statement about how confident we are that we have captured the *population* parameter, which is the true slope of the regression line. In this example, the true slope gives the *average* increase in cocaine usage based on a 1 percent increase in cannabis usage.

5. **(C)** The confidence interval gives plausible values for how much the resting metabolic rate will increase for each 1 kg increase in body mass or will decrease for each 1 kg decrease in body mass. The claim is not supported by the interval, since the interval does not contain the value 20.

6. **(D)** 1.68 is the slope of the regression line, and waist size is the independent variable.

7. **(E)** The slope of the regression line is -0.163, which is the middle of the confidence interval. The correlation coefficient has the same sign as the slope.

8. **(C)** In the second column, headed by "Coef," are the y-intercept, 0.2336, which is usually labeled "Constant" or "Intercept," and the slope, 0.008051, which is labeled with the independent x-variable, the predictor, Salary. The dependent variable is a "predicted" value, signified with a "hat" as in $\hat{y}$ or written out as in "Predicted Batting Average."

9. **(D)** The confidence interval for the slope takes the form $b \pm t^*_{n-2} \times SE(b)$. The sample slope is $b = 0.008051$, the standard error of the sample slope is $SE(b) = 0.002825$, and the critical t-score is $t^*_{25-2} = \text{invT}(0.975, 23) = 2.0687$.

10. **(A)** The t-statistic for inference on the slope is given in the T-column in the row that also gives the sample slope and the standard error of the sample slope. Note that this can also be calculated by $t = \frac{0.008051 - 0}{0.002825} = 2.850$.

11. **(A)** The P-value for inference on the slope is given in the P-column in the row that also gives the sample slope and the standard error of the sample slope. However, unless otherwise stated, computer output gives the P-value for a two-tailed test, $H_a: \beta \neq 0$. With our one-tailed test, $H_a: \beta > 0$, the P-value is half this value. Thus, $P = (0.5)(0.0091) \approx 0.0045$.

Free-Response Questions

1. (a) Assuming that all conditions for regression are met, the y-intercept and slope of the equation are found in the Coeff column of the computer printout.

$$\widehat{\text{Selling price}} = 0.890 + 1.029 \,(\text{Assessed value})$$

Both the selling price and the assessed value are in \$1,000s.

(b) *Parameter:* Let β represent the slope of the population regression line for predicting selling price from assessed value.
Procedure: t-interval for the true slope of a regression model.
Checks: It is given that all conditions are met.
Mechanics: From the printout, the standard error of the slope is $s_b = 0.08192$. With $df = 18$ and 0.005 in each tail, the critical t-values are ± 2.878. The 99% confidence interval of the true slope is

$$b \pm t s_b = 1.029 \pm 2.878(0.08192) = 1.029 \pm 0.236$$

Conclusion in context: We are 99% confident that the true slope of the regression line of selling price as a function of assessed value is between 0.79 and 1.27. In the context of the data, we are 99% confident that for every \$1 increase in assessed value, the average increase in selling price is between \$0.79 and \$1.27 (or for every \$1,000s increase in assessed value, the average increase in selling price is between \$790 and \$1,270).

(c) The entire interval, (0.79, 1.27), is positive. This shows evidence of a positive association between assessed value and selling price.

2. (a) Each additional year in age of teenagers is associated with an average of 0.4577 more texts per waking hour.

(b) The scatterplot of texts per hour versus age should be roughly linear. There should be no apparent pattern in the residual plot, and a histogram of the residuals should be approximately normal.

(c) *Parameter:* Let β represent the slope of the population regression line for predicting texts per hour from age.
Hypotheses: $H_0: \beta = 0$, $H_a: \beta \neq 0$
Procedure: A test of significance for the true slope of the regression line.
Checks: We are given that all conditions for inference are met.
Mechanics: The computer printout gives that $t = 2.45$ and $P = 0.016$.
Conclusion in context with linkage to the P-value: With this small of a P-value, $0.016 < 0.05$, there is sufficient evidence to reject H_0; that is, there is convincing evidence of a linear relationship between average texts per hour and age for teenagers ages 13–17.

(d) R-Sq = 5.8%, so even though there is evidence of a linear relationship between average texts per hour and age for teenagers ages 13–17, only 5.8% of the variability in average texts per hour is explained by this regression model (or "is accounted for by the variation in age").

Quiz 24 (Pages 309–312)

Multiple-Choice

1. **(C)** A confidence interval with all positive values gives evidence for a positive association, a confidence interval with all negative values gives evidence for a negative association, and a confidence interval with both positive and negative values gives no evidence of an association.

2. **(E)** We need to check that the distribution of the residuals is approximately normal, or at least is unimodal and roughly symmetric with no strong skewness or outliers. Ideally, this check should be made for the residuals at each possible value of x, but we very rarely have enough observations to check for normality at each x-value. So, we look at a histogram, dotplot, stemplot, or normal probability plot of all the residuals.

3. **(C)** The correlation coefficient can take any value between -1 and $+1$; however, the P-value is a probability that can only take values between 0 and $+1$.

4. **(A)** The confidence interval is a statement about how confident we are that we have captured the *population* parameter, which is the true slope of the regression line. In this example, the true slope gives the *average* change in battery life based on a one cd/m^2 increase in screen brightness.

5. **(E)** When the P-value is greater than the significance level α, there is *not* sufficient evidence to reject the null hypothesis; that is, there is *not* convincing evidence of the alternative hypothesis. In this context, there is not convincing evidence that there is a linear relationship between how long customers stay at buffets and how large the tips they leave.

6. **(D)** With $H_0: \beta = 0$ and $H_a: \beta > 0$, $t = \dfrac{b - 0}{SE(b)} = \dfrac{0.6013 - 0}{0.2488}$.

7. **(D)** In the second column, headed by Coef, are the y-intercept, 0.7737, which is usually labeled "Constant" or "Intercept," and the slope, 0.4341, which is labeled with the independent x-variable, the predictor, Packs. The dependent variable is a "predicted" value, signified with a "hat" as in $\hat{y}$ or written out as in "Predicted Cox Ratio."

8. **(C)** The spread around the regression line, that is, the differences between the predicted scores (the estimates) and the actual scores, is measured with the residual standard deviation. This is usually simply labeled "S" in computer output.

9. **(E)** The confidence interval for the slope takes the form $b \pm t^*_{n-2} \times SE(b)$. The sample slope is $b = 0.4341$, the standard error of the sample slope is $SE(b) = 0.07675$, and the critical t-score is $t^*_{7-2} = \mathtt{invT(0.95,5)} = 2.015$.

10. **(A)** The P-value for inference on the slope is given in the P-column in the row that also gives the sample slope and the standard error of the sample slope. However, unless otherwise stated, computer output gives the P-value for a two-tailed test, $H_a: \beta \neq 0$. With our one-tailed test, $H_a: \beta > 0$, the P-value is half this value. Thus, $P = (0.5)(0.0024) = 0.0012$.

11. **(E)** The P-value is a conditional probability. It is the probability of a sample with as extreme as, or more extreme than, the data obtained given that the null hypothesis is true. In this case, the P-value is the probability of getting a random sample of 7 smokers that yields a least squares regression line with a slope of 0.4341 or greater if there is no linear relationship between smoking levels (packs per day) and risk of dementia (Cox hazard ratio).

Free-Response Questions

1. (a) Predicted self-reported life satisfaction = 1.415 + 0.054(Household income in $1,000)

   (b) 86.2% of the variation in self-reported life satisfaction is accountable by this linear model (or is accountable by the variation in household income).

   (c) The standard deviation of the residuals is 0.977. That is, 0.977 gives a measure of how the points are spread around the regression line.

   (d) With $H_0: \beta = 0$ and $H_a: \beta > 0$, $t = \dfrac{b - 0}{SE(b)} = \dfrac{0.054 - 0}{0.00358} = 15.1$.

   (e) *Parameter:* Let β represent the slope of the population regression line for predicting self-reported life satisfaction from household income.

   Procedure: Confidence interval for a population slope.

   Checks: We are given that all conditions for inference are met.

   Calculation: With $df = 50 - 2 = 48$, the critical t-scores are $\texttt{±invT(0.95, 48)} = \pm 1.677$.
   Then $0.054 \pm 1.677(0.00358) = 0.054 \pm 0.006$ gives $(0.048, 0.060)$.

   Conclusion in context: We are 90 percent confident that the true slope of the least squares regression line linking self-reported life satisfaction and household income in $1,000 is between 0.048 and 0.060. That is, we are 90 percent confident that for each additional thousand dollars in household income, the self-reported life satisfaction goes up between 0.048 and 0.060, on average.

2. (a) Scatterplot is approximately linear:

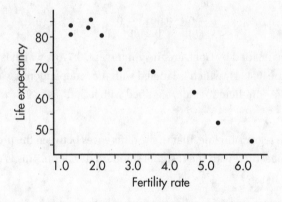

   (b) Predicted life expectancy = 95.35 − 7.864(fertility rate)

   (c) *Hypotheses:* $H_0: \beta = 0$, $H_a: \beta \neq 0$, where β is the true slope of the regression line that relates average fertility rate to women's life expectancy.

   (d) $P = 2\,[\text{tcdf}(-1000, -12.53, 9)] = 0.0000$. With a P-value this small, $0.0000 < 0.05$, there is sufficient evidence to reject H_0; that is, there is convincing evidence of a linear relationship between average fertility rate (children per woman) and life expectancy (women).

Quiz 25 (Pages 320–321)

1. *Procedure:* Chi-square test for homogeneity ("homogeneity" rather than "independence" because there are two independent samples)

   Checks: Independent random samples; all expected counts at least 5; $n \leq 0.10N$

   Hypotheses: H_0: There are no differences between the true distributions of reasons for trying out for varsity sports (social, health, or status) for men and women.

   H_a: There is at least one difference between the true distributions of reasons for trying out for varsity sports (social, health, or status) for men and women.

2. *Procedure:* One-sample z-interval for population proportion p
   Parameter: Let p represent the true proportion of the population of AP Statistics students who graduate in the top ten percent of their senior class.
   Checks: Random sample; at least 10 successes and 10 failures, that is, $n\hat{p} \geq 10$ and $n(1-\hat{p}) \geq 10$; $n \leq 0.10N$

3. *Procedure:* One-sample t-interval for population mean μ
   Parameter: Let μ represent the true mean number of books read by all high school students during their four years in secondary school.
   Checks: Random sample; population is normal or sample is large, at least 30, or sample distribution is unimodal and symmetric; $n \leq 0.10N$

4. *Procedure:* One-sample z-test for a population proportion p
   Parameter: Let p represent the true proportion of the population of racially motivated hate crimes that are motivated by anti-Black bias.
   Checks: Random sample; $np_0 \geq 10$ and $n(1-p_0) \geq 10$; $n \leq 0.10N$
   Hypotheses: H_0: $p = 0.70$, H_a: $p > 0.70$

5. *Procedure:* Chi-square test for goodness-of-fit
   Checks: Random sample; all expected counts at least 5; $n \leq 0.10N$
   Hypotheses: H_0: In cases of identity theft, 30% of the victims use their mother's maiden name for their banking password, 25% use their pet's name, 20% use "password," and 25% use something else.
   H_a: In cases of identity theft, the overall distribution of banking passwords used is different from 30% mother's maiden name, 25% pet's name, 20% "password," and 25% something else.

6. *Procedure:* One-sample t-test for population mean μ
   Parameter: Let μ represent the true mean speed of all fastballs thrown by a particular pitcher when participating in 10-minute yoga sessions.
   Checks: Random sample; population is normal or sample is large, at least 30, or sample distribution is unimodal and symmetric
   Hypotheses: H_0: $\mu = 95$, H_a: $\mu > 95$

7. *Procedure:* Linear regression t-test for slope β
   Parameter: Let β represent the true slope of the regression line relating school performance (GPA) to family income
   Checks: Random sample; scatterplot is roughly linear; no pattern in the residual plot; distribution of residuals is approximately normal; $n \leq 0.10N$
   Hypotheses: H_0: $\beta = 0$, H_a: $\beta \neq 0$. (The impact could be positive or negative.)

8. *Procedure:* Two-sample z-interval for difference in population proportions, $p_1 - p_2$
   Parameters: Let p_1 represent the true proportion of the population of patients with warts who are cured with a treatment of cryotherapy. Let p_2 represent the true proportion of the population of patients with warts who are cured with a treatment of duct tape occlusion.
   Checks: Independent random samples; at least 10 successes and 10 failures in each sample, that is, $n_1\hat{p}_1 \geq 10$, $n_1(1-\hat{p}_1) \geq 10$, $n_2\hat{p}_2 \geq 10$, $n_2(1-\hat{p}_2) \geq 10$; $n_1 \leq 0.10N_1$ and $n_2 \leq 0.10N_2$

9. *Procedure:* Two-sample t-test for difference in population means, $\mu_1 - \mu_2$
   Parameters: Let μ_1 represent the true mean cholesterol level (measured in milligrams per deciliter) for all people living in "Western" countries. Let μ_2 represent the true mean cholesterol level (measured in milligrams per deciliter) for all people living in "non-Western" countries.
   Checks: Independent random samples; for each sample, population is normal or sample is large, at least 30, or sample distribution is unimodal and symmetric; and for each sample $n \leq 0.10N$
   Hypotheses: H_0: $\mu_1 - \mu_2 = 0$, H_a: $\mu_1 - \mu_2 > 0$

10. *Procedure:* Linear regression t-interval for slope β

    Parameter: Let β represent the true slope of the regression line relating average increase in CO_2 emissions (in tons per person) to income level (in GDP per capita) among all countries of the world.

    Checks: Random sample; scatterplot is roughly linear; no pattern in the residual plot; distribution of residuals is approximately normal; $n \leq 0.10N$

11. *Procedure:* Chi-square test for independence

    Checks: Random sample; all expected counts at least 5; $n \leq 0.10N$

    Hypotheses: H_0: There is no association between race/ethnicity and whether or not an adult is unbanked.

    H_a: There is an association between race/ethnicity and whether or not an adult is unbanked.

12. *Procedure:* Two-sample z-test for difference in population proportions, $p_1 - p_2$

    Parameters: Let p_1 represent the true proportion of the population of at-risk patients receiving rivaroxaban who still have strokes. Let p_2 represent the true proportion of the population of at-risk patients receiving warfarin who still have strokes.

    Checks: Independent random samples; $n_1\hat{p}_C \geq 10$, $n_1(1 - \hat{p}_C) \geq 10$, $n_2\hat{p}_C \geq 10$, and $n_2(1 - \hat{p}_C) \geq 10$, where $\hat{p}_C = \dfrac{x_1 + x_2}{n_1 + n_2}$; $n_1 \leq 0.10N_1$ and $n_2 \leq 0.10N_2$

    Hypotheses: $H_0: p_1 - p_2 = 0$, $H_a: p_1 - p_2 \neq 0$

13. *Procedure:* Matched pair t-test for mean difference in population means, μ_{diff}

    Parameter: Let μ_{diff} represent the true mean difference (with pill − with placebo) in malaria episodes for the population of adults when they take the new pill and when they take the placebo.

    Checks: Random sample; set of differences is from a normal population or set of differences is large, at least 30, or distribution of set of differences is unimodal and symmetric

    Hypotheses: $H_0: \mu_{diff} = 0$, $H_a: \mu_{diff} < 0$

14. *Procedure:* Two-sample t-interval for difference in population means, $\mu_1 - \mu_2$

    Parameters: Let μ_1 represent the true mean infant mortality rate in the population of live births in the U.S. (mean infant deaths for every 1,000 live births).

    Let μ_2 represent the true mean infant mortality rate in the population of live births in Japan (mean infant deaths for every 1,000 live births)

    Checks: Independent random samples; for each sample, population is normal or sample is large, at least 30, or sample distribution is unimodal and symmetric; for each sample, $n \leq 0.10N$

15. *Procedure:* Matched pair t-interval for mean difference in population means, μ_{diff}

    Parameter: Let μ_{diff} represent the true mean difference (second score − first score) for the population of students taking a national standardized exam twice.

    Checks: Random sample; set of differences is from a normal population or set of differences is large, at least 30, or distribution of set of differences is unimodal and symmetric; $n \leq 0.10N$

16. *Procedure:* One-sample z-interval for true population mean μ. (Note that we use a normal distribution rather than a t-distribution because the population standard deviation is given.)

    Parameter: Let μ represent the mean waist size of the population of male high school math teachers.

    Checks: Random sample; population is normal or sample is large, at least 30, or sample distribution is unimodal and symmetric; $n \leq 0.10N$

Quiz 26 (Pages 322–323)

1. *Procedure:* Two-sample t-test for difference in population means, $\mu_1 - \mu_2$
   Parameters: Let μ_1 represent the true mean number of hours per week that first-year students study. Let μ_2 represent the true mean number of hours per week that sophomores study.
   Checks: Independent random samples; for each sample, population is normal or sample is large, at least 30, or sample distribution is unimodal and symmetric; for each sample, $n \leq 0.10N$
   Hypotheses: $H_0: \mu_1 - \mu_2 = 0$, $H_a: \mu_1 - \mu_2 > 0$

2. *Procedure:* One-sample z-interval for population proportion p
   Parameter: Let p represent the true proportion of the population of students who would report cheating by other students.
   Checks: Random sample; at least 10 successes and 10 failures, that is, $n\hat{p} \geq 10$ and $n(1 - \hat{p}) \geq 10$; $n \leq 0.10N$

3. *Procedure:* Chi-square test for independence
   Checks: Random sample; all expected counts at least 5; $n \leq 0.10N$
   Hypotheses: H_0: There is no association between blood pressure level (low, average, or high) and personality type (high-strung or easygoing).
   H_a: There is an association between blood pressure level (low, average, or high) and personality type (high-strung or easygoing).

4. *Procedure:* One-sample z-test for a population proportion p
   Parameter: Let p represent the true proportion of 3-point shot attempts successfully made by the population of Division 1 college basketball players this year.
   Checks: Random sample; $np_0 \geq 10$ and $n(1 - p_0) \geq 10$; $n \leq 0.10N$
   Hypotheses: $H_0: p = 0.34$, $H_a: p < 0.34$

5. *Procedure:* Matched pair t-test for mean difference in population means μ_{diff}
   Parameter: Let μ_{diff} represent the true mean difference (after program − before program) in the number of push-ups a clinic member can complete in 90 seconds.
   Checks: Random sample; set of differences is from a normal population or set of differences is large, at least 30, or distribution of set of differences is unimodal and symmetric
   Hypotheses: $H_0: \mu_{\text{diff}} = 0$, $H_a: \mu_{\text{diff}} > 0$

6. *Procedure:* One-sample t-test for population mean μ
   Parameter: Let μ represent the true mean body temperature in the population.
   Checks: Random sample; population is normal or sample is large, at least 30, or sample distribution is unimodal and symmetric; $n \leq 0.10N$
   Hypotheses: $H_0: \mu = 98.6$, $H_a: \mu < 98.6$

7. *Procedure:* Matched pair t-interval for mean difference in population means μ_{diff}
   Parameter: Let μ_{diff} represent the true mean difference in resting heart rates before and after an extensive exercise program for teenagers.
   Checks: Random sample; set of differences is from a normal population or set of differences is large, at least 30, or distribution of set of differences is unimodal and symmetric

8. *Procedure:* Two-sample z-interval for difference in population proportions, $p_1 - p_2$
   Parameters: Let p_1 represent the true proportion of the population of job applications with stereotypically "White" names who receive callbacks. Let p_2 represent the true proportion of the population of job applications with stereotypically "Black" names who receive callbacks.
   Checks: Independent random samples; at least 10 successes and 10 failures in each sample, that is,
   $n_1\hat{p}_1 \geq 10$, $n_1(1 - \hat{p}_1) \geq 10$, $n_2\hat{p}_2 \geq 10$, $n_2(1 - \hat{p}_2) \geq 10$

9. *Procedure:* Chi-square test for homogeneity

   Checks: Independent random samples; all expected counts at least 5; $n \leq 0.10N$

   Hypotheses: H_0: There is no difference in the true distributions of highest school levels attained between Whites, Blacks, Asians, and Hispanics (non-White).

   H_a: There is a difference in the true distributions of highest school level attained between Whites, Blacks, Asians, and Hispanics (non-White).

10. *Procedure:* Linear regression t-interval for slope β

    Parameter: Let β represent the true slope of the regression line relating selling price to miles noted on the odometer for all used cars of a particular model.

    Checks: Random sample; scatterplot is roughly linear; no pattern in the residual plot; distribution of residuals is approximately normal; $n \leq 0.10N$

11. *Procedure:* One-sample t-interval for population mean μ

    Parameter: Let μ represent the true mean pause length in the population of telemarketing calls.

    Checks: Random sample; population is normal or sample is large, at least 30, or sample distribution is unimodal and symmetric; $n \leq 0.10N$

12. *Procedure:* Two-sample z-test for difference in population proportions, $p_1 - p_2$

    Parameters: Let p_1 represent the true proportion of the population of White students who report that, at times, they feel unsafe to go to school. Let p_2 represent the true proportion of the population of non-White students who report that, at times, they feel unsafe to go to school.

    Checks: Independent random samples; $n_1\hat{p}_C \geq 10$, $n_1(1 - \hat{p}_C) \geq 10$, $n_2\hat{p}_C \geq 10$, and $n_2(1 - \hat{p}_C) \geq 10$, where $\hat{p}_C = \dfrac{x_1 + x_2}{n_1 + n_2}$; $n_1 \leq 0.10N_1$ and $n_2 \leq 0.10N_2$

    Hypotheses: H_0: $p_1 - p_2 = 0$, H_a: $p_1 - p_2 < 0$

13. *Procedure:* Chi-square test for goodness-of-fit

    Checks: Random sample; all expected counts at least 5; $n \leq 0.10N$

    Hypotheses: H_0: The U.S. incarcerated adult population by race/ethnicity is 39 percent White (non-Hispanic), 19 percent Hispanic, 40 percent Black, and 2 percent other.

    H_a: The U.S. incarcerated adult population by race/ethnicity is something different than 39 percent White (non-Hispanic), 19 percent Hispanic, 40 percent Black, and 2 percent other.

14. *Procedure:* Two-sample t-interval for difference in population means, $\mu_1 - \mu_2$

    Parameters: Let μ_1 represent the true mean weight of the population of elephants living in captivity. Let μ_2 represent the true mean weight of the population of elephants living in the wild.

    Checks: Independent random samples; for each sample, population is normal or sample is large, at least 30, or sample distribution is unimodal and symmetric; for each sample, $n \leq 0.10N$

15. *Procedure:* Linear regression t-test for slope β

    Parameter: Let β represent the true slope of the regression line relating pulse rate to quantity of caffeine consumed in a sitting.

    Checks: Random sample; scatterplot is roughly linear; no pattern in the residual plot; distribution of residuals is approximately normal; $n \leq 0.10N$

    Hypotheses: H_0: $\beta = 0$, H_a: $\beta > 0$. (Interest is in an increase in pulse rate.)

16. *Procedure:* One-sample z-test for population mean μ. (Note that we use a normal distribution rather than a t-distribution because the population standard deviation is given.)

Parameter: Let μ represent the true mean hourly wage paid for summer work to the population of high school students.

Checks: Random sample; population is normal or sample is large, at least 30, or sample distribution is unimodal and symmetric; $n \leq 0.10N$

Hypotheses: $H_0: \mu = 10.50$, $H_a: \mu < 10.50$

Quiz 27 (Pages 326–334)

1. (a)

Region	Population (in millions)	% of World Population	Wealth (GDP) (in billions)	% of World GDP
Africa	1,341	17.2	2,995	2.8
Asia	4,641	59.5	40,248	38.1
Oceania	43	0.6	2,002	1.9
Europe	748	9.6	24,881	23.6
U.S./Canada	369	4.7	31,319	29.7
Latin America	654	8.4	4,068	3.9
World Total	7,796	100.0	105,513	100.0

(b) The segmented bar chart treats each bar as a whole and divides each bar proportionally into segments corresponding to the percentage in each group.

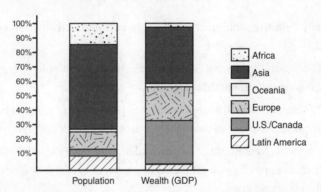

The segmented bar chart shows that both Europe and the U.S./Canada have a hugely disproportionate share of the wealth of the world when compared with their populations, while Asia, Africa, and Latin America have a low share of the wealth of the world when compared with their populations.

(c) The slopes give each region's wealth in dollars per person. The U.S./Canada region has the steepest slope and thus the highest $/person, followed by Oceania, Europe, Asia, Latin America, and Africa.

(d) There are many possible answers. For example, switching to GPI takes into account industrial pollution, which is likely much higher in the U.S./Canada than in the islands of the South Pacific. With GDP per capita as the index, the U.S./Canada ranked slightly higher than Oceania. With CPI per capita as the index, the value for Oceania might go up and the value for U.S./Canada might go down sufficiently for the ranking to be reversed.

2. (a) Fifteen percent of all U.S. Black men are likely to go to prison for the first time by age 25.

    (b) No U.S. White males are likely to go to prison for the first time between ages 60 and 70.

    (c) Age 30. The graph indicates that 16% of all U.S. Hispanic males are likely to go to prison, and half of them (corresponding to 8% on the vertical axis) will go to prison for the first time by age 30.

    (d) Answers should explain how addressing the disproportionate suspensions among students of color should lower the initial slopes of the Black and Hispanic graphs. An answer might also conclude that dismantling the school-to-prison pipeline would move the x-intercept for all 3 curves to the right, that is, raise the ages for first admissions to prison for all young people.

3. (a) Predicted total SAT $= 903.36 + 0.00116(\text{Income})$

    (b) Every \$1,000 of additional family income is associated with an average of 1.16 points higher on a student's total SAT score.

    (c) The coefficient of determination $R^2 = 94.0\%$. That is, 94.0% of the variation in total SAT scores is explained by a linear relationship between total SAT and Family Income.

    (d) Predicted Total SAT $= 903.36 + 0.00116(63,500) = 977.02$

    Residual $=$ Actual $-$ Predicted $= 970 - 977.02 = -7.02$; that is, the actual total SAT score of this student is 7.02 *less* than what was predicted by the family income.

    (e) $S = 20.936$ is the standard deviation of the residuals. In other words, it is a measure of how the points are vertically spread around the regression line.

    (f) Correlation does not imply causation. A possible confounding variable, for example, could be instructional money spent per student. Leveling per student expenditures across school districts might lead to higher SAT scores in economically disadvantaged areas.

4. (a) There is a moderate, negative, nonlinear relationship between GNI and percentage of GNI spent on development aid.

    (b) While the United States has the greatest GNI of any country in the sample, it spends one of the lowest percentages of GNI on development aid of any country in the sample.

    (c) Because the United States GNI is so large, even a small percentage of it would be large. Thus, the United States will show the highest, or one of the highest, actual amounts spent on development aid.

    (d) While Belgium, Denmark, Netherlands, Norway, and Sweden have the lowest GNIs compared with the other countries in this sample, they invest the highest percent of GNI in soft power (development aid) compared with the other countries.

    (e) There is a moderate to strong, negative, linear relationship between ln(GNI) and ln(percentage of GNI spent on development aid).

    (f) The scatterplot of ln(GNI) versus ln(percentage of GNI spent on development aid) shows a linear form and there are no leftover patterns in the residual plot.

    (g) $r = -\sqrt{0.600} = -0.7746$. (Note that r is negative as it takes the same sign as the slope.)

    (h) For each one unit increase in the natural logarithm of GNI, the natural logarithm of the percent of GNI spent on development aid *decreases* on average by 0.4179.

    (i) $\ln(\text{percentage GNI}) = 2.4135 + (-0.4179)\ln(\text{GNI})$

    (j) $\ln(\text{percentage GNI}) = 2.4135 + (-0.4179)\ln(23,400) = 2.4135 - 0.4179(10.0605) = -1.7908$ and so percentage GNI $= e^{-1.7908} = 0.167$. The predicted percentage of GNI spent on development aid for the U.S. is 0.167%, which is an overestimate of the actual value of 0.15%.

5. (a)

UN		SOHR		
		Reported Deaths	Unreported Deaths	*Totals*
	Reported Deaths	2,635	992	3,627
	Unreported Deaths	425	$D - 4,052$	$D - 3,627$
	Totals	3,060	$D - 3,060$	D

(b) Relative frequencies should be approximately equal if there is independence. The expected cell formula $\text{Exp} = \dfrac{(\text{Row sum}) \times (\text{Column sum})}{N}$ gives $2{,}635 = \dfrac{(3{,}627)(3{,}060)}{D}$. (If you look at rows, you get instead $\dfrac{2{,}635}{3{,}627} = \dfrac{425}{D - 3{,}627}$. If you look at columns, you get $\dfrac{2{,}635}{3{,}060} = \dfrac{992}{D - 3{,}060}$.) Solving any of these equations gives $D = 4{,}212$.

(c) *Convenience samples* are based on choosing individuals who are easy to reach and tend to produce data highly unrepresentative of the entire population. In this case, only very readily obtainable reports of deaths are included, violating the assumption of a random sample of deaths for that month.

(d) Both samples will tend to report the same large casualty events, thus violating the assumption of independence.

(e) The calculated value of D is almost certainly an underestimate. As shown in parts (c) and (d), the reported deaths tend to be convenience samples of more widely reported deaths and with different samples tending to report the same deaths.

6. (a) This is an example of cluster sampling, which allows for a quicker, less expensive method of sampling than a simple random sample. It assumes that each cluster, in this case each county, is representative of the whole state.

(b) Number the counties 1 through 67. Use a random number generator to pick two unique integers between 1 and 67 (ignoring repeats). The two counties to be used will consist of the two counties with numbers corresponding to the two unique numbers.

(c) The purpose of randomization in picking the 90 jurors is to obtain a sample that is representative of all eligible White jurors in the state so that any conclusions can be generalized to all eligible White jurors in the state.

(d) Random assignment here is used to minimize the effect of possible confounding variables. For example, if self-selection is allowed, perhaps some White jurors are uncomfortable with people of color and would choose to be on the all-White juries. Because it is likely that these same jurors are more likely biased against a defendant of color, it might be impossible to say whether any conclusions can be made based on the racial makeup of a jury.

7. (a) "Should books with depictions of slavery be banned from school libraries?"

(b) There are many possible answers. For example, "By restricting information and discouraging freedom of thought, censors undermine one of the primary functions of education: teaching students how to think for themselves. Should books with depictions of slavery be banned from school libraries?" This wording bias will lead to an underestimate of the true proportion of parents in favor of banning books with depictions of slavery.

(c) There are many possible answers. For example, "It is potentially dangerous to expose schoolchildren to some controversial topics. Should books with depictions of slavery be banned from school libraries?" This wording bias will lead to an overestimate of the true proportion of parents in favor of banning books with depictions of slavery.

(d) $P(\text{liberal} \cap \text{favor banning}) = P(\text{liberal}) \times P(\text{favor banning} \mid \text{liberal}) = (0.26)(0.09) = 0.0234$

(e) $P(\text{favor banning}) = P(\text{liberal} \cap \text{favor banning}) + P(\text{moderate} \cap \text{favor banning}) +$
$P(\text{conservative} \cap \text{favor banning}) = (0.26)(0.09) + (0.38)(0.12) + (0.36)(0.15) = 0.123$

(f) $P(\text{liberal} \mid \text{favor banning}) = \dfrac{P(\text{liberal} \cap \text{favor banning})}{P(\text{favor banning})} = \dfrac{0.0234}{0.123} = 0.190$

8. (a) $P(\text{Blast zone}) = \dfrac{5.5}{38.8} = 0.14175$ and $P(\text{Latino} \cap \text{Blast zone}) = P(\text{Blast zone})P(\text{Latino} \mid \text{Blast zone}) =$
$(0.14175)(0.75) = 0.10631$

(b) $P(\text{Blast zone} \mid \text{Latino}) = \dfrac{P(\text{Latino} \cap \text{Blast zone})}{P(\text{Latino})} = \dfrac{0.10631}{0.390} = 0.27259$

(c) Is $P(\text{Blast zone} \mid \text{Latino}) = P(\text{Blast zone})$? No, $0.27259 \neq 0.14175$. So, in California, being Latino and living within a blast zone are not independent events.

(d) This is a binomial with $n = 5$ and $p = 0.14175$. Then $P(X \geq 2) = 1 - [(0.85825)^5 + 5(0.14175)(0.85825)^4]$
$= 0.14979$. Calculator software gives $1 - \texttt{binomcdf(5, 0.14175, 1)} = 0.14979$.

(e) Another possible explanatory variable is poverty. It could be that blast zones tend to be associated with low-income housing and that a higher percentage of people of color than White people are low income.

9. (a) Roughly the same percentage of victims were White and were non-White. Most murders with White victims were committed by White offenders, and most murders with non-White victims were committed by non-White offenders.

(b) $P(\text{White victim} \mid \text{execution}) = \dfrac{P(\text{White victim} \cap \text{execution})}{P(\text{execution})}$, which leads to $0.76 =$
$\dfrac{P(\text{White victim} \cap \text{execution})}{0.012}$ and $P(\text{White victim} \cap \text{execution}) = (0.76)(0.012) = 0.00912$

$P(\text{non-White victim} \mid \text{execution}) = \dfrac{P(\text{non-White victim} \cap \text{execution})}{P(\text{execution})}$, which leads to $0.24 =$
$\dfrac{P(\text{non-White victim} \cap \text{execution})}{0.012}$ and $P(\text{non-White victim} \cap \text{execution}) = (0.24)(0.012) = 0.00288$

(c) $P(\text{execution} \mid \text{White victim}) = \dfrac{P(\text{White victim} \cap \text{execution})}{P(\text{White victim})} = \dfrac{0.00912}{0.505} = 0.0181$

$P(\text{execution} \mid \text{non-White victim}) = \dfrac{P(\text{non-White victim} \cap \text{execution})}{P(\text{non-White victim})} = \dfrac{0.00288}{0.495} = 0.00582$

(d) $\dfrac{0.0181}{0.00582} = 3.11$, so when the defendant is executed, it is more than three times as likely that the victim was White than non-White.

(e) This is a geometric distribution with $p = 0.012$. The mean is $\dfrac{1}{p} = \dfrac{1}{0.012} = 83.3$. The mean number of cases sampled until one is found involving the execution of the defendant is 83.3 cases.

10. (a) $P(\text{criminal}) = 0.0088.$ $P(\text{innocent}) = 1 - 0.0088 = 0.9912.$

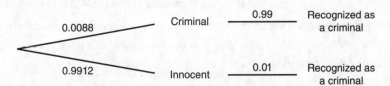

$P(\text{criminal} \cap \text{identified as criminal}) = (0.0088)(0.99) = 0.008712$

$P(\text{innocent} \cap \text{identified as criminal}) = (0.9912)(0.01) = 0.009912$

$P(\text{identified as criminal}) = 0.008712 + 0.009912 = 0.018624$

  (b) $P(\text{criminal} \mid \text{identified as criminal}) = \dfrac{P(\text{criminal} \cap \text{identified as criminal})}{P(\text{identified as criminal})} = \dfrac{0.008712}{0.018624} = 0.468$

Alternatively, a table of counts can be constructed for a hypothetical population. In this case, let's assume that the original given percentages hold exactly for a population of size 1,000,000.

	Identified as criminal	Identified as innocent	Total
Criminal			
Innocent			
Total			1,000,000

Numbers in parentheses indicate the order in which the cells can be calculated.

	Identified as criminal	Identified as innocent	Total
Criminal	(3) 8,712	(4) 88	(1) 8,800
Innocent	(5) 9,912	(6) 981,288	(2) 991,200
Total	(7) 18,624	(8) 981,376	1,000,000

Then, the conditional probability in which we are interested can be calculated straight from the first column of the table: $P(\text{criminal} \mid \text{identified as criminal}) = \dfrac{8,712}{18,624} = 0.468.$

  (c) There is a high risk of police misidentifying and acting against innocent people. By using this "99 percent accurate" facial recognition software, the probability that someone identified as criminal is actually a criminal is only 0.468.

  (d) Yes, a serious civil rights concern is that facial recognition is creating de facto algorithm-based racial profiling.

11. (a) This study looked backward by examining existing data and, therefore, was a retrospective study. Since no treatment was applied, that is, no one was told to use this forensic evidence, this study was not an experiment.

  (b) This is a binomial distribution with $n = 2,500$ and $p = \dfrac{257}{268} = 0.959.$

Thus, $E(X) = np = 2,500(0.959) = 2,397.4.$

The standard deviation $= \sqrt{np(1 - p)} = \sqrt{2,500(0.959)(0.041)} = 9.915.$

(c) The probability of a case resulting in the death penalty, given it involves flawed forensic testimony, is $\frac{32}{257} = 0.1245$. Thus, the probability of the case not resulting in the death penalty is $1 - 0.1245 = 0.8755$. The probabilities of a death penalty occurring on the first or second case examined, respectively, are 0.1245 and $(0.8755)(0.1245) = 0.1090$. So, the probability that the first case with a resulting death penalty doesn't occur until at least the third case examined is

$$1 - (0.1245 + 0.1090) = 0.7665$$

[Or recognize that this is a geometric distribution with $p = 0.1245$. Then $P(X \geq 3) = 1 - P(X \leq 2) = 1 - \text{geometcdf}(0.1245, 2) = 1 - 0.2335 = 0.7665.$]

(d) "The misapplication of forensic science" and "eyewitness misidentification" are not mutually exclusive because, if so, their intersection would be the empty set. Then the probability of their union would be $0.50 + 0.75 - 0 = 1.25 > 1$. It is possible that they are independent because, if so, the probability of their intersection would be $(0.50)(0.75)$ and the probability of their union would be

$$0.50 + 0.75 - (0.5)(0.75) = 0.875 \leq 1$$

12. (a) We have a normal distribution with mean 3,287 and standard deviation 250. Then $P(X > 3,500) = P\left(z > \frac{3,500 - 3,287}{250}\right) = 0.1971.$

(b) We have a binomial distribution with $n = 5$ and $p = 0.1971$. Then $P(X \geq 3) = 1 - P(X \leq 2) = 1 - \text{binomcdf}(5, \ 0.1971, \ 2) = 0.05572.$

(c) We have a normal distribution with mean 3,287 and standard deviation $\frac{250}{\sqrt{5}} = 111.80$. Then $P(\bar{x} > 3,500) = P\left(z > \frac{3,500 - 3,287}{111.80}\right) = 0.0284.$

(d) Possible answers might range from average yearly proceeds (about \$35 billion) to average number of slaves in the world on any given day (21 to 29 million), from numbers of child slaves to numbers of women in sexual slavery.

13. (a) A complete answer addresses center, spread, shape, and unusual features such as outliers, and mentions context. The median age of hate site members is approximately 23. The ages go from 14 to 65 with a range of $65 - 14 = 51$ and an IQR $= 30 - 17 = 13$.

There are 6 outliers between 50 and 65. Although it is not usually possible to describe distribution shape from just a boxplot, in this case the distribution of ages does appear to be skewed to the right (to the larger values).

(b) In a skewed right distribution, the mean is usually larger than the median. So, we would conclude that the quotient $\frac{\text{Mean age}}{\text{Median age}}$ is greater than 1.

(c) If mostly the younger, computer savvy members of the hate groups are the ones who are online, the median age of all hate group members is probably greater than the median age of those self-reporting on the site, 23.

(d) This is a binomial with $n = 3$ and $p = 0.186$. Then $P(X \geq 2) = 0.0909$ or

$$3(0.186)^2(0.814) + (0.186)^3 = 0.0909$$

[Or $1 - P(x \leq 1) = 1 - \text{binomcdf}(n = 3, p = 0.186, x = 1) = 1 - 0.9091 = 0.0909.$]

Quiz 28 (Pages 335–341)

1. (a) $0.19 + 0.17 - 0.06 = 0.30$

   (b) *Parameter:* Let p represent the true proportion of the population of students in grades six through ten who are victims of moderate to frequent bullying.

   Procedure: One-sample z-interval for a population proportion

   Checks: We must assume this was a random sample, $n\hat{p} = (15,686)(0.17) = 2667 \geq 10$, $n\hat{q} = (15,686)(0.83) = 13,019 \geq 10$, and $n = 15,686$ is less than 10% of all students in grades six through ten.

   Mechanics: Calculator software gives $(0.16415, 0.1759)$.

   Conclusion in context: We are 95% confident that the proportion of all students in grades six through ten who are victims of moderate to frequent bullying is between 0.16415 and 0.1759.

   (c) Because 0.22 is not in the interval of plausible values for the population proportion, there *is* convincing evidence that the proportion of all students who are victims of moderate to frequent bullying is different from the 22% claimed in the one study.

   (d) Issues that might be addressed in planning a survey include stratification among the grade levels, anonymity, and timing (early enough in the school year to allow implementation of a plan and again later in the year to assess the effectiveness of any prevention and intervention program).

2. (a) *Parameter:* Let μ represent the true mean number of firearm suicides in the U.S. each year among the population of all years.

   Procedure: A one-sample t-interval

   Checks: We are given a random sample. The nearly normal condition seems reasonable because the sample data show no outliers or strong skewness. We assume that 7 years are less than 10% of all years under consideration. Under these conditions, the mean firearm suicides per year can be modeled by a t-distribution with $n - 1 = 7 - 1 = 6$ degrees of freedom.

   Mechanics: Calculator software gives the interval $(20408, 21850)$. [We can instead calculate
   $$22,128.6 \pm 2.447 \frac{779.6}{\sqrt{7}} = 22,128.6 \pm 721.0.]$$
   Conclusion in context: We are 95% confident that the true mean number of firearm suicides in the U.S. each year is between 21,408 and 22,850.

   (b) It could well be true (actually, it is true) that most guns in Israel are technically owned by the government (by the IDF—the Israel Defense Forces). The 6.7 guns per 100 people statistic is a significant underestimate of guns in the population.

   (c) The conversation would change from number of guns in the population to strictness of gun laws pertaining to gun ownership.

   (d) The U.S. has 120.5 guns per 100 people but only $\frac{120.5}{3} = 40.2$ gun owners per 100 people. Thus, the "firearm suicides per gun owner" rate in the U.S. will be three times as great as the "firearm suicides per gun" rate. How "rates" are calculated can lead to different conclusions.

3. (a) A cumulative frequency plot that rises quickly at first and then rises at a much slower rate later corresponds to a histogram with a large area under the curve early and much less area later. Thus, it is *skewed right* (skewed to the higher values of lead). In skewed right distributions, we expect the *mean* to be greater than the median.

   (b) These figures indicate an extremely serious lead problem in the Flint water supply. The research team found that 25% of the homes had a lead water supply level of at least 15 ppb. Because 25% > 10% and 15 ppb > 5 ppb, the percent of homes with at least 5 ppb is larger than 10%.

   (c) With H_0: $\mu = 10$ and H_a: $\mu > 10$, a Type I error (mistaken rejection of a true null hypothesis) means that we think the lead level is greater than 10 ppb when, in fact, it is 10 ppb. So, we might institute costly and unnecessary cleanup measures. A Type II error (mistaken failure to reject a false null hypothesis) means that the mean lead level is greater than 10 ppb but we don't have enough evidence to think so. So, we might fail to institute a necessary cleanup. Clearly, with people's, and especially children's, health at stake, a Type II error should be more of a concern.

4. (a) The sample proportion is in the middle of the confidence interval, so $\hat{p} = \dfrac{(0.21882 + 0.28258)}{2} = 0.2507$.

   The margin of error is $0.28258 - 0.2507 = 0.03188$. We have $1.96\sqrt{\dfrac{0.2507(1 - 0.2507)}{n}} = 0.03188$, which gives $\sqrt{n} = 26.64666$, and n is about 710.

   (b) The mosaic plot shows (1) that the sample contained roughly twice as many resumes with "White" names than "non-White" names, and (2) in the sample, a greater proportion of resumes with "White" names than "non-White" names resulted in positive responses (offered interviews).

   (c) With such a small P-value, $0.00486 < 0.05$, there is sufficient evidence to reject H_0; that is, there is convincing evidence that resumes with "White" names result in a greater proportion of positive responses (offered interviews) than resumes with "non-White" names.

   (d) Because 0 is in the confidence interval, that is, the interval contains both negative and positive plausible values, there is not convincing evidence to conclude that discrimination against African-Americans is trending higher.

5. (a) This is an example of random assignment applied to a block design. Each pair of children is a block, and in each block one child is randomly assigned to receive the gun safety "treatment" while the other child receives the car safety "treatment."

   (b) H_0: $p_G = p_C$ and H_a: $p_G > p_C$ where $p_G =$ proportion of all youths 8 to 12 years old who will tell an adult about a gun in the room after being shown a gun safety video and where $p_C =$ proportion of all youths 8 to 12 years old who will tell an adult about a gun in the room after being shown a car safety video.

   (c) The two sample proportions, $\hat{p}_G$ and $\hat{p}_C$, are 4.167 standard errors apart.

   (d) This is a one-sided test with a P-value of normalcdf(4.167, 1000) = 0.000. With this small a P-value, $0.000 < 0.01$, there is sufficient evidence to reject the null hypothesis. In other words, there is convincing evidence that the proportion of all youths 8 to 12 years old who will tell an adult about a gun in the room after being shown a gun safety video is greater than the proportion of all youths 8 to 12 years old who will tell an adult about a gun in the room after being shown a car safety video.

(e) We are 99% confident that the proportion of all youths 8 to 12 years old who will tell an adult about a gun in the room after being shown a gun safety video is between 0.09342 and 0.36676 greater than the proportion of all youths 8 to 12 years old who will tell an adult about a gun in the room after being shown a car safety video.

(f) Since the confidence interval is entirely greater than 0, the two procedures give consistent results.

(g) Younger children find authority figures in uniforms to be especially persuasive. This finding suggests that we can use law enforcement to develop interventions aimed at promoting safe behavior among young children.

(h) Perhaps parents with guns are more likely to talk to their children about gun safety than parents without guns.

6. (a) *Parameters:* Let μ_{LEAD} represent the true mean yearly criminal and legal system costs of the population of low-level drug and prostitution offenders who are participating in the LEAD program. Let $\mu_{control}$ represent the true mean yearly criminal and legal system costs of the population of low-level drug and prostitution offenders who are not participating in the LEAD program.
Hypotheses: H_0: $\mu_{LEAD} = \mu_{control}$ and H_a: $\mu_{LEAD} < \mu_{control}$.
Procedure: This is a two-sample t-test for means.
Checks: We are given samples that, by design, are assumed representative and independent. The sample sizes, 203 and 115, are large enough so that by the CLT, the distribution of sample means is approximately normal and a t-test may be run. The sample sizes are assumed to be less than 10% of the population of all low-level drug and prostitution offenders.
Mechanics: Calculator software gives $t = -33.41$ with a P-value of 0.000.
Conclusion in context with linkage to the P-value: With this small of a P-value, $0.000 < 0.05$, there is very strong evidence to reject H_0. In other words, there is convincing evidence that for those adults eligible for the LEAD program, the mean yearly criminal and legal system costs of those participating in the program are less than that for those not participating.

(b) The null hypothesis was rejected. So, the possibility is of a Type I error, that is, of rejecting a true null hypothesis. A Type I error would mean that in reality, after participating in the program, the LEAD group would have the same mean days per year in jail as the control group. However, in error, we think the LEAD group has significantly fewer mean days per year in jail than the control group. A possible consequence is that the LEAD program is expanded, thinking it will reduce the future mean days per year in jail when, in reality, it won't.

(c) It is very difficult to "choose" representative samples. Using randomization has been shown to be the most effective technique to minimize the effect of possible confounding variables.

(d) "We ran a large sample statistical study and noted that for our sample, the average yearly criminal and legal costs spent on those offenders participating in the LEAD program were significantly lower than the average spent on those not participating in the program. The difference was so great that it is highly unlikely that it could have occurred by chance."

7. (a) $\dfrac{1,710}{4,658} = 36.7\%$, $\dfrac{1,423}{4,658} = 30.5\%$, and $\dfrac{543}{4,658} = 11.7\%$

$\dfrac{64,412}{116,774} = 55.2\%$, $\dfrac{10,076}{116,774} = 8.6\%$, and $\dfrac{11,600}{116,774} = 9.9\%$

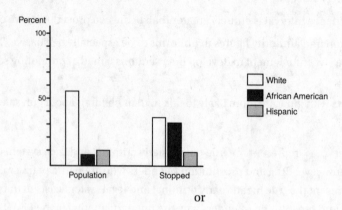

or

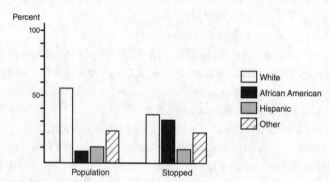

(b) *Parameter:* Let p represent the true proportion of the population of all stopped drivers who are African American.

Hypotheses: H_0: $p = 0.086$ and H_a: $p > 0.086$

Procedure: One proportion z-test

Checks: A representative (random) sample is assumed; $np_0 = 4{,}658(0.086) = 400.6 \geq 10$ and $n(1 - p_0) = 4{,}658(0.914) = 4{,}257.4 \geq 10$.

Mechanics: Software such as 1-PropZTest on the TI-84 gives $z = 53.4$ with a P-value of 0.000.

Conclusion in context with linkage to the P-value: With this low of a P-value, $0.000 < 0.001$, there is convincing evidence that African Americans are being stopped at a higher rate than what would be indicated by their population percentage.

(c) *Hypotheses:* H_0: Race/ethnicity (White, African American, Hispanic) and disposition (arrest/citation or not) are independent.

H_a: There is a relationship between race/ethnicity (White, African American, Hispanic) and disposition (arrest/citation or not).

Procedure: A chi-square test of independence

Checks: Representative (random sample assumed) and all expected cells as shown below are ≥ 5.

OBS	White	African American	Hispanic
Citation	1,058	481	237
None	652	942	306

EXP	White	African American	Hispanic
Citation	826.2	687.5	262.3
None	883.8	735.5	280.7

Mechanics: Software (such as χ^2-Test on the TI-84) gives the above expected values and $\chi^2 = 250.61$ with a *P*-value of 0.000.

Conclusion in context with linkage to the P-value: With this low of a *P*-value, $0.000 < 0.001$, there is convincing evidence of a relationship between race/ethnicity (White, African American, Hispanic) and disposition (arrest/citation or not).

8. (a) *Procedure:* Confidence interval for the slope of a regression line

   Checks: We must assume a random sample of countries (or at least a representative sample). The scatterplot looks roughly linear. The residual plot shows no pattern. The distribution of residuals is very roughly normal.

   Mechanics: From the printout, the standard error of the slope is $s_b = 0.0530$. With $df = 21 - 2 = 19$ and 0.025 in each tail, the critical *t*-values are ± 2.093. The 95% confidence interval of the true slope is $0.4404 \pm 2.093(0.0530) = 0.4404 \pm 0.1109$.

   Conclusion in context: We are 95% confident that for every 1-unit increase in the Income ratio, the mean increase in the HSP Index is between 0.3295 and 0.5513.

   (b) *Hypotheses:* H_0: $\beta = 0$ and H_a: $\beta > 0$

   Procedure: Test of significance for the slope of the regression line

   Checks: The conditions for inference are the same as those in part (a).

   Mechanics: The computer printout gives $t = 8.31$ with a *P*-value of $\frac{1}{2}(0.000) = 0.000$.

   Conclusion in context with linkage to the P-value: With this small of a *P*-value, $0.000 < 0.05$, there is sufficient evidence to reject H_0. In other words, there is convincing evidence that the higher the income ratio (that is, the higher the income inequality in a country), the greater the HSP Index (that is, the greater the health and social problems of the country).

   (c) There is strong evidence that income ratios (income inequality) have a positive linear association with HSP Indexes (health and social problems). In fact, according to the computer output, 78.41% of the variation in HSP Indexes can be explained by this linear model. However, we cannot conclude causation. We cannot say that if we work to lessen income inequality, we will reduce health and social problems in the country.

9. (a) *Hypotheses:* H_0: The population of local jails is distributed according to 34.4% convicted males, 5.0% convicted females, 53.4% unconvicted males, and 7.2% unconvicted females.

   H_a: The population of local jails is not distributed according to 34.4% convicted males, 5.0% convicted females, 53.4% unconvicted males, and 7.2% unconvicted females.

   Procedure: χ^2 goodness-of-fit test

   Checks: We are given a random sample. With a sample of size $81 + 5 + 110 + 4 = 200$, all expected cells are at least 5: 34.4% of $200 = 68.8$, 5.0% of $200 = 10$, 53.4% of $200 = 106.8$, and 7.2% of $200 = 14.4$.

   Mechanics: A calculator gives $\chi^2 = \sum \dfrac{(\text{observed} - \text{expected})^2}{\text{expected}} = 12.27$. With $df = 4 - 1 = 3$, $P = \chi^2\text{cdf}(12.27, 1000, 3) = 0.0065$.

   Conclusion in context with linkage to the P-value: With this small of a *P*-value, $0.0065 < 0.05$, there is sufficient evidence to reject H_0. In other words, there is convincing evidence that the population distribution of local jails has changed from the previous study's finding of 34.4% convicted males, 5.0% convicted females, 53.4% unconvicted males, and 7.2% unconvicted females.

(b) There are many possible ways of proportionately choosing the 500 inmates to study. We should stratify and randomly pick 44% of 500 = 220 inmates from small county jails and 56% of 500 = 280 inmates from large county jails. For example, we might use cluster sampling to pick 22 small county jails and 28 large county jails randomly. Then, in every chosen jail, number the inmates. Using a random integer generator, pick 10 unique numbers in the appropriate range corresponding to the number of inmates in each jail. The inmates with those numbers, together from all the 50 chosen jails, will form the sample to be studied.

10. (a) The sample proportion is in the middle of the interval (0.30, 0.36), so $\hat{p} = 0.33$. The margin of error is $0.36 - 0.33 = 0.03$. The critical z-scores associated with 95% confidence are ± 1.96. We have $1.96\sqrt{\frac{(0.33)(0.67)}{n}} \leq 0.03$, or $\sqrt{n} \geq \sqrt{\frac{1.96\,(0.33)(0.67)}{0.03}} = 30.72$ and $n \geq 943.7$. Conclude that the sample size was $n = 944$.

(b) With a very large P-value, there is little or no evidence to reject the null hypothesis. In other words, there is little or no evidence of an association between autism and vaccines.

(c) Given $p = 0.10$ and $n = 500$, we have $\mu_{\hat{p}} = 0.10$ and $\sigma_{\hat{p}} = \sqrt{\frac{(0.10)(0.90)}{500}} = 0.01342$. Then $P(\hat{p} > 0.12)$ $= P\left(z > \frac{0.12 - 0.10}{0.01342}\right) = P(z > 1.49) = 0.068$.

(d) $R^2 = (-0.58)^2 = 0.336$. Conclude that 33.6% of the variation in Covid-19 severity is accounted for by the linear regression model between mumps IgG titer and Covid-19 severity.

11. (a) There is a positive, linear, moderate to strong relationship between Temperature Anomaly and Time.

(b) *Parameters:* Let p_{Ind} represent the true proportion of the population of Indians who express a great deal of worry about global warming. Let p_{Chi} represent the true proportion of the population of Chinese who express a great deal of worry about global warming.
Hypotheses: H_0: $p_{\text{Ind}} = p_{\text{Chi}}$, H_a: $p_{\text{Ind}} \neq p_{\text{Chi}}$
Procedure: Hypothesis test for a difference between two proportions
Checks: Independent random samples; $n_{\text{Ind}}\hat{p}_C = (2{,}029)(0.417) = 846 \geq 10$, $n_{\text{Ind}}(1 - \hat{p}_C) = (2{,}029)(0.583) = 1{,}183 \geq 10$, $n_{\text{Chi}}\hat{p}_C = (2{,}180)(0.417) = 909 \geq 10$, and $n_{\text{Chi}}(1 - \hat{p}_C) = (2{,}180)(0.583) = 1{,}271 \geq 10$, where $\hat{p}_C = \frac{x_1 + x_2}{n_1 + n_2} = \frac{1{,}319 + 436}{2{,}029 + 2{,}180} = 0.417$; $n_{\text{Ind}} \leq 0.10N_{\text{Ind}}$ and $n_{\text{Chi}} \leq 0.10N_{\text{Chi}}$
Mechanics: $z = 29.59$, $P = 0.000$
Conclusion in context with linkage to the P-value: With this small of a P-value, $0.000 < 0.05$, there is sufficient evidence to reject H_0. In other words, there is convincing evidence of a difference between the proportion of Indians who express a great deal of worry about global warming and the proportion of Chinese who express a great deal of worry about global warming.

(c) Phrases like "climate change deniers," "corporate lobbyists," and "a dying planet" are clearly nonneutral and leading. The calculated proportion of people who express a great deal of worry about global warming will most likely be greater than the true proportion.

Quiz 29 (Pages 349–354)

1. (a) **Think:** This sounds like a straightforward algebra problem in which you name a variable, set up an equation where two proportions (two ratios) are set equal, and solve for the unknown. So, let $N =$ the whole population size, and note that 108 springboks out of N were tagged, while 12 out of the sample of 80 springboks were seen to be tagged.

Answer: $\frac{108}{N} = \frac{12}{80}$ gives $N = 720$.

(b) **Think:** That sure is a complex-looking definition of variance, but you're only asked to plug in. And note that the formula is for variance, and you're asked for the standard deviation.

Answer: $\text{var}(N) = \dfrac{108 \times 80 \times (108 - 12)(80 - 12)}{12^3} = 32,640$ and so, $\text{SD}(N) = \sqrt{32,640} = 180.67$

(c) **Think:** The sampling distribution is given to be approximately normal, so you can use a z-distribution. Even though this is not one of our standard confidence interval questions, it is still statistical inference, and, as always, you need to check conditions and give a conclusion in context, referring to the entire population.

Answer: Check conditions: Both captures are given to be random samples, $b = 12 \geq 10$ and $c - b$ $= 80 - 12 = 68 \geq 10$. A 90% z-interval has critical z-scores of ± 1.645, so the confidence interval is $720 \pm 1.645(180.67) = 720 \pm 297.2$. Thus, we are 90% confident that the total population of springboks in this wildlife preserve is between 422.8 and 1,017.2.

(d) **Think:** This is not difficult, but careful reasoning is required! If some of the marked springboks are lost, then the proportion obtained in the recapture will be smaller than it probably would have been, and if you set $\dfrac{108}{N}$ equal to a smaller number, you'll get a larger N. Now your task is to fully and clearly explain this!

Answer: We are assuming that the proportion of marked individuals within the second sample is equal to the proportion of marked individuals in the whole population. If fewer of the marked springboks are available for recapture, the proportion of marked individuals in the second sample will be smaller than it should be. Therefore, we will think that the number of originally marked individuals is a smaller proportion of the population than it really is. Thus, we will think that the population is larger than it really is.

(e) **Think:** This seems to be the same as in (d), that is, fewer correctly marked springboks around.

Answer: Again, if we think that the proportion of marked individuals in the second sample is smaller than it really is, we will think that the number of originally marked individuals is a smaller proportion of the population than it really is. Thus, we will think that the population is larger than it really is.

2. (a) **Think:** A complete answer considers shape, center, spread, and unusual features, and mentions context.

Answer:

Shape: Bimodal and roughly symmetric

Center: The center of the $\hat{w} = \dfrac{\text{Max} + \text{Min}}{2}$ calculations (where Min and Max are the minimum and maximum weights, respectively, in a set of 12 containers) is around 64.00 ounces.

Spread: From 63.91 to 64.10 (or range of 0.19 or IQR of 0.07)

(b) **Think:** Outliers are values more than $1.5 \times \text{IQR}$ below the first quartile or $1.5 \times \text{IQR}$ above the third quartile. You are given the quartiles and then can look at the dotplot as to whether any values are more than $1.5 \times \text{IQR}$ from the quartiles.

Answer: $\text{IQR} = 64.04 - 63.97 = 0.07$ and $1.5(\text{IQR}) = 0.105$

$Q_1 - 0.105 = 63.865$ and $Q_3 + 0.105 = 64.145$

Since no values are below 63.865 or above 64.145, there are no outliers.

(c) **Think:** This is a two-sided hypothesis test because you are asked whether a value varies significantly from 64, not whether it is larger or smaller than 64.

Answer: H_0: $W = 64.0$ and H_a: $W \neq 64.0$

(d) **Think:** Remember that simulation can be used to determine what values of a test statistic are likely to occur by random chance alone, assuming the null hypothesis is true. Then looking at where the test statistic falls, you can estimate a *P*-value.

Answer: For this day's sample, $\hat{w} = \dfrac{63.8 + 64.04}{2} = 63.92$, which is 0.08 away from 64.0. There are 4 values (2 above and 2 below) out of 100 this far away or farther from 64.0. With a *P*-value of 0.04, there is sufficient evidence at the 5% significance level ($0.04 < 0.05$) to reject the null hypothesis. In other words, there is convincing evidence to conclude that the machine needs adjustment.

3. (a) **Think:** The conditions for regression inference are (1) the sample must be randomly selected; (2) the scatterplot should be approximately linear; (3) there should be no apparent pattern in the residual plot; and (4) the distribution of the residuals should be approximately normal.

   Answer: The sample is given to be random, the scatterplot of Deaths vs. Cheese is roughly linear, the residual plot shows no clear pattern, and the histogram of residuals appears roughly bell-shaped (unimodal, symmetric, and without strong skewness or outliers).

   (b) **Think:** Residual = observed − predicted; you're given the observed, and you can calculate the predicted from the regression line equation. Remember that the interpretation must be in context and discuss direction, not only that the residual is of a certain size.

   Answer: The predicted number of deaths from the regression line is

   $$-2{,}977.3 + 113.13(30.1) = 427.9$$

   So, the residual = observed − predicted = $456 - 427.9 = 28.1$. The actual number of deaths was 28.1 greater than what was predicted by the regression model for the year when the per capita consumption of cheese was 30.1 pounds.

   (c) **Think:** Remember, this is a *t*-distribution with $df = n - 2$, and don't forget that confidence interval questions always involve conclusions in context.

   Answer: With $df = 10 - 2 = 8$ and 0.025 in each tail, the critical *t*-values are $\pm \text{invT}(0.975, 8) = \pm 2.306$. $b \pm t s_b = 113.13 \pm 2.306(13.56) = 113.13 \pm 31.27$. We are 95% confident that for each additional pound of cheese consumption per capita, the average increase in deaths per year from bedsheet tangling is between 81.86 and 144.40.

   (d) **Think:** Correlation does not imply causation! Most likely the two unrelated data sets follow the same time trend, resulting in correlation but no connection.

   Answer: No cause-and-effect conclusion is appropriate. Evidence of an association is not evidence of a cause-and-effect relationship. In cases like this, most likely the two unrelated data sets happen to follow the same time trend, resulting in correlation but not causation.

   (e) **Think:** The sum and thus the mean of the residuals is 0, and the SD is given by "S" in typical computer output. The residuals appear to have a roughly normal distribution, so the asked-for calculation is a straightforward normal probability!

   Answer: The sum and thus the mean of the residuals is always 0. The standard deviation of the residuals is estimated with $s = 50.0007$. With a *z*-score of $\dfrac{75}{50} = 1.5$ and a roughly normal distribution, we have $P(z > 1.5) \approx 0.067$.

4. (a) **Think:** This is a complex-looking graph right from the start! Take your time and look carefully at what the axes represent: values and their z-scores. You are only asked here for the range of CFU values, and the min and max CFU values can be easily estimated from the graph.

    Answer: Approximately $33.8 - 28.5 = 5.3$ (billion)

    (b) **Think:** There are 100 CFU values, so the fifth percentile is simply the value indicated by the fifth dot up from the minimum CFU on the graph.

    Answer: Either counting 5 dots (out of 100) or going over to the data plot from a z-score of -1.645, the 5th percentile of the data is seen to be approximately 29.75 (billion).

    (c) **Think:** Getting more complicated! You're not looking at the data here, only at that diagonal line. The 95th percentile in a normal distribution corresponds to a z-score of 1.645, which is fortunately marked on the graph.

    Answer: Going over to the slant line from a z-score of 1.645, you can see that the 95th percentile of a normal distribution with the same mean and standard deviation is seen to be approximately 34.4 (billion).

    (d) **Think:** What is the shape of a histogram? In the graph, you can see that lots of values are piled up at the upper end between 33 and 34, while at the lower end, the values are spread out from the minimum value toward the median. Remember that skewness is indicated when values are concentrated at one end and spread thinly on the other.

    Answer: Note that there are relatively few CFU values between 28 and 32, that these lower values are spread out while the values between 32 and 33 are more concentrated, and that the values between 33 and 34 are very concentrated. Conclude that the distribution is skewed left (skewed toward the lower values).

    (e) **Think:** To receive full credit for a confidence interval inference question, you must name the procedure, check conditions, find the interval, and give a conclusion in context about the whole population.

    Answer:

    Parameter: Let μ represent the true mean CFU value (in billions) of the population of probiotic capsules of this brand.
    Procedure: t-interval of a population mean.
    Checks: Random sample (given), the sample size $n = 100$ is large enough for the CLT to apply, and $n = 100$ is less than 10% of all probiotic capsules of this brand.
    Mechanics: With $\bar{x} = 32.48$, $s = 1.174$, and $n = 100$, a 95% t-interval is $(32.247, 32.713)$.
    Conclusion in context: We are 95% confident that the mean CFU value (in billions) of all probiotic capsules of this brand is between 32.247 and 32.713.

> **TIP**
>
> **Even if you have no idea how to do the first parts, a later part might be straightforward; in this case, the last part is simply a *t*-interval mean question.**

5. (a) **Think:** A Type I error is a mistaken rejection of a true null hypothesis.

    Answer: A Type I error would occur if the machine correctly puts 210 pieces into a bag but the test concludes that fewer pieces were put into the bag. A possible consequence would be that the machine is operating correctly but is shut down.

    (b) **Think:** A Type II error is a mistaken failure to reject a false null hypothesis.

    Answer: A Type II error would occur if the machine puts fewer than 210 pieces into a bag but the test fails to pick up on this. A possible consequence would be that the machine is malfunctioning, putting fewer than 210 pieces of candy into a bag; however, it is not shut down for inspection and recalibration.

(c) **Think:** Given a random variable with a normal distribution, mean μ, and standard deviation σ, the sampling distribution of the sum of n copies will have a normal distribution with mean $n \times \mu$ and standard deviation $\sqrt{n} \times \sigma$.

Answer: The sampling distribution of W is approximately normal with mean $210 \times 1 = 210$ grams and standard deviation $\sqrt{210} \times 0.03 = 0.4347$ grams.

(d) **Think:** This is a straight normal probability question, and don't forget to name the distribution and its parameters.

Answer: We have a normal distribution with mean 210 and standard deviation 0.4347. With a z-score of $\frac{209.5 - 210}{0.4347} = -1.150$, $P(X \le 209.5) = \text{normalcdf}(-1000, -1.150) = 0.0125$. [Or $\text{normalcdf}(0, 209.5, 210, 0.4347) = 0.0125$.] When the bag really does contain 210 candies, the probability that the machine will be shut down is 0.125.

(e) **Think:** If the bag contains 209 candies, the sampling distribution of W is now roughly normal with a mean of 209 and a standard deviation of $\sqrt{209} \times 0.03 = 0.4337$.

Answer: We have a sampling distribution that is roughly normal with a mean of 209 and a standard deviation of $\sqrt{209} \times 0.03 = 0.4337$. Here, 209.5 has a z-score of $\frac{209.5 - 209}{0.4337} = 1.153$, and the probability the machine will be shut down is $P(X \le 209.5) = \text{normalcdf}(-1000, 1.153) = 0.876$. [Or $\text{normalcdf}(0, 209.5, 209, 0.4337) = 0.876$.]

(f) **Think:** In (e) the null hypothesis is false, and you calculated the probability of correctly rejecting this false hypothesis. The probability of correctly rejecting a false null hypothesis, that is, not committing a Type II error, is called *power*.

Answer: The probability of correctly rejecting a false null hypothesis is called *power*.

6. (a) **Think:** You are asked for a proportionate sample and are given the proportions and the total sample size.

Answer: City: $(0.50)(2,000) = 1,000$ students
Suburbs: $(0.30)(2,000) = 600$ students
Rural: $(0.20)(2,000) = 400$ students

(b) **Think:** You are given a formula and simply asked to plug in!

Answer: $\bar{x}_{\text{overall}} = (0.5)(74.3) + (0.3)(80.4) + (0.2)(69.8) = 75.23$

(c) **Think:** Remember that the standard error of the sampling distribution of $\bar{x}$ is $\frac{s}{\sqrt{n}}$.

Answer: $SE(\bar{x}_{\text{city}}) = \frac{10.2}{\sqrt{1,000}} = 0.3226$

$SE(\bar{x}_{\text{suburb}}) = \frac{9.3}{\sqrt{600}} = 0.3797$

$SE(\bar{x}_{\text{rural}}) = \frac{12.1}{\sqrt{400}} = 0.605$

(d) **Think:** Remember that multiplying values by a constant multiplies the SD by the same constant, and when independent random variables are combined, it's the variances that add!

Answer:
$SE(\bar{x}_{\text{overall}}) = \sqrt{[(0.5)(0.3226)]^2 + [(0.3)(0.3797)]^2 + [(0.2)(0.605)]^2} = 0.4105$

(e) **Think:** You are given a statement of the procedure and are told to assume all conditions for inference are satisfied. You calculated $\hat{x}_{\text{overall}}$ in (b) and $SE(\bar{x}_{\text{overall}})$ in (d), so calculating the interval is straightforward. Don't forget a conclusion in context referring to the whole population.

Answer: The critical z-scores are $\pm\text{invNorm}(0.995) = \pm 2.576$. Then the interval is $75.23 \pm 2.576(0.4105)$ $= 75.23 \pm 1.06$, or $(74.17, 76.29)$. We are 99% confident that if all fifth graders took this mathematics exam, their mean score would be between 74.17 and 76.29.

7. (a) **Think:** Response bias occurs when the question can lead to misleading results because people don't want to be perceived as having unpopular views or looking bad to the interviewer. In this context, people might not want to admit to their dentist that they rush in brushing their teeth! Remember that when describing bias, you should always indicate in what direction the bias might distort the results.

    Answer: The patients might not want to admit to the dentist that they don't brush their teeth for as long as they should. Thus, the resulting mean brushing time from the sample would be an overestimate of the true mean brushing time.

    (b) **Think:** While the patients were selected at random, which is necessary, there is also the Normal/Large Sample condition, which says that either the population is approximately normal or the sample size is large enough $(n \geq 30)$ for the CLT to kick in. When neither of these are given, then we can look at the sample data to see if it appears reasonable to say that they come from a roughly normal population. That is, do the sample data at least look unimodal and roughly symmetric with no outliers or strong skew? Clearly this is not true here!

    Answer: We are not given that the population is roughly normal; the sample size, $n = 15$, is not large enough for the central limit theorem to apply; and the sample data are skewed left and so cannot be used to reasonably conclude that they come from a roughly normal population. Thus, it would not be appropriate to use a one-sample t-procedure to produce a confidence interval for the population mean.

    (c) **Think:** The mean of the exponential of the brushing times is given to be $\bar{x} = 99.24$, and the standard deviation is given to be $s = 44.35$. With $df = 15 - 1 = 14$, the critical t-scores are $\pm\text{invT}(0.975, 14) = \pm 2.145$. The interval is $\bar{x} \pm t^*\left(\frac{s}{\sqrt{n}}\right)$.

    Answer: $99.24 \pm 2.145\left(\frac{44.35}{\sqrt{15}}\right) = 99.24 \pm 24.56$, or $(74.68, 123.8)$. We are 95 percent confident that the true mean of the population of exponential brushing times is between 74.68 and 123.8 (10^{minutes}).

    (d) **Think:** This is a simple calculation, and you are actually given an example to mimic.

    Answer: The endpoints are $\log(74.68) = 1.87$ and $\log(123.8) = 2.09$, so the interval is 1.87 minutes to 2.09 minutes.

    (e) **Think:** Looking at Graph 2, we see that $\log \mu$ has 50 percent of the data to each side, so it must represent the median of the distribution. Also, the distribution is skewed left, so the median is greater than the mean.

    Answer: (i) The parameter $\log \mu$ is equal to the median of the population of brushing times.

    (ii) The parameter $\log \mu$ is greater than the mean of the population of brushing times. (In skewed left distributions, the median is usually greater than the mean.)

    (f) **Think:** As you saw in (e), $\log \mu$ gives the median of the population of brushing times, so the logs that are being taken in (d) are giving us a confidence interval for the median!

    Answer: The interval in (d) is a 95 percent confidence interval for the population median of the brushing times. So, we are 95 percent confident that the true median of the population of brushing times is between 1.87 minutes and 2.09 minutes.

Formulas Given on the AP Statistics Exam

I. Descriptive Statistics

$$\bar{x} = \frac{1}{n} \sum x_i = \frac{\sum x_i}{n}$$

$$s_x = \sqrt{\frac{1}{n-1} \sum (x_i - \bar{x})^2} = \sqrt{\frac{\sum (x_i - \bar{x})^2}{n-1}}$$

$$\hat{y} = a + bx \qquad\qquad\qquad \bar{y} = a + b\bar{x}$$

$$r = \frac{1}{n-1} \sum \left(\frac{x_i - \bar{x}}{s_x} \right) \left(\frac{y_i - \bar{y}}{s_y} \right) \qquad b = r\frac{s_y}{s_x}$$

II. Probability and Distributions

$$P(A \cup B) = P(A) + P(B) - P(A \cap B) \qquad P(A|B) = \frac{P(A \cap B)}{P(B)}$$

Probability Distribution	Mean	Standard Deviation
Discrete random variable X	$\mu_x = E(X) = \sum x_i \cdot P(x_i)$	$\sigma_x = \sqrt{\sum (x_i - \mu_x)^2 \cdot P(x_i)}$
If X has a **binomial** distribution with parameters n and p, then: $P(X = x) = \binom{n}{x} p^x (1-p)^{(n-x)}$, where $x = 0, 1, 2, 3, ..., n$	$\mu_x = np$	$\sigma_x = \sqrt{np(1-p)}$
If X has a **geometric** distribution with parameter p, then: $P(X = x) = (1-p)^{(x-1)} p$, where $x = 0, 1, 2, 3, ...$	$\mu_x = \frac{1}{p}$	$\sigma_x = \frac{\sqrt{1-p}}{p}$

III. Sampling Distributions and Inferential Statistics

Standardized test statistic: $\dfrac{\text{statistic} - \text{parameter}}{\text{standard error of the statistic}}$

Confidence interval: statistic $\pm$ (critical value)(standard error of statistic)

Chi-square statistic: $\chi^2 = \sum \dfrac{(\text{observed} - \text{expected})^2}{\text{expected}}$

Sampling distributions for proportions:

Random Variable	Parameters of Sampling Distribution		Standard Error * of Sample Statistic
For one proportion: $\hat{p}$	$\mu_{\hat{p}} = p$	$\sigma_{\hat{p}} = \sqrt{\dfrac{p(1-p)}{n}}$	$s_{\hat{p}} = \sqrt{\dfrac{\hat{p}(1-\hat{p})}{n}}$
For two proportions: $\hat{p}_1 - \hat{p}_2$	$\mu_{\hat{p}_1 - \hat{p}_2} = p_1 - p_2$	$\sigma_{\hat{p}_1 - \hat{p}_2} = \sqrt{\dfrac{p_1(1-p_1)}{n_1} + \dfrac{p_2(1-p_2)}{n_2}}$	$s_{\hat{p}_1 - \hat{p}_2} = \sqrt{\dfrac{\hat{p}_1(1-\hat{p}_1)}{n_1} + \dfrac{\hat{p}_2(1-\hat{p}_2)}{n_2}}$ When $p_1 = p_2$ is assumed: $s_{\hat{p}_1 - \hat{p}_2} = \sqrt{\hat{p}_c(1-\hat{p}_c)\left(\dfrac{1}{n_1} + \dfrac{1}{n_2}\right)}$ where $\hat{p}_c = \dfrac{x_1 - x_2}{n_1 - n_2}$

Sampling distributions for means:

Random Variable	Parameters of Sampling Distribution		Standard Error* of Sample Statistic
For one population: $\bar{X}$	$\mu_{\bar{x}} = \mu$	$\sigma_{\bar{x}} = \dfrac{\sigma}{\sqrt{n}}$	$s_{\bar{x}} = \dfrac{s}{\sqrt{n}}$
For two populations: $\bar{X}_1 - \bar{X}_2$	$\mu_{\bar{x}_1 - \bar{x}_2} = \mu_1 - \mu_2$	$\sigma_{\bar{x}_1 - \bar{x}_2} = \sqrt{\dfrac{\sigma_1^2}{n_1} + \dfrac{\sigma_2^2}{n_2}}$	$s_{\bar{x}_1 - \bar{x}_2} = \sqrt{\dfrac{s_1^2}{n_1} + \dfrac{s_2^2}{n_2}}$

Sampling distributions for simple linear regression:

Random Variable	Parameters of Sampling Distribution	Standard Error* of Sample Statistic
For slope: b	$\mu_b = \beta$ $\sigma_b = \dfrac{\sigma}{\sigma_x \sqrt{n}}$, where $\sigma_x = \sqrt{\dfrac{\sum (x_i - \bar{x})^2}{n}}$	$s_b = \dfrac{s}{s_x \sqrt{n-1}}$, where $s = \sqrt{\dfrac{\sum (y_i - \hat{y}_i)^2}{n-2}}$ and $s_x = \sqrt{\dfrac{\sum (x_i - \bar{x})^2}{n-1}}$

*Standard deviation is a measurement of variability from the theoretical population. Standard error is the estimate of the standard deviation. If the standard deviation of the statistic is assumed to be known, the standard deviation should be used instead of the standard error.

AP Scoring Guide

The multiple-choice and free-response sections are weighted equally.

There is no penalty for guessing in the multiple-choice section.

The investigative task counts for 25% of the free-response section. Each of the free-response questions has a possible four points.

To find your score use the following guide:

Multiple-choice section (40 questions)

Number correct × 1.25 = _____

Free-response section (5 open-ended questions plus an investigative task)

Question 1 _____ × 1.875 = _____
out of 4

Question 2 _____ × 1.875 = _____
out of 4

Question 3 _____ × 1.875 = _____
out of 4

Question 4 _____ × 1.875 = _____
out of 4

Question 5 _____ × 1.875 = _____
out of 4

Question 6 _____ × 3.125 = _____
out of 4

Total points from multiple-choice and free-response sections = _____

Conversion chart based on average cutoff scores from past exams:

Total Points	AP Score
70–100	5
55–69	4
44–54	3
30–39	2
0–29	1

In the past, roughly 15% of students scored 5, 20% scored 4, 25% scored 3, 20% scored 2, and 20% scored 1 on the AP Statistics exam. Colleges generally require a score of at least 3 for a student to receive college credit.

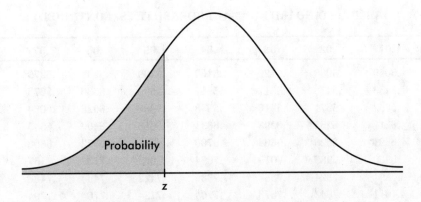

Probability

z

TABLE A: STANDARD NORMAL PROBABILITIES

z	.00	.01	.02	.03	.04	.05	.06	.07	.08	.09
−3.4	.0003	.0003	.0003	.0003	.0003	.0003	.0003	.0003	.0003	.0002
−3.3	.0005	.0005	.0005	.0004	.0004	.0004	.0004	.0004	.0004	.0003
−3.2	.0007	.0007	.0006	.0006	.0006	.0006	.0006	.0005	.0005	.0005
−3.1	.0010	.0009	.0009	.0009	.0008	.0008	.0008	.0008	.0007	.0007
−3.0	.0013	.0013	.0013	.0012	.0012	.0011	.0011	.0011	.0010	.0010
−2.9	.0019	.0018	.0018	.0017	.0016	.0016	.0015	.0015	.0014	.0014
−2.8	.0026	.0025	.0024	.0023	.0023	.0022	.0021	.0021	.0020	.0019
−2.7	.0035	.0034	.0033	.0032	.0031	.0030	.0029	.0028	.0027	.0026
−2.6	.0047	.0045	.0044	.0043	.0041	.0040	.0039	.0038	.0037	.0036
−2.5	.0062	.0060	.0059	.0057	.0055	.0054	.0052	.0051	.0049	.0048
−2.4	.0082	.0080	.0078	.0075	.0073	.0071	.0069	.0068	.0066	.0064
−2.3	.0107	.0104	.0102	.0099	.0096	.0094	.0091	.0089	.0087	.0084
−2.2	.0139	.0136	.0132	.0129	.0125	.0122	.0119	.0116	.0113	.0110
−2.1	.0179	.0174	.0170	.0166	.0162	.0158	.0154	.0150	.0146	.0143
−2.0	.0228	.0222	.0217	.0212	.0207	.0202	.0197	.0192	.0188	.0183
−1.9	.0287	.0281	.0274	.0268	.0262	.0256	.0250	.0244	.0239	.0233
−1.8	.0359	.0351	.0344	.0336	.0329	.0322	.0314	.0307	.0301	.0294
−1.7	.0446	.0436	.0427	.0418	.0409	.0401	.0392	.0384	.0375	.0367
−1.6	.0548	.0537	.0526	.0516	.0505	.0495	.0485	.0475	.0465	.0455
−1.5	.0668	.0655	.0643	.0630	.0618	.0606	.0594	.0582	.0571	.0559
−1.4	.0808	.0793	.0778	.0764	.0749	.0735	.0721	.0708	.0694	.0681
−1.3	.0968	.0951	.0934	.0918	.0901	.0885	.0869	.0853	.0838	.0823
−1.2	.1151	.1131	.1112	.1093	.1075	.1056	.1038	.1020	.1003	.0985
−1.1	.1357	.1335	.1314	.1292	.1271	.1251	.1230	.1210	.1190	.1170
−1.0	.1587	.1562	.1539	.1515	.1492	.1469	.1446	.1423	.1401	.1379
−0.9	.1841	.1814	.1788	.1762	.1736	.1711	.1685	.1660	.1635	.1611
−0.8	.2119	.2090	.2061	.2033	.2005	.1977	.1949	.1922	.1894	.1867
−0.7	.2420	.2389	.2358	.2327	.2296	.2266	.2236	.2206	.2177	.2148
−0.6	.2743	.2709	.2676	.2643	.2611	.2578	.2546	.2514	.2483	.2451
−0.5	.3085	.3050	.3015	.2981	.2946	.2912	.2877	.2843	.2810	.2776
−0.4	.3446	.3409	.3372	.3336	.3300	.3264	.3228	.3192	.3156	.3121
−0.3	.3821	.3783	.3745	.3707	.3669	.3632	.3594	.3557	.3520	.3483
−0.2	.4207	.4168	.4129	.4090	.4052	.4013	.3974	.3936	.3897	.3859
−0.1	.4602	.4562	.4522	.4483	.4443	.4404	.4364	.4325	.4286	.4247
−0.0	.5000	.4960	.4920	.4880	.4840	.4801	.4761	.4721	.4681	.4641

TABLE A: STANDARD NORMAL PROBABILITIES (CONTINUED)

z	.00	.01	.02	.03	.04	.05	.06	.07	.08	.09
0.0	.5000	.5040	.5080	.5120	.5160	.5199	.5239	.5279	.5319	.5359
0.1	.5398	.5438	.5478	.5517	.5557	.5596	.5636	.5675	.5714	.5753
0.2	.5793	.5832	.5871	.5910	.5948	.5987	.6026	.6064	.6103	.6141
0.3	.6179	.6217	.6255	.6293	.6331	.6368	.6406	.6643	.6480	.6517
0.4	.6554	.6591	.6628	.6664	.6700	.6736	.6772	.6808	.6844	.6879
0.5	.6915	.6950	.6985	.7019	.7054	.7088	.7123	.7157	.7190	.7224
0.6	.7257	.7291	.7324	.7357	.7389	.7422	.7454	.7486	.7517	.7549
0.7	.7580	.7611	.7642	.7673	.7704	.7734	.7764	.7794	.7823	.7852
0.8	.7881	.7910	.7939	.7967	.7995	.8023	.8051	.8078	.8106	.8133
0.9	.8159	.8186	.8212	.8238	.8264	.8289	.8315	.8340	.8365	.8389
1.0	.8413	.8438	.8461	.8485	.8508	.8531	.8554	.8577	.8599	.8621
1.1	.8643	.8665	.8686	.8708	.8729	.8749	.8770	.8790	.8810	.8830
1.2	.8849	.8869	.8888	.8907	.8925	.8944	.8962	.8980	.8997	.9015
1.3	.9032	.9049	.9066	.9082	.9099	.9115	.9131	.9147	.9162	.9177
1.4	.9192	.9207	.9222	.9236	.9251	.9265	.9279	.9292	.9306	.9319
1.5	.9332	.9345	.9357	.9370	.9382	.9394	.9406	.9418	.9429	.9441
1.6	.9452	.9463	.9474	.9484	.9495	.9505	.9515	.9525	.9535	.9545
1.7	.9554	.9564	.9573	.9582	.9591	.9599	.9608	.9616	.9625	.9633
1.8	.9641	.9649	.9656	.9664	.9671	.9678	.9686	.9693	.9699	.9706
1.9	.9713	.9719	.9726	.9732	.9738	.9744	.9750	.9756	.9761	.9767
2.0	.9772	.9778	.9783	.9788	.9793	.9798	.9803	.9808	.9812	.9817
2.1	.9821	.9826	.9830	.9834	.9838	.9842	.9846	.9850	.9854	.9857
2.2	.9861	.9864	.9868	.9871	.9875	.9878	.9881	.9884	.9887	.9890
2.3	.9893	.9896	.9898	.9901	.9904	.9906	.9909	.9911	.9913	.9916
2.4	.9918	.9920	.9922	.9925	.9927	.9929	.9931	.9932	.9934	.9936
2.5	.9938	.9940	.9941	.9943	.9945	.9946	.9948	.9949	.9951	.9952
2.6	.9953	.9955	.9956	.9957	.9959	.9960	.9961	.9962	.9963	.9964
2.7	.9965	.9966	.9967	.9968	.9969	.9970	.9971	.9972	.9973	.9974
2.8	.9974	.9975	.9976	.9977	.9977	.9978	.9979	.9979	.9980	.9981
2.9	.9981	.9982	.9982	.9983	.9984	.9984	.9985	.9985	.9986	.9986
3.0	.9987	.9987	.9987	.9988	.9988	.9989	.9989	.9989	.9990	.9990
3.1	.9990	.9991	.9991	.9991	.9992	.9992	.9992	.9992	.9993	.9993
3.2	.9993	.9993	.9994	.9994	.9994	.9994	.9994	.9995	.9995	.9995
3.3	.9995	.9995	.9995	.9996	.9996	.9996	.9996	.9996	.9996	.9997
3.4	.9997	.9997	.9997	.9997	.9997	.9997	.9997	.9997	.9997	.9998

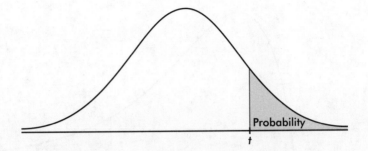

Probability

t

TABLE B: *t*-DISTRIBUTION CRITICAL VALUES

					Tail probability *p*							
df	.25	.20	.15	.10	.05	.025	.02	.01	.005	.0025	.001	.0005
1	1.000	1.376	1.963	3.078	6.314	12.71	15.89	31.82	63.66	127.3	318.3	636.6
2	.816	1.061	1.386	1.886	2.920	4.303	4.849	6.965	9.925	14.09	22.33	31.60
3	.765	.978	1.250	1.638	2.353	3.182	3.482	4.541	5.841	7.453	10.21	12.92
4	.741	.941	1.190	1.533	2.132	2.776	2.999	3.747	4.604	5.598	7.173	8.610
5	.727	.920	1.156	1.476	2.015	2.571	2.757	3.365	4.032	4.773	5.893	6.869
6	.718	.906	1.134	1.440	1.943	2.447	2.612	3.143	3.707	4.317	5.208	5.959
7	.711	.896	1.119	1.415	1.895	2.365	2.517	2.998	3.499	4.029	4.785	5.408
8	.706	.889	1.108	1.397	1.860	2.306	2.449	2.896	3.355	3.833	4.501	5.041
9	.703	.883	1.100	1.383	1.833	2.262	2.398	2.821	3.250	3.690	4.297	4.781
10	.700	.879	1.093	1.372	1.812	2.228	2.359	2.764	3.169	3.581	4.144	4.587
11	.697	.876	1.088	1.363	1.796	2.201	2.328	2.718	3.106	3.497	4.025	4.437
12	.695	.873	1.083	1.356	1.782	2.179	2.303	2.681	3.055	3.428	3.930	4.318
13	.694	.870	1.079	1.350	1.771	2.160	2.282	2.650	3.012	3.372	3.852	4.221
14	.692	.868	1.076	1.345	1.761	2.145	2.264	2.624	2.977	3.326	3.787	4.140
15	.691	.866	1.074	1.341	1.753	2.131	2.249	2.602	2.947	3.286	3.733	4.073
16	.690	.865	1.071	1.337	1.746	2.120	2.235	2.583	2.921	3.252	3.686	4.015
17	.689	.863	1.069	1.333	1.740	2.110	2.224	2.567	2.898	3.222	3.646	3.965
18	.688	.862	1.067	1.330	1.734	2.101	2.214	2.552	2.878	3.197	3.611	3.922
19	.688	.861	1.066	1.328	1.729	2.093	2.205	2.539	2.861	3.174	3.579	3.883
20	.687	.860	1.064	1.325	1.725	2.086	2.197	2.528	2.845	3.153	3.552	3.850
21	.686	.859	1.063	1.323	1.721	2.080	2.189	2.518	2.831	3.135	3.527	3.819
22	.686	.858	1.061	1.321	1.717	2.074	2.183	2.508	2.819	3.119	3.505	3.792
23	.685	.858	1.060	1.319	1.714	2.069	2.177	2.500	2.807	3.104	3.485	3.768
24	.685	.857	1.059	1.318	1.711	2.064	2.172	2.492	2.797	3.091	3.467	3.745
25	.684	.856	1.058	1.316	1.708	2.060	2.167	2.485	2.787	3.078	3.450	3.725
26	.684	.856	1.058	1.315	1.706	2.056	2.162	2.479	2.779	3.067	3.435	3.707
27	.684	.855	1.057	1.314	1.703	2.052	2.158	2.473	2.771	3.057	3.421	3.690
28	.683	.855	1.056	1.313	1.701	2.048	2.154	2.467	2.763	3.047	3.408	3.674
29	.683	.854	1.055	1.311	1.699	2.045	2.150	2.462	2.756	3.038	3.396	3.659
30	.683	.854	1.055	1.310	1.697	2.042	2.147	2.457	2.750	3.030	3.385	3.646
40	.681	.851	1.050	1.303	1.684	2.021	2.123	2.423	2.704	2.971	3.307	3.551
50	.679	.849	1.047	1.299	1.676	2.009	2.109	2.403	2.678	2.937	3.261	3.496
60	.679	.848	1.045	1.296	1.671	2.000	2.099	2.390	2.660	2.915	3.232	3.460
80	.678	.846	1.043	1.292	1.664	1.990	2.088	2.374	2.639	2.887	3.195	3.416
100	.677	.845	1.042	1.290	1.660	1.984	2.081	2.364	2.626	2.871	3.174	3.390
1000	.675	.842	1.037	1.282	1.646	1.962	2.056	2.330	2.581	2.813	3.098	3.300
∞	.674	.841	1.036	1.282	1.645	1.960	2.054	2.326	2.576	2.807	3.091	3.291
	50%	60%	70%	80%	90%	95%	96%	98%	99%	99.5%	99.8%	99.9%

Confidence level *C*

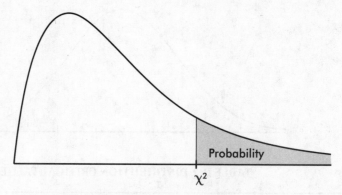

TABLE C: χ^2 CRITICAL VALUES

| df | \multicolumn{11}{c}{Tail probability p} |
	.25	.20	.15	.10	.05	.025	.02	.01	.005	.0025	.001
1	1.32	1.64	2.07	2.71	3.84	5.02	5.41	6.63	7.88	9.14	10.83
2	2.77	3.22	3.79	4.61	5.99	7.38	7.82	9.21	10.60	11.98	13.82
3	4.11	4.64	5.32	6.25	7.81	9.35	9.84	11.34	12.84	14.32	16.27
4	5.39	5.99	6.74	7.78	9.49	11.14	11.67	13.28	14.86	16.42	18.47
5	6.63	7.29	8.12	9.24	11.07	12.83	13.39	15.09	16.75	18.39	20.51
6	7.84	8.56	9.45	10.64	12.59	14.45	15.03	16.81	18.55	20.25	22.46
7	9.04	9.80	10.75	12.02	14.07	16.01	16.62	18.48	20.28	22.04	24.32
8	10.22	11.03	12.03	13.36	15.51	17.53	18.17	20.09	21.95	23.77	26.12
9	11.39	12.24	13.29	14.68	16.92	19.02	19.68	21.67	23.59	25.46	27.88
10	12.55	13.44	14.53	15.99	18.31	20.48	21.16	23.21	25.19	27.11	29.59
11	13.70	14.63	15.77	17.28	19.68	21.92	22.62	24.72	26.76	28.73	31.26
12	14.85	15.81	16.99	18.55	21.03	23.34	24.05	26.22	28.30	30.32	32.91
13	15.98	16.98	18.20	19.81	22.36	24.74	25.47	27.69	29.82	31.88	34.53
14	17.12	18.15	19.41	21.06	23.68	26.12	26.87	29.14	31.32	33.43	36.12
15	18.25	19.31	20.60	22.31	25.00	27.49	28.26	30.58	32.80	34.95	37.70
16	19.37	20.47	21.79	23.54	26.30	28.85	29.63	32.00	34.27	36.46	39.25
17	20.49	21.61	22.98	24.77	27.59	30.19	31.00	33.41	35.72	37.95	40.79
18	21.60	22.76	24.16	25.99	28.87	31.53	32.35	34.81	37.16	39.42	42.31
19	22.72	23.90	25.33	27.20	30.14	32.85	33.69	36.19	38.58	40.88	43.82
20	23.83	25.04	26.50	28.41	31.41	34.17	35.02	37.57	40.00	42.34	45.31
21	24.93	26.17	27.66	29.62	32.67	35.48	36.34	38.93	41.40	43.78	46.80
22	26.04	27.30	28.82	30.81	33.92	36.78	37.66	40.29	42.80	45.20	48.27
23	27.14	28.43	29.98	32.01	35.17	38.08	38.97	41.64	44.18	46.62	49.73
24	28.24	29.55	31.13	33.20	36.42	39.36	40.27	42.98	45.56	48.03	51.18
25	29.34	30.68	32.28	34.38	37.65	40.65	41.57	44.31	46.93	49.44	52.62
26	30.43	31.79	33.43	35.56	38.89	41.92	42.86	45.64	48.29	50.83	54.05
27	31.53	32.91	34.57	36.74	40.11	43.19	44.14	46.96	49.64	52.22	55.48
28	32.62	34.03	35.71	37.92	41.34	44.46	45.42	48.28	50.99	53.59	56.89
29	33.71	35.14	36.85	39.09	42.56	45.72	46.69	49.59	52.34	54.97	58.30
30	34.80	36.25	37.99	40.26	43.77	46.98	47.96	50.89	53.67	56.33	59.70
40	45.62	47.27	49.24	51.81	55.76	59.34	60.44	63.69	66.77	69.70	73.40
50	56.33	58.16	60.35	63.17	67.50	71.42	72.61	76.15	79.49	82.66	86.66
60	66.98	68.97	71.34	74.40	79.08	83.30	84.58	88.38	91.95	95.34	99.61
80	88.13	90.41	93.11	96.58	101.9	106.6	108.1	112.3	116.3	120.1	124.8
100	109.1	111.7	114.7	118.5	124.3	129.6	131.1	135.8	140.2	144.3	149.4

Index